Contemporary Chemistry

THE PHYSICAL SETTING

Paul S. Cohen
Saul L. Geffner

Amsco School Publications, Inc.
315 Hudson Street, New York, N.Y. 10013

Paul Cohen wishes to thank Irwin Dolkart,
master chemistry teacher, for his inspiration.

The publisher would like to thank the following teachers who acted as reviewers.

Patricia C. Koch
Irondequoit High School
Rochester, New York

Barry Kreines
Hillcrest High School
Queens, New York

Jay L. Rogoff
Hebrew Academy of
Nassau County
Uniondale, New York

Dr. Joseph L. Zawicki
Buffalo State College
Dept. of Earth Sciences and
Science Education
Buffalo, New York

Text and Cover Design: One Dot Inc.
Composition: Nesbitt Graphics Inc.
Art: Hadel Studio

Reading Passages for Chapters 5, 8, 12, 13, and 15 written by Jonathan Kolleeny, Geologist, New York State Department of Environmental Conservation.

Please visit our Web site at: ***www.amscopub.com***

When ordering this book, please specify:
769H *or* CONTEMPORARY CHEMISTRY: THE PHYSICAL SETTING, HARDBOUND

ISBN 978-0-87720-112-0

Printed in the United States of America

2 3 4 5 6 7 8 9 10 10 09 08 07

To the Student

Contemporary Chemistry: The Physical Setting is an introduction to the study of chemistry, which follows the New York State Core Curriculum. With this book, you can gain a firm understanding of the fundamental concepts of chemistry—a base from which you may confidently proceed to further studies in chemistry, or simply to a deeper appreciation of the world of science in which you live. Although the theoretical aspects of chemistry are emphasized in this book, the concepts are presented whenever possible within a laboratory-based context and applied to industrial processes and to experiences in everyday life. Thus, the author has sought a useful and realistic balance among the theoretical, descriptive, and practical aspects of chemistry.

In your previous studies of science, you may have been expected to accept information without adequate explanation of how the information was obtained. In *Contemporary Chemistry: The Physical Setting,* you will usually be given the experimental evidence that led scientists to the understandings they have. This approach will help you appreciate the part experimentation plays in chemical investigation and the growth of scientific knowledge.

The content of this book is broadly organized around three important questions:

1. What drives a chemical reaction—what makes it go?

2. What determines the rate of a reaction—how fast does it go?

3. What determines the equilibrium point of a reaction—how far will it go before equilibrium is reached?

To answer these questions, you will investigate the following general areas:

- Atomic Structure and Bonding
- Chemical Energy
- Reaction Rates and Equilibrium
- Reaction Types—Proton Transfer and Electron Transfer
- Organic Chemistry
- Nuclear Reactions

This book is designed to make learning easier for you. Many special features that stimulate interest, enrich understanding, encourage you to evaluate your progress, and enable you to review the concepts are provided. These features include:

1. ***Carefully selected, logically organized content.*** This book offers a shortened introductory chemistry course stripped of unnecessary details, which often lead to confusion.
2. ***Clear understandable presentation.*** Although you will meet many new scientific terms in this book, you will find that the language is generally clear and easy to read. Each new term is carefully defined and will soon become part of your chemistry vocabulary. The illustrations also aid in your understanding, since they, like the rest of the content, have been carefully designed to clarify concepts.
3. ***Introduction and objectives for each chapter.*** The introductory section, called Looking Ahead, is a concise description of the chapter's purpose, contents, and learning objectives. As you begin your work on each chapter, you can use the objectives as guidelines. Later you can use the objectives to pinpoint troublesome areas. The objectives will also be helpful as a way of reviewing the chapter. This section will help keep you oriented and informed about the aim of your study.
4. ***Step-by-step solutions to problems followed by practice sets.*** Problem solving is presented logically, one step at a time. Sample solutions to all types of chemistry problems are provided. These sample problems will help you approach arithmetic problems logically—a skill that should be of value in many areas of your life, not just in your study of chemistry. To enhance your newly acquired skill, you will find a practice set following most sample problems.

5. ***Taking A Closer Look.*** A feature that presents additional or in-depth information for students who wish to pursue a topic in greater detail. This section will be found in selected chapters.
6. ***End-of-chapter review questions.*** The multiple-choice questions at the end of each chapter enable you to review and assess your grasp of the chapter's content. Working out the answers to these questions is also one way to prepare for chemistry examinations you may take in the future. Some chapters are also followed by constructed-response questions, which provide practice in answering questions found in Parts B-2 and C of the Regents exam.
7. ***Chemistry Challenge.*** This section consists of SAT II-type questions as a preview for those students who wish to prepare for this test.
8. ***Readings.*** Most chapters include a reading passage that deals with an interesting topic related to that chapter. Titles such as "The Ice-Cream Diet" and "Kekulé: Lord of the Rings" are sure to grab your attention.
9. ***Glossary.*** This section contains all the **boldfaced** words found in the text along with their definitions.

The study of chemistry can be both stimulating and challenging. The authors sincerely hope that this book will increase your enjoyment of this study.

Contents

1 Matter and Energy

LOOKING AHEAD

Chemistry deals with matter, the changes matter undergoes, and the energy changes that accompany the changes in matter. In this chapter, you will study first the important characteristics of matter. You will become familiar with the standard international (SI) units used to measure these characteristics. You will then go on to learn about the forms of energy commonly involved in chemical changes. You will discover that the energy involved in a change in matter provides an important clue to the nature of the change. Finally, you will consider the three states, or phases, of matter: gases, liquids, and solids.

When you have completed this chapter, you should be able to:

- **Recognize** and **apply** standard international (SI) units of measurement.
- **Apply** Boyle's law, Charles' law, and the ideal gas law.
- **Calculate** the heat involved in phase changes and temperature changes.
- **Define** heat of vaporization; heat of fusion; vapor pressure; attractive forces.
- **Distinguish between** mass and weight; atoms and molecules; elements, compounds, and mixtures; chemical and physical properties; potential energy and kinetic energy; heat and

temperature; exothermic and endothermic changes; gases, liquids, and solids; real and ideal gases.

- **Explain** the behavior of gases according to kinetic-molecular theory.

The Study of Chemistry

As you begin your study of chemistry, you may well ask, What is chemistry? Perhaps you will be surprised to learn that there is no easy answer to that question. One answer that has been suggested is that chemistry is what the chemist does. But that is not a very helpful answer unless you already know what chemists do. So let's take a quick look at some of the work chemists around the world are doing.

Chemists in many countries are searching for drugs to fight deadly diseases, such as AIDS and cancer. Some chemists are working on ways to produce materials that can be used in place of natural foods to nourish the human body. Other chemists are doing studies that provide fundamental information about matter, such as the makeup and structure of different kinds of molecules. This kind of basic research not only furthers understanding of matter but also may lead to the development of necessary or useful products, such as drugs and synthetic fabrics. Many chemists are teachers in high schools, universities, or even in their own laboratories, where other chemists may come to learn.

This is only a small sample of the work chemists are involved in today. But it is probably large enough to give you the idea that the work of chemists is aimed toward understanding and dealing with the world and life around you.

And so it has been since the days when chemistry as a science began. Some of the earliest workers in this field were the Irish scientist Robert Boyle, who lived from 1627 to 1691; Joseph Priestley, an English clergyman and chemist, who lived and worked about 100 years after Boyle's time; and the French chemist Antoine Lavoisier, a contemporary of Priestley. These early chemists and others investigated the world around them, including such seemingly simple matter as the air they breathed. Chemists of those days were discovering what makes up the stuff of the universe. They were finding out what happens when different kinds of matter come together and interact. They were investigating the types of changes different kinds of matter can undergo. There was much to learn, even about so commonplace an event as burning. Chemistry at that time could have been defined, as it sometimes is today, as the branch of science that studies matter and the changes that matter undergoes.

But even during the early days of chemical discovery, something besides isolated pieces of knowledge about different kinds of matter began to emerge. Chemists began to see patterns in the behavior of the matter they were investigating. They began to see similarities within the great variety of things around them. From their observations, they began to formulate concepts and generalizations, or regularities, that helped to explain matter and its behavior.

Chemistry today continues in this direction. Chemists are still concerned with the composition of matter. They are equally concerned with the relationships between the properties of matter and the structure of matter. Do these relationships explain why table salt dissolves in water? Or why diamond is so hard? **Chemistry** seeks to explain how and why matter behaves as it does.

In your study of chemistry, you, too, will be concerned with the composition of matter. You will study the relationships between the properties of matter and its structure. You will study the changes that matter undergoes and the energy involved in these changes. Above all, you will gain an understanding of the unifying concepts—the big ideas—that help to explain the universe of which you are a part.

You will begin your study by considering the most basic of questions—What is matter?

What Is Matter?

Matter is the basic stuff on Earth and in the rest of the universe. It makes up your desk, your clothes, your body—the sun, moon, and stars. All these examples of matter—and any other examples you may name—have two things in common. All matter has **mass**, which is a measure of the quantity of matter. And all matter has **volume**, which is a measure of the space that matter takes up.

Mass and Weight

Mass and weight are often confused. Mass is the quantity of matter contained in a body. The mass of a body remains the same, no matter where the body is located. **Weight**, on the other hand, is a measure of the force with which gravity pulls a body toward the center of Earth. The attractive force of gravity varies with location. This means that the weight of a body changes with location, since weight depends on gravity.

The difference between mass and weight is not a new concept to you. You know that astronauts in space are said to be weightless because Earth's gravity exerts little pull on them. The mass of an astronaut in space, however, is the same as it is on Earth.

States, or Phases, of Matter

Matter commonly exists in three forms, or *states: gas, liquid,* and *solid.* At normal temperatures and pressures, air and carbon dioxide are examples of gases. Water and alcohol are examples of liquids. Iron and sulfur are examples of solids. The states of matter are also called the **phases** of matter. Thus, matter may exist in the gas phase, the liquid phase, or the solid phase.

A **gas** does not have a specific volume. Instead, it expands or contracts to fill its container. A **liquid** has a specific volume and takes the shape of its container. A **solid** is a rigid body. It has a definite volume, and its shape can be changed only by applying a force. Why gases, liquids, and solids behave as they do will be explained later.

The word "phase" can have a different meaning. Within a sample of matter, there may be some parts or regions that are different from the rest of the sample. These parts or regions are clearly separated from the other parts of the sample. A sample of matter like this is said to be made up of phases. Each of the separate parts of the sample that have the same components, or materials, and have the same characteristics, or *properties,* is one phase.

Oil and water in a container is a familiar example of a sample of matter that has two phases. Both the oil and the water are liquids. Each liquid forms a separate layer in the container. There is a sharp boundary between the two layers. Each of the liquids in the container has its own set of characteristics, or properties. The oil is one phase, the water is another phase.

Now suppose the container is shaken vigorously. The two layers will be broken up. Droplets of oil will spread throughout the water. Even now there will be boundaries between the oil and the water, and each liquid will still have its own set of properties. The sample of matter in the container will still have two phases: an oil phase and a water phase.

Dust particles suspended in air are an example of a solid phase present in gas. The many pieces of dust suspended in air make up only one phase: the solid phase.

Identifying Matter

Different kinds of matter can be described or identified by their characteristics, which you have learned are called properties. Some properties of matter are *physical properties*. Other properties are *chemical properties*. In addition to helping you recognize different kinds of matter, properties also often suggest how matter may be used.

Physical properties of matter **Physical properties** describe the qualities of a substance that can be demonstrated without changing the composition of the substance. Some physical properties describe the appearance of matter, such as its state—gas, liquid, or solid—and its color. Odor is also a physical property. Certain other physical properties describe how matter behaves. For example, the temperature at which matter boils or freezes and the solubility of matter are physical properties of this kind. Another example is **density**, which is the mass per unit of volume of a given sample of matter. Properties such as these, which can be measured and expressed by numbers, are called physical constants.

Physical constants, because they can be measured, are especially useful in identifying different kinds of matter. Pure water, for example, may be described as a colorless and odorless liquid. But many other liquids are also colorless and odorless. You have to measure the physical constants to be sure you are correct when you identify a colorless, odorless liquid as water. If, at sea level, the liquid boils at 100°C and freezes at 0°C, you know it is water.

Density and solubility are also useful physical properties, which help us identify substances. The density of water is 1.0 gram/milliliter (g/mL). Any substance with a density greater than 1.0 g/mL will sink in water, and any substance with density less than 1.0 g/mL will float. Thus it is easy to determine whether a particular substance is more or less dense than water.

Solubility in water varies widely from one substance to the other. Both salt and sugar will dissolve in water. However, you can dissolve more than five times as much sugar as salt in a given sample of water. Some common substances, such as sand and chalk, are almost insoluble in water.

Reference manuals are available that list the physical properties of thousands of different substances. Here is a sample listing for common table salt:

Name	***Formula***	***Color***	***Density***	***Melting point***	***Boiling point***	***Solubility in g/liter of water***
Sodium chloride	NaCl	Colorless	2.2 g/mL	801°C	1413°C	360 at 20°C.

Matter may undergo certain changes that cause the matter to have a different form but have no effect on the identity of the matter. A large rock, for example, may be ground or chopped into smaller pieces. The form of the rock has been changed, but each of the small pieces can still be identified as the original rock. Matter may undergo a change of state, or phase, without losing its identity. A liquid may be frozen to form a solid or boiled to form a gas. Whether liquid, solid, or gas, the matter keeps the same identity. Changes such as these, in which the form but not the identity of matter changes, are called **physical changes.**

Chemical properties of matter **Chemical properties** describe how matter behaves when it changes into another kind of matter. For example, when iron rusts, a new kind of matter forms. The properties of the new matter are quite different from the properties of the original iron. Or when some sulfur burns, a gas with a choking odor forms. This gas is new matter, different from the sulfur, which is a yellow, odorless solid. In each of these examples, the original matter—iron or sulfur—becomes a different kind of matter. Such changes are called **chemical changes.** The chemical changes that a kind of matter may undergo are examples of chemical properties.

PRACTICE

1.1 How does the density of ice compare with that of water? How do you know?

1.2 State whether each of the following illustrates a chemical property or physical property:

(a) Alcohol burns in air.

(b) Alcohol dissolves in water.

(c) Alcohol evaporates quickly at room temperature.

(d) Alcohol reacts with sulfuric acid to form ether.

(e) Helium is less dense than air.

1.3 State whether each of the following is a chemical change or a physical change:

(a) Lead melts.

(b) Oil boils.

(c) Milk turns sour.

(d) An egg becomes rotten.

(e) An apple turns brown on exposure to air.

(f) Coffee beans are ground to a powder.

(g) Hot water is passed through the coffee grounds, to prepare a cup of coffee.

The Structure of Matter—Atoms and Molecules

To understand the nature of chemical changes, you have to understand the structure of matter. Matter is made up of atoms. In some types of matter, atoms join to form larger units called molecules. The kinds of atoms present and the way they are arranged in molecules determine the particular kind of matter.

As an example, a molecule of water may be shown like this:

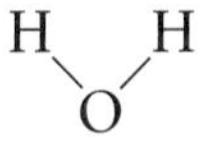

In a molecule of water, two hydrogen (H) atoms are joined to a single oxygen (O) atom. In the diagram, the lines symbolize the joining of the atoms and provide some information about how the atoms are arranged in the molecule.

The atoms in a molecule of water can be rearranged with the use of a direct electric current. The atoms are rearranged this way:

$$2\ \mathrm{H{-}O{-}H} \longrightarrow 2\ \mathrm{H{-}H} + \mathrm{O{=}O}$$

water (liquid) — hydrogen (gas) — oxygen (gas)

The electric current causes the atoms in two molecules of water to be rearranged into two molecules of hydrogen and one molecule of oxygen. Again, the lines symbolize the joining of the atoms.

The change brought about by the electric current is an example of a chemical change—one kind of matter has been changed into different matter. Notice that the total number of atoms—four of hydrogen and two of oxygen—is the same before and after the change, but the atoms are arranged in different ways. A chemical change—a change from one kind of matter into another—may be thought of as a rearrangement of atoms.

Attractive Forces in Matter

The attractive forces between atoms, **bonds**, are related to the strength of the attractive forces between smaller particles, called *protons* and *electrons,* that are within the atoms that make up the matter. Protons and electrons will be discussed later on in this chapter and in Chapter 2. Bonds and bonding will be discussed in Chapter 3.

For now, consider the bonding in a water molecule. Recall that water is symbolized as

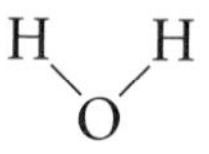

The lines that connect the letters symbolizing the atoms represent the attractive forces, or bonds, between the atoms that make up the water molecule. The angle between the bonds—in other words, the way the bonds are arranged—determines the shape, or the geometry, of the molecule.

All particles of matter, such as atoms and molecules, attract one another. The closer the particles, the stronger are the attractive forces between them. At sea level, the particles of a gas are widely separated, and the attractive forces between them are weak. For this reason, gases do not have specific volumes or shapes. On the other hand, the particles in a solid are very close to one another—the attractive forces in a solid are strong. As a result, solids have specific volumes or shapes. The attractive forces in liquids lie between the attractive forces in gases and solids, permitting a liquid to take the shape of the container. Understanding the attractive forces in matter permits understanding of changes of state, or phase changes.

Conservation of Matter

One of the chemical properties of matter is its ability either to burn or to have other matter burn in it (to support combustion). When burned, all matter behaves in the same general way. Chemists say the matter exhibits a uniform, or regular, behavior. Such behavior is commonly called a regularity, or a law, of nature.

During the 17th and 18th centuries, scientists were beginning to recognize many regularities in matter. For example, the French chemist Antoine Lavoisier determined that when matter burns, oxygen in the atmosphere is consumed. When he conducted his experiments in such a way that none of the products of burning could be lost, the total mass of matter at the beginning and at the end of the experiment was the same. In other words, Lavoisier found that when a sample of matter burns, the total mass of the products formed by burning is equal to the mass of the matter burned plus the mass of the oxygen consumed.

An example may help to make this statement clear. Suppose that 4.0 grams (g) of mercury are heated in air until all the mercury is used up. It can be shown that 4.0 grams of mercury always combine with 0.3 gram of oxygen in the air and form 4.3 grams of new matter, called mercuric oxide. This is the same as saying that the sum of the number of atoms in 4.0 grams of mercury and in 0.3 gram of oxygen is the same as the total number of atoms in 4.3 grams of mercuric oxide.

From this experiment and many similar ones, scientists formulated the **law of conservation of matter (mass).** This law states: Matter (or mass) may be neither created nor destroyed in a chemical change.

The Composition of Matter

Matter may be classified according to its physical and chemical properties. Matter may also be classified by its composition. According to this classification, matter may be an element, a compound, or a mixture.

Elements An **element** contains only one kind of atom. All matter is made up of elements, either alone or in different combinations. To date, the existence of 114 elements has been established. Of this number, 88 exist in nature and the rest have been produced, or synthesized, in the laboratory. Examples of naturally occurring elements are hydrogen, oxygen, sodium, and chlorine. Examples of elements that have been produced in laboratories are plutonium, curium, and astatine.

An **atom** is the smallest part of an element that retains the properties of the element. Every atom of any one element contains a certain number of positively charged particles, called **protons,** and an equal number of negatively charged particles, called **electrons.** Each of the atoms that make up oxygen, for example, has 8 protons and 8 electrons. Each atom of sodium has 11 protons and 11 electrons. The number of protons in the atom determines the identity of the element. Protons, electrons, and other atomic particles will be discussed in detail in Chapter 2.

A sample of matter that has the same properties throughout is said to be **homogeneous.** All samples of elements are homogeneous. In addition, any sample of a given element is exactly the same as any other sample of the element. Each sample is made up of identical atoms with the same chemical properties.

Because an element contains only one kind of atom, it cannot be broken down or changed into another element by ordinary chemical means, such as by the use of heat or an electric current. Only if the atoms of an element change—that is, the number of protons in the atoms changes—does the element become a different element. Such changes in atoms can be brought about by extraordinary means, such as by the use of an atom-smashing machine. Some elements also change into other elements through the natural process of radioactivity.

Compounds When two or more different elements combine in fixed proportions by mass, they form a **compound.** Because the elements in a compound are chemically combined, they can be separated by *chemical* means only. Water, for example, is a compound made up of hydrogen and oxygen. The percentage by mass of the two elements in the compound are 11.1 percent hydrogen and 88.9 percent oxygen. Hydrogen peroxide is another compound made up of the elements hydrogen and oxygen. In hydrogen peroxide, however, the elements combine in different proportion by mass than in water. The differences between the two compounds are shown in the table.

	Percentages of Elements by Weight	
Compound	**Hydrogen**	**Oxygen**
Water (hydrogen oxide)	11.1	88.9
Hydrogen peroxide	5.9	94.1

Still another example of a compound is mercury(II) oxide. Recall that 4 grams of mercury combine with 0.3 gram of oxygen to form 4.3 grams of mercury(II) oxide. The percentage by weight of the two elements in the compound are 93.1 percent mercury and 6.9 percent oxygen.

A compound, like an element, is homogeneous matter. In other words, any sample of a pure compound has constant properties throughout the sample. In addition, as with elements, every sample of a compound is chemically identical to any other sample of that compound. The same elements are present in the same proportion by mass in all samples of the compound. Any variety of matter that has a definite, unvarying set of properties and for which all samples have identical composition is called a **substance.** While in ordinary speech we often call any sample of matter a "substance," to a chemist, only pure elements and pure compounds are substances.

Some compounds are made up of relatively simple molecules. A molecule of water has three atoms—two of hydrogen and one of oxygen. A molecule of hydrogen peroxide has four atoms—two of hydrogen and two of oxygen. Other compounds have relatively complex molecules. Ordinary table sugar is a substance called sucrose, which contains 3 different elements, but a total of 45 atoms in each molecule.

Mixtures A mixture generally contains two or more kinds of matter in varying proportions. Therefore, mixtures are not considered true substances. For example if you were to make a mixture of table salt and sugar, you could use any proportions of salt and sugar. With an understanding of the properties of this mixture, you could also separate it into its components—salt and sugar. The components of a mixture can be separated by making use of differences in their physical properties, such as solubility and boiling and freezing points.

Some mixtures, such as table salt dissolved in water, are uniform in composition and are called **homogeneous mixtures.** Equal volumes of such a mixture have the same composition—the same number and kinds of atoms. All true solutions are homogeneous mixtures. Other mixtures, such as calamine lotion, which is zinc oxide, iron oxide, and water, and the oil and water mentioned previously, are not uniform—they are **heterogeneous mixtures.** Equal volumes of heterogeneous mixtures have different compositions. Some heterogeneous mixtures, such as orange juice, can be shaken to make them relatively uniform.

Compounds vs. mixtures A compound has a single set of properties. The properties of a mixture are the properties of each of its components and depend on the proportions in which the components are present. Take, for example, a mixture of powdered sulfur and powdered iron. Powered sulfur is yellow. Powdered iron is gray. If the proportion of sulfur in the mixture is high, the mixture will look yellow. If the proportion of iron is high, the mixture will look gray. In either case, though, the iron component of the mixture will be attracted to a magnet, just as pure iron is.

Sometimes the distinction between compounds and mixtures is not clear. An alloy, for example, may be a mixture of certain metals. Some alloys, such as brass, have specific properties—they have fixed densities, melting points, and boiling points. Such alloys are like compounds in this respect. Other alloys, such as steel, have variable properties. The densities, melting points, and boiling points of these alloys depend on the proportions in which the components are present. Such alloys are like mixtures in this respect.

Compounds always have a fixed composition. Carbon dioxide, for example, is CO_2. If you prepare carbon dioxide in the laboratory, no matter what method you use, every molecule has one carbon atom and two oxygen atoms. All carbon dioxide molecules are the same. Mixtures do not have a fixed composition. If you are preparing coffee, it can be different each time. Coffee is a mixture. Because you can vary the proportions in a mixture, an infinite number of mixtures can be formed from the same kinds of matter.

PRACTICE

1.4 When gray, powdered zinc and yellow sulfur are heated together, a flash of light is produced and an explosion takes place. The resulting solid is white. It resembles neither sulfur nor zinc. Explain these observations. What kind of matter do you think the white solid is?

1.5 Classify each of the following as element, compound, or mixture:

(a) copper (b) blue ink (c) ethyl alcohol (C_2H_5OH) (d) salt water (e) lemonade (f) iodine (g) sucrose (table sugar)

1.6 Pure aspirin, $CH_3COOC_6H_4COOH$, is a compound. Are all brands of pure aspirin the same? Explain your answer.

1.7 Which of the following are true substances? Explain your answers.

(a) water (b) cola soda (c) human blood (d) air (e) neon

Changes in Matter and Energy

Matter, you have learned, may undergo physical changes and chemical changes. Physical changes are changes in size, shape, or state (phase changes). Chemical changes result in rearrangements of atoms. Chemical changes are accompanied by changes in energy. In this section, the focus of the discussion will be on energy.

Energy is defined as the ability, or capacity, to do work. There are various forms of energy. Two forms that are often associated with changes in matter are potential energy and kinetic energy.

Potential Energy

Potential energy is the energy a body possesses because of its position with respect to another body. As the potential energy of a body decreases, its stability increases. As the potential energy increases, stability decreases. Bodies or particles that attract each other have high potential energy and low stability when they are far apart. They have low potential energy and high stability when close together.

Water molecules, like all particles, attract one another. Water molecules are farther apart in the gas state than they are in the liquid state or in the solid state. In the gas state, water molecules have higher potential energy and are less stable than water molecules in the liquid state. In turn, water molecules in the liquid state have higher potential energy and are less stable than water molecules in the solid state.

Chemical energy stored in the bonds that result from the attractive forces between particles is a form of potential energy. Stronger bonds mean greater stability. Recall that in a water molecule, two hydrogen atoms are bonded to an oxygen atom. When these bonds formed, energy was released. To change water back into its elements, the bonds must be broken—that is, energy must be absorbed. This energy may be supplied by an electric current, as was shown earlier. The relationship between energy and chemical reactions is developed further in Chapters 9 and 11.

Kinetic Energy

Kinetic energy is energy of motion. It is closely related to **temperature**, which is a measure of the average kinetic energy of a system. (A **system** is some convenient part of the environment that can be isolated for study.) Liquid water molecules have a higher average kinetic energy at 80°C than they do at 10°C.

Experiments have shown that potential energy can be changed into kinetic energy without loss. This observation is an illustration of the **law of conservation of energy.** The total amount of energy in a system is constant. Broadly viewed, this law states that under ordinary conditions energy can be neither created nor destroyed, but it can be converted from one form to another.

Heat and Temperature

Heat is a measure of the total energy in a system. Heat may be expressed in units such as calories (cal), kilocalories (kcal), joules (J), and kilojoules (kJ). The joule is the SI unit of heat.

Heat is not the same as temperature. **Temperature** is a measure of the average kinetic energy of the particles in a system. Temperature is expressed in degrees on the Fahrenheit or Celsius scales and kelvins on the Kelvin (absolute temperature) scale. The kelvin is the SI unit of temperature.

Two systems may have the same temperature but different amounts of heat. For example, 1 liter (L) and 10 liters of boiling water, both in containers at sea level, have the same temperature, 100°C. However, the larger, 10-liter sample, because it has more water, can supply more heat energy.

The following table shows the energy associated with some typical changes in matter. Note that the amount of energy involved in a chemical change is generally much larger than the amount involved in a physical change.

Change	***Kind of Change***	***Energy Change***
Coverting 1 gram of ice at 0°C to water at 0°C	Physical (Phase)	Absorbs 334 joules
Heating 1 gram of water from 0°C to water at 100°C	Physical (Temperature)	Absorbs 418 joules
Converting 1 gram of water at 100°C to steam at 100°C	Physical (Phase)	Absorbs 2260 joules
Burning 1 gram of coal	Chemical	Liberates 32,800 joules or 32.8 kilojoules
Burning 1 gram of hydrogen to form liquid water	Chemical	Liberates 143,000 joules or 143 kilojoules

Relationship Between Matter and Energy

In 1905, Albert Einstein suggested that matter and energy can be converted from one to the other—that is, matter (represented by its mass) can be converted into energy, and energy can be converted back into matter.

The mass-energy relationship can be stated mathematically by the equation $E = mc^2$. In the equation, E stands for energy, m for mass, and c for the speed of light. Einstein's theory was verified in 1932 by the British scientists John D. Cockroft and Ernest T. S. Walton. They found that when lithium is converted into helium, some small amount of mass is lost and energy is released.

The quantity of energy released was measured. It corresponded exactly to the small mass lost. As you will learn in Chapter 15, the change of a small quantity of mass into energy results in a very large quantity of energy. This was demonstrated on a vast scale with the explosion of the first atomic bomb in 1945.

As scientists came to understand that mass and energy can be changed, or converted, one into the other, the fundamental laws of mass and energy were combined. The *law of conservation of mass-energy* states: The total quantity of mass and energy in the universe is constant.

Energy and Changes of State

If you were to heat a solid, you would probably expect the temperature of the solid to increase. Most of the time, it will. Suppose, though, that you were heating a large block of ice. The ice would melt, but as long as there was still some solid left, the ice-water mixture would not get warm! The temperature of the mixture would remain at 0° while the ice melted.

The changes in temperature that occur as a substance is heated are often illustrated by graphs called "heating curves." In a heating curve, as shown in Figure 1-1, the temperature of the substance is plotted against time while the substance is heated at a constant rate.

In the interval t_1 to t_2, the solid is being warmed. Kinetic energy is increasing, as shown by the rise in temperature. At point t_2, the melting point temperature of the solid has been reached. From t_2 to t_3, the solid melts, or changes to a liquid. The heat is used to break the forces of attraction that keep the substance in a solid state, resulting in an increase in potential energy. Because

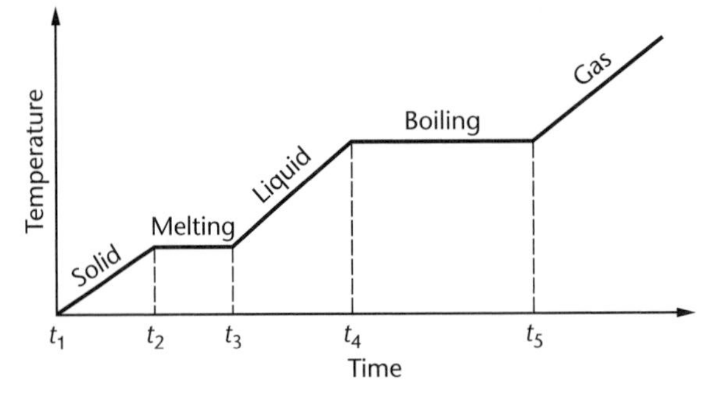

Figure 1-1 The effect of heating a solid over a period of time

the heat now increases only potential energy, the thermometer reading (temperature) remains constant. At point t_3, the substance is all liquid. Between t_3 and t_4, the liquid is being warmed. Kinetic energy again increases, and the temperature goes up. At point t_4, the boiling point temperature has been reached. In the interval from t_4 to t_5, the liquid is changing to gas. Again potential energy increases, because the forces of attraction between molecules are being broken. The temperature, which measures kinetic energy, does not change during this interval. At point t_5, the substance is all gaseous. With further heating, the kinetic energy of the molecules increases and the temperature rises.

The heat that is required to melt a given quantity of a solid is called the **heat of fusion.** The heat required to boil a quantity of liquid is called the **heat of vaporization.** Note that in Figure 1-1 the time interval during which boiling occurs is longer than the time interval during which melting occurs. Heats of vaporization are generally larger than heats of fusion. Note also, that during a phase change, at constant pressure, the temperature of a substance remains constant.

The Heating Curve of Water

Figure 1-2 shows the heating curve that results when 10.0 grams of ice, initially at −40°C, are heated at a steady rate of 840. joules per minute. You can obtain a great deal of information about a substance by examining its heating curve.

The curve levels off in two places, the melting point and the boiling point. You can see from this curve that water melts at 0°C and boils at 100°C.

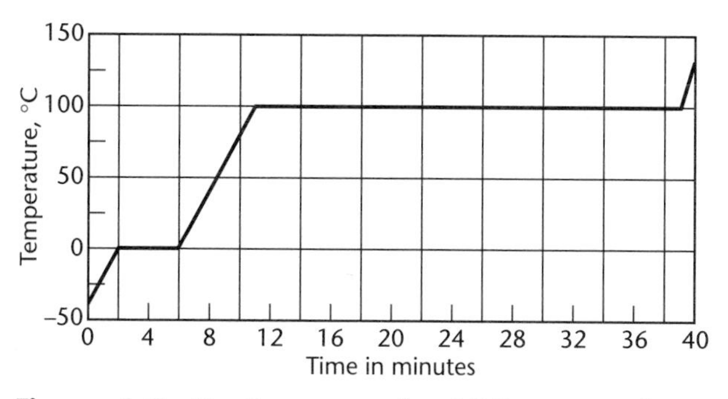

Figure 1-2 Heating curve for 10.0 grams of water heated at 840. joule/min

This curve can also be used to find the heat of fusion of ice. How long did it take to melt the ice? Since the curve levelled off at the 2-minute mark, and started to rise again at the 6-minute mark, it took 4 minutes to melt all the ice. The substance is being heated at a steady 840. joules per minute. In 4 minutes, the melting ice must have absorbed 4 × 840., or 3360 joules. It took 3360 joules to melt the 10.0 grams of ice. How many joules would it take to melt 1.0 gram of ice?

$$\frac{3360 \text{ J}}{10.0 \text{ g}} = 336 \text{ J/g}$$

The heat of fusion of ice obtained from this graph is 336 J/g.

Use the information in Figure 1-2 to find the heat of vaporization of water. Compare your result with the data in the table on page 15.

PRACTICE

1.8 Suppose you applied heat at the same rate of 840 J per minute, as shown on Figure 1-2, to 5.0 g of ice.

(a) How long would it take to melt the ice?

(b) At what temperature would the water boil?

1.9 How may a heating curve be used to determine the boiling point of a substance?

1.10 On Figure 1-2, what state is the substance in after 8 minutes of heating?

1.11 Between the 12-minute mark and the 32-minute mark in Figure 1-2, what is happening to the average kinetic energy of the substance?

Measurement of Heat

As shown in Figure 1-2 on page 17, except during a phase change, when a substance is heated, its temperature increases. The greater the amount of heat transferred to the substance, the greater the change in temperature. To calculate the exact amount of heat needed to produce a given change in temperature, two factors must be considered: the mass of the substance being heated and the nature of the substance.

If you have ever heated water in a metal pot, you are probably aware that the pot gets hot long before the water does. The amount of heat needed to change the temperature of a given mass by a given number of degrees varies enormously from substance to substance. For example, it takes 42 joules of heat to change the temperature of 10. grams of water by one kelvin, or one Celsius degree. To produce the same temperature change in 10. grams of lead requires just 1.3 joules. The quantity of heat required to change the temperature of one gram of a substance by one kelvin, or one degree Celsius, is called the **specific heat capacity**, and can be expressed in the unit *joules per gram degree,*

$$\frac{J}{g \cdot K}$$

The specific heat of water is 4.184 $\frac{J}{g \cdot K}$. (In this text, we will generally express this value to two decimals places, as 4.18 $\frac{J}{g \cdot K}$) The equation used to calculate the heat involved in a temperature change is

$$heat = mass \times change\ in\ temperature \times specific\ heat\ capacity.$$

We often use the Greek letter Δ (delta) to stand for "change in." Using q, the symbol for heat, m for mass, and C to represent the specific capacity, gives the general equation $q = mC\,\Delta T$. In water, if the heat is measured in joules, and the mass in grams, the equation becomes

$$joules = grams\ H_2O \times \Delta T \times 4.18\ J/g{\cdot}K$$

Let's say that a chemical reaction causes the temperature of 100. grams of water to increase by 4.0 K. How many joules of heat are released by the reaction? Substituting into the equation above gives us 100. g × 4.0 K × 4.18 J/g·K = 1672 joules. (Because the change in temperature has only two significant figures, our answer would be more correctly expressed as 1700 joules. The rules for correct use of significant figures are discussed in Appendix I, pages 569 to 579.)

We can use the same equation to predict the new temperature after the application of a given quantity of heat.

SAMPLE PROBLEM

PROBLEM

If 418 joules are applied to 20. g of water, initially at a temperature of 293 K, what is the final temperature of the water?

SOLUTION

Use the equation *joules* = *grams* $H_2O \times \Delta T \times 4.18$ J/g·K

$$418\text{ J} = 20\text{ g } H_2O \times \Delta T \times 4.18\text{ J/g·K}$$

Dividing both sides of the equation by the grams of water and by the specific heat, the equation becomes:

$$\frac{418\text{ J}}{20.\text{g} \times 4.18\text{ J/g·K}} = \Delta T$$

Solving the equation gives us $\Delta T = 5.0°$ K

We are not quite finished, though. The question asked not for the *change* in temperature, but for the *final* temperature. Since heat was added to the water, its temperature went up by 5 K, resulting in a final temperature of 298 K.

Heat is often measured with a device called a calorimeter. One simple type of calorimeter is shown in Figure 1-3. Basically, it consists of an insulated container in which there is a known amount of water at a known temperature. A thermometer is used to detect changes in temperature while a stirrer is used to maintain a consistent temperature throughout the liquid. Because Styrofoam™ is an excellent insulator, simple experiments are often performed in a "coffee-cup calorimeter," as shown in Figure 1-4.

When a chemical reaction takes place in a calorimeter, the energy changes that accompany the reaction will determine the temperature changes in the water. Reactions that release, or give off, heat to their surroundings are called **exothermic** reactions. Those that absorb, or remove, heat from their surroundings are called **endothermic** reactions. It is useful to think of exothermic reactions as producing heat, with endothermic reactions removing heat. Exothermic reactions will cause the temperature in the calorimeter to increase. Endothermic reactions will cause the tem-

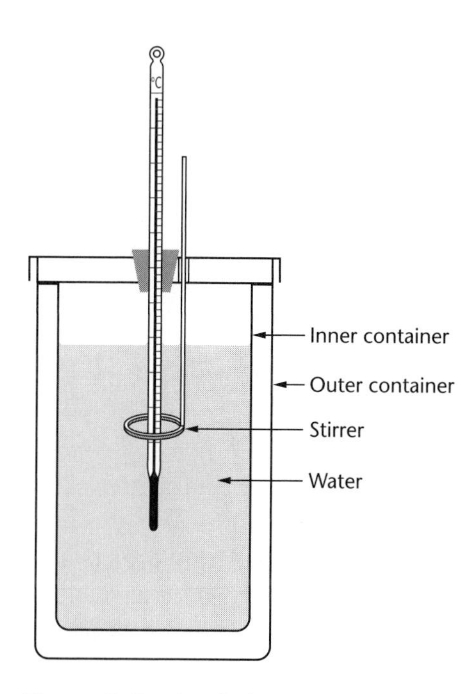

Figure 1-3 A calorimeter

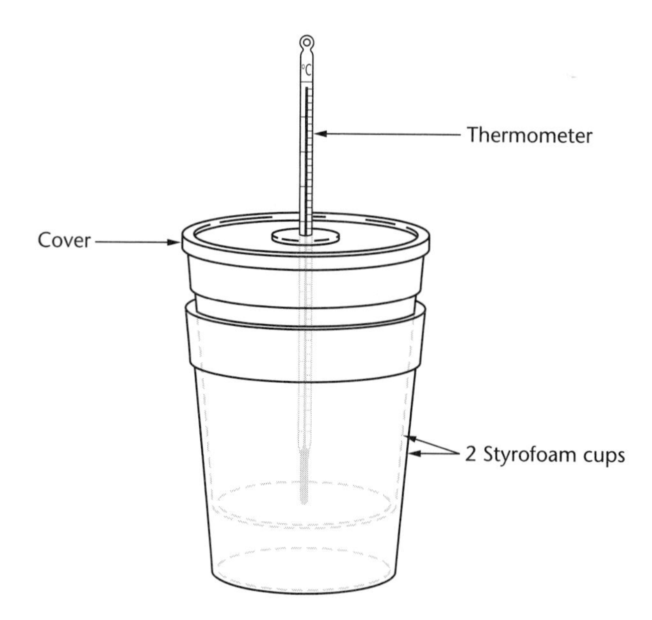

Figure 1-4 A coffee-cup calorimeter

perature to decrease. Bond formation is always exothermic. Phase changes such as freezing and condensation, which involve bond formation, are exothermic. Changes such as melting, evaporation, and boiling involve bond breaking, and are endothermic.

PRACTICE

1.12 An exothermic reaction releases 840. joules of heat to a calorimeter containing 40. g of water. Calculate the temperature change of the water.

1.13 A reaction takes place in a calorimeter containing 20. g of water at an initial temperature of 20.0°C. When the reaction is completed, the temperature of the water is 15.0°C. How many joules of heat were absorbed by the reaction?

1.14 Is the reaction in 1.13 above, exothermic or endothermic? How do you know?

1.15 How much heat must be absorbed by 50. grams of water to raise the temperature of the water from 298.0 K to 338 K?

1.16 A sample of water, upon absorbing 8360 joules of heat, became 10.K warmer. What was the mass of the water?

1.17 Ethanol is the alcohol in alcoholic beverages. When 400. joules of heat are added to a 40.0 gram sample of ethanol, initially at a temperature of 298.0 K, the temperature increases to 302.2 K. What is the specific heat of ethanol?

Other Units of Energy

Scientists have established a set of units for scientific use. These are known as SI units, from the French "Systeme International," or "International System." Some of these are listed below:

Physical Quantity	SI unit, and symbol
Mass	kilogram, kg
Length	meter, m
Temperature	kelvin, K
Energy	joule, J

However, for convenience, or sometimes just by force of habit, scientists often express quantities in units that are not SI units. In this chapter, we have already expressed masses in grams, and temperature in °C. You are probably familiar with at least two units of energy other than the joule.

Calories

Read any food label, and you will find the number of "Calories per serving." A calorie is an energy unit. Food Calories (Normally written with an upper case "C") are actually kilocalories, or thousands of calories. A calorie is equivalent to 4.18 joules. The Calorie content of a food is the quantity of energy that the food supplies to the body when it is consumed.

British Thermal Units

The British Thermal Unit, or BTU, is another unit of energy. The cooling power of an air conditioner is often given in BTU per hour. One BTU is equal to 1050 joules.

PRACTICE

The calorie is defined as the heat needed to raise the temperature of 1 gram of water by 1°C. Thus, the specific heat of water in *calories*/g·C° is exactly 1.

1.18 How many calories are needed to change the temperature of 10. grams of water from 25° to 45°C?

1.19 One glass (8 ounces) of Coca Cola™, according to the manufacturer, supplies 100 Calories. How many grams of water can be heated from its freezing point to its boiling point by the amount of energy found in a glass of Coke™?

Gases

The reasons why changes in energy accompany phase changes become clear when you consider the properties of matter in its various states. You will begin by studying the properties of gases.

Recall that molecules of a gas are separated by relatively large distances. The molecules of a gas are therefore more isolated than are the molecules of a liquid or a solid. Valuable information about individual molecules can thus be gained by studying gases.

Properties of Gases

Studies have shown that the following properties are common to all gases:

1. Gases do not have a specific shape or a specific volume. Instead, a gas takes the shape and entire volume of its container. For example, a quantity of gas in a 1-liter flask fills the flask completely. The gas has a volume of 1 liter. In a 5-liter container, the same quantity of gas has a volume of 5 liters.
2. Gases expand and contract with relative ease—that is, gases can be compressed. Many balloons can be filled from a tank of compressed helium. The volume of a balloon varies at different altitudes because the outside pressure varies. As the balloon rises, the pressure of the atmosphere decreases and the gas in the balloon expands. If the balloon continues to rise, the gas in the balloon will expand until the balloon bursts.
3. Gases at the same temperature exert pressures that are proportional to their concentrations, or number of molecules per unit of volume. The pressure in a tire increases as more molecules of air are pumped into the tire.
4. The density (mass per unit volume) of a gas is considerably less than the density of the same matter in the solid or liquid state. The volume of a given mass of gas is about 1000 times greater than the volume of the same given mass of the substance in the solid or liquid state. The volume of 28 grams of gaseous nitrogen at 0°C at sea level is about 22,400 milliliters (mL). The volume of 28 grams of solid nitrogen is only about 27.2 mL.

The Kinetic-Molecular Model

To explain the properties of gases, chemists developed the kinetic-molecular model of matter. Scientists often develop a model as a way of explaining, relating, and thinking about a scientific problem. All the facts known about the problem—observations, experimental results, and other kinds of data—are considered in developing the model. The end result is a mental image, or picture, that brings together all the available knowledge that relates to the problem.

Let us now consider the *kinetic molecular model* of matter in the gas state. This model makes certain assumptions about gases, from which attempts are made to explain the properties observed in gases. Without these assumptions, the model may offer inaccurate and untrue information. The model is based on the following assumptions:

1. Gases consist of molecules. A gas molecule may have only one atom, or it may be made up of two or more atoms. Helium and argon, for example, have only one atom per molecule. Oxygen has two atoms per molecule, and carbon dioxide has three. The attractive forces between gas molecules are negligible.
2. The total space between the molecules of a gas is very large compared with the space occupied by the molecules themselves. At normal pressures (at sea level), a volume of gas is largely empty space. This explains why gases can be readily compressed. It also explains why the molecules of one gas can readily diffuse, or mix with the molecules of other gases in the same container.
3. Gas molecules are in constant straight-line random motion. This means they move without order (see Figure 1-5). The

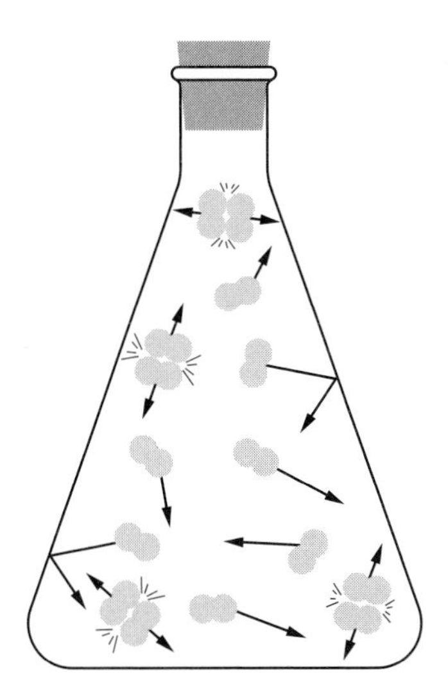

Figure 1-5 Gas molecules in random motion

random motion of gases explains why gases have no definite shape, but expand to fill the volume of any container.

4. Moving gas molecules collide with one another. These collisions are assumed to be elastic—that is, the gas molecules rebound after collision, with no net loss in energy. Gas molecules also hit the walls of the container. These impacts of gas molecules upon the walls of the container make up the pressure of the gas. The pressure is proportional to the number of impacts per unit of time and to the force of each impact. The greater the number of gas molecules in a given volume (concentration), the greater is the number of impacts and the higher the pressure of the gas.
5. The molecules of a gas move very rapidly. This explains why gases diffuse. At ordinary temperatures, the speeds of gas molecules may be higher than 1600 kilometers per hour. Such high speeds explain why gases, despite relatively small masses, may exert high pressures.

Because gas molecules move randomly, they collide with one another and with the sides of the container. During the collisions, the gas molecules exchange kinetic energies. Some molecules lose kinetic energy, and others gain kinetic energy. Thus, at a given temperature, it is unlikely that all the molecules have the same speed.

Particles in motion have kinetic energy. However, all the gas molecules in a container do not have the same speed or kinetic energy. The kinetic energy of any one molecule of a gas at a given temperature therefore cannot be described. Instead, the average kinetic energy of all the molecules is described. The average kinetic energy of the molecules of a gas is proportional to the absolute temperature (temperature on the Kelvin scale) of the gas. The higher the temperature, the higher is the average kinetic energy.

When a thermometer is used to measure the temperature of a gas, a tremendous number of molecules hits the bulb of the thermometer each second. Some of the molecules have low kinetic energy, some high. But most have kinetic energies somewhere between low and high. A graph of the kinetic energies of the molecules of a gas at some given temperature would look like the one in Figure 1-6. Points A, B, and C indicate the relative number of molecules with a given kinetic energy at three points on the curve.

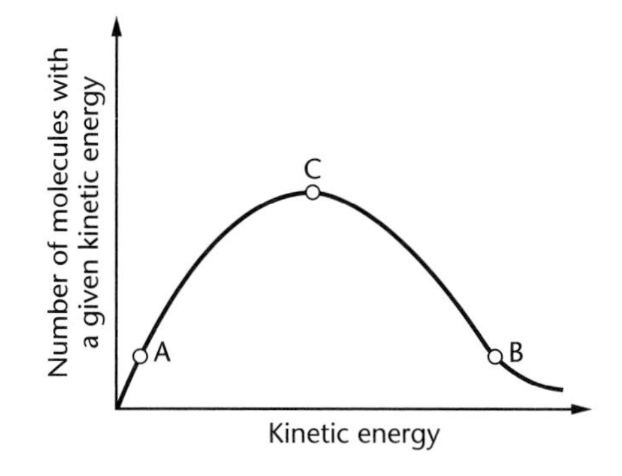

A Relatively few molecules with low kinetic energy
B Relatively few molecules with high kinetic energy
C Most molecules with kinetic energy between low and high

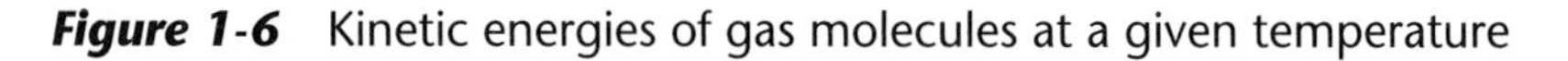

Figure 1-6 Kinetic energies of gas molecules at a given temperature

Figure 1-7 is a graph of the kinetic energies of the molecules of a gas at a different temperature. How does the temperature of the gas in Figure 1-7 compare with that in Figure 1-6?

The Gas Laws

The properties of gases were the subject of active research for many years, beginning in the 17th century. These early studies led to the formulation of the gas laws, which are among the unifying concepts of chemistry. The gas laws describe the interaction of pressure, volume, and temperature of gases.

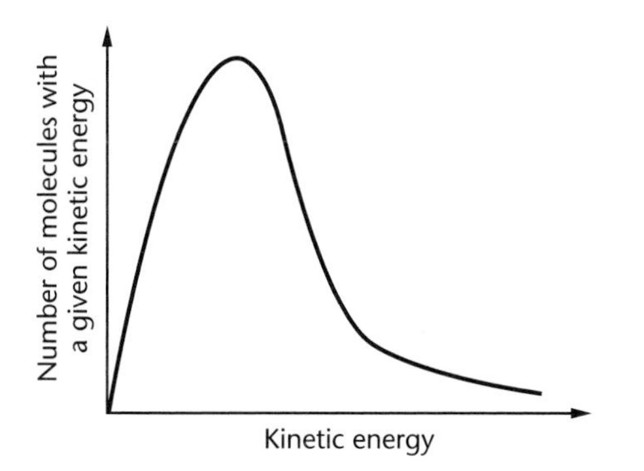

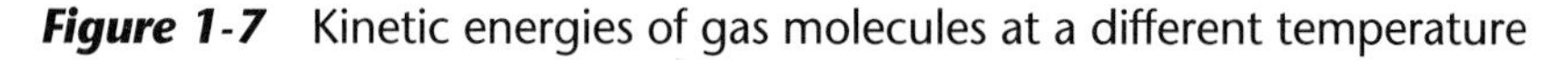

Figure 1-7 Kinetic energies of gas molecules at a different temperature

Measurements Involving Gases

Before considering the gas laws, you need some information about how these physical properties are measured.

Pressure **Pressure** is defined as force per unit area. The SI unit of force is the newton while the unit of area is the square meter. A force of one newton per square meter is called one **pascal**, the SI unit of pressure. A pascal is a very small amount of pressure. The pressure in a bicycle tire is about 500,000 pascals. The kilopascal (kPa), equal to 1000 pascals, is more commonly used. Standard atmospheric pressure is 101.3 kPa.

While it is not an SI unit, the **atmosphere** (atm) is still commonly used by chemists to measure pressure. Standard pressure is also defined as one atmosphere. A third unit of pressure is the millimeter of mercury, used because air pressure was first measured using a column of mercury; the greater the pressure, the longer the column of mercury the air could support. Standard pressure is 760 mm of mercury.

Volume The volume of a gas is defined as the amount of space in its container. Since the SI unit of length is the meter, the unit of volume is the cubic meter. However, chemists commonly express volumes using liters (L), milliliters (mL), and cubic centimeters (cm^3). Milliliters and cubic centimeters have the same volume. There are 1000 mL in 1 L, so there are also 1000 cm^3 in 1 L. Cubic centimeters are also sometimes abbreviated "cc"s, especially by medical workers.

Temperature Gas temperature is measured in kelvins, (K). The Kelvin temperature can be found by adding 273 to the Celsius temperature: $°C + 273 = K$.

In studies of gases, conditions of standard temperature and pressure are often specified. Standard temperature is 0°C (273 K) and standard pressure is one atmosphere (101.3 kPa). Standard conditions are usually indicated by the abbreviation STP.

The Kelvin temperature scale For most scientific applications, we use Kelvin, or absolute, temperature. You will recall that temperature is a measure of the average kinetic energy of a system. Does a system at 10°C have twice as much average energy as a system at 5°C? It does not! A temperature of 0°C does not correspond to a system with no kinetic energy. It just corresponds to the normal melting point of ice.

Kelvin temperature, on the other hand, is proportional to the average kinetic energy of the system. A temperature of 10 K represents twice as much energy as a temperature of 5 K. A temperature of 0 K, though impossible to achieve, would mean that the average kinetic energy of the system was zero.

A temperature of 0 K is called absolute zero. It can be approached, but never reached. Absolute zero is equivalent to a temperature of −273°C. All problems using the gas laws must be solved using the kelvin scale. Remember, to change Celsius to Kelvin, you add 273° to the Celsius temperature.

The barometer The Italian scientist Evangelista Torricelli (1608–1647) constructed a barometer using a tube of mercury (see Figure 1-8). The air at sea level can support a column of mercury 760 mm (about 30 inches) high on an average day. This was defined as standard pressure: 760 mm of mercury. In honor of Torricelli, this unit of pressure was renamed the "torr." Thus standard pressure can be expressed as 760 torr.

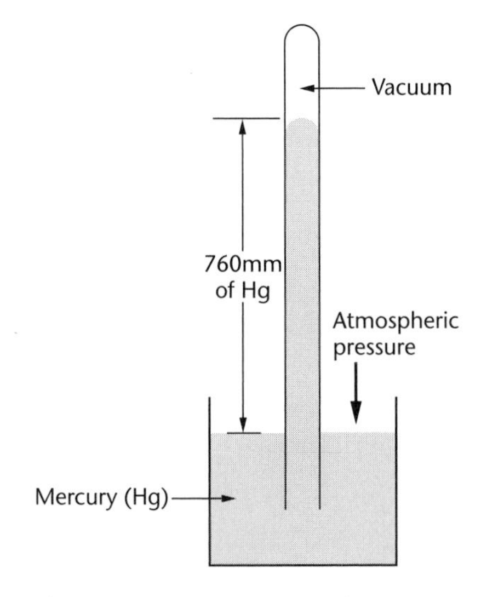

Figure 1-8 A mercury barometer. The "vacuum" above the mercury actually contains a small amount of mercury vapor

Building a Barometer: Water or Mercury?

Perhaps you have noticed that if you hold a glass under water, open side up, then turn it upside-down, and slowly lift it, as long as the glass is in contact with the water, none of the water comes out of the glass. The pressure exerted by the short column of water in the glass is far smaller than normal air pressure, so the water stays in the glass. If we use a taller glass, the water will exert a greater pressure. (If you have ever dived under water, you have experienced the pressure exerted by a column of water.) If we use a tall enough tube of glass, the pressure of the water will be greater than normal air pressure, and some water will come out of the glass until the two pressures are equal. Now, any change in air pressure will cause the water in the glass tube to move either up or down. The height of the water can thus be used to measure the air pressure. We have built a water barometer.

Before we go into the barometer business, however, there are a few additional things to consider. How long would the tube of glass need to be? Recall that the height of the mercury in the barometer (Figure 1-8) was 760 millimeters. Mercury is 13.6 times denser than water. Therefore, a column of water, in order to exert the same pressure as a column of mercury, needs to be 13.6 times higher than a column of mercury.

PRACTICE

1.20 What is the normal boiling point of water in kelvins?

1.21 Convert a temperature of 253 K to degrees Celsius.

1.22 A reaction in a calorimeter caused the temperature of a sample of water to increase by 27°C. What was the temperature change in Kelvin temperature?

1.23 What pressure in atmospheres is equivalent to a pressure of 380 mm of Hg?

The height of the water column must be $13.6 \times 760.$ millimeters $= 10,300$ millimeters, or 34 feet! Odds are, a barometer three stories tall will never be a popular household item.

Still, with enough glass tubing, a tub of water, and a large ladder, we could build and read our water barometer. Then, there would be another problem to consider. Recall that the "vacuum" above the mercury column in Figure 1-8 actually contains a small amount of mercury vapor. The vapor pressure of mercury at room temperature is 0.002 millimeter of mercury, too small to significantly affect the height of the column. However, the vapor pressure of water at room temperature is 24 millimeters of mercury, which would push the column of water down more than a foot. Since the vapor pressure changes when the temperature changes, our water column could move up and down noticeably, even when the air pressure was constant.

Mercury, with its high density, low vapor pressure, and low chemical activity, is the ONLY substance that would have permitted Torricelli and the other scientists of his time to measure air pressure accurately and practicably. It was these special properties of mercury as well that allowed Robert Boyle to use mercury in his study of the relationship between pressure and volume of gases. We can only wonder how the history of chemistry might have been affected, had mercury not been available to the scientists of the seventeenth century.

Boyle's Law

Boyle's law compares the pressure and volume of a gas when the temperature is held constant. **Boyle's law** states: At constant temperature, the volume of a fixed mass of gas varies inversely with its pressure. This relationship can be expressed mathematically as

$$V \text{ (at constant temperature)} = \frac{k}{P} \quad \text{or} \quad PV = k$$

where V = volume, P = pressure, and k = some constant number. k can be determined by experiments. This same relationship is expressed graphically in Figure 1-9, on the next page.

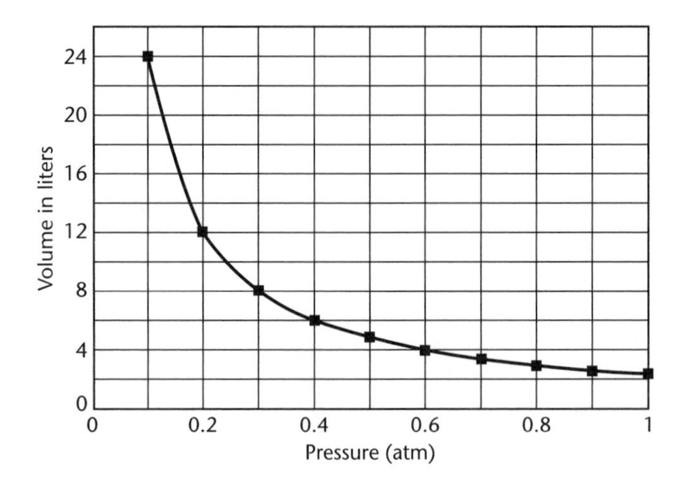

Figure 1-9 The relationship of gas volume to gas pressure

The equation $PV = k$ means that, with the proper units, the product of the pressure and volume of a fixed mass of gas has a fixed value. Notice, in Figure 1-9, if you multiply any value of P by the corresponding value of V, the product is always the same (2.4). This means that P and V are inversely related (at constant temperature). In other words, the volume of a gas increases as the pressure decreases and the volume decreases as the pressure increases as long as the temperature remains unchanged. PV at the beginning of an experiment must equal PV at the end. This relationship is often expressed with the equation:

$$P_1V_1 = P_2V_2$$

or, original pressure × original volume = new pressure × new volume. This equation is most useful in solving pressure-volume problems.

SAMPLE PROBLEM

PROBLEM

The pressure on 1.20 L of a gas at constant temperature is increased from 100. kPa to 120. kPa. What is the new volume of the gas?

SOLUTION
Set up the pressure-volume relationships according to Boyle's law. At constant temperature,

$$P_1V_1 = P_2V_2$$

This can be rewritten as

$$\underset{\text{new volume}}{V_2} = \underset{\text{original volume}}{V_1} \times \underset{\text{correction factor}}{\frac{P_1}{P_2}}$$

Substitute the given values, and solve the equation.

$$V_2 = 1.20 \text{ L} \times \frac{100. \text{ kPa}}{120. \text{ kPa}} = 1.00 \text{ L}$$

PRACTICE

1.24 A 4.0-liter sample of gas exerts a pressure of 2.0 atm. If the gas is allowed to expand to a volume of 16 L (at constant temperature), find the new pressure of the gas.

1.25 A gas has a volume of 24 mL at a pressure of 405.2 kPa. Assuming no change in temperature, what would the volume of this gas be at standard pressure?

1.26 At constant temperature, what happens to the pressure exerted by a gas if its volume is doubled?

Charles' Law

Charles' law compares the temperature and volume of a gas when the pressure is held constant. **Charles' law** states: The volume of a fixed mass of gas at constant pressure varies directly with its absolute temperature (°C + 273). This relationship can be expressed mathematically as

$$V \text{ (at constant pressure)} = kT \text{ or } \frac{V}{T} = k$$

where V = volume, k = a constant, and T = Kelvin temperature.

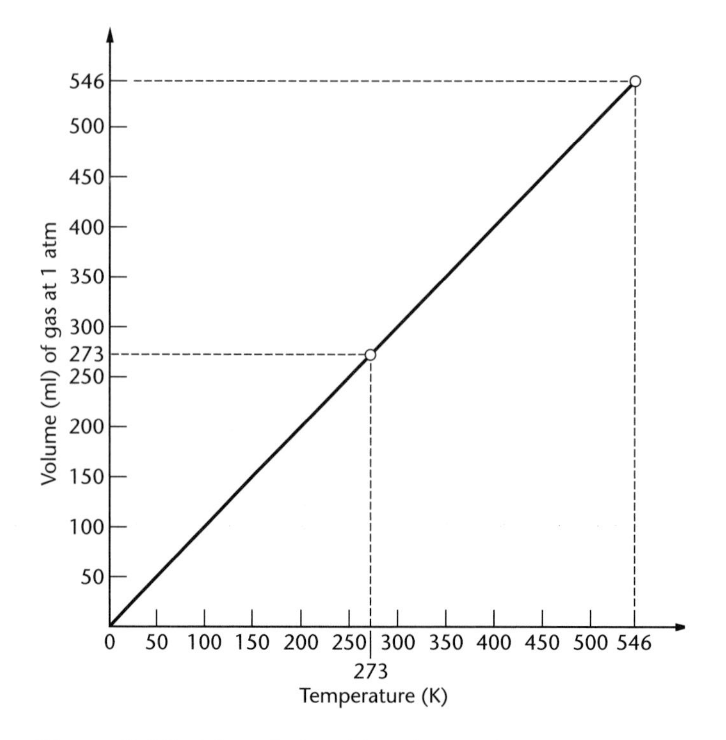

Figure 1-10 The relationship of gas volume to gas temperature

If the absolute temperature of a gas at constant pressure is doubled, the equation tells you that the volume is also doubled. The relationship between Kelvin temperature and volume can also be expressed as

$$\frac{V_1}{T_1} = \frac{V_2}{T_2}.$$

This form of the Charles' law equation is most useful in solving problems. The same relationship is expressed graphically in Figure 1-10, above. Notice that as T goes from 273 K to 546 K, V changes from 273 mL to 546 mL.

SAMPLE PROBLEM

PROBLEM

The temperature of 150 mL of a gas at constant pressure is increased from 20.°C to 40.°C. What is the new volume of the gas?

SOLUTION

Set up the temperature-volume relationships according to Charles' law. At constant pressure,

$$\frac{V_1}{T_1} = \frac{V_2}{T_2}$$

This can be rewritten as

$$\underset{\text{new volume}}{V_2} = \underset{\text{original volume}}{V_1} \times \underset{\text{correction factor}}{\frac{T_2}{T_1}}$$

Before substituting the given values recall that the temperature must be expressed in kelvins. Add 273 to the Celsius temperature

$$T_1 = 20.° + 273° = 293 \text{ K}$$

$$T_2 = 40.° + 273° = 313 \text{ K}$$

Substitute these and solve the equation.

$$V_2 = 150 \text{ mL} \times \frac{313 \text{ K}}{293 \text{ K}} = 160 \text{ mL}$$

Since the temperature of the gas was increased at constant pressure, the volume of the gas was also increased, according to Charles' law. Doubling the Celsius temperature of a gas at constant pressure does not double the volume.

PRACTICE

1.27 At a temperature of 27°C, 30. mL of a gas are collected. If the gas is heated, at constant pressure, to a temperature of 127°C, find the new volume of the gas.

1.28 A student collects 40. mL of a gas at a temperature of 364 K. The gas is cooled at constant pressure, until it reaches standard temperature. What is the new volume of the gas?

1.29 What would happen to the volume of a gas if both the Kelvin temperature and the pressure of the gas were doubled?

Gay-Lussac's Law

Gay-Lussac's law describes the relationship between the pressure and temperature of a gas when the volume is held constant. If a gas is heated in a sealed container, the particles move faster. They strike the walls of the container harder and more often, producing an increase in pressure. **Gay-Lussac's law** states that the pressure of a fixed mass of gas at constant volume varies directly with the Kelvin temperature. This can be expressed mathematically as

$$\frac{P}{T} = k \text{ (at constant volume)}$$

This is exactly the same kind of relationship as that between the temperature and the volume in Charles' law. Thus a graph of pressure versus Kelvin temperature will produce a straight line through the origin, like that shown in Figure 1-10 on page 34.

PRACTICE

1.30 A tank contains oxygen at STP. If the Kelvin temperature of the gas is doubled, what is the new pressure of the gas?

The Combined Gas Law

The three gas laws we have discussed can be combined into a single expression:

$$\frac{PV}{T} = k \quad \text{or} \quad \frac{P_1V_1}{T_1} = \frac{P_2V_2}{T_2}$$

In each of the three gas laws, one of the variables was held constant. A constant term is equal on both sides of the equation and drops out of the expression. For example, if the temperature is constant, then $T_1 = T_2$, and the expression, with the temperature omitted, is identical to the Boyle's law equation, $P_1V_1 = P_2V_2$. Similarly, if you keep the pressure constant, and omit the pressure terms from the combined gas law, the resulting equation is identical to the Charles' law equation.

The combined gas law allows us to solve problems in which the temperature, pressure and volume all change.

SAMPLE PROBLEM

PROBLEM

At a temperature of 27°C and a pressure of 50.65 kPa, 60.0 mL of a gas is collected. What would the volume of the gas be at STP?

SOLUTION

Recall that STP is defined as a temperature of 273 K and a pressure of 101.3 kPa. Set up the relationship according to the combined gas law,

$$\frac{P_1V_1}{T_1} = \frac{P_2V_2}{T_2}$$

Remember that the temperature must be in kelvins, so T_1 is 300. K, V_1 was 60.0 mL, and P_1 was 50.65 kPa. P_2 is standard pressure, 101.3 kPa, and T_2 is standard temperature, 273 K. Rearranging the equation to solve for V_2

$$V_2 = V_1 \times \frac{T_2}{T_1} \times \frac{P_1}{P_2}$$

Substituting the given values,

$$V_2 = 60.0\ \text{mL} \times \frac{273\ \text{K}}{300.\ \text{K}} \times \frac{50.65\ \text{kPa}}{101.3\ \text{kPa}}$$

$$V_2 = 27.3\ \text{mL}$$

The volume of the gas at STP would be 27.3 mL.

PRACTICE

1.31 At a temperature of 300. K and a pressure of 2.0 atm, a gas has a volume of 240 mL. If the temperature is increased to 400. K, and the new pressure is 4.0 atm, what is the new volume of the gas?

1.32 A gas has a volume of 25.0 L at a temperature of 27°C and a pressure of 1.00 atm. The gas is allowed to expand to a new volume of 50.0 L. The pressure is now 0.40 atm. What is the final temperature?

The Ideal Gas Law

Suppose you blow up a balloon to a volume of 5.0 liters. There are three different ways you could cause the volume of the air in the balloon to increase. You could decrease the pressure on the balloon. Boyle's law tells us that when the pressure decreases, the volume increases. You could heat the balloon, which Charles' law predicts will increase its volume. Or you could simply blow more air into the balloon. This introduces a new variable, the number of gas particles. Obviously, in the case of the balloon, increasing the number of gas molecules in the balloon will increase the volume of the balloon.

The ideal gas law includes the term "n," for the number of gas molecules. The equation becomes

$$\frac{PV}{nT} = R \qquad \text{or} \qquad PV = nRT$$

R is a constant, called the universal gas constant. Its value depends only on the choice of units for P, V, n, and T. Some values of R are listed on the table of physical constants in Appendix 4.

This *ideal gas law* applies to gases under conditions that allow molecules to be relatively far apart. If molecules are closer together, they have certain effects on one another because of the positive and negative particles (protons and electrons) in their atoms. These effects, as you know, are attractive forces. When molecules are far enough apart, the attractive forces among the molecules are small. Under these conditions, all gases, regardless of their composition, follow the ideal gas law.

PRACTICE

1.33 Explain the following observations about gases on the basis of the behavior of gas molecules. (You might want to refer back to the kinetic model described on pages 24 to 27)

(a) When additional air is pumped into a bicycle tire, the pressure of the air in the tire increases.

(b) When the air inside a bicycle tire gets warmer, the pressure of the air in the tire increases.

1.34 During the Olympic Ice Hockey competition the rink used was considerably larger than the rinks used in the National Hockey League. It was observed that there were far fewer collisions between the players and the boards. If we think of the players as resembling gas molecules, and the rink size represents the volume of the gas, which of the gas laws is analogous to our observation?

Avogadro's Hypothesis

One aspect of the relationships expressed by the ideal gas law was stated in 1811 by the Italian scientist Amedeo Avogadro. *Avogadro's hypothesis* states that equal volumes of all gases at the same temperature and pressure contain the same number of molecules. That is, if n_1 is the number of molecules in gas_1 and n_2 is the number of molecules in gas_2, $n_1 = n_2$. This can be verified by rewriting the ideal gas equation for each gas.

$$n_1 = \frac{P_1 V_1}{RT_1} \quad n_2 = \frac{P_2 V_2}{RT_2}$$

If $P_1 = P_2$, $V_1 = V_2$, and $T_1 = T_2$, then $n_1 = n_2$, no matter what the gases are.

Chemists use a unit called a **mole** to count the very small particles, such as atoms and molecules, with which they deal. A mole is the amount of any substance that contains 6.02×10^{23} particles (for example, atoms, molecules, or electrons). This number is known as **Avogadro's number.** (This number, like many numbers you will use in chemistry, is expressed in a system called *exponential notation*. If you are not familiar with this system, turn to Appendix 2.)

It can be shown that at 273 K and 1 atmosphere, 22.4 liters of gas contain 1 mole of molecules. This is true whether the molecules are light, like those of hydrogen, or heavier, like those of uranium hexafluoride.

Molar Gas Volume

When the ideal gas equation is solved for volume, $V = nRT/P$, it can be seen that the volume of one mole of a gas ($n = 1$) depends only on its temperature and pressure. At conditions of standard temperature and pressure (STP), the volume of 1 mole of any gas is 22.4 liters. The volume of 1 mole of a gas is called the **molar gas volume.** Since we know the molar gas volume at STP, 22.4 liters, we can use the combined gas law to find it under any other conditions of temperature and pressure. For example, at 1 atm and 25°C (298 K), the molar volume of any gas is 24.5 liters.

Diffusion of Gases

The continuous, random motion of the molecules of a gas causes the gas to diffuse in all directions. The rate at which gas molecules diffuse depends on the mass of the molecules and their temperature. Under the same conditions of temperature and pressure, light molecules diffuse more rapidly than heavy molecules. This relationship was investigated in 1833 by the Scottish scientist Thomas Graham.

SAMPLE PROBLEM

PROBLEM

A red balloon and a green balloon are both inflated to a volume of 10.0 L. The red balloon is inflated with helium gas. The green balloon is inflated with carbon dioxide gas.

Both balloons are at the same temperature and pressure. When released, the red balloon rises to the ceiling. The green balloon falls to the floor.

Compare the gases in the two balloons for each of the following variables.

(a) Number of molecules of gas in the balloon.

(b) Average kinetic energy of the gas in the balloon.

(c) Rate of diffusion of the gas in the balloon.

(d) Density of the gas in the balloon.

SOLUTION

(a) The number of molecules of gas is the same in both balloons. Avogadro's hypothesis states that equal volumes of all gases at the same temperature and pressure contain the same number of molecules.

(b) The average kinetic energies are the same in both balloons. Temperature is a measure of average kinetic energy. Since both gases are at the same temperature, they must both have the same average kinetic energies.

(c) The helium gas, in the red balloon, diffuses faster. Graham's law states that light molecules diffuse faster than heavy molecules. The gas in the balloon that floats is evidently lighter than the gas in the balloon that sinks.

(d) The gas in the red balloon, helium, since it floats in air must be less dense than air. The carbon dioxide gas is more dense than air, so the green balloon sinks. The gas in the green balloon must be denser than the gas in the red balloon. (In fact, carbon dioxide gas is 11 times more dense than helium gas.)

Dalton's Law of Partial Pressures

Suppose a metal container holds oxygen gas at a pressure of 3.0 atmospheres. Neon gas is then pumped into the container until the total pressure is 5.0 atmospheres. The gases do not react with each other, and the pressure exerted by each gas is independent of the other gas. Since the oxygen pressure was 3.0 atmospheres before the neon was added, it is still 3.0 atmospheres after the neon was added. If the total pressure has become 5.0 atmospheres, we would conclude that the pressure of the neon must be 2.0 atmospheres.

The pressure of each gas in a mixture of different gases is called the **partial pressure** of the gas. In the system described above, the partial pressure of the oxygen is 3.0 atmospheres, and the partial pressure of the neon is 2.0 atmospheres. The total pressure, 5.0 atmospheres, is the sum of the *partial* pressures. The English scientist John Dalton summarized these observations in 1805. **Dalton's law of partial pressures** states: At constant temperature, the pressure of a mixture of gases, which do not react, equals the sum of the partial pressures of the gases in the mixture. Expressed mathematically,

$$P = P_1 + P_2 + P_3 + \ldots$$

where P = total pressure and $P_1, P_2, P_3, \ldots$ = partial pressures.

A metal tank contains oxygen, at a partial pressure of 3.0 atmospheres, and neon at a partial pressure of 2.0 atmospheres. Since both gases are in the same tank, they must have the same volume, and we can also assume they are at the same temperature. The difference in pressure can only be due to a difference in the number of molecules. Since oxygen exerts 3/5 of the pressure, 3/5 of the molecules must be oxygen, and the other 2/5 must be neon. Dalton's Law includes a second equation, which shows the relationship between the number of molecules, or moles of each gas, and its partial pressure. If a container contains 2 gases, and we call them gas "A" and gas "B", then

$$\frac{P_A}{P_B} = \frac{moles\ A}{moles\ B}$$

In the tank described above, suppose we find that there are 9.0 moles of oxygen gas. How many moles of neon are there? The equation above becomes $3.0/2.0 = 9.0/x$, where x = moles of neon. Solving the equation, we find that there are 6.0 moles of neon.

SAMPLE PROBLEM

PROBLEM

A 10-liter tank contains nitrogen gas at a pressure of 1.0 atm. Oxygen is pumped into the tank until the total pressure in the tank is 4.0 atm. What is the partial pressure of the oxygen gas in the tank? The gases do not react, and the temperature is constant.

SOLUTION
Dalton's law of partial pressures states that the pressure of a mixture of gases equals the sum of the partial pressures of the gases in the mixture. In this case, then,

$$P_{\text{(total)}} = P_{\text{(nitrogen)}} + P_{\text{(oxygen)}}$$

Substituting the given values and solving the equation:

$$4.0 \text{ atm} = 1.0 \text{ atm} + P_{\text{(oxygen)}}$$

$$P_{\text{(oxygen)}} = 3.0 \text{ atm}$$

The partial pressure of the oxygen is 3.0 atm.

PRACTICE

1.35 A tank of gas with a total pressure of 12.0 atm contains a mixture of oxygen, nitrogen, and argon. If the partial pressure of the nitrogen is 10.0 atm, and the partial pressure of the oxygen is 1.5 atm, what is the partial pressure of the argon gas?

1.36 In problem 1.35, above, if the tank is found to contain 3.0 moles of argon gas, how many moles of oxygen gas must there be in the tank?

Deviation from the Gas Laws

Recall that the ideal gas law predicts the behavior of gases correctly only when there are relatively large distances between gas molecules. This is true of all the gas laws. When gas molecules are close together, deviations from the gas laws occur. For example, because of attractions between molecules that are close together, volume decreases more with increased pressure than Boyle's law predicts.

The strengths of attractions between molecules vary from one gas to another. Generally speaking, larger molecules have stronger attractions for each other than do smaller molecules. Since attractions between molecules cause deviations from the gas laws, the weaker the attractions, the more ideal the gas.

At very high temperatures, the gas molecules are moving extremely fast, and the attractions between them are very weak, too weak to cause any deviations from ideal gas behavior. However, there is another possible source of deviations from ideal behavior.

One of our assumptions about gases was that the total space between the molecules is very large compared with the space occupied by the molecules themselves. We assume, for example, that a 10-liter container of a gas contains 10 liters of available space. The space taken up by the gas particles themselves is so small compared with 10 liters that we ignore it. At very high pressures, however, there is such a high concentration of gas in the container that the particles take up a significant portion of the space. How much space would depend on the size of the molecules and the number of molecules. At very high pressures, the particle size can no longer be ignored. Since the actual amount of free space becomes significantly less than the volume of the container, the pressures become greater than those predicted by the gas laws.

The major causes of deviations from the gas laws are attractions between molecules and the space occupied by the molecules. These are most significant at low temperatures, high pressures, or both. Hence gases deviate most from the gas laws at low temperatures, high pressures, or both. Gases are most ideal at high temperatures and low pressures. Gases such as hydrogen, oxygen, nitrogen, and carbon dioxide show little or no deviation from the gas laws at standard pressure and temperature. Under those conditions, these common gases may be considered ideal gases.

Liquids

When the attractive forces between molecules are sufficiently strong, the attractions cause a substance to be a liquid at ordinary temperatures and pressures. For example, the attractive forces between oxygen molecules are relatively weak, and so oxygen is a gas at 25°C and 1 atm. The attractive forces between water molecules, however, are strong enough to cause this compound to be a liquid at 25°C and 1 atm. Molecules in a liquid do not have the same freedom of motion as do molecules in a gas. As a result, the properties of liquids are different from the properties of gases.

Properties of Liquids

The general properties of liquids may be summarized as follows:

1. Liquids have clearly defined boundaries and take the shape of their containers. This means that liquids have a definite volume. Unlike a gas, a given volume of a liquid cannot fill every container.
2. Liquids, like gases, are able to flow. Both liquids and gases are *fluids.*
3. Increasing the pressure on a liquid has practically no effect on its volume. Liquids are virtually incompressible.
4. Increasing the temperature of a liquid produces only small increases in its volume. All liquids do not expand at the same rate for the same increase in temperature. Mercury is an exception. Mercury expands uniformly on heating and contracts uniformly on cooling. This property of mercury explains its use in thermometers.
5. Liquids resist flow in varying degrees, a property called *viscosity.* This property depends on the sizes and shapes of the molecules and on the attractive forces between them. Where the attractive forces are small and the molecules are small, the viscosity is low. Liquids with low viscosity flow easily, as in the case of alcohol and ether. Where the attractive forces are large and the molecules are large and complex, the viscosity is high. Liquids with high viscosity flow with difficulty, as in the cases of glycerine and heavy oils. As the temperature rises, the increasing kinetic energy of the liquid molecules weakens the attractive forces. Resistance to flow therefore decreases with increasing temperature in most liquids.
6. The surface of a liquid behaves like a stretched membrane on which objects denser than the liquid may be supported. This property is called *surface tension.* It results from the unequal distribution of attractive forces at the surface of the liquid (see Figure 1-11 on page 46). The surface molecules are

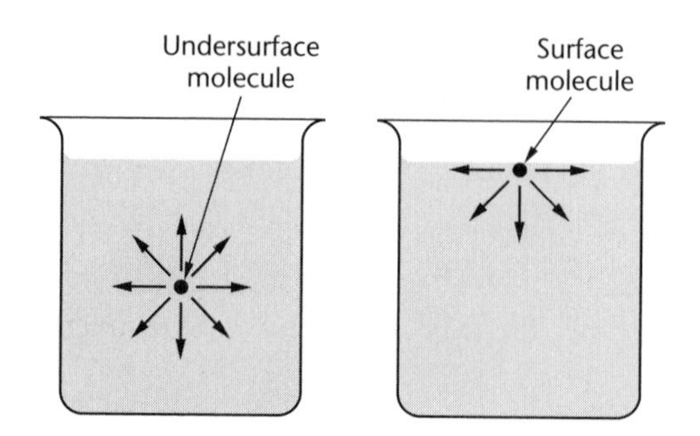

Figure 1-11 Surface tension

attracted by other molecules of the liquid in all directions except from the top. The net effect of this uneven distribution of forces is to pull the surface molecules together.

Surface tension is the reason liquid drops have a spherical shape and why some insects can walk on the surface of water. The surface tension of water can be decreased through the addition of soap to the water. With increasing temperature, the attractive forces between the molecules of a liquid tend to weaken. Surface tension thus decreases with increasing temperature.

7. Liquids evaporate spontaneously—that is, they change into vapor at any temperature.

Evaporation and Vapor Pressure

Recall that not all the molecules of a gas at any given temperature have the same kinetic energy. This is true of the molecules of a liquid also.

Those molecules of a liquid that have higher kinetic energies move more rapidly than the molecules with lower kinetic energies. The molecules with higher kinetic energies may escape in the form of vapor from the surface of the liquid. This phenomenon, called *evaporation,* or *vaporization,* may occur at any temperature. As the temperature increases, more molecules gain greater kinetic energy and more of them escape to form vapor. Because the molecules with higher kinetic energies escape during evaporation, the remaining liquid has molecules with lower kinetic energies. The temperature of a liquid therefore drops during evaporation; evaporation is an endothermic process.

When a liquid is placed in a closed container, gas molecules escape from the surface of the liquid and are found in the space above it. This vapor exerts pressure against the walls of the container. The pressure is called **vapor pressure.** Eventually, the rate at which molecules change from liquid to vapor and escape (evaporation) equals the rate at which molecules change from vapor to liquid (a process called *condensation*) and return. A condition known as *liquid-vapor equilibrium,* or *vapor pressure equilibrium,* results (see Figure 1-12).

The vapor pressure of a liquid depends on forces between molecules within the liquid and the average kinetic energy, or temperature, of the liquid. Stronger forces between molecules result in the liquid having a lower vapor pressure. Increasing the temperature raises the vapor pressure. The table on the next page shows the vapor pressure of water at various temperatures.

When the vapor pressure of a liquid equals the outside pressure pushing down on the liquid, the liquid boils. As shown in the table, water boils at 100°C when the outside, or atmospheric, pressure equals 101.3 kilopascals (1 atm). If the atmospheric pressure were only 2.33 kPa, water would boil at a temperature of 20°C. At a pressure higher than 1 atm, water boils at a temperature higher than 100°C, a principle utilized in pressure cookers. The *boiling-point temperature* of a liquid is that temperature at

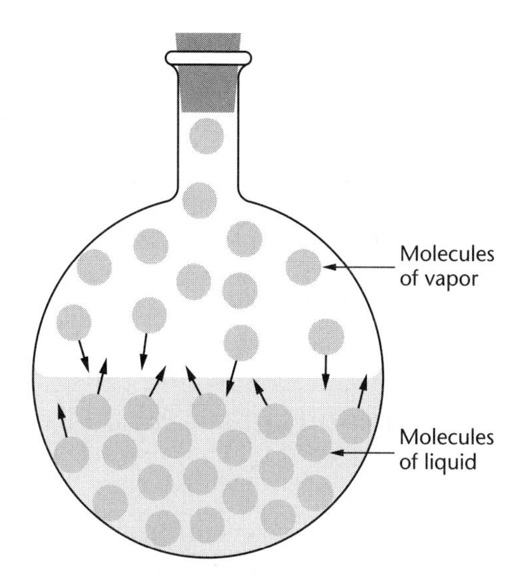

Figure 1-12 Liquid-vapor equilibrium

Temperature (°C)	Vapor Pressure (kilopascals)	Temperature (°C)	Vapor Pressure (kilopascals)
0	0.61	50	12.3
5	0.87	60	19.8
10	1.23	70	31.1
20	2.33	80	47.3
30	4.20	90	70.0
40	7.32	100	101.3

which its vapor pressure equals the external, or atmospheric, pressure acting on the liquid.

Vapor Pressure Curves

The rate of evaporation, boiling point, and vapor pressure of liquids depend upon the strength of the attractive forces between the molecules of the substances. These attractive forces are called intermolecular attractions.

Liquids with weak intermolecular attractions will evaporate quickly. They will have high vapor pressures and low boiling points. You can study these relationships by looking at the vapor pressure curves of a few different liquids (Figure 1-13).

You can see that for all four liquids the vapor pressure increases as the temperature increases. Recall that a liquid will boil when its vapor pressure equals the prevailing atmospheric pressure. Since standard atmospheric pressure is 101.3 kilopascals (kPa), the temperature at which the vapor pressure of a liquid reaches 101.3 kPa is called its **normal boiling point.** Using Figure 1-13, you can find the normal boiling points of the four liquids listed. You can see, for example, that the vapor pressure of acetone reaches 101.3 kPa at a temperature of about 56°C. Therefore, 56°C is the normal boiling point of acetone. Use the vapor pressure curve of water to find its normal boiling point. The answer, as you already know, is 100°C.

Using the vapor pressure curves, you can also find the boiling points of the liquids listed at pressures other than 101.3 kPa. For example, you can see that at a pressure of 70. kPa, acetone would boil at a temperature of 45°C.

The attractive forces of these liquids can be compared as well. The liquid with the strongest intermolecular attractions should have the lowest vapor pressure and the highest boiling point. The

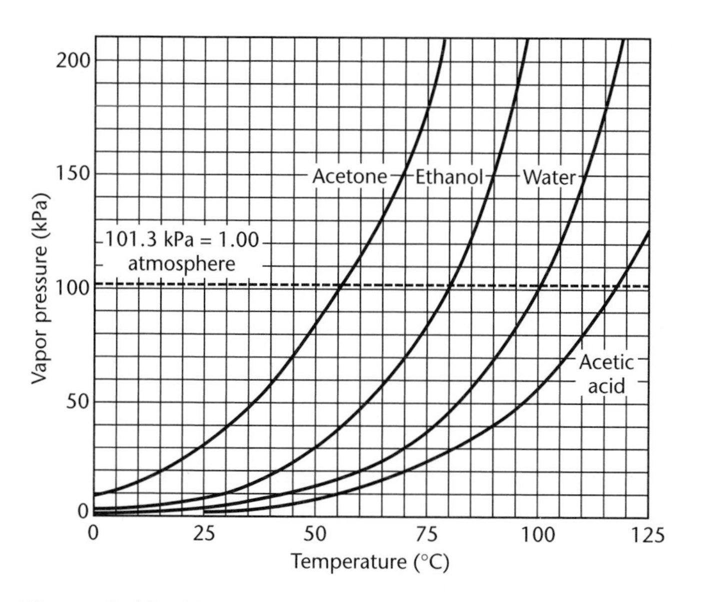

Figure 1-13 Vapor pressure curves for four liquids

strong attractions tend to prevent evaporation and keep the substance in the liquid state. Acetic acid has the strongest intermolecular attractions of the liquids shown.

PRACTICE

Base your answers on Figure 1-13.

1.37 At room temperature, which of the four liquids evaporates most rapidly?

1.38 What is the normal boiling point of acetic acid?

1.39 What is the external pressure, if ethanol is boiling at a temperature of 70°C?

1.40 Which of these liquids has the WEAKEST intermolecular attractions?

1.41 How would you expect the surface tension in acetic acid to compare with that of water? Explain your prediction.

1.42 Acetone is often used in nail polish removers. Its users eventually notice that when spilled on the hands, acetone dries very quickly, and feels quite cold. Explain these observations on the basis of the intermolecular attractions in acetone.

Heat of Vaporization

When heat is applied to a liquid that is at its boiling temperature, the potential energy of the molecules increases. The average kinetic energy of the molecules (temperature) remains constant. At its boiling point, the heat energy required to change a fixed amount of liquid to gas is known as the *heat of vaporization* of the liquid. Heat of vaporization is often expressed in joules per gram. The heat of vaporization of water at its boiling point is 2260 J/g. Heat of vaporization may also be expressed in kilojoules per mole.

Freezing, or Solidification

The average kinetic energy of the molecules of a gas is sufficiently great to overcome the attractive forces between the molecules. As a gas is cooled, the average kinetic energy of its molecules is decreased. At some point, the average kinetic energy is no longer sufficient to overcome the attractive forces between the molecules. As a result, the gas becomes a liquid—condensation occurs. Continued cooling of the liquid further decreases the kinetic energy of the molecules. Eventually, the liquid becomes a solid—freezing, or solidification, takes place.

At atmospheric pressure, some substances change directly from solid to gas without going through the liquid phase. This direct change from solid to gas is known as **sublimation.** A common substance that behaves in this way at or near atmospheric pressure is carbon dioxide. At room temperature and pressure, gaseous carbon dioxide sublimes from a piece of dry ice, which is solid carbon dioxide. At much higher pressures (more than five times atmospheric pressure), carbon dioxide follows the more usual sequence of changes in state—that is, from solid to liquid to gas.

Solids

How do solids differ from liquids and gases? The kinetic energy of molecules is greatest in the gaseous state. Since the molecules are widely spaced, the potential energy of gaseous molecules is also greatest. This is true because work was done and energy was consumed in order to separate the molecules.

In liquids, the kinetic energy of molecules is lower because heat is released when the liquid is formed from the gas. A decrease in kinetic energy also results in a decrease in potential energy because the molecules have come closer to one another.

Solids have the least potential energy because energy has been released in all the changes from the gaseous to the solid state. In a solid, the particles are arranged in the least random (most ordered) fashion.

Attractive Forces in Solids

A solid forms when the average kinetic energy of the particles of a substance is low enough that particles are held in fixed positions. The particles in a solid vibrate but do not move significantly from their positions. Forces of attraction between particles in solids vary from very weak to very strong.

The average kinetic energy, or temperature, at which forces between particles are just strong enough to hold the particles in a relatively fixed position is known as the *melting-point temperature.* If heat is added to a solid at its melting-point temperature, the solid will change to a liquid. If heat is removed from a liquid at the melting-point temperature, the liquid will change to a solid. This is the same as the *freezing-point temperature.*

The melting points of solids vary greatly, in accordance with variations in attractive forces. The melting points of solids are therefore a fair indication of the strength of the attractive forces within them. The table below shows the melting points of a few solids.

Solid	***Melting Point (°C)***
Oxygen	−218.4
Water	0
Potassium	62.3
Sodium chloride	801
Iron	1535

Clearly, the forces between molecules in solid oxygen are weak. The forces between particles in sodium chloride are strong, and those between particles in iron are stronger still.

Solids can be classified according to the kinds of attractive forces within them. The different kinds of solids and the forces within them will be considered in Chapter 3.

Fusion and Crystallization

The change of a liquid to a solid is called solidification, or freezing. The opposite change, from a solid to a liquid, is called *melting,* or *fusion.* Both of these changes, as you have learned, occur at the melting-point temperature. For water, solidification and fusion occur at 0°C at sea level (1 atm).

The heat energy required to change a fixed amount of a solid to a liquid at the melting-point temperature is called the *heat of fusion.* For water, the heat of fusion is 334 joules per gram. Heat of fusion may also be expressed in kilojoules per mole.

In the reverse process, heat energy is released when a liquid changes to a solid. This energy is usually called the *heat of crystallization.*

Properties of Solids

As with the properties of gases and liquids, the properties of solids are related to their internal structures. The general properties of solids may be summarized as follows:

1. Solids have a specific volume (unlike gases) and a definite shape (unlike gases and liquids).
2. Solids do not flow. Instead, they are rigid bodies that retain their shape unless subjected to a considerable force.
3. Solids are considerably denser than gases but in most cases are only slightly denser than liquids.
4. Increasing the pressure on a solid has very little effect on its volume. Thus solids are virtually incompressible.

PRACTICE

In solving these problems, recall that the heat of fusion of water is 334 J/g, the heat of vaporization is 2260 J/g, and the specific heat of water is 4.18 J/g·K.

1.43 How much heat is needed to melt 5.00 grams of ice at its melting point?

1.44 How much heat is needed to boil 10.0 grams of water at its boiling point?

1.45 How much heat is released when 5.00 grams of water freezes?

1.46 What is the total amount of heat required to melt 2.00 grams of ice, heat it to its normal boiling point, and then boil the entire sample?

TAKING A CLOSER LOOK

The Phase Diagram

The phase at which a particular substance exists depends on temperature and external pressure. For example, at standard pressure water boils at 100°C. At that pressure and temperatures below 100°C, water is a liquid, while at that pressure and at temperatures above 100°C water is a gas. When the external pressure changes, the boiling point changes. Using the vapor pressure table on page 48, you can see that if the external pressure is 2.33 kPa, water will boil at a temperature of 20°C. At that pressure, water will be a gas at temperatures above 20°C and a liquid at temperatures below 20°C. (Of course, the water will still freeze if the temperature is low enough.)

The **phase diagram** is a graph that shows the phases present at various temperatures and pressures. Figure 1-14 shows the

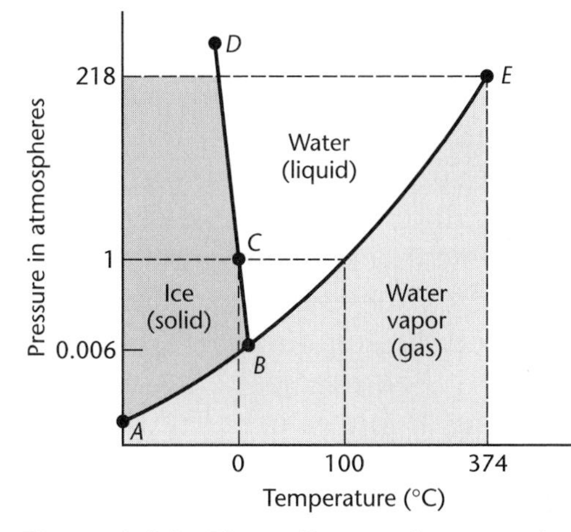

Figure 1-14 Phase diagram for water (not drawn to scale)

phase diagram of water. (Note that the phase diagram is not drawn to scale; in order to show the boundaries between the phases clearly, neither axis shows a regular interval)

On the phase diagram, line B-E marks the boundary between the liquid and gas phases. Any point on that line represents a boiling point. You can see, for example, that at a pressure of 1 atm, the boiling point is 100°C. As the pressure increases, the boiling point increases, up to a boiling point of 374°C at a pressure of 218 atm. Here, however, the curve stops. At temperatures above 374°C, water cannot be turned into a liquid no matter how much pressure is applied. Above this temperature, water exists in the gas phase only.

The **critical temperature** of a substance is defined as the temperature above which that substance cannot exist in the liquid phase; 374°C is the critical temperature of water. At the critical temperature, the attractive forces between the molecules have become too weak to form a liquid. The heat of vaporization, which is 2260 J/g for water at 100°C, drops to zero at the critical temperature. The surface tension at the critical temperature is also zero.

As the pressure decreases, the boiling point decreases. Eventually, the boiling point line intersects the melting point line (B-D) at point B. At a temperature of 0°C, the vapor pressure of water is 0.006 atm (0.61 kPa). If the external pressure is lowered to 0.006 atm, water will boil at 0°C. Under these conditions, we would actually see ice water boiling. All three phases would be present at the same time. These are the conditions at point B on our phase diagram. Point B is called the **triple point.** The triple point is defined as the conditions of temperature and pressure at which a given substance can exist in all three phases at the same time.

At pressures below the pressure at the triple point, there is only one boundary line present, line A-B. This line separates the solid and gas phases. Thus at these low pressures, no liquid phase exists. The solid turns directly into the gas by sublimation.

At standard pressure, the solid-liquid boundary line, line B-D, crosses the 1 atm line at point C. You can see that the temperature is 0°C, the normal melting point of ice. The line slopes slightly to the left (negatively). As pressure increases, the melting point decreases slightly. This is true for water but for very few other substances. Water is unusual in that it expands when it freezes. Ice floats on water because its greater volume gives it a lower density than that of water. An increase in pressure favors a decrease in volume. In the case of water, an increase in pressure favors the production of the liquid, because it has a smaller volume than the solid. Ice skates exert enough pressure on the ice to cause the ice beneath the blades to melt.

In most substances, the molecules are closer together in the solid state than in the liquid state. The solid has a smaller volume than the liquid and will sink in the liquid. If you melt wax in a test tube, you will see that the solid wax falls to the bottom of the test tube. For most substances, the solid-liquid line in the phase diagram will slope to the right (positively).

Figure 1-15 shows the phase diagram for carbon dioxide. Note that at 1 atm, only the solid and gas phases exist. Solid carbon dioxide (dry ice) will sublime under ordinary conditions of temperature and pressure. Liquid carbon dioxide can exist only if the pressure is raised above 5.1 atm, the pressure at the triple point. Note also how the solid-liquid line slopes to the right (positively). In this respect, carbon dioxide resembles the vast majority of known substances.

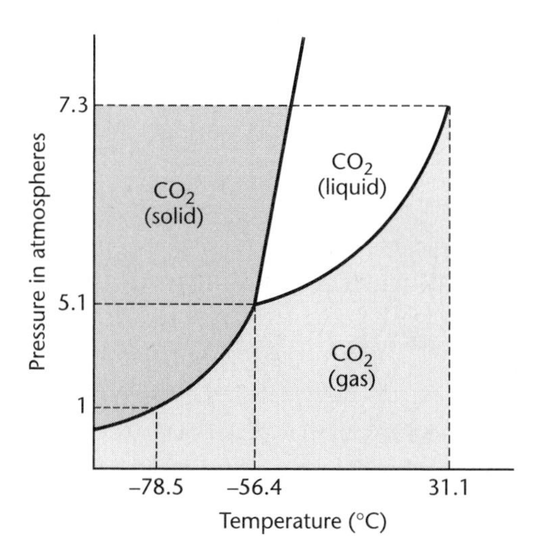

Figure 1-15 Phase diagram for carbon dioxide

The Ideal Gas Law

Recall that the ideal gas equation is

$$PV = nRT$$

In this equation, n is the number of molecules. We usually express the number of molecules in moles. A mole is a certain number of objects, just the way a dozen is a certain number of objects. A mole is 6.02×10^{23} particles.

You know the volume of a mole of gas at STP is 22.4 liters. You can use this information to find the value of R, the universal gas constant. Rearranging the equation to solve for R you get

$$R = \frac{PV}{nT}$$

When you express standard pressure as 1 atm, standard temperature as 273 K, and n as 1 mole, the equation becomes:

$$R = \frac{1\ \text{atm}(22.4\ \text{L})}{1\ \text{mole}(273\ \text{K})}$$

Solving this equation gives us $R = 0.082\ \text{L} \cdot \text{atm/mole} \cdot \text{K}$. You can use this equation to find the volume of any amount of an ideal gas under any conditions of temperature and pressure.

SAMPLE PROBLEM

PROBLEM

What is the volume of 2.0 moles of oxygen at 25°C and a pressure of 1.5 atm?

SOLUTION

Rearranging the ideal gas equation to solve for volume gives

$$V = nRT/P$$

Substituting $n = 2.0$ moles, $R = 0.082\ \text{L} \cdot \text{atm/mole} \cdot \text{K}$, $T = 298$ K and $P = 1.5$ atm, gives you

$$V = \frac{2.0\ \text{moles} \times 0.082\ \text{L} \cdot \text{atm}/\text{mole} \cdot \text{K} \times 298\ \text{K}}{1.5\ \text{atm}}$$

$$V = 33\ \text{L}\ (32.6)$$

PRACTICE

1.47 How many moles of helium gas are there in a 10.0 L balloon, inflated to a pressure of 1.00 atm at a temperature of 300. K?

1.48 The number of moles of gas in a container is doubled, the temperature is doubled, and the volume is made four times greater (quadrupled). What has happened to the pressure of the gas in the container?

CHAPTER REVIEW

The following questions will help you check your understanding of the material presented in the chapter.
Write the number of the word or expression that best completes the statement or answers the question.

1. The average kinetic energies of the molecules in two gas samples can best be compared by measuring their (1) temperatures (2) volumes (3) pressures (4) densities.
2. At a pressure of one atmosphere, the temperature of a mixture of ice and water is (1) 25 K (2) 32 K (3) 273 K (4) 298 K.
3. A gas sample occupies 10.0 mL at 1.0 atm of pressure. If the volume of the gas changes to 20.0 mL and the temperature remains the same, the pressure will be (1) 0.25 atm (2) 0.50 atm (3) 1.0 atm (4) 2.0 atm.
4. Under which conditions does the behavior of a gas deviate most from the ideal gas behavior? (1) high temperature and high pressure (2) high temperature and low pressure (3) low temperature and high pressure (4) low temperature and low pressure
5. As the pressure on an enclosed liquid is decreased, the boiling point of the liquid (1) decreases (2) increases (3) remains the same.

6. Which is an exothermic, physical change? (1) paper burns (2) ice melts (3) alcohol freezes (4) acetone boils
7. At constant pressure, the volume of a mole of any ideal gas varies (1) directly with the radius of the gas molecules (2) inversely with the radius of the gas molecules (3) directly with the Kelvin (absolute) temperature (4) inversely with the Kelvin (absolute) temperature.
8. When 5 g of a substance are burned in a calorimeter, 3 kilojoules of energy are released. The energy released per gram of substance is (1) 600 joules (2) 1700 joules (3) 5000 joules (4) 15,000 joules
9. Which will be found in a closed vessel partially filled with pure water at 25°C and 1.0 atm? (1) water vapor only (2) liquid water only (3) water vapor and liquid water (4) liquid water and solid water (ice)
10. 83.6 joules of heat are added to 2 g of water at a temperature of 10°C, the resulting temperature of the water will be (1) 5°C (2) 20°C (3) 30°C (4) 40°C.
11. A liquid that evaporates rapidly at room temperature most likely has a high (1) vapor pressure (2) boiling point (3) melting point (4) attraction between molecules.

Base your answers to questions 12 through 16 on the graph below, which shows the relationship between the vapor pressure and the temperature of four liquids.

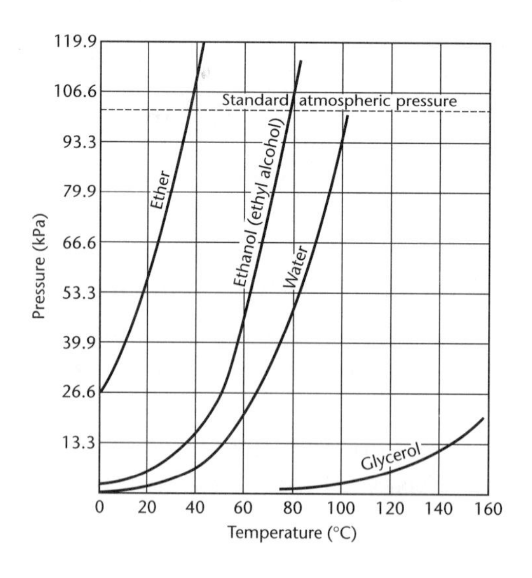

12. The vapor pressure of which liquid will increase the most if its temperature is changed from 20°C to 40°C? (1) ether (2) ethanol (3) water (4) glycerol

13. When the pressure exerted on the surface of glycerol equals that of 13.3 kiloPascales, its boiling point will be (1) 25°C (2) 140°C (3) 145°C (4) 160°C.

14. The liquid whose molecules exert the strongest attractive force on one another is (1) glycerol (2) alcohol (3) water (4) ether.

15. Water boils at a temperature of 80°C when the pressure on its surface equals (1) 39.9 kPa (2) 49.3 kPa (3) 53.3 kPa (4) 101.3 kPa.

16. The normal boiling point of ether is (1) 38°C (2) 42°C (3) 100°C (4) 760°C.

17. When you splash room-temperature aftershave on your face, you notice that it feels colder than would room-temperature water. You could logically conclude that (1) aftershave has a higher boiling point than water (2) aftershave has a higher vapor pressure than water (3) aftershave has stronger intermolecular attractions than water (4) aftershave has a slower rate of evaporation than water

18. A liquid boils at the lowest temperature at a pressure of (1) 1 atm (2) 2 atm (3) 50 kPa (4) 101.3 kPa

19. A liquid's freezing point is −38°C and its boiling point is 357°C. What is the number of Kelvin degrees between the boiling point and the freezing point of the liquid? (1) 319 (2) 395 (3) 592 (4) 668

20. At constant pressure, the volume of a gas will increase when its temperature is changed from 10°C to (1) 263 K (2) 273 K (3) 283 K (4) 293 K.

21. The temperature of a sample of water in the liquid phase is changed from 15°C to 20°C by the addition of 2100 J. What is the mass of the water? (1) 10 g (2) 50 g (3) 100 g (4) 5000 g

22. How many joules are required to change 100.0 g of ice at 0°C to water at 0°C? (Heat of fusion of water = 334 J/g) (1) 3.34 J (2) 3.34 kJ (3) 33.4 KJ (4) 33.4 J

23. Oxygen and helium gas are mixed to a total pressure of 12 atm. If the partial pressure of the oxygen is 4.0 atm, what is

the partial pressure of the helium? (1) 3.0 atm (2) 4.0 atm (3) 8.0 atm (4) 48 atm.

24. Sixty milliliters of a gas are collected at 60°C. If the temperature is reduced to 30°C and the pressure remains constant, the new volume of the gas will be equal to

(1) $60 \text{ ml} \times \frac{30}{60}$ (3) $60 \text{ ml} \times \frac{303}{333}$

(2) $60 \text{ ml} \times \frac{60}{30}$ (4) $60 \text{ ml} \times \frac{333}{303}$

25. If 12 L of a gas are collected at 300. K and a pressure of 2.0 atm, and the temperature is increased to 600. K while the pressure is decreased to 1.0 atm, what is the new volume of gas? (1) 6 L (2) 12 L (3) 24 L (4) 48 L

26. Which is a chemical property of water? (1) It freezes. (2) It decomposes. (3) It evaporates. (4) It boils.

27. Which is the most common property of all liquids? (1) definite shape (2) definite volume (3) high compressibility (4) high vapor pressure

28. As the average kinetic energy of the molecules of a sample increases, the temperature of the sample (1) decreases (2) increases (3) remains the same.

29. At constant pressure, which curve best shows the relationship between the volume of an ideal gas and its absolute temperature? (1) A (2) B (3) C (4) D

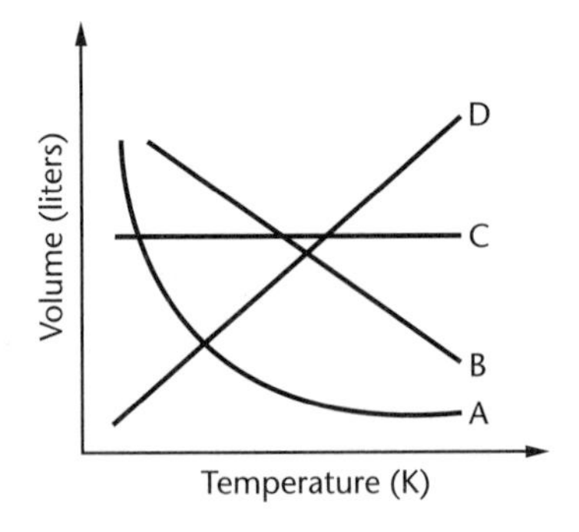

30. Which temperature equals +20 K? (1) −253°C (2) −293°C (3) +253°C (4) +293°C

31. A sample of hydrogen has a volume of 100. mL at a pressure of 50.0 kPa. If the temperature is constant and the pressure is raised to 100. kPa, the volume will be (1) 25.0 mL. (2) 50.0 mL (3) 100. mL (4) 200. mL

32. As the temperature of a liquid increases, the vapor pressure of the liquid (1) decreases (2) increases (3) remains the same.

33. As the temperature of a gas increases, at constant volume, the pressure exerted by the gas (1) decreases (2) increases (3) remains the same.

34. As the volume of a gas increases, at constant temperature, the pressure exerted by the gas (1) decreases (2) increases (3) remains the same.

35. At the same temperature and pressure, 5.0 L of nitrogen gas contains the same number of molecules as (1) 22.4 L of oxygen gas (2) 5.0 L of carbon dioxide gas (3) 10.0 L of hydrogen gas (4) 15.0 L of argon gas.

36. As water boils at constant pressure, the average kinetic energy of the water molecules (1) decreases (2) increases (3) remains the same.

37. Which is an example of an endothermic physical change? (1) A piece of magnesium burns in air. (2) Water vapor condenses. (3) Ice melts. (4) Gaseous carbon dioxide becomes solid dry ice.

38. When a piece of cotton is dipped into acetone, and the cotton is rubbed against a chalk board, the streak of acetone dries very quickly, much more quickly than a streak of water at the same temperature. From this you can conclude that acetone (1) has weaker intermolecular attractions than water (2) evaporates more slowly than water (3) has a lower vapor pressure than water (4) has a higher boiling point than water.

39–40. Base your answers on the information given about the following four gases:

Gas	***Hydrogen***	***Chlorine***	***Sulfur dioxide***	***Ammonia***
Normal boiling point	20 K	238 K	263 K	240 K

39. Which of these gases has the strongest intermolecular attractions? (1) hydrogen (2) chlorine (3) sulfur dioxide (4) ammonia

40. Which gas would behave most like an ideal gas at standard temperature and pressure? (1) hydrogen (2) chlorine (3) sulfur dioxide (4) ammonia

CONSTRUCTED RESPONSE

1. A 10.0 liter metal tank contains oxygen gas at a pressure of 12.0 atm. Half of the oxygen gas is allowed to escape from the tank. What is the volume of the oxygen gas remaining in the tank? What is the pressure of the oxygen gas remaining in the tank? Explain your answers.
2. Mercury liquid has an extremely low vapor pressure. Based on this information, make predictions about its forces of attraction, rate of evaporation, and boiling point.
3. (a) A student performs a chemical reaction in a calorimeter containing 200. g of water. The temperature of the water goes up from 30.°C to 80.°C. How much heat did the reaction produce?

   (b) A second student performs the same chemical reaction, using the same amount of chemical, but only 100 g of water, also initially at 30°C. Assuming that the reaction produces the same amount of heat as in (a) above, what should be the temperature change of the water? In fact, the temperature goes up only 70°C. Why?

CHEMISTRY CHALLENGE

The following questions will provide practice in answering SAT II-type questions.

For each question in this section, one or more of the responses given is correct. Decide which of the responses is (are) correct.

Then choose

(a) if only I is correct;
(b) if only II is correct;
(c) if only I and II are correct;
(d) if only II and III are correct;
(e) if I, II, and III are correct.

Summary: Choice	a	b	c	d	e
True statements:	I only	II only	I & II	II & III	I, II, & III

1. Samples of helium gas and carbon dioxide gas, at the same temperature, pressure, and volume, also have the same
I. number of molecules.
II. average kinetic energy.
III. number of atoms.

2. Conditions at standard temperature and pressure include
I. $T = 273$ K.
II. $P = 760$ torr.
III. $P = 1$ atm.

3. The volume of a gas varies
I. directly with the Celsius temperature.
II. directly with the number of molecules.
III. inversely with the pressure.

4. A pure, solid substance is heated at a steady rate, from an initial temperature of 20°C. The temperature goes up steadily until it reaches 52°C. It remains at 52°C for several minutes. The temperature then increases steadily to 170°C, where it once again remains constant for several minutes. Correct conclusions from these data include
I. The substance melts at 52°C.
II. The substance boils at 170°C.
III. At a temperature of 110°C the substance is a liquid.

5. Correct statements about ideal gases include
I. The molecules do not collide with each other.
II. There are negligible attractions between the molecules.
III. The size of the molecules is insignificant compared with the amount of space between them.

6. Phases that may exist at pressures below the triple point pressure are
I. solid. II. gas. III. liquid.

7. A chemist would consider salt water
   I. a substance.
   II. a mixture.
   III. homogeneous.
8. Endothermic physical changes include
   I. melting. II. boiling. III. sublimation.
9. A liquid with relatively weak intermolecular attractions would have
   I. a relatively high vapor pressure.
   II. a relatively high boiling point.
   III. a relatively low rate of evaporation.
10. An increase in the temperature of a gas (at constant pressure) will change its
    I. mass. II. density. III. volume.
11. The formation of chemical bonds is associated with
    I. absorption of heat.
    II. decreased potential energy.
    III. endothermic change.
12. An increase in the temperature of a sample of liquid mercury will cause an increase in its
    I. surface tension.
    II. vapor pressure.
    III. volume.

2 Atomic Structure

LOOKING AHEAD

This chapter deals with the structure of the atom, particularly with the subatomic particle called the electron. Today's model of the atom—the quantum mechanical model—evolved from the work of many scientists and provides a basis for understanding the behavior of elements and the nature of bonds between atoms.

When you have completed this chapter, you should be able to:

- **Describe** the general structure of the atom.
- **Define** nucleon, atomic number, atomic mass, isotope, gram-atomic mass, mass number, orbital, sublevel, ionization energy, electron affinity, and electronegativity.
- **Explain** some of the early significant experiments that led to our present understanding of the atom.
- **Relate** the experimental results from atomic emission spectroscopy to Bohr's model for electrons in atoms.
- **State** two key differences between the Bohr model and the quantum mechanical model.
- **Diagram** the electron configuration of an atom.
- **Determine** the number of valence electrons of an atom on the basis of electron configuration.

Early Models of the Atom

The idea that matter is composed of tiny, indivisible particles was proposed by Democritus, a Greek philosopher, about 2500 years ago. In fact, the word "atom" comes from a Greek word meaning "indivisible." However, the ancient concept of an atom was very different from the modern one. Democritus believed that all atoms were exactly the same. What made one substance different from another was the way the atoms were linked together. The Greek philosophers did not do experiments to test their beliefs, and the existence of atoms remained an unproved and unpopular concept until just about 200 years ago. In 1803, on the basis of a great deal of experimental evidence then available to him, an Englishman named John Dalton proposed the first modern atomic theory.

Dalton's theory included the following statements:

1. Elements are composed of tiny, indivisible particles, called atoms.
2. All atoms within a given element are identical.
3. Atoms of different elements are different.
4. During a chemical reaction, atoms can be neither created nor destroyed; they are simply rearranged.

As we learn more about atoms, we will see that the first two of these statements are not entirely accurate. However, Dalton's atomic theory gave scientists a far better view of the nature of matter than had been proposed previously. Dalton is often called "The Father of Atomic Theory."

Dalton believed that no particles smaller than atoms existed. Evidence that there are smaller, or *subatomic* particles came from the study of electricity.

Static Electricity

We have all occasionally experienced the mild electric shock that results from walking across a carpet and then touching a metal object. If you pull a woolen sweater over your head on a cold, dry day, you may hear a crackling sound and, in a dark room, may even see the sparks produced by static electricity. In the 1750s, Benjamin Franklin studied the charge that develops when various objects are rubbed against wool or silk. He concluded that there are just two types of charge, which he called positive (+) and negative (−). Objects with like charges repel each other, while objects with opposite charges attract each other. The cause of these static charges, however, remained a mystery for another hundred years.

The Cathode Ray Tube

In the nineteenth century, scientists discovered how to generate large amounts of electricity. They observed that when most of the air is pumped out of a glass tube, the remaining gas will glow when a sufficient amount of electricity is applied to it. Chemists discovered that by placing different gases in the tubes, they could produce different colors of light; the orange glow of neon lights is produced this way. Certain solids, such as zinc sulfide, also will glow in a vacuum tube when electricity is passed through them. By coating a metal with zinc sulfide, and using a narrow slit to focus the resulting beam of light, the scientists were able to conclude that the beam was moving from the negative terminal to the positive terminal. The negative terminal of a battery is called a cathode, and so the beam of light became known as a cathode ray. Cathode ray tubes of the type shown in Figure 2-1 were used to determine the nature of these cathode rays.

The electron On the basis of his experiments with cathode ray tubes, the British scientist J.J. Thomson concluded that the cathode

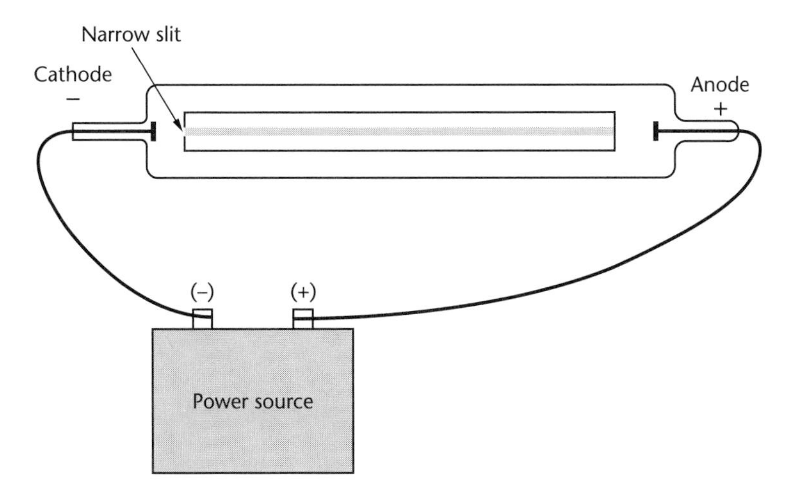

Figure 2-1 Cathode ray tube

ray was actually a stream of negatively charged particles. He named these particles electrons. Thomson found that all elements contained identical electrons, and later experiments indicated that the mass of an electron is far smaller (more than 1800 times) than the mass of the lightest atom. Since atoms are neutral, and contain negative electrons, they must also contain a source of positive charge.

The nucleus Thomson had found a negative subatomic particle, the electron. With no positive particle having been discovered, he proposed what became known as the "plum pudding" model of the atom. In this model, negatively charged electrons are imbedded in a positively charged "pudding" like small pieces of fruit. (See Figure 2-2) In 1909, a team of scientists led by Ernest Rutherford designed an experiment to test this model.

Rutherford's experiment made use of polonium, a radioactive element that gives off positively charged alpha particles. A stream of these particles was allowed to strike a very thin sheet of gold or copper. Behind the metal foil, there was a fluorescent screen.

A diagram of the effects Rutherford and his co-workers observed is shown in Figure 2-3. Most of the alpha particles passed through the metal foil with very little interference. Each time one of these alpha particles hit the fluorescent screen, a flash of light was given off. But a few particles did not pass directly through the foil. They were deflected, or turned aside from their paths. Most of the deflections were at small angles, but some were at relatively large angles. About one alpha particle in 20,000 was deflected through an angle greater than 90°. In effect, some of the

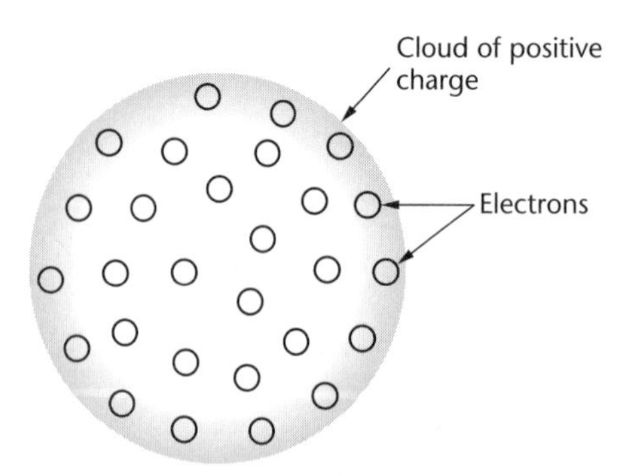

Figure 2-2 The "Plum Pudding" model

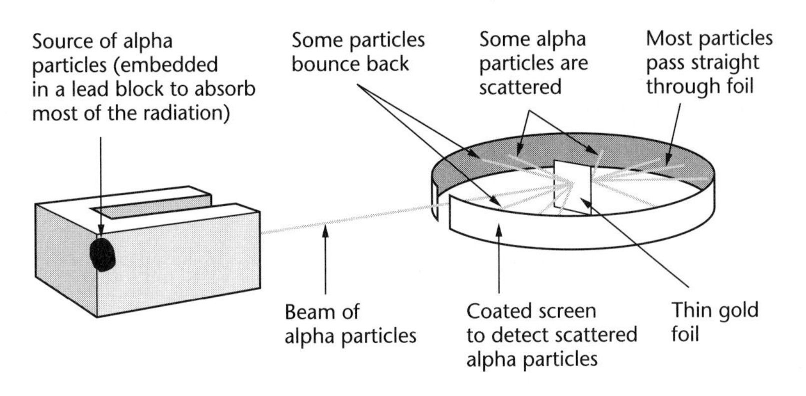

Figure 2-3 Rutherford's experiment

particles seemed to move back toward the polonium from which they had been given off. Rutherford and his co-workers reacted with surprise when they observed these large deflections. Rutherford is quoted as saying, "It was almost as incredible as if you fired a 15-inch shell at a piece of tissue paper and it came back and hit you."

To better understand Rutherford's experiment, visualize yourself blindfolded, standing in front of a fence. You do not know what kind of fence it is, so to find out, you start throwing small pebbles at the fence. Most of the time you hear nothing at all. Occasionally, you hear a faint "clink" as the pebble seems to contact something on its way through the fence. And then, on very rare occasions, the pebble strikes something solidly, and bounces back at you. What would you expect the fence to look like? (See Figure 2-4)

Rutherford concluded that like our fence, atoms consist mostly of empty space. However, they also contain a tiny, massive center, which he called the nucleus. The nucleus is positively

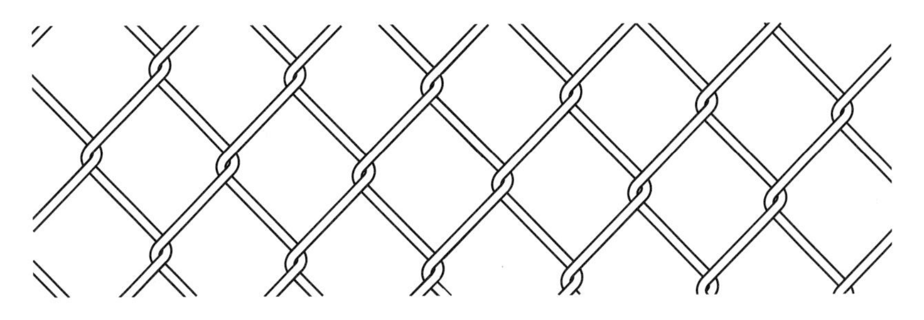

Figure 2-4 One type of fence

charged, and although extremely small, it contains most of the mass of the atom.

The experiments of Thomson, Rutherford, and many others led to our modern view of the general structure of the atom.

The General Structure of an Atom

An atom, you will recall, is the smallest part of an element that retains the properties of the element. Atoms, in turn, are made up of smaller parts, or particles. These fundamental particles of an atom include protons, electrons, and neutrons. There are other atomic particles also, but they will not be discussed in this chapter.

The proton is a particle with a positive electrical charge. It is assigned a unit charge of +1. The mass of a proton is almost exactly the mass of a hydrogen atom.

The electron is a negatively charged particle with a unit charge of −1. The electron is the smallest of the three fundamental particles. It has a mass only 1/1837 the mass of a proton.

The **neutron** has no charge—it is an electrically neutral particle. The mass of a neutron is only very slightly greater than the mass of a proton. For all practical purposes, the mass of a neutron and the mass of a proton are considered equal.

Atoms are electrically neutral. This means that the number of protons in an atom must be equal to the number of electrons. Recall, too, that the number of protons (and of electrons) in an atom is different for each of the elements. The number of protons in an atom is called the **atomic number.** An atomic number describes a particular atom and the element of which it is a part.

Together, protons and neutrons make up nearly the entire mass of an atom. The sum of these two particles—protons and neutrons—is called the **mass number** of an atom. The mass of the electron is so small that it is usually ignored when the mass of an atom is considered. This is true even when an atom contains 100 or more electrons.

Protons and neutrons are also called **nucleons.** They are located in the center of the atom, in the region called the **nucleus.** The electrons are located outside the nucleus. Each kind of atom has a specific arrangement of electrons.

The Size of Atoms

Although the nucleus may have more than 100 protons and more than 150 neutrons, the nucleus is only a very small region of the atom. An average atom may have a radius of about 1×10^{-8} centimeter (cm), one-hundred millionth of a centimeter. The radius of its nucleus is about 1×10^{-13} cm (one-ten trillionth of a centimeter). Thus the radius of the whole atom is 10^5 (one hundred thousand) times greater than the radius of the nucleus of the atom.

Both the atom and the nucleus can be thought of as spheres. The volume of the whole atom can be compared with the volume of the nucleus alone, because the volume of a sphere is proportional to the cube of its radius. The ratio of the volume of an atom to the volume of its nucleus is

$$\frac{(1 \times 10^{-8}\ \text{cm})^3}{(1 \times 10^{-13}\ \text{cm})^3} = \frac{1 \times 10^{-24}\ \text{cm}^3}{1 \times 10^{-39}\ \text{cm}^3} = 1 \times 10^{15}$$

The volume of an atom is about 10^{15} (one quadrillion) times larger than the volume of its nucleus. Nearly the entire mass of an atom is packed into the tiny nucleus.

Atomic Number

Rutherford and his co-workers repeated the "gold leaf" experiment many times, using different metals to deflect alpha particles. They concluded that each element has a characteristic number of positive charges in its nucleus. This conclusion was later confirmed by the English physicist Henry Moseley. Moseley used the term atomic number to refer to the number of positive charges in the nucleus of an atom. Since each proton carries one positive charge, the atomic number represents the number of protons in the nucleus of an atom. In a neutral atom, the atomic number also equals the number of electrons outside the nucleus, since the charge of the electrons balances the positive charge of the nucleus.

The atomic number of an element never changes. If the number of protons in a nucleus changes, the element changes into another element with a different atomic number. As long as the number of protons does not change, the element remains the same.

Atomic Mass

As you know, atoms are very small in size. The actual mass of any atom is also very small. For example, the mass in grams of an oxygen atom is approximately 2.67×10^{-23} gram. To avoid using such small numbers, an arbitrary scale for the mass of elements has been developed. On the scale, the mass of the elements is given in units called **atomic mass units.** One atomic mass unit (amu) is equal to $\frac{1}{12}$ the mass of an atom of the most abundant form of carbon. This form of carbon (carbon-12) is assigned a relative atomic mass of exactly 12 amu. Therefore, an atom of an element with exactly twice the mass of a carbon-12 atom has a mass of 24 amu, and one with half the mass of carbon-12 has a mass of 6 amu. You should remember that atomic mass units are relative, or comparative, masses, the standard being carbon-12.

One atomic mass unit is very close to the mass of a proton or neutron. In practice, therefore, both the neutron and the proton are given a mass of 1 amu. Thus it is possible to describe the mass of an atom, or the mass number, as the sum of the protons and neutrons in its nucleus.

Molecular Mass

The mass of a molecule (the **molecular mass**) is the sum of the atomic masses of all the atoms in the molecule. Thus the molecular mass of hydrogen is 2 amu, because a molecule of hydrogen consists of two atoms, each of which has an atomic mass of 1 amu. The molecular mass of water (H_2O) is 18 amu.

$$\mathrm{H} = 2 \times 1\ \mathrm{amu} = 2$$

$$\mathrm{O} = 1 \times 16\ \mathrm{amu} = \underline{16}$$

$$\mathrm{H_2O} = 18\ \mathrm{amu}$$

Since atomic mass units are relative masses, molecular masses are relative also.

PRACTICE

2.1 How did Thomson's view of the atom differ from that of Democritus?

2.2 When alpha particles are shot at a thin gold leaf, only one particle in about 20,000 bounces back. How would Rutherford explain this observation?

2.3 What is the atomic number of a neutral atom that contains 14 electrons? Explain your answer.

2.4 A certain atom contains 6 protons, 8 neutrons, and 6 electrons. How many and which of these particles are in the nucleus of this atom?

Atomic Mass and Isotopes

Look up the atomic mass of chlorine in a reference table, such as the one in Appendix 5. Do you find that chlorine has atomic mass 35.453? Why is the atomic mass of chlorine a fractional number instead of a whole number, such as 35 or 36?

Recall that Dalton believed that all atoms of the same element are identical. This turns out not to be true. Although all atoms of the same element have the same number of protons, there can be varying numbers of neutrons. All chlorine atoms have 17 protons. However, some chlorine atoms have 18 neutrons, while others have 20 neutrons. Different forms of the same element are called **isotopes.** Isotopes share the same number of protons, but have a different number of neutrons. Since neutrons have mass, but no charge, the number of neutrons affects the mass of an atom, but has no effect at all on its chemical properties.

The atomic mass of 35.453 is the *average* mass of a chlorine atom. It is equal to the weighted average of the masses of its isotopes. Suppose, for example, that the form of chlorine with 18 neutrons has a mass of 35 amu, while the form with 20 neutrons has a mass of 37 amu. If 75.4% of the chlorine atoms have a mass of 35 amu, and the rest of the atoms have a mass of 37 amu, we can calculate the mass of an average chlorine atom.

Finding the atomic mass from the % of each isotope Follow these steps to determine the average atomic mass of an atom from the percent of each isotope in the element:

1. Multiply the mass of each isotope by its decimal percent.
2. Add all of the results from step 1.

If the 75.4% of the chlorine has a mass of 35 amu, then 24.6% has a mass of 37 amu. Calculation reveals that 75.4% of 35 is 26.39 and 24.6% of 37 is 9.10. Adding the two results,

26.39 + 9.10 = 35.49, which is approximately equal to the listed atomic mass of chlorine.

The **atomic mass** of each naturally occurring element, as listed in reference tables, actually represents the *weighted average* mass of the atoms of the isotopes of that element.

Mass Number

Isotopes of an element are chemically identical to each other. The different number of neutrons produces a different mass in each isotope. To distinguish one isotope from another, chemists use the mass number of the isotope. Mass number is defined as the sum of the number of protons and the number of neutrons. One isotope of carbon is carbon-12. The 12 is the mass number of this isotope. Since a carbon atom always has 6 protons (its atomic number is six), carbon-12 must also have 6 neutrons.

$$\textit{mass number} = p + n$$

where p is the number of protons, and n is the number of neutrons. You will recall that protons and neutrons are both called *nucleons* because they are found in the nucleus. The mass number, then, is equal to the number of nucleons.

When an element is represented with a symbol, the mass number is usually written as a superscript before the symbol. Carbon-12, for example, would be written ^{12}C. The atomic number is usually written as a subscript before the symbol. Since carbon has 6 protons, the symbol for carbon-12 could be written $^{12}_{6}C$. The more common isotope of chlorine would be $^{35}_{17}Cl$. You can determine the number of neutrons in $^{35}_{17}Cl$ by subtracting the atomic number, 17, from the mass number, 35. There are 18 neutrons in an atom of $^{35}_{17}Cl$. The number of neutrons in an isotope always equals the mass number minus the atomic number.

PRACTICE

2.5 The element magnesium consists primarily of three isotopes: ^{24}Mg, ^{25}Mg, and ^{26}Mg. The composition of naturally occurring magnesium is 79% ^{24}Mg, 10% ^{25}Mg, and

11% ^{26}Mg. Find the atomic mass of Mg. (Assume that the mass of each isotope is equal to its mass number)

2.6 For each of the following atoms, indicate the number of protons, neutrons, electrons, and nucleons:

(a) $^{39}_{19}K$ (b) $^{14}_{6}C$ (c) $^{40}_{18}Ar$

Atomic Mass Versus Atomic Weight

As you know, mass and weight are different. Mass is an unchanging property of an object. Weight depends on the gravitational attraction exerted on an object in any given environment. In other words, weight changes with the location of an object. An object will weigh slightly more in a valley than it will on a mountaintop.

Historically, chemists sometimes used the term *weight* when they should have used *mass*. As a result, many terms used in chemistry still include weight. This can be very confusing, especially to beginning chemistry students. Since chemistry deals with changes in matter and since mass represents the quantity of matter, in this book *mass* will be used instead of *weight*. Also, *mass* terms instead of *weight* terms will be used whenever possible.

Gram-Atomic Mass

The **gram-atomic mass** of an element is the atomic mass of the element expressed in grams. Thus the gram-atomic mass of sodium is 23.0 grams.

The number of atoms in the gram-atomic mass of each of the elements is the same. By experiments, it can be shown that this number is 6.02×10^{23}. You will recognize this number as Avogadro's number.

The mass of a single atom can be computed from its gram-atomic mass and the number of atoms in a gram-atomic mass. For example, 6.02×10^{23} atoms of sodium have a mass of 23.0 grams. The mass in grams of one sodium atom is

$$\frac{23.0 \text{ grams}}{6.02 \times 10^{23} \text{atoms}} = 3.82 \times 10^{-23} \text{ gram/atom}$$

Hydrogen has an atomic mass of 1.008 atomic mass units (amu), or a gram-atomic mass of 1.008 grams. What is the mass of a hydrogen atom? Since 1.008 grams of hydrogen contain 6.02×10^{23} atoms, one hydrogen atom has a mass of

$$\frac{1.008 \text{ grams}}{6.02 \times 10^{23} \text{atoms}} = 1.67 \times 10^{-24} \text{ gram/atom}$$

This means that 1.008 amu, or approximately 1 amu, has a mass of 1.67×10^{-24} gram.

Electrons in Atoms

So far, you have a picture of the atom as a tiny nucleus with protons and neutrons and a vast surrounding space where the electrons are located. Let us now concentrate on the electrons and their arrangement in the "empty space" of the atom. This study will lead to the modern concept of the atom.

The current model of atoms, including the arrangement of electrons, was developed between 1900 and 1930. Many models, based on the work of Niels Bohr, Erwin Schrödinger, and many others, led to the modern model. The story of how scientists arrived at today's concept of the atom is a good example of the way models are modified when new evidence does not agree with the accepted model.

Emission Spectroscopy

Much of what we know about the atom and its electrons was learned in the early 1900s from studies of the energy given off by excited atoms. The atoms of elements are normally in a **ground state**, the lowest energy state. However, when gaseous elements are heated or subjected to an electrical discharge, the electrons in the gaseous atoms absorb energy. The atoms are then said to be in an **excited state**. Excited atoms give off energy in the form of electromagnetic radiation.

Electromagnetic radiation is energy that travels through space as waves at a speed of 3×10^8 meters per second (m/s). The waves can be pictured as pulses that vibrate in a regularly repeating pattern (see Figure 2-5).

Waves are characterized by wavelength (λ) and frequency (f). **Wavelength** is the distance between peaks or between troughs in

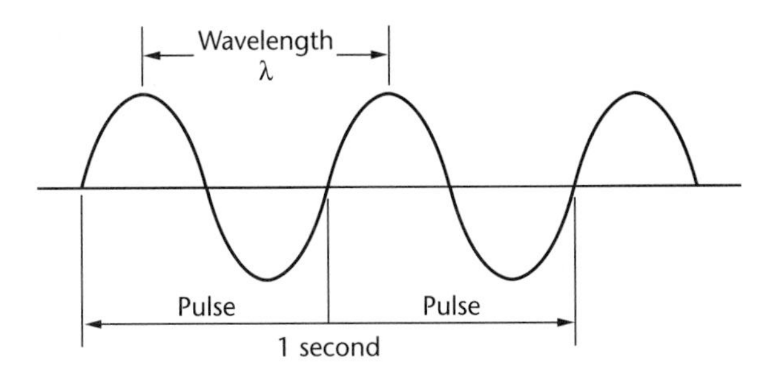

Figure 2-5 An electromagnetic wave

the wave. **Frequency** is the number of complete pulses, or cycles, that pass a given point in one second. (In Figure 2-5, two pulses are marked off. If they occur in 1 second, the frequency is 2 vibrations per second.) Wavelength and frequency vary inversely; the greater the frequency of a wave, the shorter its wavelength.

Electromagnetic radiation covers a broad range, or *spectrum,* and includes waves from 10 kilometers in length to those with wavelengths that are a millionth of a millionth of a meter long. As the table below shows, ultraviolet radiation, visible light, infrared radiation (heat), and radio waves are included in the electromagnetic spectrum. The study of electromagnetic radiation is called **spectroscopy. Emission spectroscopy** is the study of the energy—the electromagnetic radiation—emitted by excited atoms.

The amount of energy emitted by excited atoms is related to the frequency of the waves. This relationship of frequency to amount of energy was established as a result of the work of Max Planck, a German physicist. It is expressed by the formula $E = hf$,

Electromagnetic Radiation

Decreasing wavelength ↓			Increasing frequency ↓
	Radio (including radar and television)		
	Infrared (heat)		
	Red	Visible light	
	Orange	Visible light	
	Yellow	Visible light	
	Green	Visible light	
	Blue	Visible light	
	Violet	Visible light	
	Ultraviolet		
	X rays		
	Gamma		
	Cosmic		

where E is energy, h is a number called Planck's constant, and f is frequency. From the formula, you know that as f increases, E also increases. Thus the higher the frequency of the waves, the greater is the energy given off.

Sunlight, as you probably know, is made up of light with different wavelengths. In the visible spectrum, red light has the longest wavelength and the lowest frequency, violet the shortest wavelength and the highest frequency. The different wavelengths are separated when sunlight passes through a prism, because each different wavelength is bent through a different angle. If the separated wavelengths fall onto a white surface, a rainbow—a spectrum of colors from red to violet—is seen. The colors are not separated from one another but blend together. This is called a *continuous spectrum* (see Figure 2-6). The continuous spectrum contains every wavelength and every frequency of light waves, from red to violet. Since the energy can be found from the frequency, the spectrum must also contain every possible amount of energy, from the high-energy violet waves, down to the lower energy red waves.

When the radiation from an excited atom is directed through a prism, this radiation, too, is separated into its various wavelengths. If the separated wavelengths are allowed to fall on a white screen, each wavelength in the visible spectrum leaves a colored line on the screen. (Wavelengths in the invisible part of the spectrum can be detected by using a photographic plate instead of a screen.) Figure 2-7 shows the spectrum formed by an

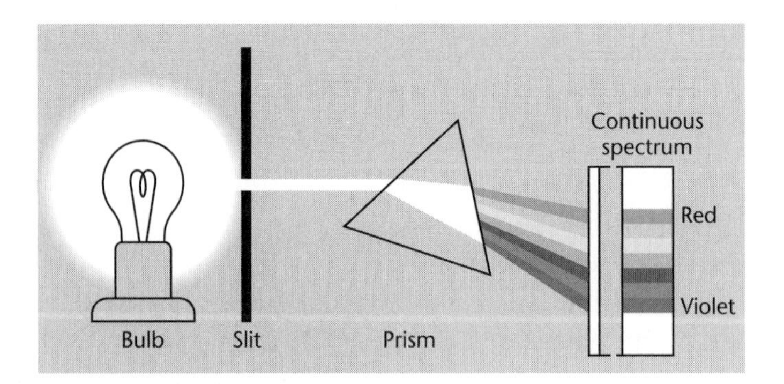

Figure 2-6 Continuous spectrum

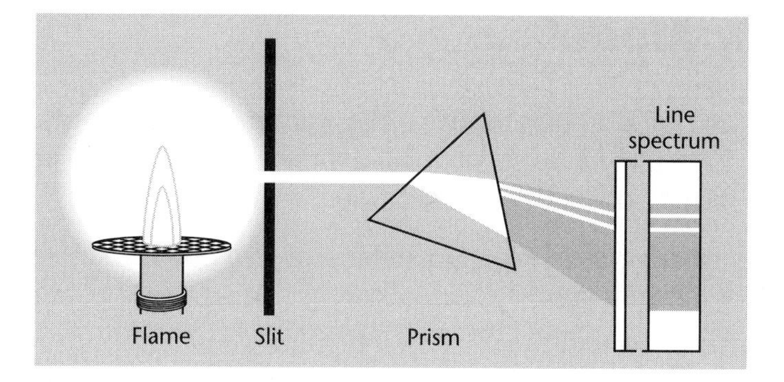

Figure 2-7 Line spectrum

element that emits radiation of two different wavelengths in the visible spectrum. A spectrum such as this one is called a *line spectrum.* A line spectrum is *discontinuous*—the lines appear at some locations but not at others. Each line represents a particular frequency of light, which means it also indicates a particular energy of light. Excited atoms must emit light of only certain energies. We say that the energy produced by the excited atom is **quantized.** It can have only certain values within a given range.

The spectroscopic studies of the early 1900s showed that each element has a different line spectrum. Thus, for a given element, a specific number of lines can be observed at specific positions on either a screen or a photographic plate. In effect, each element has its own spectroscopic "fingerprint." These observations led to a new concept of the atom—the Bohr model.

PRACTICE

2.7 Which color of visible light contains the smallest amount of energy?

2.8 A student, in explaining the difference between "continuous" and "quantized" says: "A piano is quantized, while a violin is continuous." Do you think this is a good analogy? Explain your answer.

The Bohr Model

From the results of spectroscopic studies of the hydrogen atom, Niels Bohr, a Danish physicist, proposed a new model of the atom in 1913. The main points of Bohr's model are as follows:

1. Electrons revolve around the nucleus of the atom in specific orbits, or energy levels.
2. An atom has several orbits, each representing a specific energy level. The energy levels of an atom are very much like the steps of a ladder. Only specific energy values (steps) can exist—there are none between. Electrons cannot exist at any position except one of the energy levels. The energy levels are represented by 1, 2, 3, 4, . . . , or by E_1, E_2, . . . , E_n, or by *K*, *L*, *M*, *N*, ... (The subscript *n* is the symbol for any whole number.)
3. When a hydrogen atom is not excited—that is, the atom is in its ground state—its electrons are in orbits close to the nucleus. The electrons are at the lowest energy level, E_1.
4. If a hydrogen atom receives energy—that is, if the atom is excited—an electron is displaced farther away from the nucleus to one of the higher energy levels, E_2, E_3, E_4, . . . , E_n.
5. An atom emits energy when an electron falls from a higher energy level to a lower energy level in one sudden drop, or transition. The energy released is electromagnetic radiation.
6. The frequency (f) of the radiation emitted depends on the difference between the higher and lower energy levels involved in the transition.

$$E_{\text{higher}} - E_{\text{lower}} = hf$$

The Bohr model explained the discontinuous line spectra (plural of *spectrum*) that had been observed in spectroscopic studies of excited hydrogen atoms. Let's see how.

When an electron is close to the nucleus of the atom, its potential energy is low. When the electron is farther away from the nucleus, its potential energy is comparatively high. Now recall that an increase in potential energy is always accompanied by absorption of energy. Conversely, a decrease in potential energy is always accompanied by release of energy. Thus an electron of an excited atom absorbs energy in the form of heat or electricity when it moves to a higher energy level. When the electron returns to the ground state, it emits energy. Each wave of the emitted radiation has a specific hf value. This is because the transition of the electron can occur only between certain energy levels. (Remember the steps of a ladder.) Hence the energy given off is quantized.

A diagram of the Bohr model for the simplest of all elements, hydrogen, is shown in Figure 2-8. The diagram shows six of the possible energy levels the atom may have.

The hydrogen atom has only one electron. Ordinarily, the electron is at the lowest possible potential energy, in energy level E_1. If energy is absorbed, the electron may be pushed away from the nucleus to a higher energy level, E_2 or E_3, for example. The potential energy of the electron will then be higher. When the electron returns to the lower level, E_1, its potential energy relative to the nucleus decreases and energy is emitted.

As the diagram shows, some changes, or transitions, between energy levels result in emission of energy in the visible region of the electromagnetic spectrum. These emissions appear as lines of different colors. Other transitions result in release of energy outside the visible region, such as in the ultraviolet region. You cannot see these radiations, but you can bend them in a prism and observe them as black lines on special photographic plates.

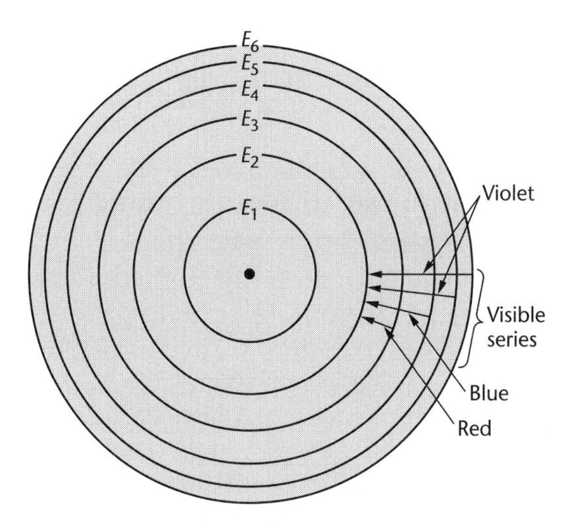

Figure 2-8 Bohr model of the hydrogen atom

In summary, the importance of the Bohr model lay in the fact that it interpreted experimental evidence—discontinuous line spectra—in terms of energy levels for electrons. According to the model, electrons may exist at some energy levels but not at others. In other words, some transitions in the potential energy of electrons are possible, but others are not.

The Quantum Mechanical Model

The Bohr model of the atom successfully explained the lines in the hydrogen spectrum. But when scientists tried to extend the model to explain atoms with more than one electron, they failed. In addition, Bohr's model made some assumptions that were not borne out experimentally. For example, Bohr assumed that the orbiting electron does not approach the nucleus. Clearly, a new or revised model was needed.

The revised model is called the *quantum mechanical,* or *wave mechanical, model.* The words *quantum* and *quantized* are used for variables, such as the energy of an electron, that may have certain values only.

The two main features of the quantum mechanical model (and differences from the Bohr model) are as follows:

1. The electron is treated mathematically as a wave in the quantum mechanical model. In the Bohr model, the electron is treated as a particle. In fact, the electron has properties of both particles and waves.

The French physicist Louis de Broglie predicted the existence of matter waves in 1924. Today we speak of the "duality of matter." Matter has some wave characteristics, just as waves have some matter characteristics. The smaller the particle being studied, the more important the wave characteristic. If you are studying the movement of a baseball, you can predict its path very accurately by just treating the ball as matter. If you are studying the movement of an electron, however, you must consider the behavior of waves.

2. In the Bohr model, the energy of the electron is described in terms of a definite orbit, or pathway. In the quantum mechanical model, the energy is described in terms of the probability

of locating the electron in a region of space outside the nucleus. The electron can be very close to the nucleus or very far away. However, the probability of the electron being a certain distance from the nucleus most of the time is high.

In 1927, Werner Heisenberg, who had studied under Niels Bohr, proposed what came to be known as the **Heisenberg uncertainty principle.** He stated that it is impossible to measure exactly both the position and the momentum of an object simultaneously, and proposed an equation that computes the minimum amount of error in such measurements. The error is extremely small, but it becomes significant when applied to objects as small as electrons. We cannot say exactly where an electron will be, but we can determine where it is most likely to be.

You can see this by studying Figure 2-9. It shows the probability of finding the hydrogen electron (called a 1*s* electron) at different distances from the nucleus. The dark region of the cloud is the region in space where the probability of finding the electron is greatest. The radius (*r*) encloses a region within which the electron can be found about 90 percent of the time. At some large distance from the nucleus, the probability of finding the electron approaches zero. At this distance, the electron has become completely separated from the nucleus. This region is represented by the outer edge of the diagram in Figure 2-9.

In the quantum mechanical model the word **orbital** is used to describe the shape of the region where an electron may be. For *s* electrons, the region is spherical, as shown in Figure 2-9.

Quantum Numbers

The mathematics of the quantum mechanical model makes it possible to describe the energy level of an electron using four

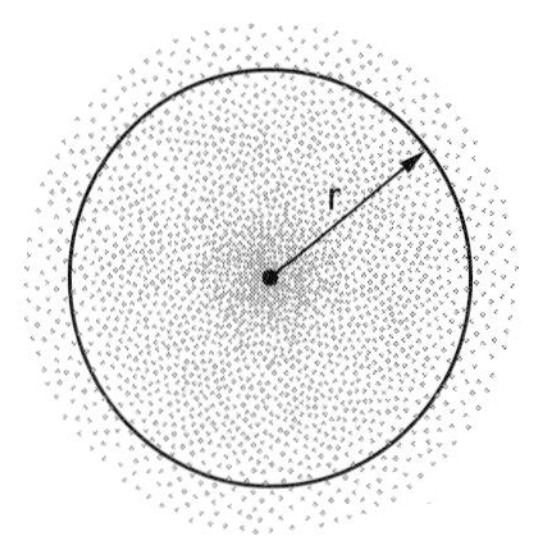

Figure 2-9 Electron density distribution of a hydrogen electron.

quantum numbers. The first number is the principal quantum number, or principal energy level, *n;* the second quantum number, or sublevel, *l;* the third quantum number, or orbital, *m;* and a number describing a characteristic called *electron spin.* In this study, you will be concerned chiefly with the principal quantum number.

No two electrons in an atom can have an identical set of four quantum numbers. This is in accordance with the exclusion principle, developed in 1925 by the Austrian-American physicist Wolfgang Pauli. The **Pauli exclusion principle** states that no more than two electrons can occupy the same orbital in an atom. It also states that the two electrons must have opposite spins.

Let us now consider each of the quantum numbers separately.

The principal quantum number The principal quantum number, *n,* describes the most probable distance of the electron from the nucleus and has whole number values—1, 2, 3, 4, ... These energy levels are, in a way, similar to the *K, L, M, N* shells of the Bohr model and are called the principal energy levels.

The maximum number of electrons in any principal energy level n is $2n^2$. Thus in the first energy level, there can be a maximum of 2 electrons; in the second, 8; in the third, 18; and so on.

If we know the electron capacities of the principal energy levels, and we assume that the lower energy level fills first, we can easily predict the electron distributions for the first 18 elements. (To go beyond 18, as we shall see, additional information is needed.)

SAMPLE PROBLEM

PROBLEM

Draw the structure of the atom with the symbol $^{31}_{15}P$. Show the contents of the nucleus, and the electrons in their energy levels.

SOLUTION

The atomic number 15 indicates that a neutral atom of phosphorus has 15 protons and 15 electrons. The number of neutrons is found by subtracting the atomic number from the mass number (see page 74), so there are 16 neutrons.

The 15 electrons must be distributed as follows: The first energy level can hold 2 electrons. The second can hold 8. That leaves 5 electrons in the third energy level, which is only partially filled. We might draw the structure as follows:

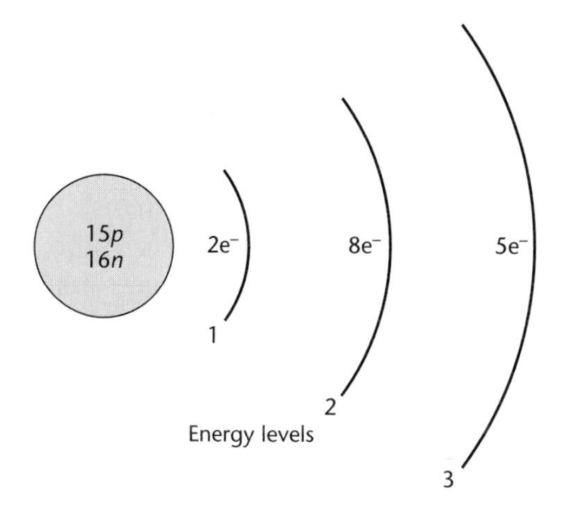

PRACTICE

2.9 Draw structures for the following atoms, showing the protons and neutrons in the nucleus, and the electrons in the energy levels:

(a) $^{27}_{13}\text{Al}$ (b) $^{35}_{17}\text{Cl}$ (c) $^{40}_{18}\text{Ar}$

2.10 How many electrons would there be in the second principal energy level of a silicon atom?

Sublevel As scientists continued to study the spectra of atoms, they found that many of the single lines of the spectra were actually made up of fine, closely spaced lines. The origin of all these lines could not be explained by transitions, or jumps, of electrons between the principal energy levels. This led to the view that the principal energy levels are divided into sublevels. The energy values of the sublevels differ by only small amounts.

Sublevels are described by the second quantum number, l. The number of sublevels in any principal energy level is equal to the principal quantum number, n. Thus, when $n = 1$, the first principal energy level has one sublevel: the *s* sublevel. The two electrons that may be in the first principal energy level are called 1*s* electrons.

The second principal energy level ($n = 2$) has two sublevels. The first sublevel is again called the *s* sublevel. The second sublevel is called the *p* sublevel. The electrons in the first sublevel are 2*s* electrons. The electrons in the second sublevel are 2*p* electrons. There may be two 2*s* electrons in the first sublevel. There may be up to six 2*p* electrons in the second sublevel. Thus the second principal energy level can have a maximum of 8 electrons.

The third principal energy level ($n = 3$) is split into three sublevels: 3*s*, 3*p*, and 3*d*. There may be a total of 18 electrons. The fourth principal energy level ($n = 4$) follows the pattern that has been developing. There are four sublevels in the fourth principal energy level: 4*s*, 4*p*, 4*d*, and 4*f*. There may be a total of 32 electrons.

The arrangement of electrons in principal energy levels and sublevels is shown in the table below.

Principal Energy Level	Principal Quantum Number (*n*)	Maximum Number of Electrons in Each Sublevel				Total Number of Electrons
		s	*p*	*d*	*f*	
1	1	2				2
2	2	2	6			8
3	3	2	6	10		18
4	4	2	6	10	14	32

Orbital Careful analysis of the lines of spectra show that sublevels are made up of orbitals. An orbital is a region of space around a nucleus in which the probability of finding an electron is high. Orbitals represent the third quantum number, *m*. Each sublevel has a specific number of orbitals. There is one orbital in an *s* sublevel, three orbitals in a *p* sublevel, five orbitals in a *d* sublevel, and seven orbitals in an *f* sublevel.

Each sublevel and its orbitals are related to the shape of the region that an electron may occupy. Any *s* sublevel is spherical, as shown in Figure 2-9 on page 83. Sublevels described as *p* are shaped like a dumbbell, as shown in Figure 2-10. The three orbitals in a *p* sublevel are oriented along three axes: *x*, *y*, and *z*.

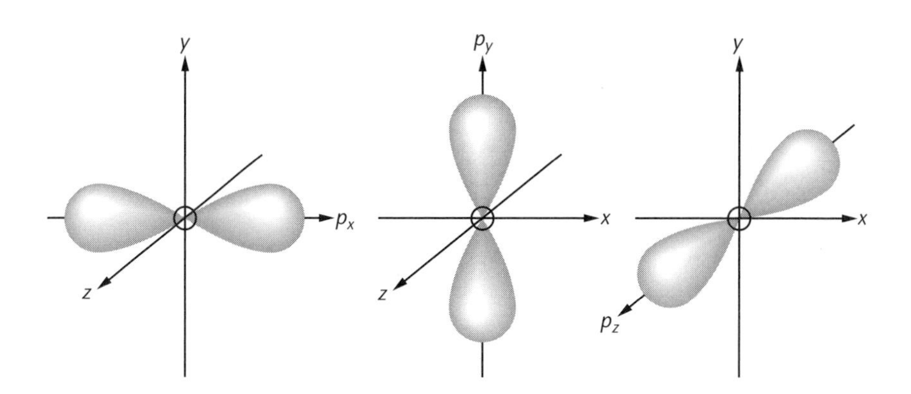

Figure 2-10 The shape of orbitals in a *p* sublevel

The shapes of *d* and *f* sublevels are difficult to represent with a simple drawing and need not concern you.

It is important to realize that an orbital does not represent the path of an electron; it represents a region in which the electron can be found 90% of the time. To illustrate this concept, suppose that while in the field, the New York Yankees' shortstop, Derek Jeter, wore shoes that made a mark wherever he went. At the end of a few innings, as shown in Figure 2-11 on page 88, there would be marks all over the left side of the infield, and part of the outfield, but there would clearly be a region where Mr. Jeter could be found the vast majority of the time. This region would be Derek Jeter's orbital, but we prefer to call it "shortstop."*

Electron spin Spin is the fourth quantum number. An electron acts as if it were spinning on an axis. When a negatively charged object spins, it creates a magnetic field. If it spins counter clockwise, the north pole of the magnetic field is "up" while if it spins clockwise, the north pole is "down." For this reason, the spin is generally represented in terms of the resulting magnetic field—either up or down.

A maximum of two electrons can occupy one orbital. The two electrons can have the same first three quantum numbers, but in accordance with the Pauli exclusion principle, the fourth quantum number must be different. Thus the two electrons in an orbital must have opposite spins.

Within a given sublevel in a free atom, the energies of orbitals are equal.

*The author wishes to make it clear that he is NOT a Yankees fan.

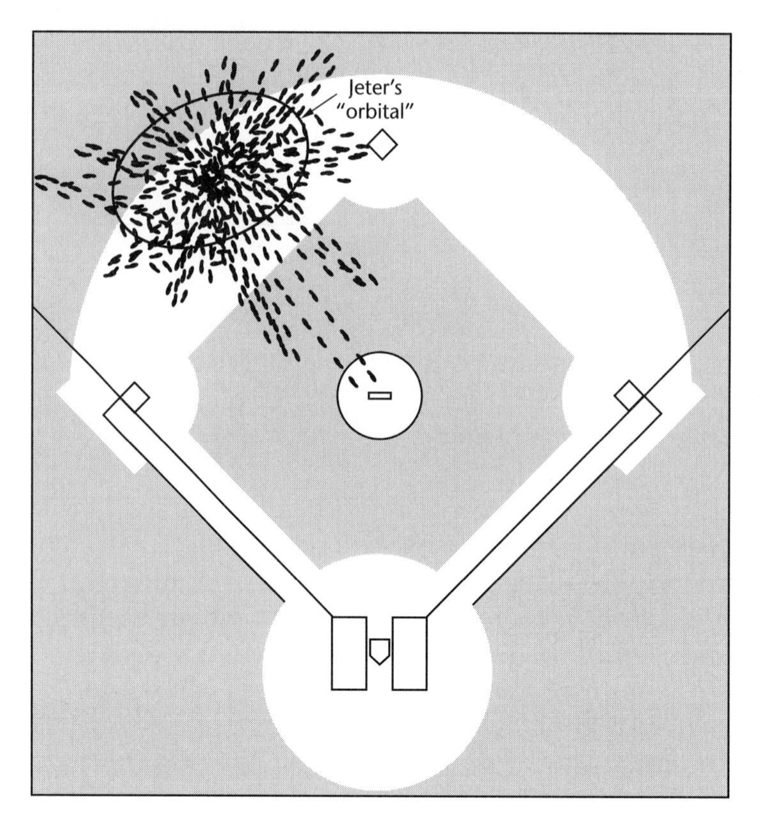

Figure 2-11 A baseball player's "orbital"

Orbitals and their electrons can be shown in various ways. In this book, an orbital will be represented as □. An orbital with one electron will be represented as [↑] or [↓], depending on the spin of the electron. An orbital with two electrons will be represented as [↑↓]. Some examples are shown in the following chart.

1s	***1s***	***1s***	***2p***
[↑]	[↓]	[↑↓]	[↑][ ][ ]
One electron in 1*s* orbital	One electron with opposite spin in 1*s* orbital	Two electrons with opposite spins in filled 1*s* orbital	Three orbitals in 2*p* sublevel. A total of six electrons can be contained in this sublevel, but only one is present here.

Quantum Numbers—A Summary

Some of the information derived from quantum numbers is summarized in the following table. You will find it useful to refer to this table as you go on with your study.

Principal Energy Level (n)	Sublevels (l)	Number of Orbitals (m) per Sublevel	Total Number of Electrons per Sublevel	
1	*s* (spherical)	1	2	
2	*s*	1	2	8
	p (dumbbell)	3	6	
3	*s*	1	2	18
	p	3	6	
	d	5	10	
4	*s*	1	2	32
	p	3	6	
	d	5	10	
	f	7	14	

Electron Configurations

Electrons normally occupy the lowest energy orbital available. Figure 2-12 on page 90, shows the order in which the orbitals are filled. Beginning at the bottom of the diagram, the 1*s* orbital is at the lowest energy level. It can receive two electrons. When this orbital is filled, an electron goes to the orbital with the next lowest energy level, 2*s*. This energy level also can receive two electrons. When this orbital is filled, the orbital with the next higher energy level will be occupied. The *p* orbitals have the next higher energy values. A total of six electrons can be added to complete this energy level.

Generally, the orbitals increase in energy value in the order *s, p, d,* and *f*. Notice, however, that the 4*s* orbital is at a slightly lower energy level than the 3*d* orbitals. Also, the 5*s* orbital is lower than the 4*d* orbitals. In other words, the *s* orbital of a

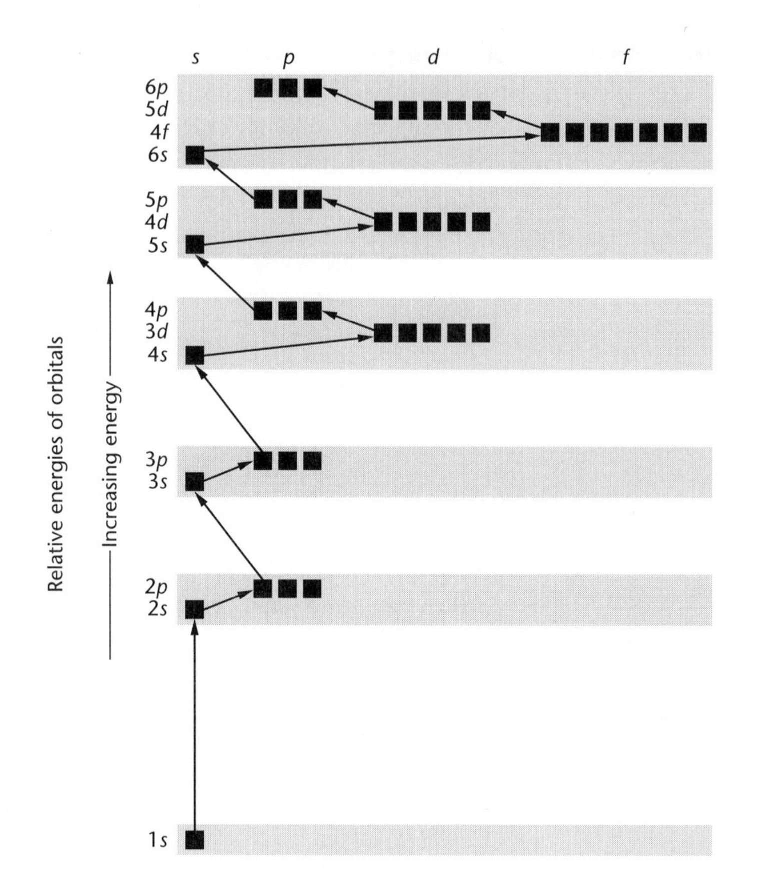

Figure 2-12 Order of filling orbits

higher principal energy level may be occupied before the *d* orbitals of a lower principal energy level.

Hund's Rule

A *p* sublevel can hold up to six electrons—two in each of the three *p* orbitals ↑↓|↑↓|↑↓. Suppose that a particular *p* sublevel has only two electrons in it. How are they arranged? Here are four possibilities:

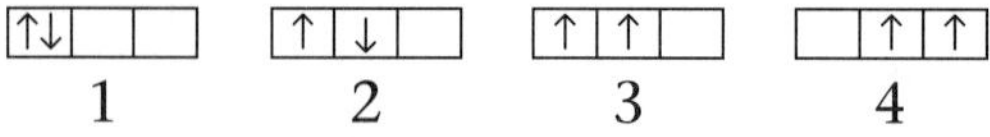

In structure 1, both electrons are in the same orbital. Electrons in the same orbital are called *paired electrons*. In structures 2, 3, and 4, the electrons are in separate orbitals. A single electron in an orbital is called an *unpaired electron.*

Hund's rule predicts the way electrons will fill the orbitals within a given sublevel. It states that:

1. No orbital in a sublevel may contain two electrons unless all of the orbitals in that sublevel contain at least one electron. In other words, the orbitals fill one electron at a time.
2. All unpaired electrons on a given atom have the same spin.

We can see that structure 1 is incorrect, because it has two electrons in the same orbital while other orbitals in the sublevel are empty. Structure 2 is also incorrect, because the unpaired electrons do not have the same spin. Structures 3 and 4 are both correct. However, it is customary to fill the orbitals from left to right when illustrating the electron configuration of an atom. Therefore, structure 3 is the one that would normally be used to illustrate a *p* sublevel containing two electrons.

Electron Configuration of the Elements

Let us now fill in electrons at the proper energy levels for the first 21 elements. We will take the elements in order of increasing atomic number and make sure that each successive electron enters the lowest energy level available. As you work with this section, it will be helpful for you to refer to the Periodic Table in Appendix 5. You will notice that the elements are arranged in the table in order of atomic number. Also, the elements are placed in certain horizontal rows, also called **periods**, and in vertical columns, also called **groups** or **families.** The periods are numbered from 1 to 7, and the groups from 1 to 18. (Older versions of the table used Roman numerals to number the groups, sometimes followed by a letter A or B.) You will understand the significance of the periods and groups as you go on. The Periodic Table will also be discussed at length in Chapter 5.

The first principal energy level In hydrogen, the single electron enters the first principal energy level (n = 1), which has only an *s* sublevel. The *s* sublevel contains one orbital. The electron configuration of hydrogen can be shown this way:

1*s*

H [↑] (1)

Helium, atomic number 2, has two electrons in the 1*s* sublevel. These electrons fill the 1*s* orbital. The electron configuration is shown as

1*s*

He [↑↓] (2)

Because two electrons are the maximum number that can occupy any orbital, helium has a completely filled 1*s* orbital. In addition, two electrons are the maximum number that can occupy the first principal energy level. Helium has therefore also completed the first principal energy level.

The horizontal arrangement of hydrogen and helium make up the first row, or period, of the Periodic Table.

The second principal energy level Lithium, atomic number 3, first fills the sublevel of lowest energy, the 1*s* sublevel, with two electrons. The third electron enters the next lowest energy level available—the 2*s* sublevel of the second principal energy level ($n = 2$). This energy level has two sublevels, *s* and *p*. Beryllium, atomic number 4, completes the 2*s* sublevel. In boron, atomic number 5, the fifth electron enters the *p* sublevel, which contains three *p* orbitals. The electron configurations of lithium, beryllium, and boron can be shown as

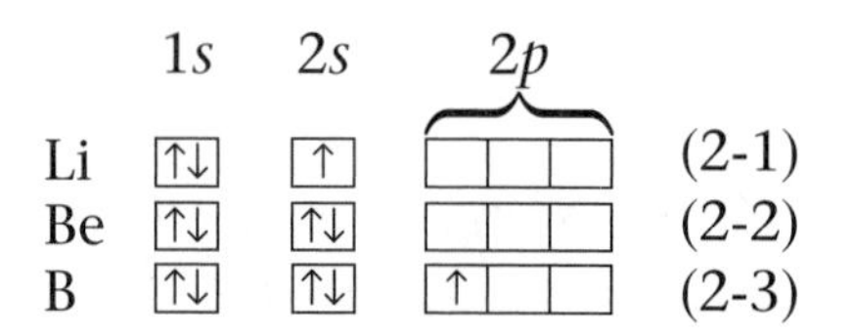

In carbon, atomic number 6, the sixth electron enters the second *p* orbital instead of the first *p* orbital, as predicted by Hund's rule. This is due to the tendency of particles with similar charges to repel one another and thus to move as far apart as possible.

1*s* 2*s* 2*p*

C [↑↓] [↑↓] [↑|↑|] (2-4)

In the same manner, electrons are added to nitrogen, oxygen, fluorine, and neon until the second principal energy level is complete. The electron configurations are given below.

	1*s*	2*s*	2*p*	
N	↑↓	↑↓	↑ ↑ ↑	(2-5)
O	↑↓	↑↓	↑↓ ↑ ↑	(2-6)
F	↑↓	↑↓	↑↓ ↑↓ ↑	(2-7)
Ne	↑↓	↑↓	↑↓ ↑↓ ↑↓	(2-8)

Neon fills the second principal energy level. The second row of the Periodic Table contains eight elements.

The third principal energy level The eleventh electron in sodium must enter the third principal energy level in the 3*s* sublevel.

	1*s*	2*s*	2*p*	3*s*	3*p*	
Na	↑↓	↑↓	↑↓ ↑↓ ↑↓	↑		(2-8-1)

The electron configurations of the next seven elements can be completed in the same way that the configurations for the elements in the previous principal energy level are completed.

	1*s*	2*s*	2*p*	3*s*	3*p*	
Mg	↑↓	↑↓	↑↓ ↑↓ ↑↓	↑↓		(2-8-2)
Al	↑↓	↑↓	↑↓ ↑↓ ↑↓	↑↓	↑	(2-8-3)
Si	↑↓	↑↓	↑↓ ↑↓ ↑↓	↑↓	↑ ↑	(2-8-4)
P	↑↓	↑↓	↑↓ ↑↓ ↑↓	↑↓	↑ ↑ ↑	(2-8-5)
S	↑↓	↑↓	↑↓ ↑↓ ↑↓	↑↓	↑↓ ↑ ↑	(2-8-6)
Cl	↑↓	↑↓	↑↓ ↑↓ ↑↓	↑↓	↑↓ ↑↓ ↑	(2-8-7)
Ar	↑↓	↑↓	↑↓ ↑↓ ↑↓	↑↓	↑↓ ↑↓ ↑↓	(2-8-8)

If you compare the electron structures of the elements in the vertical columns of the Periodic Table, you will see the beginning of similarities. In lithium and sodium, for example, the *s* sublevel is half-filled. Lithium and sodium are the first two elements of Group 1—the alkali metal family. Other common members of this family include potassium, rubidium, and cesium. The presence of the half-filled *s* sublevel accounts for similar properties in this group of elements. (Hydrogen also has a half-filled *s* sublevel. However, for reasons that will be discussed later, hydrogen does not belong to the alkali metal family.)

Transition elements The maximum number of electrons in the third principal energy level is 18. This level contains a *d* sublevel with five *d* orbitals that have a total of ten electrons. When these ten electrons are added to the eight electrons in the 3*s* and 3*p* orbitals, the maximum number of 18 electrons ($n = 3$) is reached.

The nineteenth electron of potassium might be expected to enter the 3*d* sublevel. However, if you check Figure 2-12 on page 90, you will notice that the 4*s* orbital is lower in energy than the 3*d* orbital. The nineteenth electron of potassium therefore enters the 4*s* orbital. The twentieth electron of calcium completes this orbital.

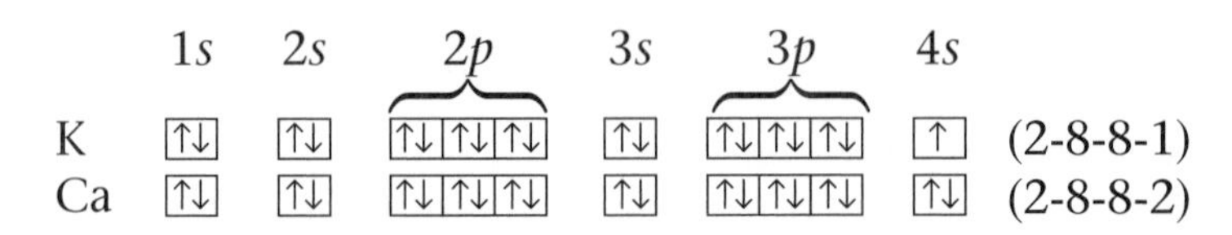

The next most energetic sublevel is 3*d*. The twenty-first electron of scandium enters the 3*d* sublevel. In terms of principal energy levels only, we can say that the electron configuration of scandium is 2-8-9-2.

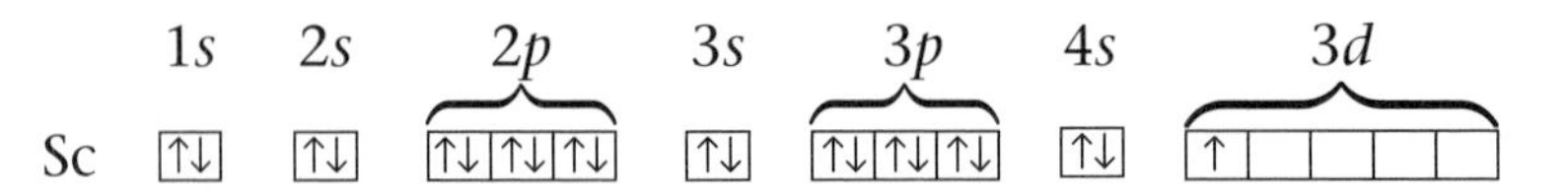

The next nine elements (titanium through zinc) complete this sublevel, which is in the third principal energy level. When this sublevel is complete, electrons enter the 4*p* sublevel. Elements that complete inner, or lower, energy levels before completing outer, or higher, energy levels are called **transition elements.** Transition elements are discussed again in Chapter 5.

The third row of the Periodic Table has eight elements. However, the third principal energy level has 18 elements. Ten of these elements (scandium through zinc) are transition elements. They are in the fourth row of the Periodic Table.

The Periodic Table in Appendix 5 gives the electron configurations of most of the elements in terms of their principal energy levels.

As you learned from Bohr theory, electrons may absorb energy and jump to higher energy levels. These higher energy levels are called excited states. When an element is in an excited state, it will not have the electron structure shown in this Periodic Table.

The structures in the Periodic Table are all for elements in their ground states.

Using the reference table to fill the orbitals Our Periodic Table shows the electron configurations only in terms of the shells, or principal energy levels. From this information it is easy to also determine the orbital configurations. We need to remember that the electron capacities of the *s, p, d* and *f* sublevels are 2, 6, 10 and 14, respectively. For example, consider the electron configuration of a Ca atom. Our table gives us the configuration "2-8-8-2". There are 2 electrons in the first shell, 8 in the second, 8 in the third, and 2 in the fourth. When a shell contains only 2 electrons, only the *s* sublevel can be involved. A shell with 8 electrons must contain 2 electrons in the *s* sublevel, and 6 electrons in the *p*. Filling in the orbitals gives us the structure shown on page 94. Now examine the configuration of an iron atom, element number 26. The Periodic Table lists the configuration as "2-8-14-2." What orbitals are used by the 14 electrons in the third shell? The *s* holds 2 electrons, and the *p* 6, which leaves 6 electrons in the 5 orbitals of the *d* sublevel. The correct configuration of a ground state iron atom shows that it contains 4 unpaired electrons.

1*s*	2*s*	2*p*	3*s*	3*p*	4*s*	3*d*
↑↓	↑↓	↑↓ ↑↓ ↑↓	↑↓	↑↓ ↑↓ ↑↓	↑↓	↑↓ ↑ ↑ ↑ ↑

PRACTICE

2.11 Using arrows to represent electrons, as shown in this chapter, draw the orbital configurations of the following atoms in the ground state:

(a) N (b) Si (c) Mn (d) Zn

2.12 How many unpaired electrons are there in each of the following elements, in the ground state?

(a) $_{8}O$ (b) $_{15}P$ (c) $_{18}Ar$ (d) $_{24}Cr$

2.13 Each of the four elements on page 96 is shown with an *incorrect* electron configuration. In each case, indicate why the configuration is incorrect, and draw the correct configuration.

	1s	2s	2p	3s
(a) $_6C$	[↑↓]	[↑↓]	[↑↓][ ][ ]	
(b) $_9F$	[↑↓]	[↑↓]	[↑↓][↑↓][↑↓]	
(c) $_{12}Mg$	[↑↓]	[↑↓]	[↑↓][↑↓][↑↓]	[↑↑]
(d) $_7N$	[↑↓]	[↑↓]	[↑][↓][↑]	

The Notation System of Showing Electron Configuration

You have been working with a pictorial way of showing the number of electrons in an atom and the distribution of the electrons in orbitals. By this method, the one electron in hydrogen can be shown as

	1s
H	[↑]

and the eight electrons in oxygen can be shown as

	1s	2s	2p
O	[↑↓]	[↑↓]	[↑↓][↑][↑]

Electron configuration can also be shown by a notation system. By this method, the hydrogen electron is represented by $1s^1$. The electrons in oxygen are represented by $1s^22s^22p^4$. In both cases, the coefficients refer to the principal energy level (the first quantum number).

The exponents, or superscripts, indicate the number of electrons at the specified sublevels.

The order of notation is important. Refer again to the elements of the fourth row. Potassium has one electron in the fourth principal energy level and none in the 3*d* sublevel. Calcium has two electrons in the fourth principal energy level and none in the $3d$ sublevel. The electron configuration of potassium is $1s^22s^22p^63s^23p^64s^1$. Calcium is $1s^22s^22p^63s^23p^64s^2$.

Beginning with scandium, electrons are added to the $3d$ sublevel.

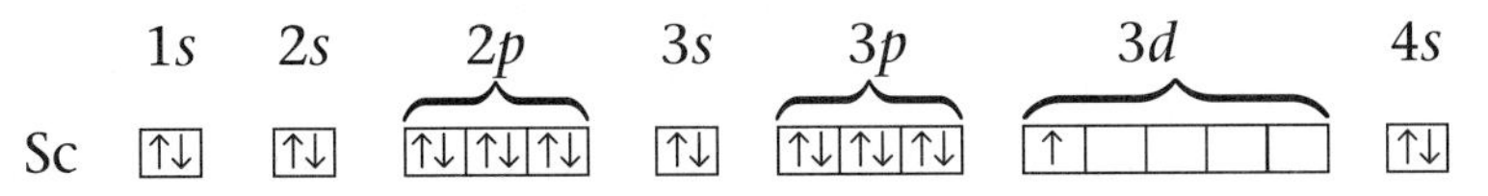

The electron configuration of scandium would be written $1s^22s^22p^63s^23p^63d^14s^2$. Notice that the $3d$ sublevel is written before the $4s$, even though the $4s$ filled first. (See the configuration of calcium, above.) It is customary when writing electron configurations to keep the sublevels in the same principal energy level together. Doing so makes it easier to view the contents of each principal energy level. In scandium, you can see that the first principal energy level has 2 electrons, the second has 8, the third has 9 and the fourth has 2.

A shortcut method of notation is sometimes used. Iron, one of the transition elements, has the electron configuration $1s^22s^2$ $2p^63s^23p^63d^64s^2$. The electron configuration of iron can also be shown as [Ar]$3d^64s^2$. [Ar] represents the electron configuration of the element argon: $1s^22s^22p^63s^23p^6$. When [Ar] is used to stand for the configuration of the first eighteen electrons of iron, only the eight electrons at the outer energy levels of the iron atom need be shown.

Argon is one of the so-called **noble gases.** The outermost principal energy levels of the noble gases have their full complement of electrons. This means that electrons from other atoms cannot enter these levels. As a result, the noble gases do not readily form chemical bonds with other elements. The other noble gases are helium, neon, krypton, xenon, and radon. The noble gases form Group 18 of the Periodic Table. (They were also called *inert gases* because of their limited chemical activity.)

With their filled energy levels, all the noble gases can be used as shortcuts for writing electron notations. Thus magnesium can be shown as [Ne]$3s^2$ and potassium as [Ar]$4s^1$.

Valence Electrons

The electrons in the outer energy levels of atoms are those that are involved in the formation of chemical bonds. As you will later learn, in Chapter 3, atoms may be bonded by losing, gaining, or sharing electrons.

The electrons of an atom that are likely to become involved in the formation of bonds are called the **valence electrons.** In all the elements of Groups 1, 2, 13–18, that is, in all elements except the transition elements, Groups 3–12, the valence electrons are all the electrons in the outermost principal energy level that contains electrons. Take, for example, an element with the electron configuration $1s^22s^22p^63s^23p^2$. This element contains four valence electrons, because there are four electrons in the third principal energy level. For elements in Groups 1, 2, 13–18, the number of valence electrons is also the same as the number in the units place of the group number of the element. Thus the elements in Group 1 have the valence electron, and those in Group 15, have five valence electrons.

Estimating the number of valence electrons for the transition elements is more difficult and does not need to be studied in this course. It is worth noting, however, that a transition element may have valence electrons in more than one principal energy level. For example, the electron configuration of an iron atom is $[Ar]3d^64s^2$. When iron forms one type of bond, three electrons are involved: two $4s$ electrons and one $3d$ electron.

SAMPLE PROBLEM

PROBLEM

The electron configuration for a phosphorus (P) atom is $1s^22s^22p^63s^23p^3$. Indicate for this atom the number of

(a) completely filled principal energy levels

(b) completely filled sublevels

(c) completely filled orbitals

(d) unpaired electrons

SOLUTION

(a) The first and second principal energy levels are filled, with two and eight electrons respectively. The third principal energy level has five electrons in it. It is not completely filled, since it can hold 18 electrons. There are two completely filled principal energy levels.

(b) The 1*s*, 2*s*, 2*p*, and 3*s* are completely filled. The 3*p* is only half filled. There are four completely filled sublevels.

(c) An *s* sublevel contains one orbital; a *p* sublevel contains three orbitals. The total number of completely filled orbitals is six, as shown in the diagram below.

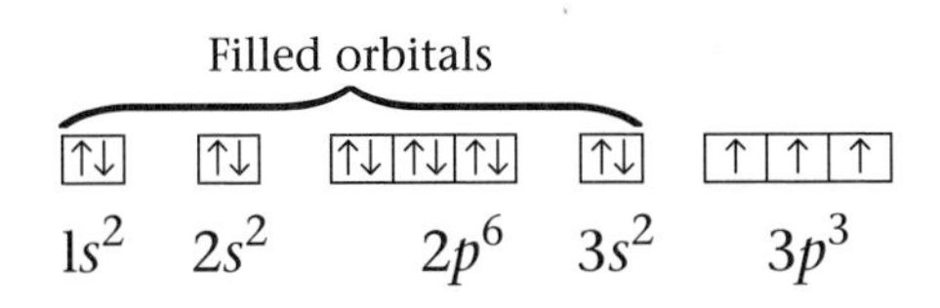

(d) As you can see in the diagram above, there are three unpaired electrons in a phosphorus atom.

PRACTICE

2.14 For an oxygen atom in the ground state, find the number of (a) unpaired electrons (b) completely filled principal energy levels (c) completely filled orbitals.

2.15 A certain atom in an excited state has the electron configuration $1s^2 2s^1 2p^3$.

(a) While the atom is in this state, how many of the electrons are unpaired?

(b) What is the electron configuration of this atom in the ground state?

Ionization Energy

All atoms are electrically neutral. They have an equal number of positively charged protons in the nucleus and negatively charged electrons outside the nucleus. Recall, though, that one way atoms are bonded is by the loss of one or more electrons. The loss of an electron from a neutral atom results in the formation of a positively charged particle, called a **positive ion.**

Electrons are held in the atom by the attractive force of the positively charged nucleus. Energy must therefore be expended in removing an electron from the atom to form a positive ion. The energy required to remove the most loosely held electron from an isolated, neutral, gaseous atom is the **ionization energy.**

Let us represent a neutral, gaseous atom of element X as X^0. An electron removed from X^0 to form the positive ion X^+ involves the first ionization energy. Another electron removed from X^{1+} to form X^{2+} ion involves the second ionization energy. The removal of another electron from X^{2+} ion to form X^{3+} ion involves the third ionization energy.

As successive electrons are removed, the original particle becomes more positively charged and attracts electrons more strongly. As a result, it becomes progressively harder to remove electrons as the charge on the ion increases. Ionization energies therefore increase as each additional electron is removed.

The amount of energy needed to remove the most loosely held electron from a gaseous atom depends on the following conditions:

1. The charge of the nucleus. As the nuclear charge (atomic number) increases, the force of attraction between the nucleus and the electrons increases.
2. The distance from the nucleus to the outermost energy level that has electrons. This distance is called the *radius* of the atom. As the number of occupied principal energy levels increases, the radius of the atom increases. Because the electron is farther from the nucleus, it is held more loosely, and less energy is required to remove it from the atom.
3. The screening, or shielding, effect of the electrons of the inner energy levels. This effect tends to reduce the force exerted by the nucleus on the electrons in the outer energy level.
4. The sublevel of the outer electrons. Generally, *s* electrons are more tightly held to the nucleus (they are at a lower energy level) than are *p* electrons.

Ionization energy is generally expressed in kilojoules per mole. For example, it requires 520 kJ to remove one mole of electrons from one mole of lithium atoms.

The following table lists the first ionization energies for a number of elements. The elements are arranged in the order in which they appear in the Periodic Table.

The same ionization energies are plotted on the graph shown in Figure 2-13. The change of ionization energy as atomic numbers increase is repeated regularly, as you can see in the graph.

Atomic Number	Element	Ionization Energy (kJ/mole)	Atomic Number	Element	Ionization Energy (kJ/mole)
1	H	1312	11	Na	496
2	He	2372	12	Mg	736
3	Li	520	13	Al	578
4	Be	900	14	Si	787
5	B	801	15	P	1012
6	C	1086	16	S	1000
7	N	1402	17	Cl	1251
8	O	1314	18	Ar	1521
9	F	1681	19	K	419
10	Ne	2081	20	Ca	590

Such a change is called *periodic*. Look at the elements of Period 2 (lithium, Li, to neon, Ne) and Period 3 (sodium, Na, to argon, Ar). The ionization energies for the elements between lithium and neon generally increase. Then there is a large drop to sodium (and the ionization energy for sodium is lower than the ionization energy for lithium). Between sodium and argon, the ionization energies again generally increase. Then there is another large drop to potassium, K (and the ionization energy for potassium is lower than the ionization energy for sodium). The general trend is an increase of ionization energy within a period (lithium to neon) and a decrease within a group (lithium to potassium). These observations agree with the conditions listed previously. (Some irregularities in these trends can be observed. Ionization energy decreases from beryllium, Be, to boron, B; from nitrogen, N, to oxygen, O; from magnesium, Mg, to aluminum, Al; and

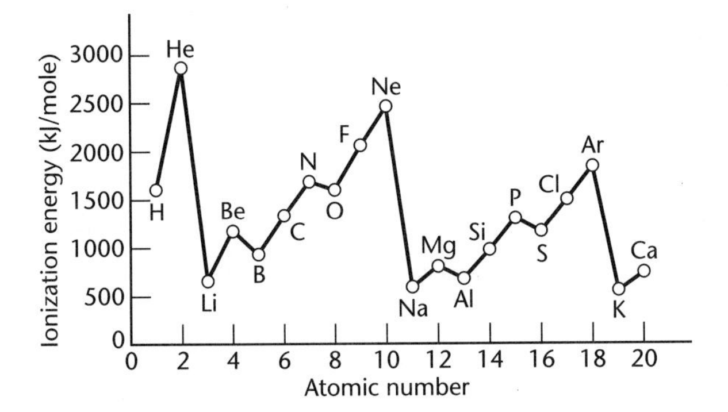

Figure 2-13 Trends in ionization energies

from phosphorus, P, to sulfur, S. The irregularities can be explained by a more detailed study of the electron structure in each pair of atoms.)

Ionization energy is a valuable measure of how firmly the valence electrons are held in a given atom. This information, in turn, helps to explain the formation of chemical bonds between atoms.

Electron Affinity and Electronegativity

Atoms can be bonded by losing, gaining, or sharing electrons. When an atom loses an electron and becomes a positively charged ion, energy is absorbed. The amount of energy needed is the ionization energy. The opposite process may also occur. An atom may receive an electron and become a negatively charged ion. In this process, energy generally is released. The amount of energy released is called **electron affinity.**

The ability of atoms in molecules to attract the electrons that bond the atoms also can be compared. This ability, which the American chemist Linus Pauling (1901–1994), has called "the power of an atom in a molecule to attract electrons to itself," is called **electronegativity.**

Chemists have set up a scale of the relative electronegativities of the elements. Electronegativity values of some elements are given in Figure 2-14. Of the elements shown, fluorine, F, has the greatest attraction for electrons. Fluorine is therefore given the highest electronegativity value (4.0). Further study of the electronegativity values in the table shows:

1. In every horizontal row, the electronegativity value is lowest for the elements in Group 1 (the alkali metals).
2. The electronegativity values increase in a regular manner from the alkali metals to the elements in Group 17 (the halogen elements).
3. In every vertical group of the Periodic Table (with some minor exceptions), the electronegativity values decrease from the top to the bottom of the group. The most electronegative element is at the top of Group 17.
4. The least electronegative element (excluding the noble gases) is at the bottom of Group 1.

ELECTRONEGATIVITIES OF SOME ELEMENTS
(on the arbitrary Pauling scale)

1	2	13	14	15	16	17
H 2.1						
Li 1.0	Be 1.6	B 2.0	C 2.6	N 3.0	O 3.4	F 4.0
Na 0.9	Mg 1.3	Al 1.6	Si 1.9	P 2.2	S 2.6	Cl 3.2
K 0.8	Ca 1.0	Ga 1.8	Ge 2.0	As 2.2	Se 2.6	Br 3.0
Rb 0.8	Sr 1.0					I 2.7
Cs 0.8	Ba 0.9					

Figure 2-14 Partial Periodic Table showing electronegativities

5. Except for some irregularities in the values for ionization energies, ionization energies and electronegativities follow similar trends. If you think about it, you will see why this is so. Atoms of elements that have low ionization energies exert a weak attractive force on their own outer electrons and also on the outer electrons of other atoms.

Ionization energies, electron affinities, and electronegativities are examples of some periodic properties of the elements. These properties are discussed fully in Chapter 5, along with other periodic properties.

TAKING A CLOSER LOOK

Order of Fill

The order in which electrons go into the available sublevels is called the order of fill. There are rules that predict the order of fill for most elements. You can use these rules to predict an electron configuration for an element from its atomic number alone, without consulting a reference table.

The sublevels fill in the following order: 1*s*, 2*s*, 2*p*, 3*s*, 3*p*, 4*s*, 3*d*, 4*p*, 5*s*, 4*d*, 5*p*, 6*s*, 4*f*, 5*d*, 6*p*, 7*s*, 5*f*, 6*d*. Instead of memorizing the order above, you may prefer to use a chart like this one:

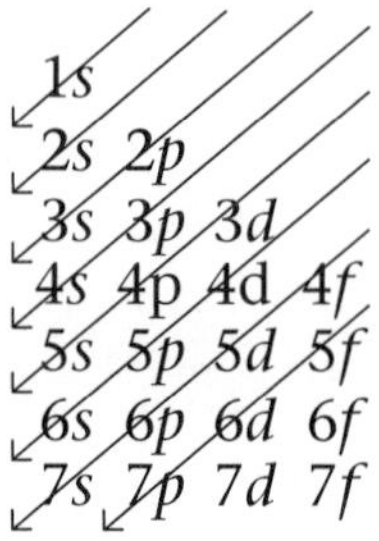

1s
2s 2p
3s 3p 3d
4s 4p 4d 4f
5s 5p 5d 5f
6s 6p 6d 6f
7s 7p 7d 7f

This chart simply lists all of the sublevels in each of the seven principal energy levels, one under another. By drawing diagonal lines, as shown, you obtain the correct order of fill. To obtain the electron configuration for an element, you simply fill the sublevels, in the order shown, until you run out of electrons. Remember that an *s* sublevel can hold 2 electrons, a *p* can hold 6, a *d* can hold 10, and an *f* can hold 14 electrons.

SAMPLE PROBLEM

PROBLEM

What is the probable electron configuration of an arsenic atom (atomic number = 33)?

SOLUTION

You need to fill in 33 electrons. Following the order of fill, begin with $1s^2 2s^2 2p^6 3s^2 3p^6$. Thus far you have used up 18 electrons. The next sublevel to fill is the 4*s*, followed by the 3*d*. If you fill the 4*s* with 2 electrons, there are 10 electrons left for the 3*d* and 3 electrons for the 4*p*, giving us the correct configuration: $1s^2 2s^2 2p^6 3s^2 3p^6 3d^{10} 4s^2 4p^3$ (Note that you write the 3*d* before the 4*s*, even though the 4*s* fills first. Recall the custom of keeping sublevels in the same principal energy level together when you write these configurations.)

PRACTICE

2.16 Without using a reference table, write the electron configurations for elements with the following atomic numbers.
(a) 25 (b) 40 (c) 29

Exceptions to the Order of Fill

If you tried Practice 2.16, the structure you obtained for element 29, (Cu) was probably $1s^22s^22p^63s^23p^63d^94s^2$. However, the structure copper is $3d^{10}4s^1$ instead of $3d^94s^2$. Copper is one of several exceptions to the order of fill. Let us examine why some of these exceptions occur.

As you move farther out from the nucleus, the energy differences between the sublevels become smaller and smaller. The 4*s* and 3*d,* for example, are very close in energy. The 4*s* generally fills before the 3*d*. However, there is extra stability when *p, d,* or *f* sublevels are exactly filled and exactly half filled. By putting 10 electrons into the 3*d* sublevel it is filled completely, giving it added stability and making it just lower in energy than the 4*s*.

You can see a similar occurrence in chromium, which has the configuration $[Ar]3d^54s^1$. Here the 3*d* sublevel is exactly half filled. The extra stability causes an electron to prefer the 3*d* sublevel to the 4*s*. There are many other exceptions to the order of fill. Not all of them have been completely explained.

Trends in Ionization Energy

The trends in ionization energy are shown in Figure 2-13 on page 101. You can see that as you move from left to right, across a period, the ionization energy generally increases. In Period 2, from lithium, Li, to neon, Ne, you see this trend clearly. However, you can see two exceptions to the upward trend. The ionization energy dips between beryllium Be, and boron, B, and then dips again between nitrogen, N, and oxygen, O. Notice that similar dips occur in every period, not just in Period 2. Why?

Beryllium has the configuration $1s^22s^2$. the configuration of B is $1s^22s^22p^1$. Since boron has one additional proton pulling on its electrons, you might expect its ionization energy to be higher than that of beryllium. However, the electron being removed from the boron is a $2p$ electron, while the electron removed from a beryllium atom would be a $2s$ electron. A $2p$ electron is screened by the $2s$ electrons, and thus, lost more easily.

The dip between nitrogen and oxygen can be explained using a principle already mentioned, the extra stability of half-filled sublevels. The configuration of nitrogen is $1s^22s^22p^3$. The p sublevel is half filled. In oxygen, which is $1s^22s^22p^4$ the p is no longer exactly half filled, and it is easier to remove an electron from it.

Electron Configurations of Ions

Elements may gain or lose electrons to form charged particles called ions. The electron configurations of these ions are not listed on most reference tables, because they can generally be determined from the configurations of the atoms.

When atoms lose electrons they form positive ions. These ions are positively charged because, having lost electrons, they now contain more positive protons than negative electrons. The electrons lost are from the highest principal energy level of the atom. For example, in the transition elements, the outer s electrons are lost before the inner d electrons. Consider the structure of magnesium (Mg), element number 12, $1s^22s^22p^63s^2$. To form a 2+ ion, it must lose two electrons. These electrons come from the last principal energy level. Taking two electrons from the $3s$ sublevel leaves a configuration for the Mg^{2+} ion of $1s^22s^22p^6$. This configuration is identical to that of a neutral atom of neon. Particles with the same number of electrons are said to be *isoelectronic*. Thus the Mg^{2+} ion is isoelectronic to a neon atom.

Because of the loss of electrons, positive ions are always smaller than their neutral atoms. In the case of the Mg^{2+} ion, an entire principal energy level has been lost. Because it now has only two principal energy levels, the Mg^{2+} ion is much smaller than the magnesium atom (less than half as large).

The sodium ion, Na^+, has exactly the same electron configuration as the Mg^{2+} ion. Sodium atoms have 11 electrons, so the 1+ ion has 10. Magnesium atoms have 12 electrons, so the 2+ ion has 10. How would a Na^+ ion and a Mg^{2+} ion compare in size? They have exactly the same number of electrons and exactly the same electron configurations. However, magnesium has 12 protons, while sodium has 11 protons. Since there is a greater positive charge pulling on the electrons in the magnesium ion, it is smaller than the sodium ion. Generally speaking, the larger the positive charge in the nucleus of an isoelectronic ion, the smaller the ion.

Negative ions have more electrons than protons. The O^{2-} ion, for example, has 10 electrons. Oxygen has an atomic number of 8. When oxygen becomes a 2− ion it gains two electrons. The O^{2-} ion, then, has the configuration $1s^22s^22p^6$. This makes the O^{2-} ion isoelectronic to the Na^+ ion and the Mg^{2+} ion discussed above. However, negative ions are larger than their neutral atoms, because the additional electrons repel each other. The O^{2-} ion is larger than Na^+ or Mg^{2+} and larger than neutral Ne, which also has 10 electrons. Generally speaking, the more negative the isoelectronic ion, the larger its radius.

PRACTICE

2.17 Titanium has an atomic number of 22. The Ti^{2+} ion has two unpaired electrons.

(a) How many electrons are there on this ion?

(b) What is the electron configuration of a Ti^{2+} ion?

2.18 What is the electron configuration of an Fe^{3+} ion? Iron is atomic number 26. (Hint: the correct structure has five unpaired electrons.)

CHAPTER REVIEW

The following questions will help you check your understanding of the material presented in the chapter.

Data required for answering questions in this chapter will be found in Table S and in the Periodic Table in Appendix 5.

1. Which is the correct orbital notation for the electrons in the second principal energy level of a beryllium atom in the ground state?

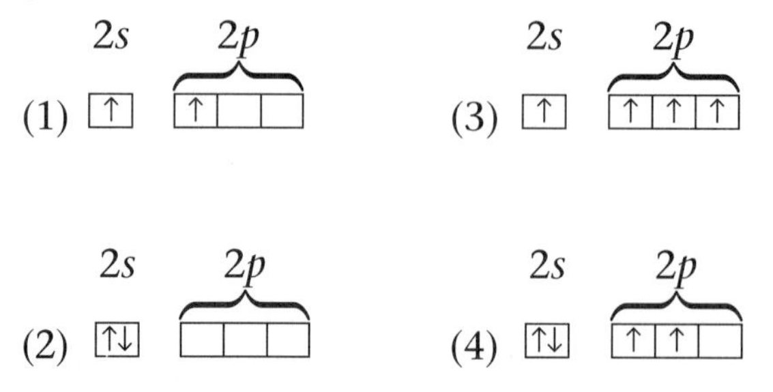

2. An atom with the electron configuration $1s^2 2s^2 2p^6 3s^2 3p^6 4s^2$ has an incomplete (1) second principal energy level (2) 2*s* sublevel (3) third principal energy level (4) 3*s* sublevel.
3. How many electrons are in a neutral atom of $^{7}_{3}Li$? (1) 10 (2) 7 (3) 3 (4) 2
4. When an aluminum atom loses three electrons it would form an ion with a charge of (1) 1+ (2) 1− (3) 3+ (4) 3−
5. Which set of particles is arranged in order of increasing mass? (1) H_2, H, H^+ (2) H^+, H, H_2 (3) H_2, H^+, H (4) H, H^+, H_2
6. Isotopes of the same element do not have the same (1) number of electrons (2) atomic number (3) mass number (4) electron configuration.
7. How many moles of helium contain the same number of molecules as 4 moles of neon? (1) 20 (2) 10 (3) 8 (4) 4
8. Which correctly represents an atom of neon containing 11 neutrons? (1) $^{11}_{10}Ne$ (2) $^{20}_{11}Ne$ (3) $^{21}_{10}Ne$ (4) $^{21}_{11}Ne$
9. A Mg^{2+} ion has the same electron configuration as (1) Na^0 (2) Ar^0 (3) F^- (4) Ca^{2+}
10. An atom of which element in the ground state contains electrons with a principal quantum number (*n*) of 4? (1) Kr (2) Ar (3) Ne (4) He
11. Which energy level transition represents the greatest absorption of energy? (1) 1*s* to 3*p* (2) 2*p* to 3*s* (3) 3*s* to 3*p* (4) 3*s* to 4*s*
12. Which is the electron configuration of the element having the highest ionization energy? (1) 2-3 (2) 2-4 (3) 2-6 (4) 2-7

13. Energy in the form of light is emitted when an electron in a hydrogen atom moves from a 3*s* sublevel to a (1) 4*s* (2) 3*p* (3) 3*d* (4) 2*s*

14. Which element requires the least amount of energy to remove its most loosely bound electron? (1) Li (2) Mg (3) K (4) Ca

15. The isotopes $^{1}_{1}H$ and $^{2}_{1}H$ have the same (1) atomic number (2) mass number (3) number of neutrons (4) density.

16. What is the total charge on an ion that contains 10 electrons, 13 protons, and 15 neutrons? (1) 1− (2) 1+ (3) 3− (4) 3+

17. In which way does an Na^{+} ion differ from an Na^{0} atom? (1) atomic number (2) mass number (3) number of electrons (4) nuclear charge

18. What is the maximum number of electrons that can occupy the 3*d* sublevel? (1) 6 (2) 8 (3) 10 (4) 14

19. An element in its ground state has 5 valence electrons. Which is the correct distribution of these electrons?

20. What is the number of orbitals in a *p* sublevel? (1) 7 (2) 6 (3) 3 (4) 1

21. Which element in the ground state has a valence electron in a *p* subshell? (1) Na (2) Mg (3) Al (4) Be

22. The maximum number of electrons possible in any principal energy level (principal quantum number = n) is equal to (1) n (2) $2n$ (3) n^2 (4) $2n^2$

23. Which represents the electron configuration of an isotope of oxygen in the ground state? (1) $1s^22s^22p^1$ (2) $1s^22s^22p^2$ (3) $1s^22s^22p^3$ (4) $1s^22s^22p^4$

24. As the elements in Period 3 are considered in order of increasing atomic number, the number of principal energy levels

in each successive element (1) decreases (2) increases (3) remains the same.

25. Which is the electron configuration for a neutral atom in the ground state? (1) $1s^22s^23s^1$ (2) $1s^22s^22p^43s^1$ (3) $1s^2 2s^22p^63p^1$ (4) $1s^22s^22p^63s^1$

26. What is the number of sublevels in the third principal energy level? (1) 1 (2) 2 (3) 3 (4) 4

27. An ion has the electron configuration $1s^22s^22p^63s^23p^6$ and a charge of 1−. The number of protons in its nucleus is (1) 16 (2) 17 (3) 18 (4) 19.

Base your answers to questions 28 through 30 on the following electron configuration of a neutral atom:

$$1s^22s^12p^3$$

28. How many protons are in the nucleus of this atom? (1) 6 (2) 5 (3) 3 (4) 2

29. How many principal energy levels are in this electron structure? (1) 1 (2) 2 (3) 3 (4) 4

30. How many incomplete orbitals are indicated by this electron configuration? (1) 1 (2) 2 (3) 3 (4) 4

31. When the aluminum atom is in the ground state, how many orbitals contain only one electron? (1) 1 (2) 2 (3) 3 (4) 13

32. Which orbital contains the valence electrons of a calcium atom? (1) 1*s* (2) 2*s* (3) 3*s* (4) 4*s*

33. The number of neutrons in ${}^{3}_{1}H$ is (1) 1 (2) 2 (3) 3 (4) 4.

34–38 Base your answers on the ground state electron configurations of four elements given below. In each case choose the element that best fits the description. (1) 2-7 (2) 2-4 (3) 2-8-1 (4) 2-8-2

34. The element with the fewest valence electrons

35. The element with the greatest electronegativity

36. The element with the lowest ionization energy

37. This element contains no unpaired electrons.

38. This element contains five completely filled orbitals.

CONSTRUCTED RESPONSE

1. A certain element has an extremely low ionization energy. Predict how this element would compare with most other elements in:
   (a) electronegativity
   (b) number of principal energy levels
   (c) number of valence electrons.
   Explain each of your answers.
2. In the Periodic Table, you can see that Period 2 consists of eight elements. (Li-Ne) The sublevels that fill in this period are the 2*s* and the 2*p*. For Periods 3, 5, and 6, identify
   (a) the number of elements in the period, and
   (b) the sublevels that fill in the period.
3. Bohr's theory of the atom contained the following assertions:
   (a) Electrons move in fixed, circular orbits around the nucleus.
   (b) The energy of the electron is quantized.
   (c) Electrons emit energy when they fall from higher to lower energy levels.

   Which of these assertions has been changed in modern atomic theory? How has it been changed?
4. Based on the electron configurations given on the Periodic Table in Appendix 5, which 4 elements in Groups 3 to 10 have no unpaired electrons?

CHEMISTRY CHALLENGE

The following questions will provide practice in answering SAT II-type questions.

Use the Periodic Table in Appendix 5 to help you.

1. The S^{2-} ion has the same number of electrons as (1) Ne (2) Na^+ (3) Ca^{2+} (4) K
2. Of the following ions, the one with the smallest ionic radius is (1) O^{2-} (2) F^- (3) N^{3-} (4) Na^+
3. An ion with the electron configuration $1s^22s^22p^63s^23p^6$ has a charge of 2−. The number of protons in this ion is (1) 18 (2) 20 (3) 16 (4) 22
4. The sublevel filling in elements 39–48 is the (1) 3*d* (2) 4*d* (3) 4*f* (4) 5*f*
5. How many unpaired electrons are there on an element with the electron configuration $[Ar]3d^64s^2$? (1) 6 (2) 5 (3) 4 (4) 2
6. What is the next sublevel to fill after the 4*d*? (1) 4*p* (2) 3*s* (3) 5*p* (4) 5*s*
7. As an atom gains electrons, its radius (1) decreases (2) increases (3) remains the same
8. A V^{2+} ion loses three electrons. The new ion formed would have the symbol (1) V^{1-} (2) V^{5+} (3) Ca^{2+} (4) Ca^{1-}
9. Which element in Period 4 has six unpaired electrons in the ground state? (1) Cr (2) Mn (3) Fe (4) Ar
10. Elements that have low ionization energies (1) tend to have small atomic radii (2) tend to have low electronegativities (3) tend to form negative ions (4) tend to be found on the right side of the Periodic Table.

Heisenberg Might Have Slept Here

Werner Heisenberg is known best for his "uncertainty principle," which says that there is no way we can measure with complete certainty the behavior of a physical object. The uncertainty principle helped Heisenberg win the Nobel Prize in 1932. In Heisenberg's case, though, the uncertainty principle applies to his life as much as it does to his science.

Heisenberg headed the German atomic energy project during World War II. His objective was to develop an atomic bomb for Germany. If he had succeeded, the course of history might have been changed. We are uncertain as to why the German project failed. Some historians believe that Heisenberg deliberately sabotaged the project to prevent Nazi Germany from acquiring so powerful a weapon. Others insist that Heisenberg simply miscalculated the amount of fissionable material needed to build a nuclear weapon; as a loyal and patriotic German, he would have built the bomb had he been able to. Heisenberg's biography, written by David Cassidy, is entitled *Uncertainty: The Life and Science of Werner Heisenberg.*

Playwright Michael Frayn found the uncertainty surrounding Heisenberg so fascinating that he wrote the play *Copenhagen* about Heisenberg's 1941 meeting with the Danish physicist Niels Bohr. Before the war the two men had been good friends. However, by 1941 Germany had invaded and occupied Denmark, which strained their friendship. In the play, Frayn presents three versions of what might have happened during that meeting. *Copenhagen* had successful runs both in London and New York, and it has recently been made into a movie. The playwright believes that the uncertainty principle is as applicable to human thinking as it is to the movement of electrons.

Once, while driving through New Jersey, I passed a car with the bumper sticker that said, "Heisenberg Might Have Slept Here." I laughed out loud.

3 Bonding

LOOKING AHEAD

The binding forces between atoms, ions, and molecules explain a great deal about chemical reactions. Chemical bonds also relate to structure, which, in turn, explains chemical properties. Although you will study many kinds of bonds, only one force in nature accounts for the formation of all chemical bonds. This force is the attraction between positively charged and negatively charged matter.

When you have completed this chapter, you should be able to:

- **Describe** bonding in molecules, metals, ionic compounds, and network solids, and associate properties with different types of bonds.
- **Discuss** factors that influence intermolecular attractions.
- **Determine** the bonding in a particle, given the components, a property, or a table of electronegativities.
- **Distinguish** between single, double, and triple covalent bonds.
- **Predict** the shapes of the molecules formed when the central bonding atom is B, Al, C, N, P, O, or S.
- **Relate** potential energy to stability; exothermic and endothermic reactions to changes in potential energy; molecular shape to polarity; and the bonds within a molecule to its polarity.
- **Diagram** the covalent bonding structures in simple molecules.

Formation of Chemical Bonds

When two hydrogen atoms come close enough together, they unite and form a diatomic hydrogen molecule. However, when two helium atoms come close together, they do not unite and form a helium molecule. They remain as separate atoms. Sodium atoms unite with chlorine atoms and form a stable compound—sodium chloride, or table salt. Silver atoms and copper atoms do not unite. What are the factors that determine whether or not atoms will unite? What are the different ways in which atoms can unite?

When atoms unite, attractive forces tend to pull the atoms together. These attractive forces are called **chemical bonds**, often referred to in shortened form as bonds. When a chemical bond forms, energy is released. When a chemical bond breaks, energy is absorbed. Thus when two atoms are held together by a chemical bond, the atoms are at a lower energy condition than when they are separated.

It is a principle of science that systems at low potential energy levels are more stable than systems at high potential energy levels. A ball rolls unaided downhill but not uphill. This is because the ball has less potential energy and greater stability on the bottom of the hill than it has on top of the hill. *Chemical changes*—that is, bond-making and bond-breaking—*occur if the change leads to a lower energy condition and hence to a more stable structure.* This observation suggests that chemical changes proceed in the direction that liberates energy. Or, to put it another way, nature favors exothermic reactions. (This is only part of the total concept. You will learn more about the forces that determine the direction of chemical reactions in Chapter 11.)

Forces Between Atoms

The forces that tend to establish chemical bonds between two atoms result from the interactions between the protons and electrons in the atoms. These forces include

1. Repulsions between the electrons of the two atoms.
2. Repulsions between the nuclei of the two atoms.
3. Attractions between the nucleus of one atom and the electrons of the other atom.

A chemical bond results when force 3 is greater than the sum of forces 1 and 2. Or, a chemical bond between two atoms results when the forces of attraction between the nuclei and the electrons of opposite atoms are greater than the repulsions between the two electron systems and between the two nuclei. Thus a chemical bond may be characterized as the force resulting from the simultaneous attraction of two nuclei for one or more pairs of electrons.

Sizes of Atoms and Molecules

Bond formation occurs when one atom approaches another, if the net forces of attraction are greater than the net forces of repulsion. The distance between the bonded atoms is often called the *bond length.* Figure 3-1 shows a model of two bonded gaseous iodine atoms. The distance D_C represents the distance between the nuclei of the two bonded atoms. This bond length is found to be 264 picometers. One picometer, abbreviated pm, is 10^{-12} meter. One half of the bond length, labeled R_C, is called the *covalent atomic radius*. It is found to be 132 pm. This is the radius of a bonded gaseous iodine atom.

Atomic sizes are, at best, only approximations, since single atoms cannot be isolated. Further, the boundary of an atom is uncertain—electrons may appear at any distance from the nucleus, although the probability is greater nearer the nucleus. In addition, the presence of other atoms affects the electron distribution in a given atom. Atomic sizes become meaningful only if measured in the same manner and only if used for comparison. Then

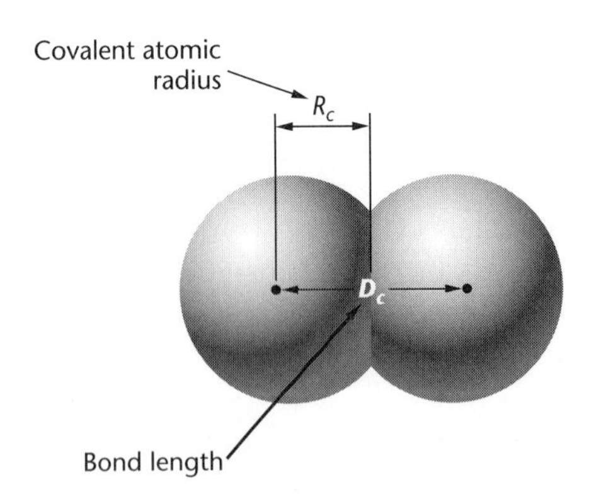

Figure 3-1 Covalent radius

atomic sizes offer a means of relating physical and chemical properties of atoms and the particles in which they appear.

Bond Energy and Stability

Consider two hydrogen atoms coming together. As the atoms approach each other, forces of attraction and repulsion come into play. The electron of the first atom "feels" the attraction of the positive nucleus of the other atom. The electron of the second atom "feels" the attraction of the positive nucleus of the first atom. At the same time, the two electrons repel each other, and the two nuclei repel each other. At some point, the forces of attraction balance the forces of repulsion. At this point, a system of lower potential energy, or greater stability, is reached. This situation is shown in Figure 3-2.

Distance *d* is the distance between the two atoms where the potential energy is lowest. At this position, a stable bond can form between the two atoms. If the atoms come closer, the positive nuclei begin to repel each other strongly. The electrons also begin to repel each other strongly. A sharp increase in potential energy results. This increase means that the forces of repulsion are prevailing.

The distance between the atoms corresponding to the lowest potential energy can be represented as the distance *d* between the two nuclei in the hydrogen molecule (Figure 3-3). Since the

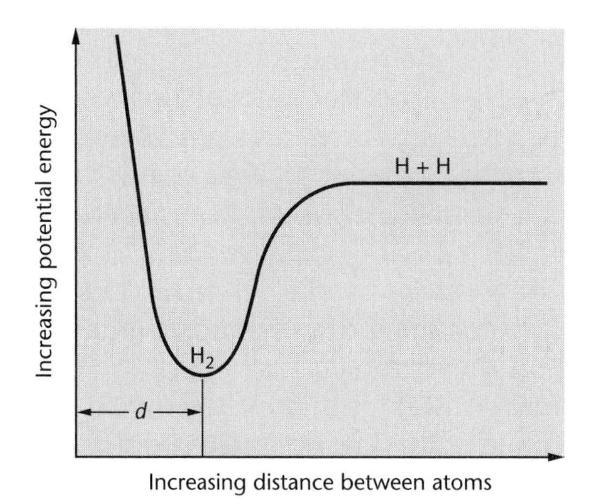

Figure 3-2 Potential energy diagram for the formation of a hydrogen molecule

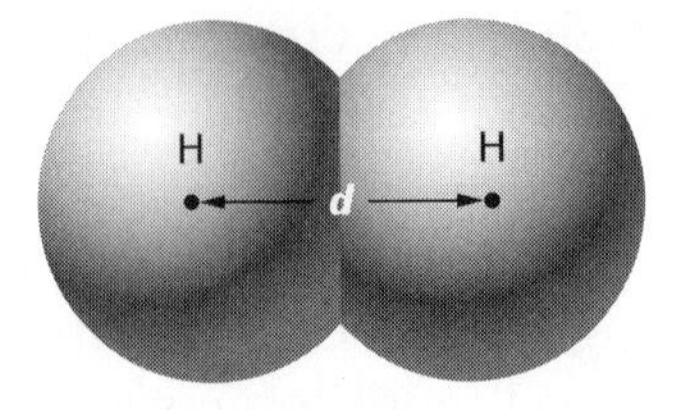

Figure 3-3 Interatomic distance in a hydrogen molecule (lowest potential energy)

electrons are in motion, this distance is not fixed. It is an *average* distance. The distance is found to be 74 pm. This bond is extremely strong—435 kJ/mole are required to break it. Conversely, 435 kJ/mole are released when the bond forms.

Bonds Between Atoms

The attractive forces between negatively and positively charged matter account for the formation of all chemical bonds. The simultaneous attraction of electrons from one atom to the positively charged nucleus of another atom constitutes the attractive force. Several kinds of bonds between atoms form because of these interactions.

The Covalent Bond

The most common type of bond between atoms is formed by the sharing of one or more pairs of electrons. This type of bond is called a **covalent bond**. One, two, or three pairs of electrons may be shared, resulting in one, two, or three covalent bonds. Each atom usually contributes one electron for a given shared pair. In some instances, a single atom may contribute both electrons of a pair.

The bonding capacity of an atom can be related to the number of unpaired valence electrons the atom contains. Some examples are given in the table on the next page.

Notice that carbon is shown with only one electron in the 2*s* sublevel. Carbon in the ground state is represented as follows:

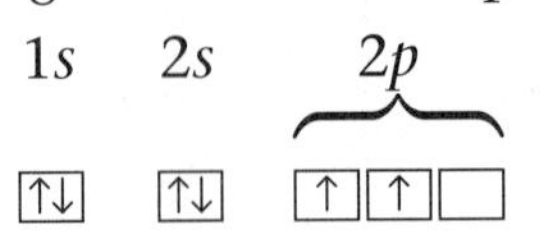

Element	Valence Electrons 1s	2s	2p	Bonding Capacity (number of unpaired valence electrons)
Hydrogen	[↑]			1
Carbon		[↑]	[↑][↑][↑]	4
Nitrogen		[↑↓]	[↑][↑][↑]	3
Oxygen		[↑↓]	[↑↓][↑][↑]	2
Fluorine		[↑↓]	[↑↓][↑↓][↑]	1
Neon		[↑↓]	[↑↓][↑↓][↑↓]	0

This structure shows two unpaired electrons, which would permit two covalent bonds, as in a compound like

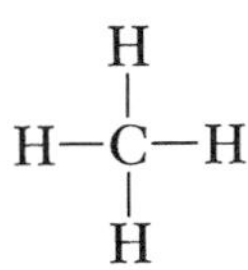

Such compounds have been prepared, but they are very unstable. However, compounds such as

```
     H
     |
  H—C—H
     |
     H
```

are easily prepared. This suggests another structure for a carbon atom.

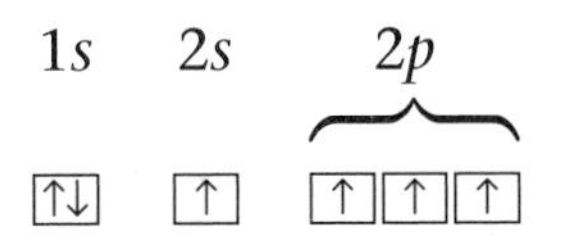

To attain this structure, an electron from 2*s* is promoted to the higher 2*p* sublevel. The energy required to effect this change is small. In exchange for the loss of energy, the carbon atom gains additional bonding capacity. In fact, in most carbon compounds, carbon atoms have *four* covalent bonds.

Covalent bonding can be shown conveniently by diagramming the valence electrons. In the examples that follow, the first diagram represents the atoms of a hydrogen molecule, H_2. The second represents a molecule of hydrogen fluoride, HF.

H [↑] 1s
H [↓] 1s

F [↑↓] 2s [↑↓][↑↓][↑] 2p
H [↓] 1s

Another common way of showing covalent bonds is by an electron-dot diagram. In this type of diagram, the symbol for the element is used to stand for the *kernel* of the atom, which consists of the nucleus and all the electrons of the atom except the valence electrons. The valence electrons are shown as dots around the symbol. For example, the electron-dot diagram for hydrogen is H·. The electron-dot diagram for fluorine is

$$\cdot\ddot{\underset{\cdot\cdot}{\text{F}}}:.$$

The electron-dot diagram showing the covalent bonding of the atoms of a hydrogen molecule is H:H. Each atom in a hydrogen molecule has one valence electron.

The electron-dot diagram for a molecule of hydrogen fluoride is

$$\text{H}:\ddot{\underset{\cdot\cdot}{\text{F}}}:.$$

The fluorine atom has 7 valence electrons, one of which it contributes to the covalent bond. The hydrogen atom contributes 1 electron to the bond. These electron-dot diagrams are also known as "Lewis structures," named after the American chemist Gilbert N. Lewis.

In a typical covalent bond, each atom can attain, through sharing, the same number of electrons in its outer principal energy level as has the nearest noble gas—that is, either 2 or 8 electrons. For example, through sharing, each H atom of H_2 has attained 2 electrons, as in the noble gas He. In HF, the F atom has attained, through sharing, 8 electrons, as in Ne.

Hydrogen, with its single principal energy level, is the only element that can attain, through sharing, a noble-gas structure with just two valence electrons. For all of the other elements, covalent bonds generally result in a total of 8 valence electrons.

We can state that elements generally bond so as to attain a structure containing 8 valence electrons. This statement is known as the **octet rule**.

Polar and Nonpolar Covalent Bonds

In the previous chapter, you learned about the electronegativity scale. Electronegativity, you will recall, measures the relative attraction atoms have for electrons. Consider the HF molecule, described above. Fluorine, with an electronegativity of 4.0, has a much stronger attraction for electrons than does hydrogen, at an electronegativity of 2.1. Therefore, when hydrogen and fluorine share electrons, they do not share them equally.

Fluorine pulls the electron pair more strongly, and so acquires more than half of the negative charge of the two electrons. Thus the fluorine atom becomes slightly negative. The hydrogen atom, with its smaller share of the electron pair, becomes slightly positive. The bond, as shown in Figure 3-4, is positive on one side and negative on the other. It is called a **polar covalent bond**. Any covalent bond between different atoms will have some polarity. The greater the difference in electronegativity between the two atoms, the more polar the bond.

Now consider the bond between the hydrogen atoms in the hydrogen molecule, H_2. Since both atoms have exactly the same attraction for the electrons, the bond is a **nonpolar covalent bond**. No charge develops on either side of the bond. The bond in diatomic elements, such as F_2 and Cl_2, is always nonpolar covalent.

Multiple Covalent Bonds

Following the rule that elements generally attain through covalent bonding the same number of electrons as the nearest noble gas, some atoms share two or even three pairs of electrons.

$$\overset{\delta^+}{H}-\overset{\delta^-}{F}$$

Figure 3-4 The Greek letter delta (δ) is used to show partial charges resulting from unequal sharing of electrons

This generally results in the element acquiring 8 valence electrons, in accordance with the octet rule. In the examples that follow, each line between atoms represents a pair of shared electrons—a covalent bond. The electron-dot diagrams for the molecules are also given.

Ethylene, C_2H_4

$$\begin{array}{ccc} H & & H \\ | & & | \\ C & = & C \\ | & & | \\ H & & H \end{array} \quad \text{or} \quad \begin{array}{ccc} H & & H \\ \cdot\cdot & & \cdot\cdot \\ C & :: & C \\ \cdot\cdot & & \cdot\cdot \\ H & & H \end{array}$$

Acetylene, C_2H_2 $\quad H{-}C{\equiv}C{-}H \quad$ or $\quad H{:}C{:::}C{:}H$

Note how in both molecules, the carbon atoms form the number of bonds necessary to provide each carbon atom with an octet, 8 valence electrons.

Dot Structures

By applying the octet rule, you can draw correct dot structures for many molecules. Let us consider the structure of SO_2, the gas produced when sulfur is burned in oxygen. Since a dot structure shows all valence electrons, you first need to know how many such electrons are found in each atom. By consulting the Periodic Table (in Appendix 5) you can see that both sulfur and oxygen have 6 valence electrons. Now try to draw the dot structure. Figure 3-5 A shows the correct number of valence electrons. However, there are only 6 electrons around the sulfur, which violates the octet rule. By using one double bond, as shown in Figure 3-5 B, you can achieve 8 valence electrons around the sulfur and around both oxygens. Figure 3-5 B, then, is the preferred dot structure of SO_2.

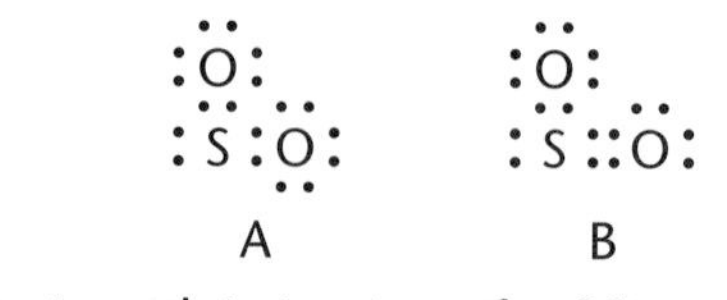

Figure 3-5 Electron-dot structures for SO_2

PRACTICE

3.1 Draw dot structures of the following molecules, which obey the octet rule. Remember that hydrogen, with only one energy level, will acquire 2 valence electrons.

(a) NH_3 (b) H_2CO (c) SO_3 (d) BrCl (e) O_3 (f) N_2

3.2 Arrange the following bonds, from most highly polar to least polar. Refer to the table of electronegativities on page 103.

(a) H—I, (b) Br—Cl, (c) N—I, (d) H—O

3.3 Hydrogen cyanide, the gas used in the gas chambers in some states in the United States, has the structure

$$H{-}C{\equiv}N$$

What is the total number of electrons represented by the lines connecting the C to the N?

3.4 What atom would form a nonpolar covalent bond with a fluorine atom?

Coordinate Covalent Bonds

Let us consider another way in which covalent bonds may form. A particle (an atom or an ion) that can accept a pair of electrons unites with a particle that can donate a pair of electrons. A coordinate covalent bond forms between the two particles. In a **coordinate covalent bond,** both electrons in a shared pair are supplied by a single atom. In an ordinary covalent bond, each atom supplies one electron to form the shared pair.

The hydrogen ion (H^+) is an example of a particle that can accept a pair of electrons. The hydrogen ion has a positive charge, and its outer electron level is completely unoccupied. This means that it has room for two electrons to attain the helium structure (He**:**).

Nitrogen, in the compound ammonia (NH_3), has an unshared pair of electrons. The electron-dot diagram of ammonia is

$$\begin{array}{c} H \\ \cdot\cdot \\ H{:}N{:}H \\ \cdot\cdot \end{array}$$

pair of electrons is not involved in bond formation
ound. This pair of electrons, called a **lone pair**, can be
a coordinate covalent bond. H^+, with its completely
n level, can accept these electrons, as follows:

$$\begin{array}{c}\mathrm{H}\\ \mathrm{H{:}\underset{\cdot\cdot}{\overset{\cdot\cdot}{N}}{:}H}\end{array} + \mathrm{H^+} \longrightarrow \left[\begin{array}{c}\mathrm{H}\\ \mathrm{H{:}\underset{\cdot\cdot}{\overset{\cdot\cdot}{N}}{:}H}\\ \mathrm{H}\end{array}\right]^+$$

$[NH_4]^+$ is called the ammonium ion.

Water has two lone pairs of electrons and can form a coordinate covalent bond with H^+, as shown:

$$\begin{array}{c}\mathrm{{:}\underset{\cdot\cdot}{\overset{\cdot\cdot}{O}}{:}H}\\ \mathrm{H}\end{array} + \mathrm{H^+} \longrightarrow \left[\begin{array}{c}\mathrm{H{:}\underset{\cdot\cdot}{\overset{\cdot\cdot}{O}}{:}H}\\ \mathrm{H}\end{array}\right]^+$$

$[H_3O]^+$ is called the hydronium ion. The hydronium ion can also be written as $H^+(H_2O)$ and as $H^+(aq)$, in which (*aq*) represents water.

Note that in the structures shown for the NH_4^+ and the H_3O^+ ions, while you know that one of the bonds is coordinate covalent, you cannot tell which one it is. Once a coordinate covalent bond forms, it is indistinguishable from an ordinary covalent bond. All four bonds on the ammonium ion are the same length and have the same bond energy.

Ions such as these, which contain two or more atoms, are called **polyatomic ions**. A list of some of the more common polyatomic ions appears in Appendix 4. Most polyatomic ions contain coordinate covalent bonds.

Metallic Bonds

The atoms of metals have few—usually no more than three—valence electrons. These valence electrons are free to move throughout the solid, producing positively charged metallic ions. Such electrons are called mobile, or nonlocalized, electrons. Thus the particles in a metal are positive ions surrounded by mobile electrons that can drift from one atom to another. You can think of **metallic bonds** as a sea of mobile valence electrons surrounding relatively stationary positive ions.

Metallic bonds are strong. The strength of any bond depends on the strength of the attractive forces between positively and negatively charged particles. The strong bonds of metals result from the attractions of all the positive ions for the electrons surrounding them. This tends to produce a rigid solid with a definite shape. Considerable energy is required to overcome these attractive forces—that is, to liquefy a solid. Thus metals, as a rule, have high melting points. The mobile electrons account for the luster of metals and for their ability to conduct heat. The mobile electrons also account for the ability of metals to conduct electricity. (An electric current is simply the flow of electrons through a wire.) More about the properties of solids will be discussed later.

Ionic Bonds

Recall that electronegativity is "the power of an atom in a molecule to attract electrons to itself." When a chemical bond forms between elements that have greatly differing electronegativity, the electrons are displaced completely and a transfer of electrons takes place. As a result, positive and negative ions are formed. Such a bond is called an **ionic bond**.

Sodium chloride (NaCl), or common table salt, is an ionically bonded compound. Sodium has a low ionization energy and low electronegativity (0.9). Chlorine has a high ionization energy and also high electronegativity (3.2). The valence electron of sodium transfers to chlorine. Sodium, Na^+, and chlorine, Cl^-, ions are formed.

$$\mathrm{Na}\cdot \curvearrowright :\!\ddot{\underset{\cdot\cdot}{\mathrm{Cl}}}\!:$$

Enormously large numbers of ions (about 6×10^{23} for less than 60 grams of NaCl) attract one another in a three-dimensional pattern called a lattice, which forms a crystal of NaCl. A small section of the ionic solid NaCl is shown in Figure 3-6.

The diagram shows that the radius of Na^+ is small compared with that of Cl^-. Positively charged ions are smaller than the same neutral atom. The excess positive charge of an ion draws electrons closer to the nucleus, shrinking the ion. Negative ions are larger than the same neutral atom. The increased electron-electron repulsions of the ion expand the ion.

A crystal of sodium chloride is composed of ions situated at the corners of a cube, with equally constructed cubes extending in all directions. As you can see in Figure 3-6, there are no sodium chloride molecules present—no specific Cl^- "belongs" to a specific Na^+. The electron-dot diagrams of ionic compounds are therefore different from those of molecular compounds. Usually, the symbols are placed in brackets. No electrons are shown on the positive ions. The valence electrons of the negative ion and the electrons gained from the positive ion are shown on the negative ion.

$$[\mathrm{Na}]^+ \; [:\ddot{\underset{\cdot\cdot}{\mathrm{Cl}}}:]^-$$

The strength of the attractions between the large numbers of positive and negative ions in an ionic compound is such that the compounds are solids at ordinary conditions. (Recall that metals are generally solids.) Most ionic solids have a high melting point. The melting point of NaCl, for example, is 801°C.

Ionic solids do not conduct an electric current, because the ions do not have enough mobility. When the solid is fused, or melted, the ions become more mobile, and the liquid conducts electricity.

Many ionic compounds are soluble in water. Since water is a polar compound, electrical interactions weaken the attractive forces in ionic compounds. This breaks the ionic lattice, and the compound dissolves. The resulting solution has mobile ions and is a strong conductor of electricity.

How can you predict the formation of ionic bonds? The table of electronegativities is helpful. You have already used it to distinguish nonpolar covalent bonds, which form between atoms

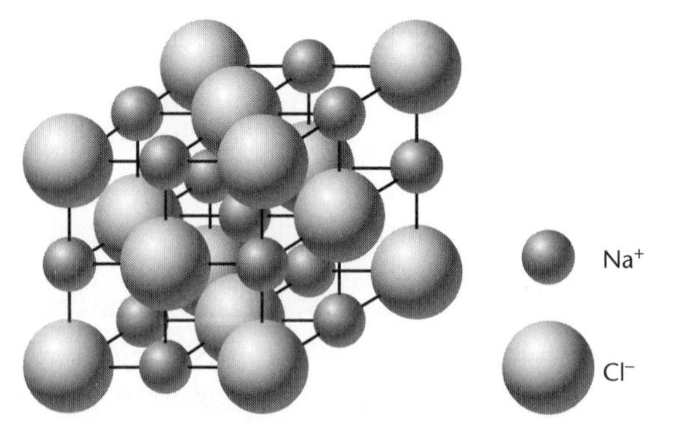

Figure 3-6 A portion of the ionic compound NaCl

of identical electronegativity, from polar covalent bonds, which form when atoms have different electronegativities. The element with the higher electronegativity pulls electrons harder, and so acquires a greater share of the electron pair and a slight negative charge. The greater the difference in electronegativity, the more unequal the sharing and the more polar the bond.

When the difference in electronegativity reaches 1.7 or more, however, the sharing is so unequal that the electrons essentially belong to the more electronegative element. That element thus acquires a full negative charge, and the bond is considered an ionic bond. The general trends in bond formation are shown in the table below:

Electronegativity Difference	***0***	***>0, <1.7***	***>1.7***
Bond type	Nonpolar covalent	Polar covalent	Ionic
Example	O_2	HCl	NaCl

There are some important exceptions to these rules, however. The hydrogen ion, H^+, is too small to occupy a site in an ionic crystal. Therefore, HF is polar covalent, even though the electronegativity difference is greater than 1.7. Metal hydrides, such as NaH and CaH_2 are generally ionic, even though the electronegativity difference is less than 1.7. Compounds containing metals, especially those of Groups 1 and 2 in the Periodic Table, are almost always ionic. Compounds that contain only nonmetals are usually covalent.

Chemists frequently refer to the ionic character of a bond. The greater the difference in electronegativity, the greater the ionic character of the bond. You may recall that in Practice 3.2 you were asked to determine which bond from a group of bonds was the most highly polar. Asking which bond has the greatest ionic character means the same thing.

PRACTICE

3.5 Based on the information in Figure 2-14 on page 103, state whether each of the following compounds is ionic or covalent.

(a) BrCl (b) HCl (c) KCl (d) $CaBr_2$

3.6 Sodium metal bonds readily to each of the elements in Group 17 of the Periodic Table. Of the bonds formed between sodium and the Group 17 elements, which would have the greatest ionic character?

Network Solids—Macromolecules

Certain elements, such as carbon and silicon, can form very hard solids with high melting points. Diamond and graphite are two common solid forms of carbon. Diamond is the hardest form of carbon. Its structure is a crystal—that is, diamond has a regularly repeating arrangement of atoms. (In a crystal of NaCl, the Na^+ ions and Cl^- ions form a regularly repeating arrangement.) In the diamond crystal, an atom of carbon forms four equally spaced covalent bonds with four neighboring carbon atoms situated at the corners of a tetrahedron.

Carbon atoms are represented by spheres in the diagram of a diamond crystal in Figure 3-7. Each carbon atom is surrounded by four other carbon atoms equally spaced to form a three-dimensional structure. Each of these four atoms is, in turn, surrounded by four other atoms. Just as there are no molecules of NaCl in a crystal of NaCl, there are no molecules of C_2, C_4, or any other small number of C atoms in a diamond crystal. Instead, all the atoms in the crystal are joined in one large network. Such a struc-

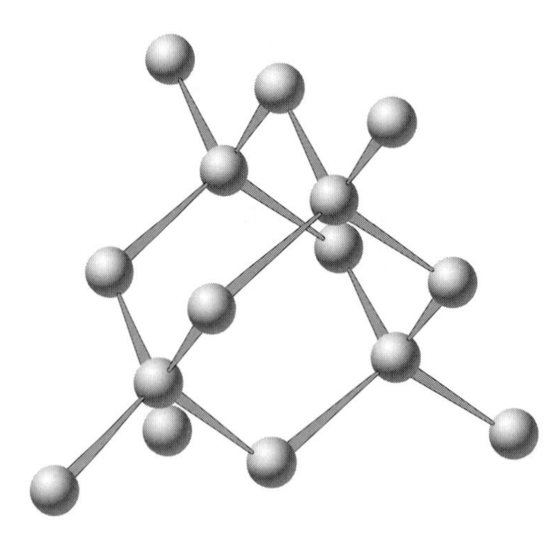

Figure 3-7 Part of a diamond crystal

ture is called a **network solid**, or a **macromolecule**. The covalent bonds in such solids are called **network bonds**.

The structure of diamond explains some of its properties. Diamond is an extremely poor conductor of electricity, because all electrons are firmly held in covalent bonds. There are no unoccupied orbitals, so electrons cannot move from one carbon atom to another. Such electrons are called localized. Diamond is very hard and has a high melting point (3500°C). These properties are due to the strong forces acting on the small carbon atoms that are held together by covalent bonds from all directions.

Graphite, another crystalline form of carbon, conducts electricity and is relatively soft. Like diamond, the carbon atoms in graphite form a network solid in which the atoms are covalently bonded. Why, then, do diamond and graphite have such different properties?

Both diamond and graphite have four valence electrons. In diamond, each carbon atom is bonded to four neighboring carbon atoms in a three-dimensional structure, a tetrahedron. All the valence electrons in a carbon atom are used up. In graphite, each carbon atom is covalently bonded to three other carbon atoms in a flat, or planar, structure (Figure 3-8). As in diamond, the bonds in graphite are equally spaced. The fourth valence electron, however, is not localized—it does not belong to a specific bond. Instead, this electron is mobile and spends an equal amount

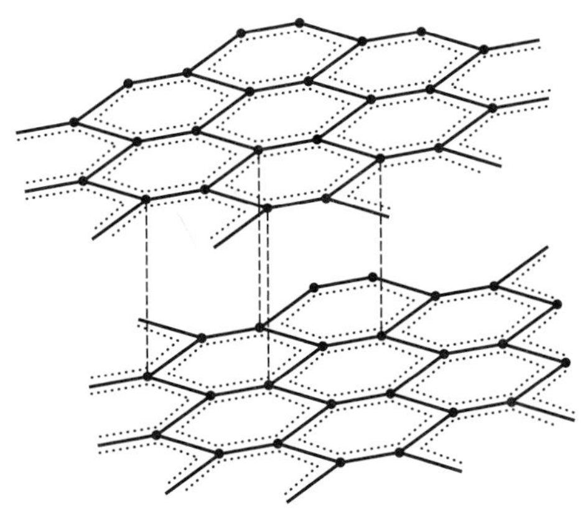

Figure 3-8 Part of a graphite crystal. The dotted lines represent the mobile electrons.

of time between the neighboring carbon atoms. Because of its mobile electrons, graphite conducts electricity.

The planar graphite molecules form giant layers of sheets that are bonded to one another by weak forces, called dispersion forces, which will be discussed later. These sheets, because they are weakly bonded, can slide over one another. This accounts for the softness of graphite and its lubricating properties.

Silicon carbide (SiC) is an interesting compound. It is extremely hard and has a high melting point. It has a very regular crystalline structure. It is made up of equal numbers of silicon and carbon atoms. Both of these elements are in Group 14 of the Periodic Table. They form four strong covalent bonds that are equally spaced. Silicon carbide exists as a network solid that is similar to diamond except that half of its atoms are silicon.

Sand, which is silica (SiO_2), is another example of a network solid. Many other natural compounds of silicon, such as mica, are network solids.

The Shapes and Polarities of Molecules

When a rubber rod is rubbed against fur, it acquires a negative charge. If the rod is then held next to a thin stream of water, the stream bends towards the rod. Since a negatively charged object attracts the water, we might expect a positively charged object to repel it. Yet when a positively charged rod, such as a glass rod stroked with silk, is brought near the stream of water, as in

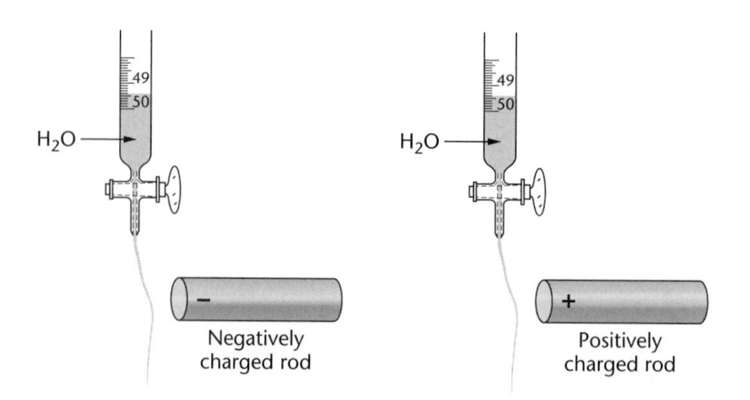

Figure 3-9 A stream of water is attracted to a charged rod

Figure 3-9, it also attracts the water. Why are water molecules attracted by both positively and negatively charged objects?

The dot structure of water is shown this way:

$$^{\delta+}\mathrm{H}:\ddot{\underset{\cdot\cdot}{\mathrm{O}}}:^{\delta-}$$
$$\mathrm{H}^{\delta+}$$

The oxygen, which has a much higher electronegativity than hydrogen, becomes slightly negative (δ^-), and the hydrogen becomes slightly positive (δ^+). Thus the side of the molecule containing the hydrogens is positively charged, and the opposite side of the molecule is negatively charged. Molecules that contain two, oppositely charged centers of charge are called **dipoles**, or **polar molecules**. When a negatively charged rod is brought near the stream of water, the water molecules spin so that their positive side faces the rod. The molecules are then attracted by the rod. Similarly, a positively charged rod causes the water molecules to face their negative side to the rod, which attracts the stream.

The polarity of molecules determines many of their physical and chemical properties. Polarity depends both on the shape of the molecule, and the kinds of bonds within the molecule.

The Shapes of Molecules

Pairs of electrons in a molecule repel each other. They therefore tend to be as far from each other as possible within the limits of a regular geometric structure. The valence shell electron pair repulsion theory (often abbreviated VSEPR) states that because electron pairs repel, molecules take shapes that keep the valence electron pairs as far apart as possible. This theory makes it possible to predict the shapes of molecules as shown in the table below. The electron pairs include all electrons, both bonded and nonbonded (lone pairs), around the central atom of the molecule. Let us examine a few examples.

Number of Electron Pairs	***Shape of Molecule***
1 or 2	Linear
3	Planar triangular
4	Tetrahedral

Beryllium fluoride (BeF_2) When two electron pairs surround a central atom (beryllium), a straight line, or linear, structure allows the maximum repulsion of electrons. The structure may be shown in either of the following ways:

$$:\ddot{\underset{..}{F}}:Be:\ddot{\underset{..}{F}}: \qquad\qquad F—Be—F$$

Boron trifluoride (BF_3) Three electron pairs surround the central boron atom. The geometric structure that can accommodate the electrons in the molecule is planar triangular. (Note both BeF_2 and BF_3 are exceptions to the octet rule.)

Methane (CH_4) Four electron pairs surround the central carbon atom. The molecule has a tetrahedral structure.

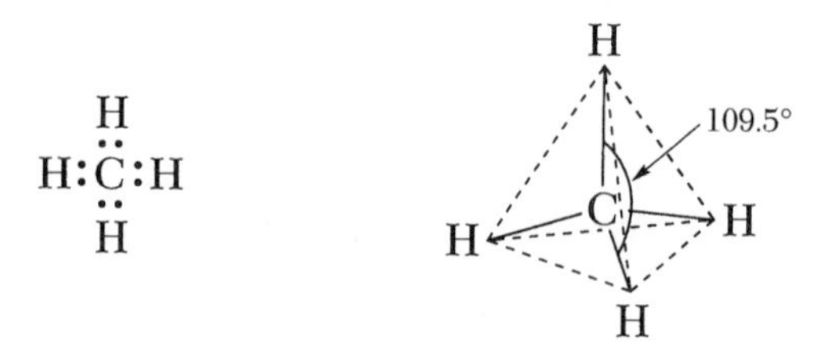

The next two examples are somewhat special cases of important and common molecules.

Ammonia (NH_3) Four electron pairs (as in methane) surround the central nitrogen atom three bonded pairs and one nonbonded (lone) pair. The electron pairs are arranged in a tetrahedral structure. The nitrogen and the three hydrogens together form a triangular pyramid, as shown in the diagram.

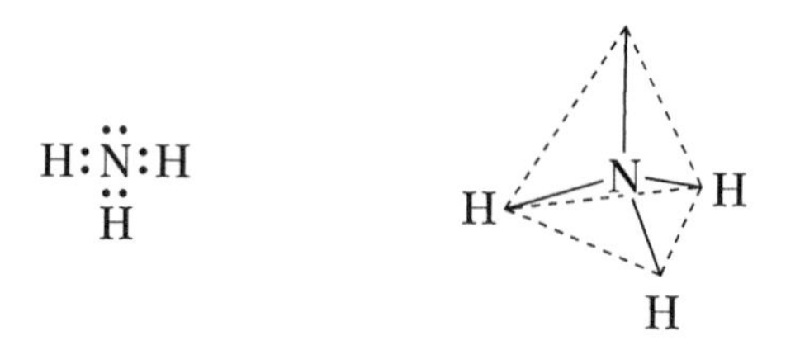

With the nonbonded electrons omitted, the shape of an ammonia molecule can be represented as follows:

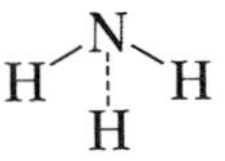

$$\begin{array}{c} \text{N} \\ \text{H} \diagup \;\vdots\; \diagdown \text{H} \\ \text{H} \end{array}$$

The dashed line here indicates a three-dimensional shape, with the bond directed away from you, into the paper.

Water (H.₂O) Four electron pairs surround the central oxygen atom—two bonded pairs and two nonbonded pairs. If the nonbonded electrons are included, the shape of the water molecule is tetrahedral, similar to the molecules of methane and ammonia.

The normal bond angle in a regular tetrahedron is about 109.5°, as shown in the diagram of CH_4. In water, this angle is distorted, because nonbonded electron pairs repel each other more than do bonded electron pairs. The bond angle in H_2O is approximately 105. A two-dimensional sketch of a water molecule can be shown as follows:

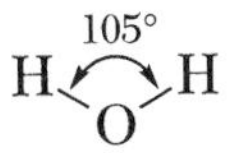

$$\begin{array}{c} 105^\circ \\ \text{H} \diagdown \;\; \diagup \text{H} \\ \text{O} \end{array}$$

In describing the shapes of molecules, the lone pairs are often omitted. Thus water is described as being bent, or angular, and ammonia is said to be pyramidal.

Polar and Nonpolar Molecules

In this section, the shape of the overall molecule, not only the bonds within it, is considered. To be polar or a dipole, a molecule must satisfy two conditions:

1. The molecule must have at least one polar bond (bond with ionic character).
2. The shape of the molecule must be such as to permit a net displacement of charge (one end becoming negative; the other, positive). An example of a polar molecule is H_2O.

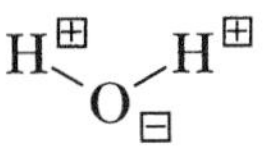

$$\begin{array}{c} \text{H}^{\boxplus} \diagdown \;\; \diagup \text{H}^{\boxplus} \\ \text{O}_{\boxminus} \end{array}$$

The water molecule is polar because the bonds between oxygen and hydrogen are polar (have ionic character), and the asymmetric (unsymmetrical) shape of the molecule permits a net displacement of charge. The hydrogen end is partially positive, and the oxygen end is partially negative. If the chemical symbols were left out, a diagram representing the molecule would look like a diagram of a magnet.

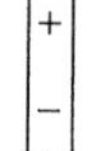

An example of a nonpolar molecule is CO_2.

$$\overset{\ominus}{O}{=}\overset{\oplus}{C}{=}\overset{\ominus}{O}$$

The bonds between carbon and oxygen are polar covalent (the electronegativity of C is 2.6 and that of O is 3.4). The O atoms are partially negative, and the C atom is partially positive. However, the linear shape does not permit a net displacement of charge. Instead, the charges cancel, producing a net charge of zero. When the chemical symbols are left out, a diagram of the molecule does not look like a diagram of a magnet. In CO_2, although the *bonds* are polar, the *molecule* is nonpolar.

− + −

Molecules of diatomic elements, such as N_2, O_2, and Cl_2, are nonpolar. There cannot be any difference in electronegativity between the bonded atoms, since they are the same. This means that there is no net displacement of charge and no polar bonds. The conditions for molecular polarity are not met.

If the atoms of a diatomic molecule are different, as in HF, the difference in electronegativity permits displacement of charge. Such molecules are polar.

When the four atoms bonded to a C atom are identical, the molecule is nonpolar. Its symmetrical shape does not permit a net displacement of charge. The two-dimensional diagram of CCl_4 shows that carbon tetrachloride is nonpolar.

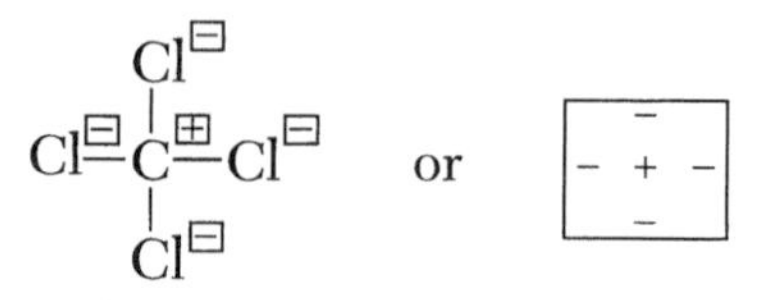

When the four atoms bonded to C are not all the same, the molecule will most likely be polar. Thus the two-dimensional diagram of CH_3Cl is

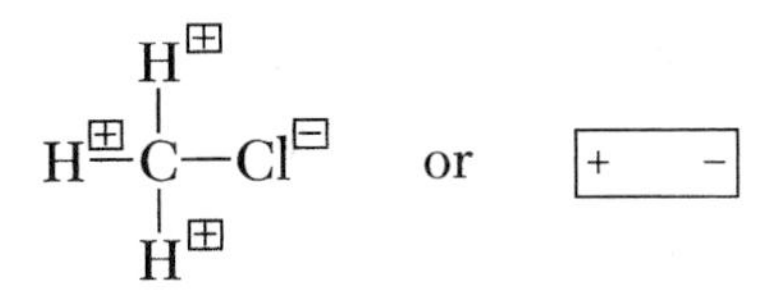

In summary, the presence of polar bonds does not necessarily make the overall molecule polar. How the bonds relate to one another—that is, the net effect of the bonds—determines the polarity of a particular molecule.

Shapes of Common Molecules

By learning the shapes of some common molecules, you can predict the shapes of a great many others, and determine whether these molecules are polar or nonpolar. In the chart below, the letter "X" is used to represent *any* nonmetallic atom.

Polar molecules		**Nonpolar molecules**	
Formula	**Shape**	**Formula**	**Shape**
HX	Linear	X_2	Linear
H_2O	Bent or angular	CO_2	Linear
NH_3	Pyramidal	CX_4	Tetrahedral

Using this chart, you can see, for example, that CI_4 would be tetrahedral and nonpolar, while HBr would be linear and polar. You can also use the chart to figure out the shape of H_2S. Sulfur is in the same group on the Periodic Table as oxygen. Therefore, it is likely that H_2S is similar in structure to H_2O. You can predict that it should be bent and polar covalent.

PRACTICE

3.7 Predict the shapes of the following molecules and indicate whether they are polar covalent or nonpolar covalent. Explain your predictions.

(a) I_2 (b) HCl (c) PH_3 (d) $SiCl_4$ (e) CS_2

3.8 Figure 3-9 on page 129 shows how a thin stream of water is attracted by a charged rod. Which of the substances listed in Practice 3.7, when in the liquid state, would behave in a manner similar to that of water? Explain your answer.

Attractive Forces Between Molecules

So far, you have considered bonds between atoms or ions—covalent bonds within a molecule, ionic bonds, metallic bonds, and network bonds. In this section, you will be concerned with attractive forces between molecules.

Van der Waals Forces

One of our assumptions about ideal gases is that the molecules do not attract one another. However, at sufficiently low temperatures and high pressures, any gas can be liquified. Therefore, there must be some attractions between these molecules. Attractions between molecules, in general, are known as van der Waals forces, named for the Dutch chemist, Johannes van der Waals. Chemists have identified two types of van der Waals forces, dispersion forces, and dipole-dipole attractions.

Dispersion forces The elements helium, hydrogen, oxygen, and nitrogen are gases at ordinary conditions. When gases such as these are cooled sufficiently, they liquefy. If cooling continues, they become solids. In the liquid state, the molecules attract one another enough to overcome the random motion associated with their kinetic energies. In the solid state, the molecules have the least random motion. The molecules of these gases are nonpolar. (Why?) What, then, is the nature of the attractive forces in these gas molecules?

In the solid state, the melting points of the simple gaseous elements and of many gaseous compounds are very low, usually well below 0°C. Their boiling points are also very low. In other words, with only slight increases in temperature, the molecules separate enough for the solids to become liquids and for the liquids to become gases. The forces that act to hold the molecules in the solid state and in the liquid state must therefore be very weak.

The attractive forces, or bonds, that hold together nonpolar molecules are called **dispersion forces**. Melting and boiling points

Gas	Atomic Radius (Å)	Melting Point (°C)	Boiling Point (°C)
^{2}He	0.93	−271.9 (26 atm)	−268.9
^{10}Ne	1.12	−248.7	−245.9
^{18}Ar	1.54	−189.3	−185.7
^{36}Kr	1.69	−157	−152.9
^{54}Xe	1.90	−111.5	−107.1

are a measure of the strength of these forces. It is believed that dispersion forces result from the interaction of the electron clouds of neighboring molecules. These interactions may temporarily produce oppositely charged regions in neighboring molecules, resulting in weak attractive forces.

If dispersion forces result from interactions of electrons, the dispersion forces must become stronger as the number of electrons in related molecules increases. The table above shows the noble gases in order of increasing number of electrons. Notice the increase in melting and boiling points.

Dispersion forces result from a temporary polarity of molecules. The attractions between the molecules are, as in all chemical bonds, the attractions between positively and negatively charged bodies. Although dispersion forces are weak compared with other chemical bonds, they may be strong enough to cause an element or compound to be liquid or solid under room-temperature conditions. The following table shows the first four elements in Group 17 (the halogen family).

With increasing numbers of electrons, the sizes of the halogen molecules increase. The increased numbers of electrons in the molecules result in increased electron interactions. With the resulting increase in dispersion forces, a halogen, such as iodine, exists in the solid state at room temperature and pressure.

Molecule	Total Number of Electrons	State at Room Temperature and Pressure
F_2	18	Gas
Cl_2	34	Gas
Br_2	70	Liquid
I_2	106	Solid

The general rule that dispersion forces increase with increasing molecular size holds true only if the elements being compared are related, as are the halogens. This rule and its limitations apply to compounds as well as to elements. For example, the rule holds true for simple hydrocarbons, compounds made up of hydrogen and carbon. The simple hydrocarbons form a related series that ranges from CH_4 (gas) through C_8H_{18} (liquid) to $C_{20}H_{42}$ (solid).

Dipole-dipole forces Dipole-dipole forces are the attractive forces, or bonds, that occur between polar molecules. The positive end of one polar molecule attracts the negative end of an adjacent polar molecule. These forces tend to cause the melting points and boiling points of polar compounds to be higher than those of nonpolar compounds of similar size.

Hydrogen Bonds

Hydrogen reacts with highly electronegative nonmetals, such as fluorine, nitrogen, and oxygen, to form hydrogen compounds with unusually high boiling points. The boiling points of such compounds are shown in Figure 3-10. In the diagram, a dashed line connects the hydrogen compounds formed with the elements of Group 17, the halogens. The solid line connects the hydrogen compounds formed with the elements of Group 16.

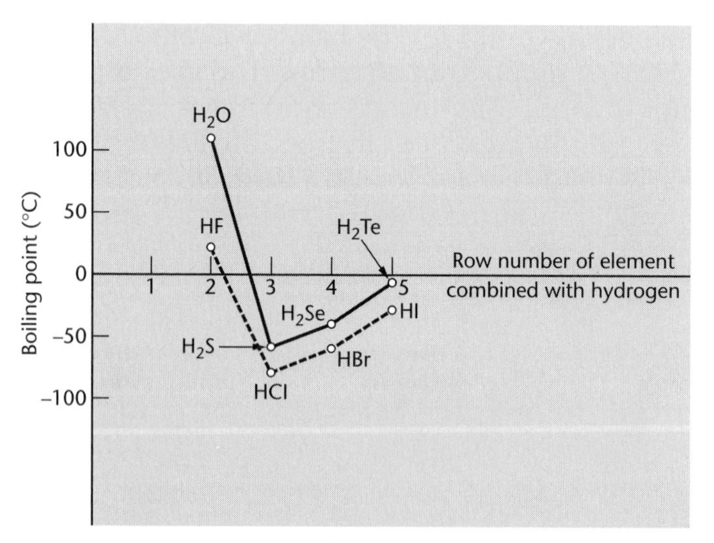

Figure 3-10 Trends in boiling points (Groups 16 and 17)

Notice how high the boiling point of HF is, compared with the boiling points of the other hydrogen-halogen compounds. Also notice the high boiling point of H_2O as compared with the boiling points of the other compounds formed with hydrogen and the elements in Group 16. Such behavior suggests that liquid HF and liquid H_2O are not made up of simple molecules. Instead, each liquid consists of a number of molecules joined together in a chain by hydrogen atoms. A **hydrogen bond** is an attractive force between molecules that occurs when an atom of hydrogen acts as a bridge between two small, highly electronegative atoms, such as the oxygens in H_2O or the fluorines in HF. The bond tends to create chains of weakly linked molecules. Before these complex molecules can be brought to the boiling point, the chains of molecules must be broken down into simpler molecules. This requires an expenditure of energy. Then energy is used to boil the simpler molecules. The net effect is an unusually high boiling point for liquid HF or liquid H_2O.

Why does a hydrogen bond form between molecules in liquid HF? HF has exceptionally strong polar bonds. The H region of the molecule is positively charged, and the F region is negatively charged. An attractive force exists between the H atom in one HF molecule and the F atom in another HF molecule. The chain of HF molecules formed can be shown as

$$F—H\cdots F—H\cdots F—H$$

In this formula, the solid line is the bond in the parent HF molecule. The dotted line is the hydrogen bond. H_2O, another polar covalent compound, is also bonded by hydrogen atoms, as shown in Figure 3-11.

If H_2O molecules were not hydrogen-bonded, it is likely that H_2O would be a gas at room temperature. Hydrogen bonds unite

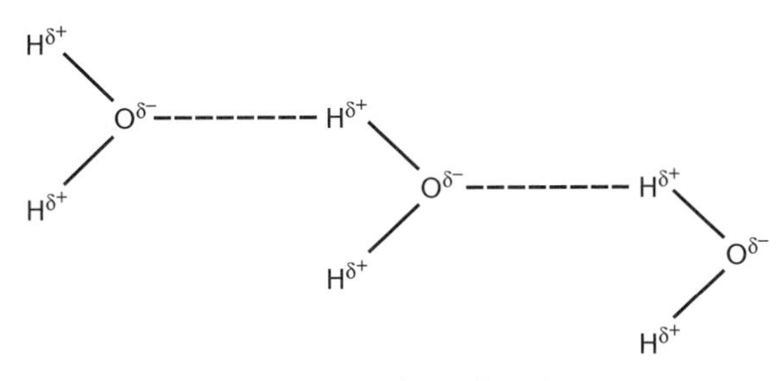

Figure 3-11 Hydrogen bonding in water

H_2O molecules and hold them together in the liquid and solid states. Without hydrogen bonding, life as we recognize it probably could not exist because there would be no liquid water at room temperature. Many life processes cannot go on in the absence of water.

Nitrogen has a lower electronegativity than oxygen. As a result, the hydrogen bonds between NH_3 molecules, while significant, are weaker than those between H_2O molecules. Hydrogen bonds are significant *only* when molecules contain H—F, H—O, or H—N bonds.

Hydrogen bonds play a vital role in the formation of giant molecules, such as proteins and DNA (the hereditary material of living things). In addition, hydrogen bonds strengthen certain plastics.

Summary of Molecular Bonds

1. Nonpolar molecules are attracted to each other by van der Waals forces called dispersion forces.
2. If the molecule is somewhat polar to begin with, the van der Waals attractions are strengthened. The attractions are called dipole-dipole forces.
3. When hydrogen atoms are bonded to a small, highly electronegative element, such as fluorine, oxygen, or nitrogen, added intermolecular attractions, called hydrogen bonds, result.

Dispersion forces, dipole-dipole forces, and hydrogen bonds are weaker than covalent, metallic, or ionic bonds.

The elements and compounds that exhibit van der Waals forces are called molecular substances, as contrasted with ionic, metallic, or network substances. Gases that are liquified or solidified are examples of molecular substances. There is little attractive force between the widely separated molecules of gases, and the bonding between atoms is covalent. Because van der Waals forces and hydrogen bonds are relatively weak, molecular substances tend to have relatively low melting and boiling point temperatures.

PRACTICE

3.9 State which of the following pairs of molecules has the higher boiling point. Explain your answer.

(a) HF or HCl (b) CH_4 or C_3H_8 (c) C_3H_8 or C_2H_5OH

3.10 Octane, C_8H_{18}, is a large molecule with a boiling point of 125.8°C. Yet KF has a far higher boiling point, 1500°C, than does octane. Explain this observation.

3.11 Glycerol, which has the formula $C_3H_5(OH)_3$, boils at 290°C. Propanol, C_3H_7OH, boils at 97°. Account for this large difference in boiling point.

Review of All Bonds

The table on the next page is provided as a summary of bonding. The table shows only the outstanding properties of each kind of bond and a few examples of the substances in which each type occurs.

Bonding and the Solid State

The bonding forces you have learned about are responsible for the formation of four distinctly different types of solids. These differ in their physical properties and in the types of bonding forces that form them.

Ionic bonding results in the formation of **ionic solids**. Positive and negative ions attract each other, forming an ionic crystal. These solids have high melting points. Because the ions are held firmly in place, the solids do not conduct electricity in the solid state. When they are melted or dissolved in water, however, the ions are free to move, and these substances become excellent conductors of electricity. Sodium chloride is a familiar example of an ionic solid.

Covalent bonding often results in the formation of neutral molecules. Covalent molecules are attracted to each other by intermolecular attractions. Because these attractions are weak, these **molecular solids** have low melting and boiling points. They do not conduct electricity in the solid or in the liquid state. Ice, dry ice (solid CO_2), and iodine are common molecular solids.

Enormous chains of covalent bonds result in the formation of network solids. These generally contain carbon or silicon, and include diamond, graphite, and silicon dioxide. Because the cova-

Bonds Between Atoms

Bond	***Properties***	***Examples***
Covalent	Moderately strong Energy in range of 400 kJ/mole	Hydrogen and chlorine gases Macromolecules, such as graphite and diamond
Metallic	Moderately strong to strong Solids conduct electricity	Sodium, copper, silver
Ionic	Very strong Solids do not conduct electricity but liquids do Solids often form compounds that are water soluble at room temperature and pressure	Sodium chloride, lithium fluoride

Bonds Between Molecules

Van der Waals	Relatively weak	
Dispersion forces		Nonpolar molecular substances, such as liquified gases; solids with low melting point temperatures, such as solid CO_2
Dipole-dipole forces		Polar molecular substances such as HCl
Hydrogen bonds	Somewhat stronger than van der Waals forces Account for strong attractions in polar molecules that contain H and a highly electronegative element	Hydrogen fluoride, water ammonia

lent bonds extend throughout the network solid, these substances have exceptionally high melting points. They are also extremely hard, and they are often used as abrasives in industry.

Pure metals form **metallic solids**. The positively charged kernels are attracted to a sea of mobile electrons. These substances are excellent conductors of electricity in the solid and liquid states. Metallic solids also exhibit luster, malleability, and ductility. *Luster* refers to the characteristic shininess associated with metals. *Malleability* is the ability to be hammered into thin sheets without becoming brittle. Aluminum foil is an application of the malleability of aluminum metal. *Ductility* is the ability to be stretched out into thin wires without breaking. The tungsten wire in a light bulb must be extremely thin in order to produce the heat needed to make it glow brightly. All of the metallic elements form metallic solids when pure. Different metals can form mixtures called alloys, such as brass and bronze. These are metallic solids as well.

PRACTICE

3.12 Some physical properties of five substances are listed on the chart below.

Substance	*State*	*Melting Point (°C)*	*Boiling Point (°C)*	*Electrical Conductivity*
A	Liquid	−39	357	Excellent conductor in solid and liquid states
B	Liquid	−130	36	Poor conductor in all states
C	Solid	765	812	Poor conductor in solid state; good conductor in liquid state
D	Solid	3500	4200	Poor conductor in all states
E	Solid	842	1240	Excellent conductor in solid and liquid states

The five substances are calcium (Ca), mercury (Hg), pentane (C_5H_{12}), calcium bromide ($CaBr_2$), and diamond (C).

Match each substance with its letter (A, B, C, D, or E) in the chart on page 143, and explain your answer.

TAKING A CLOSER LOOK

Hybrid Orbitals

In explaining how carbon forms four bonds, you saw how one of the 2*s* electrons moves to the 2*p* orbitals, resulting in the formation of four half-filled orbitals. Carbon can then bond to four hydrogen atoms to form methane, CH_4. The four bonds are equivalent to one another and equidistant from one another, in a tetrahedral arrangement. However, if one of the bonding electrons is in an *s* sublevel, and the other three are in the *p* sublevel, you would not expect to get four identical bonds. The theory of *hybridization* provides an explanation.

The one *s* and three *p* orbitals rearrange themselves to form four identical orbitals. These are called sp^3 hybrids. The electrons in these orbitals get as far away from each other as possible, which is a tetrahedral arrangement. Each of the four sp^3 orbitals is then available to bond to a hydrogen atom, forming the tetrahedral substance CH_4. Carbon forms sp^3 hybrid orbitals whenever it forms four single bonds. Figure 3-12 shows how the sp^3 orbitals form in a carbon atom.

Four sp^3 hybrids

Figure 3-12 Hybridization of the orbitals in a carbon atom

In BCl_3, there are only three pairs of electrons around the boron atom. Once again, all of the bonds are identical. Since only three orbitals are needed, however, the boron atom forms sp^2 hybrid orbitals. The three sp^2 orbitals are as far away from each other as possible, which is 120°, in a triangular arrangement.

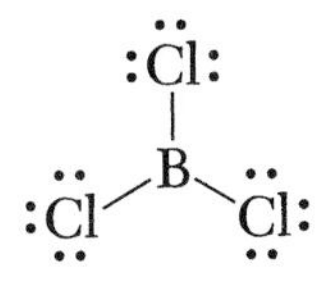

BeI_2 contains only two pairs of electrons around the Be atom. Thus only two hybrid orbitals are formed. They are called *sp* orbitals, and are as far away from each other as possible, at 180°, in a linear arrangement.

:I–Be–I:

Double and Triple Bonds

As you know, the N_2 molecule contains a triple bond. Its dot structure is:

:N⋮⋮N:

In the ground state, nitrogen has the electron configuration $1s^2 2s^2 2p^3$. The three bonds must be between electrons in *p* orbitals. The three *p* orbitals are arranged at 90° to each other. It is easy to see how a single bond can form, as shown in Figure 3-13.

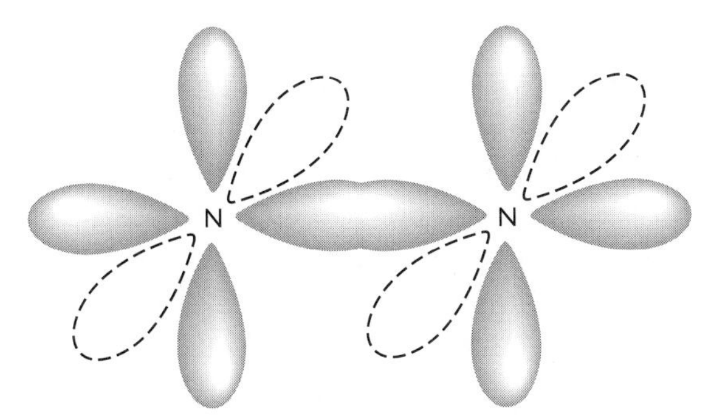

Figure 3-13 A single N—N bond. The dotted lines represent *p* orbitals coming in and out of the page.

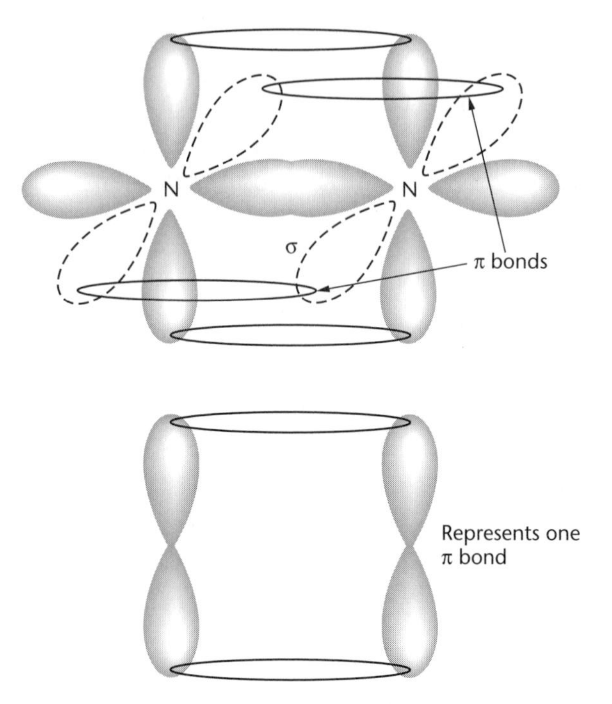

Figure 3-14 Sigma and pi bonds

How do the other two bonds form? The remaining *p* orbitals are parallel to each other. They bond sideways, above and below the molecule, as shown in Figure 3-14. These sideways bonds between parallel *p* orbitals are called pi (π) bonds. Bonds between orbitals that lie directly between the two nuclei are called sigma (σ) bonds. A triple bond consists of one sigma and two pi bonds. A double bond consists of one sigma and one pi bond. The formation of pi bonds pulls the nuclei closer together. Double and triple bonds are therefore shorter than single bonds.

Carbon dioxide has the structure O═C═O. The carbon must have one sigma bond and one pi bond with each oxygen. Suppose the pi bond on the left is above and below the molecule. Since *p* orbitals are always at 90° to each other, the pi bond on the right must be above and below the plane of the paper. The electrons in the two sigma bonds must be in hybrid orbitals. Since there are only two such bonds, they are at 180° to each other, and the carbon must form two *sp* hybrid orbitals.

The hybridization around a carbon atom depends upon the number of sigma and pi bonds the carbon forms. When the carbon forms four sigma bonds, as in methane, the hybridization is sp^3, and the molecule is tetrahedral. When the carbon forms one double bond and two singles, which implies three sigma bonds and one pi bond, the hybridization is sp^2, and the molecule is triangular. Formaldehyde, HCHO, has a triangular geometry, as shown below.

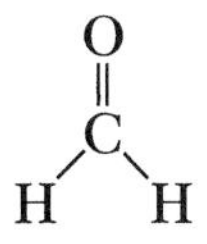

When a carbon atom forms two double bonds, or a triple and a single bond, there are two pi bonds and two sigma bonds. The hybridization is then *sp,* and the molecule is linear. Ethyne, C_2H_2, has a linear geometry, as shown.

$$H{-}C{\equiv}C{-}H$$

Resonance

When you draw a dot structure for the SO_3 molecule, it is necessary to form a double bond in order to achieve an octet around the sulfur atom. Our dot structure might look like this.

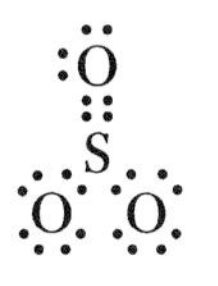

You would expect one of the bonds, the one that includes the pi bond, to be shorter than the others. Yet experiments have shown that all of the bonds are the same length. The sulfur is sp^2 hybridized, and the molecule is an equilateral triangle. If all of the bonds are equivalent, then where is the pi bond? The explanation is that the pi bond is shared equally by all three oxygen atoms. The *p* orbital on the sulfur is parallel to *p* orbitals on each oxygen atom. Thus it bonds to all of them equally. When a pi bond is delocalized, that is, spread over two or more sites, we say that there is *resonance,* and we call the substance a *resonance hybrid.*

The structure showing the double bond in one place, shown above, is called a *contributing structure.* However, it is incorrect to think that the double bond is sometimes on one oxygen, and sometimes on another. It is always shared by all three. You might represent a sulfur trioxide molecule like this:

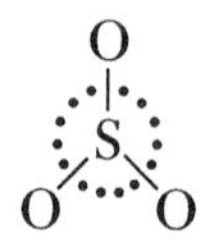

Polyatomic Ions

You are already familiar with the chart of polyatomic ions in Appendix 4. Let us briefly examine the structures of these ions. You have seen how a hydrogen ion, H^+, can form a coordinate covalent bond with ammonia, NH_3, to form the ammonium ion, NH_4^+. To draw the dot structure of a polyatomic ion, we subtract one valence electron for each positive charge, or add one for each negative charge. Nitrogen has five valence electrons, and each hydrogen one. Therefore, if we subtract one electron to account for the positive charge of the ammonium ion, that leaves eight electrons. The dot structure looks like this:

$$\left[\begin{array}{c} \mathrm{H} \\ \mathrm{H}:\ddot{\underset{\cdot\cdot}{\mathrm{N}}}:\mathrm{H} \\ \mathrm{H} \end{array}\right]^{+}$$

What type of hybridization would you expect around the nitrogen atom?

Most of the polyatomic ions are negatively charged, which means they have extra electrons. Consider the sulfate ion, SO_4^{2-}. Sulfur has 6 valence electrons, but since the ion has a 2− charge, you add 2 more electrons to the sulfur, making a total of 8. It is then easy to draw the dot structure.

$$\left[\begin{array}{c} :\ddot{\mathrm{O}}: \\ :\underset{\cdot\cdot}{\ddot{\mathrm{O}}}:\underset{\cdot\cdot}{\ddot{\mathrm{S}}}:\underset{\cdot\cdot}{\ddot{\mathrm{O}}}: \\ :\underset{\cdot\cdot}{\mathrm{O}}: \end{array}\right]^{2-}$$

The sulfur is sp^3 hybridized, and the ion is tetrahedral.

If we remove one oxygen atom, we are left with the sulfite ion, SO_3^{2-}. Now the sulfur has three bonds plus one lone pair of electrons. The ammonia molecule, which has been discussed previously, also had three bonds and one lone pair around the central atom, and it had a pyramidal shape. The sulfite ion has a pyramidal shape for the same reasons. The hybridization is sp^3, but there is now a lone pair. Note that while there was a pi bond and resonance in the SO_3 molecule, there is no pi bond and no resonance in the SO_3^{2-} ion.

```
  ..  ..  ..     2−
[ :O: S :O: ]
  ..  ..  ..
     :O:
      ..
```

PRACTICE

3.13 Draw the PO_4^{3-} ion. Which of the ions discussed in this section does it closely resemble?

3.14 Draw the SO_2 molecule. What type of hybridization would you expect around the sulfur atom? Is there resonance?

3.15 The carbonate ion, CO_3^{2-}, and the nitrate ion, NO_3^-, have the same structure, hybridization, and geometry. They also resemble the SO_3 molecule, shown previously. Draw the two ions. Why do they have identical structures?

3.16 From the list of polyatomic ions in Appendix 4, choose an ion whose structure would most closely resemble that of the sulfite ion, SO_3^{2-}, and draw its dot structure.

CHAPTER REVIEW

The following questions will help you check your understanding of the material presented in the chapter.

1. According to the table of electronegativity values that is given on page 103, which pair of elements forms a compound with the greatest ionic character? (1) H and F (2) Na and Cl (3) Ca and O (4) Cs and N

2. A Ba^{2+} ion differs from a Ba^{0} atom in that the ion has (1) more electrons (2) more protons (3) fewer electrons (4) fewer protons.
3. Which property best accounts for the conductivity of metals? (1) the protons in metallic crystals (2) the malleability of most metals (3) the filled inner electron shells of most metals (4) the free electrons in metallic crystals
4. Which is a nonpolar covalent substance? (1) CCl_4 (2) NH_3 (3) H_2O (4) KCl
5. Which can form a coordinate covalent bond?

$$(1)\ \begin{array}{c} \mathrm{H} \\ \mathrm{H}:\ddot{\underset{\cdot\cdot}{\mathrm{C}}}:\mathrm{H} \\ \mathrm{H} \end{array} \qquad (2)\ \mathrm{H}:\mathrm{H} \qquad (3)\ \begin{array}{c} \mathrm{H} \\ \mathrm{H}:\ddot{\underset{\cdot\cdot}{\mathrm{Si}}}:\mathrm{H} \\ \mathrm{H} \end{array} \qquad (4)\ \begin{array}{c} \\ \mathrm{H}:\ddot{\underset{\cdot\cdot}{\mathrm{O}}}: \\ \mathrm{H} \end{array}$$

For each of questions 6 and 7 select the number of the substance, chosen from the following table, that best answers that question.

Substance	***Melting Point (K)***	***Boiling Point (K)***
(1) Sodium chloride	1074	1686
(2) Helium	1	4
(3) Diamond	3773	4473
(4) Water	273	373

6. Which substance has molecular forces of attraction due mainly to dispersion forces?
7. Which substance forms a molecular solid made up of polar molecules?
8. As the molecular mass of the compounds of a related series of simple hydrocarbons increases, the boiling point (1) decreases (2) increases (3) remains the same.
9. A certain solid, when it is in the liquid state or dissolved in water, will conduct electricity. In the solid state it will not conduct electricity. This solid must contain (1) ionic bonds (2) metallic bonds (3) covalent bonds (4) coordinate bonds.
10. Which type of bond is most likely to be formed between phosphorus and chlorine? (1) nonpolar covalent (2) polar covalent (3) ionic (4) network

11. Which is an example of a dipole? (1) N_2 (2) H_2 (3) CH_4 (4) NH_3
12. Which molecule has a triple covalent bond? (1) F_2 (2) O_2 (3) N_2 (4) H_2
13. Which best explains why a methane (CH_4) molecule is nonpolar? (1) Each carbon-hydrogen bond is polar. (2) Carbon and hydrogen are both nonmetals. (3) Methane is a compound. (4) The methane molecule is symmetrical.
14. A solid substance is soft, has a low melting point, and is a poor conductor of electricity. The substance is most likely (1) an ionic solid (2) a network solid (3) a metallic solid (4) a molecular solid.
15. Which pair of elements forms a bond with the least ionic character? (1) P—Cl (2) Br—Cl (3) H—Cl (4) O—Br
16. Which molecule is the most polar? (1) H_2O (2) H_2S (3) H_2Se (4) HI
17. Which compound, in the liquid state, most readily forms hydrogen bonds between its molecules? (1) HF (2) HCl (3) HBr (4) HI
18. Which type of bonding involves positive ions immersed in a sea of mobile electrons? (1) ionic (2) nonpolar covalent (3) polar covalent (4) metallic
19. Hydrogen forms a negative ion when it combines with sodium to form NaH. This is primarily because hydrogen (1) loses an electron to sodium (2) has a greater attraction for electrons than sodium has (3) is a larger atom than sodium (4) has a smaller ionization energy than sodium.
20. Which is the formula of a nonpolar molecule containing nonpolar bonds? (1) CO_2 (2) H_2 (3) NH_3 (4) H_2O
21. Chlorine would be most likely to form a covalent bond with (1) barium (2) potassium (3) zinc (4) nitrogen.
22. Which compound has the lowest boiling point at standard pressure? (1) NaI (2) HI (3) MgI_2 (4) AlI_3
23. Dipole-dipole attractive forces are strongest between molecules of (1) H_2 (2) CH_4 (3) H_2O (4) CO_2.
24. Which compound, in the liquid phase, conducts electricity best? (1) H_2O (2) H_2S (3) NH_3 (4) NaCl
25. Which type of bond is formed when a hydrogen ion (H^+) reacts with an ammonia molecule (NH_3)? (1) a coordinate

covalent bond (2) a nonpolar covalent bond (3) a metallic bond (4) an ionic bond

26. The forces of attraction that exist between hydrogen molecules in liquid hydrogen are due to (1) ionic bonds (2) hydrogen bonds (3) molecule-ion forces (4) dispersion forces.

27. Which compound is a network solid? (1) SiO_2 (2) Na_2O (3) H_2O (4) CO_2

28. Which is the correct electron-dot formula for the ammonia molecule?

(1) $H\cdot\overset{\overset{H}{\cdot}}{N}\cdot H$ (2) $H:\underset{\cdot\cdot}{\overset{\overset{H}{\cdot\cdot}}{N}}:H$ (3) $H:\overset{\overset{H}{\cdot\cdot}}{N}:H$ (4) $H:\underset{\underset{H}{\cdot\cdot}}{\overset{\overset{H}{\cdot\cdot}}{N}}:H$

29. The NCl_3 molecule is pyramidal, while BCl_3 is triangular. It is thus most likely that (1) Both molecules are polar. (2) The NCl_3 is polar, while the BCl_3 is nonpolar. (3) The NCl_3 is nonpolar, while the BCl_3 is polar. (4) Both molecules are nonpolar.

For questions 30-35 choose the substance from the list below that best fits the description. Choices may be used more than once or not at all.

(1) BaO (2) HF (3) CO_2 (4) H_2 (5) BrF

30. A polar covalent molecule that exhibits strong hydrogen bonding
31. A molecule containing nonpolar covalent bonds
32. A nonpolar molecule containing polar covalent bonds
33. An ionic substance
34. A molecule containing double bonds
35. A polar molecule that forms no hydrogen bonds

CONSTRUCTED RESPONSE

1. Explain each of the following observations on the basis of bonding forces:
   (a) Solid Mg conducts electricity, while solid MgO does not.
   (b) Salt conducts electricity when melted, while sugar, ($C_{12}H_{22}O_{11}$) does not.

(c) Br_2 is a liquid at room temperature, while I_2 is a solid.

(d) Water has a much higher boiling point than that of H_2S.

2. Dichloromethane has the formula CH_2Cl_2. Draw a dot structure of the molecule and state whether it is a dipole. What is the actual shape of the molecule?

CHEMISTRY CHALLENGE

The following questions will provide practice in answering SAT II-type questions.

Each question below consists of a statement and a reason. Select

(a) if both the statement and reason are true and the reason is a correct explanation of the statement;

(b) if both the statement and reason are true, but the reason is NOT a correct explanation of the statement;

(c) if the statement is true but the reason is false;

(d) if the statement is false but the reason is true;

(e) if both the statement and reason are false.

Example:

Statement	*Reason*
Water boils at 100°C.	Water is a small molecule, containing only 10 electrons.

The right answer would be (*b*). Both the statement and the reason are true. However, the reason water boils at 100° is due to hydrogen bonds. Small molecules generally have lower boiling points.

Statement	*Reason*
1. CaO conducts electricity when melted.	CaO is a polar molecule.
2. Xe has a higher boiling point than Kr.	Xe has stronger van der Waals forces than Kr.

Statement	*Reason*
3. Tungsten (W) has a very high melting point.	Network solids have very high melting points.
4. Water has a linear shape.	There are two lone pairs on the oxygen atom in water.
5. Gold conducts electricity.	Metallic bonds provide a sea of mobile electrons.
6. The bond in HF has more ionic character than the bond in HCl.	There is a greater difference in electronegativity between H and F than between H and Cl.
7. H_2 has an extremely low boiling point.	There are hydrogen bonds formed between the H_2 molecules.
8. The Cl_2 molecule contains double bonds.	Cl_2 obeys the octet rule.
9. Diamond is the hardest substance known.	Diamond is a network solid.
10. CCl_4 has a square planar shape.	The farthest apart the four bonds can get from each other is 90°.

When Eight Is Not Enough

We have seen that nonmetals generally form covalent bonds so as to attain an octet, a set of eight valence electrons. Using the octet rule, we can easily draw the dot structure of SF_2, sulfur difluoride.

:S:F:
:F:

Based on this structure, we would predict that this molecule should have a bent shape, like that of water. In addition, chemists have prepared SF_4 and SF_6. Let us look first at SF_6. Our dot structure gives us 12 electrons around the sulfur. Chemists call this an "extended octet." The six pairs of electrons form a shape called an octahedron.

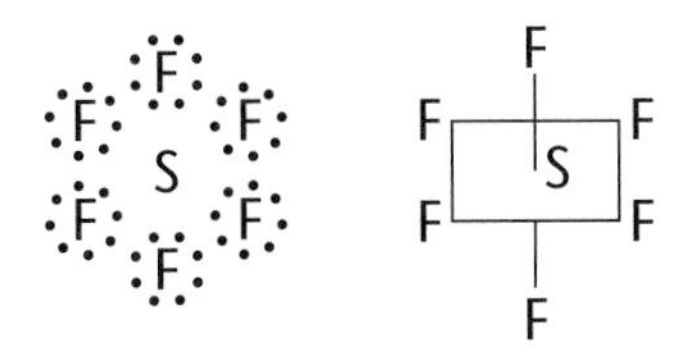

In SF_4 , our dot structure shows 5 pairs of electrons around the sulfur, with 4 bonds and one lone pair. The unusual shape that results is called a "seesaw" or a "saw horse." If you find the shape difficult to visualize, try this: Hold your arms out in front of you at about a 120° angle from each other. Now if sulfur were in your heart, you would have a fluorine in your head, a fluorine at your feet, and a fluorine in each hand. The lone pair of electrons would be directly behind you.

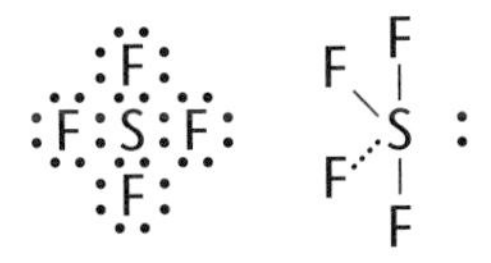

Until just 40 years ago, chemists believed that no compounds of Group 18 elements could be prepared, since these elements already have complete octets. However, if sulfur can end up with 12 valence electrons, why not xenon? In 1962, XeF_4, the first compound of a Group 18 element, was prepared. The dot structure of xenon tetrafluoride shows six pairs of electrons around the central Xe atom, exactly as in SF_6. However, there are only four bonds formed, so there are two lone pairs around the Xe atom. The resulting molecule is in the shape of a square with the lone pairs above and below the square.

4 Formulas and Equations

LOOKING AHEAD

Chemical formulas and equations are brief, convenient ways of expressing chemical information. They are, in effect, the language of chemistry. Learning to write formulas and equations will enable you to communicate ideas more easily and more exactly, not only in the field of chemistry but also in many other scientific fields.

When you have completed this chapter, you should be able to:

- **Write** chemical formulas for ionic substances, based on the charges of the ions.
- **Write** chemical formulas for molecular substances, based on a system of numerical prefixes.
- **Assign** oxidation numbers.
- **Correctly name** chemical compounds.
- **Write** and **balance** a chemical equation.
- **Calculate** molecular and molar mass.

Chemical Formulas

You learned about the symbols of elements in Chapter 2. By now, you probably know and use the symbols of many common elements. When elements exist in combination—that is, in a compound—the symbols of the elements involved are combined into a **formula.** Every chemical formula provides important information about the compound it represents.

The Information in Formulas

A chemical formula tells what elements are present in a combination. In the compound with the formula $CaCl_2$, for example, the Ca tells us calcium is part of the combination; the Cl_2 tells us two chlorines are present. (The small "2" is called a subscript.) Therefore, a chemical formula also tells how much of each element is present. In other words, the formula states the quantitative composition of the compound.

This quantitative composition can be stated even more specifically. Recall that the mole is the counting unit of chemistry. One mole equals 6.02×10^{23} particles—atoms, ions, or molecules. Calcium chloride is an ionic compound. Therefore, the formula $CaCl_2$ states that 1 mole of $CaCl_2$ contains 1 mole of Ca^{2+} ions and 2 moles of Cl^- ions.

Writing Chemical Formulas

Given the name of a compound, you need to be able to determine the formula of that compound. Compounds are named according to rules that enable you to do this. Ideally, there should be only one correct name for each formula. Unfortunately, some common compounds have names that were given to them long before chemists designed a system of naming compounds. Water and ammonia are examples of compounds generally called by their "common names."

Binary Compounds

Binary compounds contain two different elements. These compounds are named by first stating the metal followed by the nonmetal with the ending "ide." Common examples are sodium chloride (NaCl) and calcium oxide (CaO). If the binary compound consists of two nonmetals, then the nonmetal with the lower electronegativity is named first. Thus the compound CO_2 is called carbon dioxide, and not dioxygen carbide. Remembering

ıpounds are neutral will help you write correct formu-
ary ionic compounds. The positive charge of the metal
:xactly balanced by the negative charge of the nonmetal.
did chemists arrive at the formula for calcium chloride,
gı. above as $CaCl_2$? Calcium ions have a charge of 2+. Chloride ions have a charge of 1−. For calcium chloride to be neutral, there must be two chlorides for every calcium. Obviously, to write the formula of an ionic substance, it is necessary to know the charges of the ions that make up the substance. While these often can be determined from the electron configurations of the atoms, the charges of the most common ions should be committed to memory (see table below).

Charges of Some Common Ions

1+	2+	3+	1−	2−	3−
Sodium Na^+	Magnesium Mg^{2+}	Aluminum Al^{3+}	Fluoride F^-	Oxide O^{2-}	Nitride N^{3-}
Potassium K^+	Calcium Ca^{2+}		Chloride Cl^-	Sulfide S^{2-}	
Hydrogen H^+	Barium Ba^{2+}		Bromide Br^-		
Silver Ag^+	Zinc Zn^{2+}		Iodide I^-		

Using the charges of these ions, you can write formulas for the compounds containing these ions.

What is the formula of aluminum oxide? You can see that the aluminum ion is 3+, while the oxide ion is 2−. What combination of aluminum and oxide will result in a charge of zero? You need two aluminums for every three oxides, resulting in the formula Al_2O_3. Note that the total charge on two aluminum ions is 6+, and the total charge on three oxide ions is 6−, so that the compound is neutral. Students frequently arrive at the correct formula by using the "crisscross method."

Writing formulas by the crisscross method Follow these steps:

1. Write the symbol for each ion, including its charge.
2. Crisscross the charges. The charge of the nonmetal becomes the subscript of the metal and vice versa. Do not include the

sign of the charge (the + or −) in your subscripts, and do not write 1 as a subscript. (When no number is written, the subscript 1 is understood.)

3. Reduce these subscripts to lowest terms.
4. When the ions have equal but opposite charges, the subscripts are usually dropped. If in step 1 you get $K^{1+}Cl^{1-}$, for example, in step 2 you will get K_1Cl_1. The formula is correctly written KCl. In like manner, $Mg^{2+}O^{2-}$ becomes MgO.

SAMPLE PROBLEM

PROBLEM

What is the formula of sodium oxide?

SOLUTION

Steps: 1. The ions are $Na^{1+}O^{2-}$

2. Crisscrossing,

$$Na^{1+} \times O^{2-}$$

gives us Na_2O. Remember, we do not write 1 as a subscript.

3. The subscripts are already in lowest terms.

PRACTICE

4.1 Write the formulas for the following substances:

(a) barium oxide (b) calcium iodide (c) aluminum sulfide (d) sodium nitride (e) potassium sulfide.

Elements That Form Multiple Ions

The metals listed in the table on page 158 commonly form just one ion. Calcium is 2+ in virtually all of its compounds, and sodium ions are always 1+. There are other metals, however, particularly the transition metals (Groups 3–12 in the Periodic Table), that can form at least two different ions. One such metal is iron. Iron forms ions of 2+ and 3+. Thus iron can form two different compounds with the chloride ion: $FeCl_2$ and $FeCl_3$. To avoid confusion, the compounds are given different names. $FeCl_2$ is called iron(II) chloride, and $FeCl_3$ is called iron(III) chloride. The Roman numeral indicates the charge of the positive ion. These Roman numerals are used only when an atom forms ions with two or more charges. Thus ZnS is called zinc sulfide, because zinc is always 2+, while NiS is called nickel(II) sulfide, because nickel can also form a 3+ ion.

SAMPLE PROBLEM

PROBLEM

What is the formula for tin (IV) oxide?

SOLUTION

The Roman numeral tells you that tin is 4+. The oxide ion, you will remember, is 2−. Crisscrossing gives us Sn_2O_4, which in lowest terms becomes SnO_2. The correct formula for tin (IV) oxide is SnO_2.

PRACTICE

4.2 Write correct formulas for the following substances:

(a) iron (III) oxide (b) cobalt (II) chloride (c) lead (IV) sulfide (d) nickel (II) oxide

Polyatomic Ions

In Chapter 3 you examined the structure of the ammonium ion, NH_4^+. The ammonium ion is a polyatomic ion, an ion containing two or more atoms. The atoms within the ion are covalently bonded, but the entire ion, since it has a positive charge, can form ionic bonds with negative ions. Most polyatomic ions are negatively charged. A list of these ions appears in Appendix 4.

A shorter list of the most common polyatomic ions, their fo[illegible] las, and their charges appears in the table below. Note that a p[illegible] atomic ion whose name ends in "ite" generally has one few[illegible] oxygen atom than a polyatomic ion of the same element whos[illegible] name ends in "ate."

To write formulas for compounds containing polyatomic ions follow the same rules you used for binary compounds. Treat the polyatomic ion as a single unit. The formula of the polyatomic ion never changes. For example, consider the formula of aluminum nitrate. First, write the ions and their charges. $Al^{3+}NO_3^-$. Now, crisscross the charges. This means the nitrate gets the subscript 3. This subscript applies to the entire nitrate ion. Therefore you write the ion in parentheses, and the correct formula would be written $Al(NO_3)_3$. What is the formula for iron(III) sulfate? You know that iron must be 3+, from the Roman numeral, and sulfate is listed below as 2−. Crisscrossing the charges gives us $Fe_2(SO_4)_3$, the correct formula for iron(III) sulfate. The formula for iron(II) sulfate, on the other hand, would be $FeSO_4$. Note that you do not use parentheses here, because only *one* sulfate is needed. Parentheses are used only to show the presence of two or more of the same polyatomic ion.

Charges on Some Polyatomic Ions

1+	1−	2−	3−
Ammonium NH_4^+	Hydroxide OH^-	Sulfate SO_4^{2-}	Phosphate PO_4^{3-}
Hydronium H_3O^+	Nitrate NO_3^-	Sulfite SO_3^{2-}	
	Nitrite NO_2^-	Carbonate CO_3^{2-}	
	Acetate $C_2H_3O_2^-$	Chromate CrO_4^{2-}	

PRACTICE

4.3 Write formulas for the following substances:

(a) aluminum phosphate (b) sodium carbonate (c) barium nitrate (d) nickel(III) hydroxide (e) ammonium sulfate e following substances:

)$_4)_2$ (b) $Ca(OH)_2$ (c) NH_4Cl (d) Na_2S
(f) Na_2SO_4 (g) $FeSO_4$

ecular and Empirical Formulas

The formulas you have written thus far have been for ionic compounds. The formula expresses in lowest terms the relationship between the ions in the compound. As you saw in Chapter 3, however, bonds between atoms often are covalent and do not result in the formation of ions. Covalent bonds result in the formation of neutral molecules. The formula of a molecule states the exact composition of the molecule, and the formula is frequently not in lowest terms.

The glucose molecule has the formula $C_6H_{12}O_6$. This formula indicates that each molecule contains 6 carbon atoms, 12 hydrogen atoms, and 6 oxygen atoms. $C_6H_{12}O_6$ is the molecular formula of glucose. If this formula is written in lowest terms, it becomes CH_2O. A formula that expresses the smallest whole number ratio among its elements is called an **empirical formula**. The empirical formula of glucose is CH_2O. Many different molecules can have the same empirical formula. For example, the formula of acetic acid can be written $HC_2H_3O_2$. A molecule of acetic acid, then, contains 2 carbons, 4 hydrogens, and 2 oxygens. In lowest terms, that comes out to CH_2O, which is the same as the empirical formula of glucose.

As was the case with ionic compounds, molecular compounds are named so as to suggest the correct formula. Several different methods are currently in use.

Prefix names You probably know that CO_2 is called carbon dioxide. The prefix "di" is used to indicate the number of oxygen atoms. The prefixes used in this system are: *mon(o)-* indicating one; *di-*, two; *tri-*, three; *tetr(a)-*, four; *pent(a)-*, five; and *hex(a)-*, six. (The vowels in parentheses are often dropped when preceding another vowel.) Some compounds named this way include:

carbon monoxide, CO	sulfur dioxide, SO_2
dinitrogen trioxide, N_2O_3	sulfur trioxide, SO_3
diarsenic pentoxide, As_2O_5	carbon tetrachloride, CCl_4

It is easy to determine the formulas from the names of these compounds.

PRACTICE

4.5 Give the formulas of the following compounds:
(a) silicon tetrabromide (b) dinitrogen monoxide
(c) carbon disulfide (d) phosphorus trichloride

The Stock system You will recall that iron can form two different compounds with chlorine, $FeCl_2$ and $FeCl_3$. In the names, a Roman numeral indicates the charge of the iron in the compound. Thus the first is called iron (II) chloride, and the second, iron (III) chloride. This system of using Roman numerals to indicate the charge of the positive ions is called the Stock system.

Oxidation Numbers

Molecular substances do not contain ions. To use the Stock system, you must understand a new term, the oxidation number. The **oxidation number**, or **oxidation state**, is the charge an atom would acquire if all of its bonds were treated as ionic bonds. In assigning oxidation numbers, you treat all shared electrons as if they were taken by the atom with the higher electronegativity. For example, consider the dot structure of water.

$$:\ddot{O}:H$$
$$\dot{H}$$

Since oxygen has a higher electronegativity than does hydrogen, all of the shared electrons are assigned to the oxygen. This gives oxygen eight valence electrons, a gain of two, for an oxidation state of -2. Each hydrogen is assigned no electrons, a loss of one. Each hydrogen atom has an oxidation state of $+1$. Note: In oxidation numbers, the sign of the charge is written before the value.

Generally, it is not necessary to look at the structure of a molecule to determine the oxidation states of the elements. The sum of the oxidation states in any molecule must equal zero. If you know the common oxidation states of a few elements, you can use them to determine the oxidation states of the other elements in a compound. For metals, the oxidation state is the same as the ionic charge.

To find the oxidation states of elements from chemical formulas, follow these rules:

1. The sum of the oxidation states in any substance is zero.
2. The sum of the oxidation states in any ion is equal to the charge of that ion.
3. The oxidation state of a metal is the same as the charge of the metal ion.
4. Several nonmetals normally show only one oxidation state when they are negative. (In a formula, the element with the negative oxidation state is almost always written last.) These negative oxidation states are the same as the charges of the negative ions, listed in the table on page 158. When nonmetals are bonded to other nonmetals, remember that the less electronegative nonmetal will have a positive oxidation state.

SAMPLE PROBLEMS

PROBLEM

1. What is the oxidation state of the chlorine in $KClO_3$?

SOLUTION

The sum of the oxidation states must be 0. Potassium is always +1. Oxygen is −2, but there are three oxygens, for a total of −6. For the sum to equal zero, the Cl must be +5, [(+1) + (+5) + (−6) = 0]. The oxidation state of the Cl is +5. In the formula below, the oxidation state of each element is written above it, and the total oxidation state for all the atoms of that element is written below.

$$\begin{array}{ccc} +1 & +5 & -2 \\ \mathrm{K} & \mathrm{Cl} & \mathrm{O_3} \\ +1 & +5 & -6 = 0 \end{array}$$

PROBLEM

2. What is the oxidation state of the sulfur in the sulfate ion, SO_4^{2-}?

SOLUTION

Recall that the sum of the oxidation states for an ion must equal the charge of the ion, in this case, −2. Oxygen is −2, but there are 4 oxygens, for a total of −8. Therefore the sulfur must be +6. (+6) + (−8) = −2. The oxidation state of the sulfur is +6.

$$\begin{array}{ccc} +6 & -2 & 2- \\ \mathrm{S} & \mathrm{O_4} & \\ +6 & -8 & = 2- \end{array}$$

To determine oxidation states of elements that are part of a polyatomic ion, it is almost always easier to consider the polyatomic ion separately. For example, what is the oxidation state of nitrogen in the compound $Al(NO_3)_3$? This compound contains 1 Al, 3 Ns and 9 Os. You could find the oxidation state of the nitrogen, since you know that aluminum is always 3+, and oxygen is 2−. However, it is easier to consider the nitrate ion separately. The nitrate ion is NO_3^-. Three oxygens give a total of 3 × 2− = 6−. For the sum of the oxidation states to equal the 1− charge of the ion, the nitrogen must be 5+.

$$\begin{array}{ccc} +5 & -2 & -1 \\ \mathrm{N} & \mathrm{O_3} & \\ +5 & -6 & = -1 \end{array}$$

PRACTICE

4.6 Find the oxidation number of the underlined element in each compound:

(a) $K\underline{Mn}O_4$ (b) $Ca\underline{C}O_3$ (c) $\underline{Fe}SO_4$ (d) $H_2\underline{C}_2O_4$

4.7 Find the oxidation number of the underlined element in each of the following ions:

(a) $\underline{N}O_2^-$ (b) $\underline{P}O_4^{3-}$ (c) $\underline{Mn}O_4^-$ (d) $\underline{Cr}_2O_7^{2-}$

Applying the Stock System to Molecular Substances

Molecular substances may be named using the Stock system, which uses Roman numerals to indicate the positive oxidation state of an element. In this way, carbon dioxide becomes carbon (IV) oxide, and sulfur trioxide becomes sulfur (VI) oxide. Similarly, ionic substances are sometimes given prefix names. Na_3PO_4 is often called trisodium phosphate. Many chemists prefer that there be just one correct name for each substance. The general rules they have proposed are:

1. Use Stock names for compounds of metals and compounds of hydrogen.
2. Use prefix names for compounds that contain nonmetals only.

If these rules are obeyed, Na_3PO_4 is called sodium phosphate, and SO_3 is sulfur trioxide.

PRACTICE

4.8 Apply the rules above to give the correct name for each of the following substances.

(a) NO (b) N_2O_3 (c) Fe_2O_3 (d) CO (e) NiO
(f) $ZnCl_2$ (g) SCl_2

Other Systems of Naming Compounds

The Stock system has been in common use for only 40 years. Before that, metal ions with different charges were assigned separate names. For example, Fe^{2+} was called ferrous, while Fe^{3+} was called ferric. When there are only two different ions, the one with the lower charge is given the ending "ous" while the one with the higher charge ends in "ic." In this system, the Latin names for the elements generally are used. Here is a table of some of the more common names and their symbols.

Ferrous Fe^{2+}	Cuprous Cu^{+}
Ferric Fe^{3+}	Cupric Cu^{2+}
Stannous Sn^{2+}	Mercurous Hg_2^{2+}
Stannic Sn^{4+}	Mercuric Hg^{2+}
Plumbous Pb^{2+}	Aurous Au^{+}
Plumbic Pb^{4+}	*Auric Au^{3+}

*In the James Bond film "Goldfinger," Mr. Goldfinger's first name is Auric. Ian Fleming obviously knew some chemistry!

Using this system, the compound $FeCl_2$ is called ferrous chloride, while $FeCl_3$ is called ferric chloride.

There is a specific system for naming acids. It is discussed in Chapter 12. There is also a special method of naming organic compounds, which will be discussed in Chapter 14.

Using Formulas to Find Mass

The mass of a molecule is equal to the sum of the masses of the atoms that make up the molecule. Since you know the formula for water is H_2O, you can find the mass of a water molecule. The mass of two hydrogen atoms is 2×1 amu, or 2 amu. A single oxygen atom has a mass of 16 amu. (The atomic masses of the elements are listed in Appendix 5.) The molecular mass of water is 2 amu + 16 amu, or 18 amu.

The mass of one mole of a substance is equal to its molecular mass expressed in grams. The mass of 1 mole of water, then, is 18 grams. The molecular mass expressed in grams is sometimes called the **gram molecular mass.**

To find the mass of NaCl you would simi
of its elements to arrive at 58 amu. However
stance, does not consist of molecules. The 5
mass of the formula, NaCl, and is called the
mole of NaCl has a mass of 58 grams. The for
in grams is often called the **gram formula m**

Since both the gram formula mass and the gram molecular mass represent the mass of one mole of the substance, we use the term **molar mass to** apply to both. The molar mass is the mass of one mole of a substance, and it is equal to its formula mass in grams.

SAMPLE PROBLEMS

PROBLEM

1. What is the molar mass of aluminum nitrate?

SOLUTION

First, you need to know that the correct formula for aluminum nitrate is $Al(NO_3)_3$.

Now add up the masses of all of the atoms in the formula. In 1 mole of aluminum nitrate there is 1 mole of Al, 3 moles of N, and 9 moles of O. Adding up the masses:

$$1\ Al = 1 \times 27 = 27\ grams$$
$$3\ N = 3 \times 14 = 42\ grams$$
$$9\ O = 9 \times 16 = \underline{144}\ grams$$
$$Total\ mass = 213\ grams$$

The molar mass is always expressed in grams, or grams per mole.

ALTERNATE SOLUTION

You could have found the molar mass of aluminum nitrate by treating the nitrate ion as a single unit. A nitrate ion, NO_3^- has a molar mass of 62 grams (the mass of one N, 14, + the mass of three O's, 48). The molar mass of $Al(NO_3)_3$ is then

$$1\ Al = 1 \times 27 = 27\ grams$$
$$3\ NO_3 = 3 \times 62 = \underline{186}\ grams$$
$$Total\ mass = 213\ grams$$

PROBLEM

2. What is the molar mass of $BaCl_2$?

SOLUTION

If we round off the atomic masses to the nearest whole number, we get $137 + 2 \times 35 = 207$ grams. However, if we round off the atomic masses to the nearest tenth, we get $137.3 + 2 \times 35.5 = 208.3$. Our molar mass, to the nearest whole number, is now 208 grams. Obviously, 208 grams is the more accurate answer. When finding molar masses it is a good idea to include at least one more decimal place in your atomic masses than you need in your final answer.

PRACTICE

4.9 Find the molar mass to the nearest whole number of each of the following compounds:

(a) NH_4Cl (b) $Fe_2(SO_4)_3$ (c) sodium sulfate (d) calcium carbonate (e) $(NH_4)_2Cr_2O_7$

Chemical Equations

When magnesium burns in air, magnesium combines with oxygen, and a compound called magnesium oxide is formed. A chemist may express this reaction with a word equation,

$$\text{magnesium} + \text{oxygen} \longrightarrow \text{magnesium oxide.}$$

This equation is read "magnesium plus oxygen yields magnesium oxide." The material to the left of the "yields" sign, arrow ($\longrightarrow$), is called the **reactant(s)**, and the material to right of the arrow is called the **product(s)**.

Chemists find it more effective to express a chemical reaction using the formulas of the substances involved. The word equation above becomes:

$$Mg + O_2 \longrightarrow MgO$$

This equation, called a **skeleton equation**, shows the reactants and products, but does not indicate in what proportion they are reacting. If you examine the skeleton equation closely, you can see there are two atoms of oxygen on the left and only one atom of oxygen on the right.

The total mass of all the atoms on the left is 24 + 32 = 56 amu, while the mass on the right is only 24 + 16 = 40 amu. If the reaction actually occurred this way it would defy the law of conservation of mass. Chemists write equations to show the proportions in which materials react and products are formed, in accordance with the law of conservation of mass. These are called **balanced equations.** In a properly balanced equation, there must be the same number of atoms of each element on the left side of the equation as on the right side.

To balance an equation, you first examine the skeleton equation to see if anything is unbalanced. In the equation

$$Mg + O_2 \longrightarrow MgO,$$

you can see that there are two oxygens on the left, so you must have two oxygens on the right. To have two oxygens, two MgO must be formed. Once you have two MgO, you now need two Mg on the left. The balanced equation is

$$2\ Mg + O_2 \longrightarrow 2\ MgO.$$

You might be tempted to balance the equation this way:

$$Mg + O_2 \longrightarrow MgO_2$$

This is balanced, but it is incorrect, because the formula of magnesium oxide is MgO, not MgO_2. You cannot write MgO_2 because, if it exists, it would be a completely different substance, and it would not be formed by this reaction. When you are balancing equations you cannot change the formulas of the substances in the reaction!

Balancing Equations

There are some special techniques for balancing difficult equations, which will be discussed in Chapter 13. Most simple chemical equations can be balanced by inspection. Follow these steps:

1. Write the correct formula for each substance in the reaction. Separate the reactants from the products with the "yields" sign and each substance from the next with a "+."

2. Check the number of atoms of each element on both sides of the equation. If the same polyatomic ion appears on both sides of the equation, it can be treated as a single unit.
3. Determine which elements and/or ions are present in unequal quantities on the two sides of the equation. Place large numbers, called coefficients, in front of the substances, as needed, until each element and/or ion is present in equal quantities on both sides of the equation.
4. Make sure that your coefficients are in the lowest possible ratio. Where no coefficient is written, it is understood that the coefficient is one.

SAMPLE PROBLEMS

PROBLEM

1. Iron rusts in oxygen to form iron (III) oxide; write the balanced equation.

SOLUTION

Writing the correct formulas gives us

$$Fe + O_2 \longrightarrow Fe_2O_3$$

On the left there is 1 atom of Fe and 2 atoms of O, while on the right there are 2 atoms of Fe and 3 atoms of O. To make the oxygens equal on both sides, put the coefficient 3 in front of the O_2, and a 2 in front of the Fe_2O_3. Now there are 6 oxygen atoms on both sides. There are also now 4 iron atoms on the right, in the 2 Fe_2O_3, so we need a 4 in front of the Fe on the left. The balanced equation is

$$4\ Fe + 3\ O_2 \longrightarrow 2\ Fe_2O_3$$

PROBLEM

2. Balance the equation $Na + H_2O \longrightarrow H_2 + NaOH$

SOLUTION

In this case, the correct formulas are already given, so you move directly to the next step. On the left, there are 1 Na, 2 Hs, and 1 O. On the right, there are 3 Hs, 1 Na, and 1 O. (Note: An atom may appear in more than one substance on the same side of the reaction, as hydrogen, H, does here. They must be added together to obtain the total number of atoms of that element.) The only unbalanced element is the hydrogen. Since on the left all of the hydrogen is in H_2O, water, you will always have an even number of hydrogens on the left. You can get an even number of hydrogens on the right, by putting a 2 in front of the NaOH. Thus far you have

$$Na + H_2O \longrightarrow 2\ NaOH + H_2$$

You can balance the sodium, Na, by placing a 2 in front of the Na on the left. You can balance the H, by putting a 2 in front of the H_2O (giving us 4 Hs on each side).

$$2\ Na + 2\ H_2O \longrightarrow 2\ NaOH + H_2$$

Checking the oxygens, we see that there are now 2 Os on each side of the equation. The equation is balanced.

PRACTICE

4.10 Balance each of the following chemical equations:

(a) $Zn + HCl \rightarrow ZnCl_2 + H_2$

(b) $FeCl_3 + NaOH \longrightarrow Fe(OH)_3 + NaCl$

(c) $HgO \longrightarrow Hg_2O + O_2$

(d) Silver + sulfur $\longrightarrow$ silver sulfide

(e) Chlorine + aluminum bromide $\longrightarrow$ bromine + aluminum chloride

Types of Chemical Reactions

To help understand and predict the course of chemical reactions, chemists classify these reactions into categories. Most of the simple reactions you have studied thus far belong to one of four categories of chemical reactions.

Combination Reactions

You have learned that magnesium burns in oxygen gas to produce magnesium oxide.

$$2\ Mg + O_2 \longrightarrow 2\ MgO$$

In this reaction, two simple substances combine to form a single substance. Reactions in which two or more substances combine to form one substance are called **combination reactions.** They are also known as **synthesis** or **composition reactions.** The reactants in a combination reaction need not be elements. The reaction between calcium oxide and carbon dioxide to form calcium carbonate is a combination reaction.

$$CaO + CO_2 \longrightarrow CaCO_3$$

Decomposition Reactions

Decomposition, or **analysis, reactions** are the exact opposite of combination reactions. In a decomposition, a single substance produces two or more simpler substances. When potassium chlorate is heated, it produces oxygen and potassium chloride.

$$2\ KClO_3 \longrightarrow 2\ KCl + 3\ O_2$$

This is an example of a decomposition reaction. Hydrogen peroxide, H_2O_2, decomposes slowly on exposure to light.

$$2\ H_2O_2 \longrightarrow 2\ H_2O + O_2$$

Single Replacement Reactions

When a piece of zinc is placed in a solution of copper (II) sulfate, the zinc goes into solution as zinc ions, and metallic copper comes out of solution.

$$Zn + CuSO_4 \longrightarrow ZnSO_4 + Cu$$

The zinc is said to have replaced the copper, and the reaction is called a **single replacement reaction.** All single replacement reactions have the same essential format:

$$\text{element} + \text{compound} \longrightarrow \text{other element} + \text{other compound}$$

Metals replace metals, and nonmetals replace nonmetals in these reactions. However, hydrogen, which is not considered a metal, often acts as a metal in single replacement reactions. When magnesium is placed in dilute sulfuric acid (H_2SO_4), the magnesium quickly disappears, and the mixture bubbles vigorously. The bubbles indicate the formation of a gas, in this case hydrogen. An upward arrow, ↑, or (*g*), is often used in chemical equations to indicate the formation of a gas.

$$Mg + H_2SO_4 \longrightarrow MgSO_4 + H_2 \uparrow$$

This is another single replacement reaction. The magnesium has replaced the hydrogen. Elements can only replace elements that are less chemically active than they. Magnesium is more active than hydrogen, and zinc is more active than copper, so both of the reactions shown above do occur. Single replacement reactions are discussed in more detail in Chapter 13.

Double Replacement Reactions

Double replacement reactions generally occur between ionic compounds in solution. For example, when clear solutions of silver nitrate and sodium chloride are mixed, a white precipitate of silver chloride is formed, while sodium nitrate remains in solution. Compounds in solution are indicated by (*aq*).

$$AgNO_3\ (aq) + NaCl\ (aq) \longrightarrow NaNO_3\ (aq) + AgCl \downarrow$$

A **precipitate** is an insoluble product, and is often indicated in chemical equations with a downward arrow, ↓, or (*s*). In double replacement reactions, the positive ions simply "change partners." For an actual chemical change to occur, there must be some reason why the ions do not simply switch back again. When a precipitate is formed, the ions leave the solution, so the reaction cannot go backwards. Double replacement reactions occur when a precipitate is formed, a gas is formed, or a molecular substance such as water is formed. In each of these cases, ions are removed from the solution. In the reaction between sodium hydroxide and hydrochloric acid (HCl),

$$NaOH + HCl \longrightarrow NaCl + H_2O$$

water, a molecular substance, is formed. The H^+ ions have changed places with the Na^+ ions, so you can see that this is clearly a double replacement reaction.

PRACTICE

4.11 Refer to the five reactions given in Practice 4.10. Classify each of these reactions, using the reaction types described in this section.

Predicting the Products of Chemical Reactions

Thus far, you have written balanced equations, given the reactants and the products of the reaction. Using your understanding of single and double replacement reactions, you now should be able to write a complete balanced equation, based only on the reactants.

SAMPLE PROBLEMS

PROBLEM

1. Write a balanced equation for the reaction between zinc metal and copper (II) sulfate.

SOLUTION

First write the formulas of the reactants: $Zn + CuSO_4$.

Analyze the reaction: An element is reacting with a compound, so this is a single replacement reaction. The metal, zinc, must replace the other metal, copper. The products of the reaction are copper and zinc sulfate.

Write the correct formulas for the products: $Cu + ZnSO_4$

Balance the equation: $Zn + CuSO_4 \longrightarrow Cu + ZnSO_4$

Checking each element, we see that the equation is already balanced.

PROBLEM

2. Write a balanced equation for the reaction between aluminum metal and a solution of zinc chloride.

SOLUTION
The formulas of the reactants are Al and $ZnCl_2$.

An element is reacting with a compound, so this is a single replacement reaction. The aluminum is replacing the zinc, so the products are aluminum chloride and zinc.

The formulas of the products are $AlCl_3$ and Zn. (If you have forgotten how to find the correct formulas, you should review "formula writing" at the beginning of this chapter.)

The equation, $Al + ZnCl_2 \rightarrow AlCl_3 + Zn$, is not balanced, because there are not the same number of chlorines on both sides. The balanced equation is $2\ Al + 3\ ZnCl_2 \rightarrow 2\ AlCl_3 + 3\ Zn$.

PROBLEM

3. Write a balanced equation for the reaction between solutions of barium chloride and sodium sulfate.

SOLUTION
The formulas of the reactants are $BaCl_2$ and Na_2SO_4.

This is a reaction between two ionic substances in solution. It is a double replacement reaction. The ions "change partners," so the products are sodium chloride and barium sulfate.

The formulas of the products are NaCl and $BaSO_4$.

Balancing the equation $Na_2SO_4 + BaCl_2 \longrightarrow BaSO_4 + NaCl$ gives us our answer, $Na_2SO_4 + BaCl_2 \longrightarrow BaSO_4 + 2\ NaCl$.

PRACTICE

4.12 Write balanced chemical equations for each of the following chemical reactions.

(a) Magnesium reacts with a solution of nickel(II) chloride.

(b) Aluminum reacts with a solution of silver nitrate.

(c) Iron(III) chloride and silver nitrate react in solution.

(d) Hydrogen reacts with hot copper(II) oxide.

(e) A solution of barium hydroxide is added to a solution of iron(III) sulfate.

CHAPTER REVIEW

The following questions will help you check your understanding of the material presented in the chapter.

1. The correct formula for lead(IV) oxide is (1) PbO (2) Pb_2O (3) PbO_2 (4) Pb_2O_2.
2. When the equation $C_2H_6 + O_2 \longrightarrow CO_2 + H_2O$ is correctly balanced, the coefficient in front of O_2 will be (1) 7 (2) 10 (3) 3 (4) 4.
3. The correct formula for aluminum sulfate is (1) Al_2S_3 (2) Al_3S_2 (3) $Al_2(SO_4)_3$ (4) $Al_3(SO_4)_2$.
4. Given the unbalanced equation

   $$\_Al_2(CO_3)_3 + \_Ca(OH)_2 \longrightarrow \_Al(OH)_3 + \_CaCO_3$$

   the sum of the coefficients for the balanced equation is (1) 5 (2) 9 (3) 3 (4) 4.
5. The reaction $Pb(NO_3)_2 + 2\ NaI \rightarrow 2\ NaNO_3 + PbI_2 \downarrow$ is classified as a (1) combination reaction (2) decomposition reaction (3) single replacement reaction (4) double replacement reaction.
6. What is the most likely formula for the compound formed when magnesium reacts with nitrogen? (1) Mg_2N (2) Mg_3N_2 (3) Mg_2N_3 (4) Mg_5N_2
7. Element X has an electron configuration of 2-8-2. Element X will most likely form oxides with the formula (1) X_2O (2) X_2O_3 (3) XO (4) XO_2.
8. What is the molar mass of $Al(OH)_3$? (1) 46 g (2) 78 g (3) 132 g (4) 44 g

9. A sulfide with the formula X_2Y_3 will form if sulfur combines with any element in Group (1) 1 (2) 2 (3) 13 (4) 14,
10. What is the empirical formula of the compound N_2O_4? (1) NO (2) NO_2 (3) N_2O (4) N_2O_3
11. Nitrogen(IV) oxide is also called (1) nitrogen monoxide (2) nitrogen dioxide (3) dinitrogen monoxide (4) tetranitrogen dioxide
12. When the equation $Ca(ClO_3)_2 \longrightarrow CaCl_2 + O_2$ is balanced, the coefficient in front of O_2 will be (1) 1 (2) 2 (3) 3 (4) 4.
13. A certain metal, M, forms a sulfate with the formula M_2SO_4. What would be the likely formula for the nitrate formed by element M? (1) MNO (2) MNO_3 (3) $M(NO_3)_2$ (4) M_2NO_3
14. What is the oxidation state of the nitrogen atom in the compound KNO_3? (1) +1 (2) +3 (3) −3 (4) +5
15. What is the oxidation state of the Cr in the chromate ion, CrO_4^{2-}? (1) +6 (2) +2 (3) +3 (4) +8

CONSTRUCTED RESPONSE

1. Write a balanced equation for the reaction:

$$\text{calcium} + \text{water} \longrightarrow \text{calcium hydroxide} + \text{hydrogen}$$

After you have balanced the equation, find the total mass of the reactants and the total mass of the products. Show that the law of conservation of mass has been obeyed.

2. The carbonate, nitrate, sulfate, and phosphate ions listed on page 161 each contain a nonmetal in its maximum oxidation state.
   (a) Find the oxidation states of the carbon, nitrogen, sulfur and phosphorous in these ions.
   (b) How is the maximum oxidation state related to the electron configuration of each of these elements?
   (c) What would you predict should be the maximum oxidation state of chlorine?
   (d) Find the ion in Appendix 4 that shows chlorine in its maximum oxidation state.

Chemical Riddles

Your new knowledge of names and formul
understand and create a special kind of hu
dles that chemists like to share with one a
around the Bunsen burner toasting marsh
most famous riddles follows. What is this

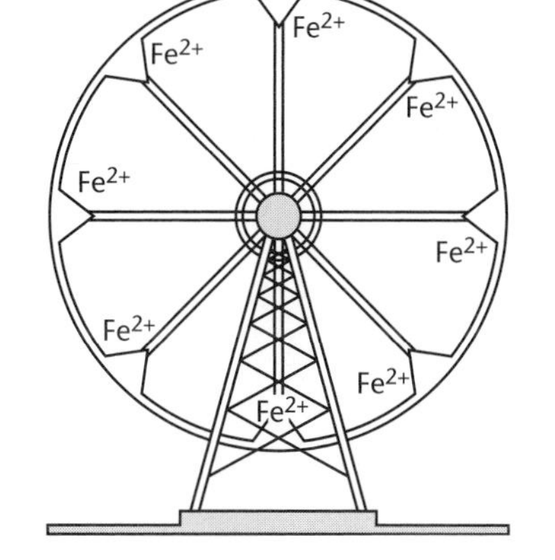

Answer: A ferrous wheel! Recall that the Fe^{2+} ion is called "ferrous." See if you can figure out the following chemical riddles. The answers are on the next page.

1. What does this represent?

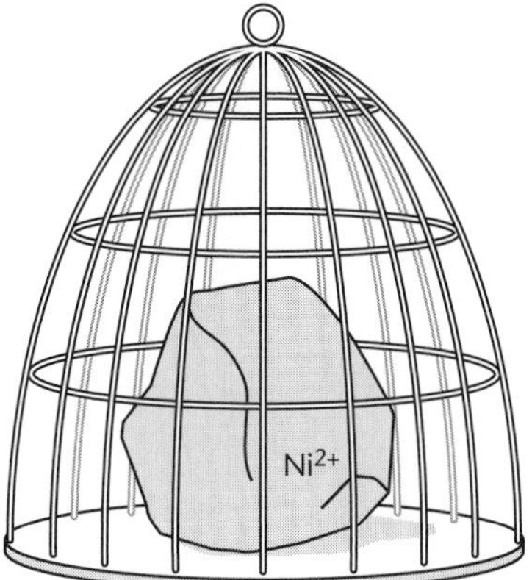

does St. Ni^{2+} represent?

hat do you get if a barium combines with two sodiums?

What is this? $HClO_2$—$HClO_2$—$HClO_2$—$HClO_2$
(Hint: $HClO_2$ is called chlorous acid.)

5. On October 12, 1492, who discovered lead?

Answers:

1. Nickelous Cage
2. Saint Nickelous
3. A BaNaNa (banana)
4. A chlorous line
5. Christopher Plumbous

5 The Periodic Table

LOOKING AHEAD

This chapter describes the organization and important features of the Periodic Table of the Elements. From the arrangement of the elements in the Periodic Table you can derive many relationships in physical and chemical properties among the elements.

When you have completed this chapter you should be able to:

- **Describe** the general organization of the Periodic Table; the approximate location of metals, nonmetals, and metalloids in the Periodic Table.
- **Correlate** changes in atomic radii with changes in ionization energy and in electronegativity; the number of valence-electrons with typical bonding behavior.
- **Define** periodicity, as it applies to the elements; metal, nonmetal, and metalloid; active metal; transition element.
- **Predict** the properties of elements in Groups 1, 2, 13–18 from the positions of the elements in the Periodic Table.

Origins of the Periodic Table

Chemists would be overwhelmed with isolated pieces of information if they did not have some way of relating the facts they know about the more than 100 known elements. The Periodic Table provides a means of organizing information so that relationships among elements can be clearly seen and understood.

Even before so many elements were known, scientists were searching for relationships among elements. Three important attempts to determine such relationships were made during the nineteenth century by the English physician William Prout, the German chemist Johann Döbereiner, and the English chemist John Newlands. All three of these scientists based their work on the model of the atom proposed in 1803 by the English scientist John Dalton.

The Dalton Model

The Dalton model proposed the following:

1. All matter consists of simple bodies (elements) and compound bodies (compounds). The smallest part of a simple body is the atom. The smallest part of a compound body is the compound atom (later called the molecule).
2. All atoms of the same element have the same properties, such as shape, size, and mass. Atoms of different elements have different properties.
3. When matter undergoes chemical change, atoms of different elements either combine or separate from one another.
4. Atoms cannot be destroyed, even during chemical change.

Prout's Hypothesis

In 1814, Prout suggested that all of the other elements are developed from hydrogen. In other words, Prout proposed that hydrogen is the fundamental element. Prout reached this conclusion because he observed that the masses of atoms are whole-number multiples of the atomic mass of hydrogen. The notion of atomic mass as the relative mass of an atom on a scale where hydrogen has a mass of one unit grew out of this idea. Later, oxygen became the standard. Later still, the concept of atomic mass unit (amu), with carbon-12 as a standard, was developed (pages 186–187).

At the time, Prout's idea seemed revolutionary. Now, it appears that Prout may have been close to the truth. Evidence from studies of radioactive changes led modern scientists to believe that all elements may be derived from hydrogen. You will learn more about radioactive changes in Chapter 15.

Döbereiner's Triads

In 1817, Döbereiner noticed that certain groups of three elements have related properties. Döbereiner called these groups *triads*. If the elements of a triad are arranged in order of increasing atomic mass, the atomic mass of the middle element is the average of the atomic masses of the other two elements. Take the triad chlorine, bromine, and iodine as an example. The atomic mass of bromine (80) is close to the average of the atomic masses of chlorine (35) and iodine (127).

The properties of the middle element of a triad are also approximately midway between the properties of the other two elements. Thus bromine is less reactive than chlorine but more reactive than iodine.

Newlands' Law of Octaves

In 1865, Newlands suggested that if the elements are arranged in order of increasing atomic mass, the first and eighth elements have related properties. Let's take an example of the first 14 elements (except hydrogen) that were known at Newlands' time. If you arrange these 14 elements in order of increasing atomic mass in columns of seven, this is what you get:

lithium	sodium
beryllium	magnesium
boron	aluminum
carbon	silicon
nitrogen	phosphorus
oxygen	sulfur
fluorine	chlorine

Just as the first note in an octave of the musical scale resembles the eighth note, so the first element—lithium—resembles the eighth element—sodium. The second element resembles the ninth; the third resembles the tenth; and so on. This relationship, because of its resemblance to the musical scale, is called the **law of octaves.** (Newlands was unaware of the existence of the elements neon and argon. If these elements are included, the relationship no longer resembles a musical scale.)

The Mendeleev-Meyer Periodic Classification

The ideas of Prout, Döbereiner, and Newlands were met with doubt and, in some cases, scorn. Yet these ideas proved to be forerunners of one of the most important schemes of organization in all of chemistry—periodic classification.

In 1869, Dmitri Mendeleev, a Russian chemist, stated that the properties of the elements are periodic functions of their atomic weights (masses). This means that the properties of elements repeat regularly and are related to the atomic weights (masses) of the elements. Mendeleev's conclusion was based on a study of the chemical properties observed in the elements then known.

At about the same time, Lothar Meyer, a German chemist, came to the same conclusion. Meyer's conclusion was based on studies of some of the physical properties of the same elements.

The observation by both chemists is called the **periodic law.** The law states that if the elements are arranged in order of increasing atomic weight (mass), the physical and chemical properties repeat themselves regularly. These repetitions are referred to as *periodic functions,* or as *periodicity.* You will later learn that these properties are periodic functions of *atomic numbers.* This means that the properties of elements are repeated when the elements are arranged in order of increasing atomic number.

All the elements known at the time of Mendeleev and Meyer were arranged in the form of a table. A portion of the table appears below.

	Groups						
Series	***I***	***II***	***III***	***IV***	***V***	***VI***	***VII***
1	H						
2	Li	Be	B	C	N	O	F
3	Na	Mg	Al	Si	P	S	Cl
4	K	Ca	□	□	As	Se	Br
	Cu	Zn	□	Ti	V	Cr	Mn
5	Rb	Sr	In	Sn	Sb	Te	I
	Ag	Cd	Y	Zr	Nb	Mo	□

□ represents an element that was predicted but had not yet been discovered.

This arrangement placed elements with similar properties in chemical families called *groups*. The groups are the vertical columns of the table, designated by Roman numerals. The properties of an element in any group are closely related to the properties of the elements directly above and below it.

Each group of the table has a subgroup written to the right of the elements in the main group. Cu (copper) and Ag (silver), for example, are in a subgroup of Group I. Elements in a subgroup are more closely related to one another than they are to the elements in the main group. Cu and Ag are more closely related to each other than they are to Li (lithium) and Na (sodium) in main Group I.

The Mendeleev-Meyer table was also based on the Dalton model of the atom. Because the Dalton model is imperfect, the table based on it is too. The reasons for the periodicity of properties are not apparent. The reasons for the subgroups cannot be explained. Finally, when certain elements, such as argon and potassium, cobalt and nickel, and tellurium and iodine, are arranged in order of increasing atomic weight (mass), their properties do not fall properly into place. In these cases, it seems that the order of the two members of a pair of elements should be reversed.

Despite the limitations of the table, it was enormously useful. The table enabled scientists to predict the existence of many elements that had not yet been discovered. These predicted, but undiscovered, elements are indicated by the symbol □ in the table.

Predictions based on the table were remarkably accurate. For example, notice that there is a symbol for a predicted element in the square directly below Si (silicon). The properties that this predicted element was expected to have are shown in the following table. The table also lists the properties that the element—germanium—did, in fact, have when it was discovered in 1886.

Property	***Prediction for Element Below Silicon***	***Observed Property of Germanium***
Atomic weight	72	72.32
Specific heat	0.073	0.076
Specific gravity	5.5	5.47
Formula of oxide	XO_2	GeO_2

In summary, the Mendeleev-Meyer table was a great step forward in the task of classifying the elements, and it led to the discovery of many new elements.

The Modern Periodic Table

The work set in motion by the Mendeleev-Meyer system of classification led to further developments during the latter half of the 19th century. By 1900 the noble gas elements, helium, neon, argon, xenon, and krypton, had been discovered and added to the table. One problem, however, was as yet unresolved. Why did a few elements, when grouped by atomic mass, fail to appear in their proper place?

This problem was finally resolved in 1913, when a 26-year-old English physicist, Henry Moseley, was able to determine the atomic number of each of the elements. When the elements were grouped by atomic number, instead of by atomic mass, every element fell into its proper group, with elements of similar properties. Thus the periodic law was revised, and the development of the modern periodic table was possible.

The **revised periodic law** states: The properties of the elements are periodic functions of their atomic number. In other words, when the elements are arranged in order of increasing atomic number, the properties of the elements repeat regularly. Notice that in this arrangement, argon (atomic number 18) precedes potassium (atomic number 19), which has a lower atomic mass than argon. The same reversal of order occurs with cobalt and nickel and with tellurium and iodine. The properties of these pairs of elements now fall into place.

Recall that the atomic number indicates the number of protons (and also of electrons) in an atom. You will see why periodicity is a function of atomic number when you consider some examples of properties that repeat regularly. First, though, you should become familiar with the basic organization of the Periodic Table.

Organization of the Periodic Table

In the revised Periodic Table, the elements are arranged in order of increasing atomic number. The elements fall into horizontal rows and vertical columns. Horizontal rows, called periods, are labeled 1, 2, 3, ... Vertical columns, called groups, or families, were labeled with Roman numerals and letters, such as IIA and IIB. The subgroups, described on page 168, became the Group B elements, while the main groups became the Group A elements.

A final change in the Periodic Table was made recently, by the International Union of Pure and Applied Chemistry (IUPAC). The A and B designations were discarded entirely, and the groups were simply numbered consecutively, from 1 to 18. The elements in Groups 1, 2, 13–17 are now often called the "representative elements," while the elements in Groups 3–12 are called "transition elements." The modern Periodic Table is shown in Appendix 5.

Atomic masses are shown above atomic numbers in the table. Atomic masses are based on the carbon-12 standard—that is, the masses are determined on a scale in which the most common isotope of carbon is assigned a mass of exactly 12.00. The mass numbers reflect the weighted average of the naturally occurring isotopes.

The electron configuration of an atom is also given for each element. In this table, the common oxidation states are also provided. This information applies to some of the periodic properties of the elements, as you will see.

In this table, as in most Periodic Tables now in use, the row of elements 58–71 (lanthanides) and the row of elements 90–103 (actinides) appear separately at the bottom of the table. This placement makes it easier to follow the regularities of all the other elements. Let's now look at examples of periodicity.

Atomic Radii

A clear example of a periodic trend is the regularly repeating decreases and increases in the atomic radii of elements when they are arranged according to increasing atomic numbers. Recall that the atomic, or covalent, radius is one half of the bond length R_C (see Figure 3-1, page 116). If you examine the atomic radii shown for each element in the Periodic Table, leaving out the transition elements, Groups 3–12 and the noble gases Group 18, you will find a regular pattern within both periods and groups. Within periods, atomic radii decrease. Within groups, atomic radii increase.

Let us examine the reasons for these trends in atomic radii. As we move left to right, within a period, the number of principal energy levels stays the same, while the number of protons and electrons increases. The attraction of the protons for the electrons is much stronger than the repulsion between the electrons within the same energy level. As the positive charge of the nucleus increases, due to additional protons, we would expect the electron cloud to be pulled closer to the nucleus. Thus, the atomic radius decreases across a period.

As we go down a group, the number of protons increases, as does the number of occupied energy levels. The increased number of protons tends to pull the electron clouds closer to the nucleus. However, each principal energy level is "screened" or "shielded" from the nucleus by the inner energy levels. Each new outer energy level is farther from the nucleus than the inner level was, so the radii generally increase as we go down a group.

The periodic increases and decreases can be shown by a graph, as in Figure 5-1. Notice that the elements at the peaks of the graph form Group 1. The elements at the lowest points of the graph form the beginning of Group 17. Within these groups, atomic radii increase as atomic numbers increase. These changes in atomic radii are consistent with trends in ionization energy

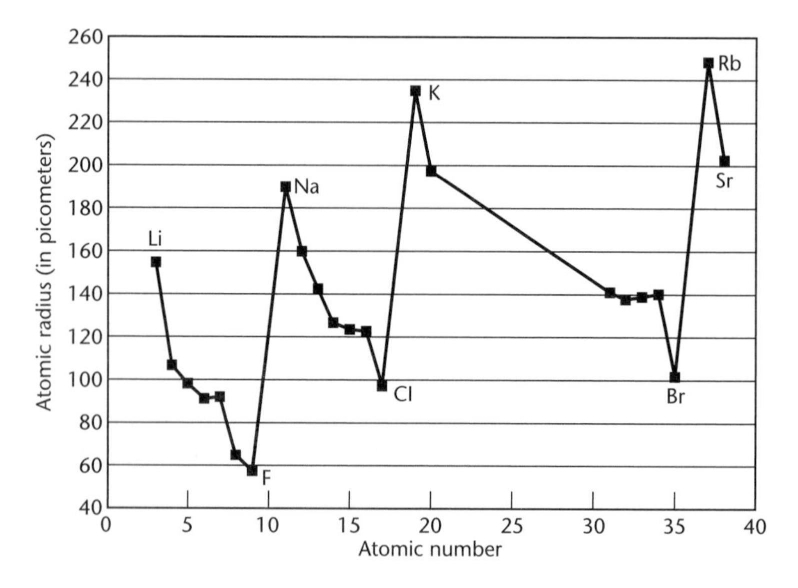

Figure 5-1 Periodic trends in atomic radii (transition elements and Group 18 omitted)

and electron affinity. They are also consistent with the change from metallic to nonmetallic character of the elements.

Ionization Energy

Recall that ionization energy is the energy required to remove the most loosely held electron from a gaseous atom. Trends in ionization energy are shown by the graph in Figure 5-2. The values of ionization energy increase and decrease regularly as atomic numbers increase.

Ionization energy generally increases across periods and decreases down groups. If you compare the graphs in Figures 5-1 and 5-2, you will see that trends in ionization energy are inversely related to trends in atomic radii. As atomic radii decrease, it becomes more difficult to remove an electron—that is, ionization energy increases. As atomic radii increase, it becomes easier to remove an electron—ionization energy decreases.

Electron Affinity and Electronegativity

Electron affinity refers to the energy released when an atom receives an electron and becomes a negative ion. Electro-

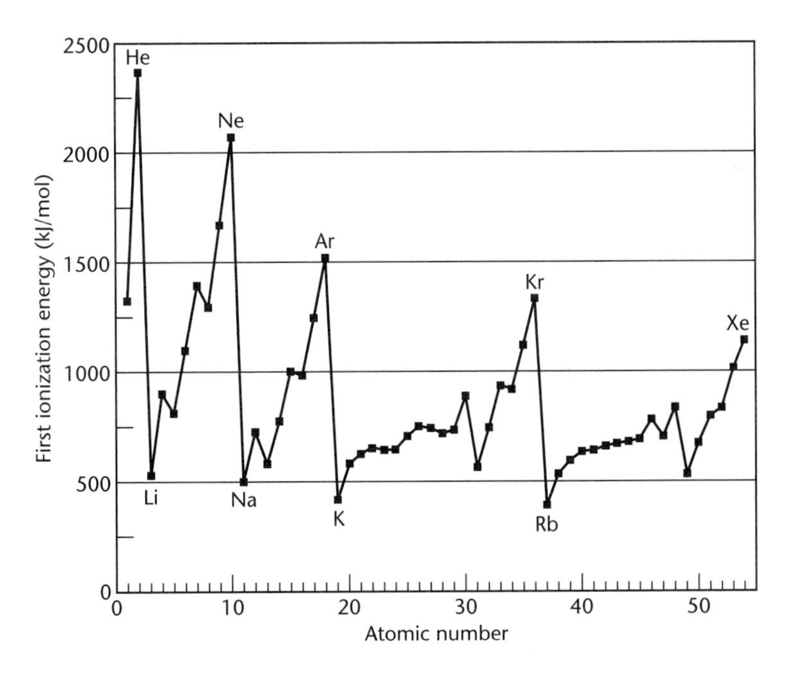

Figure 5-2 Periodic trends in ionization energy

negativity refers to the tendency of an atom to attract electrons.

Electronegativity and electron affinity are also periodic properties of atoms and can be related to the trends in atomic radii. As atomic radii decrease within a period, outer electrons are held more tightly and become harder to remove. Thus, in general, electron affinity and electronegativity increase across a period. As atomic radii increase down a group, outer electrons are held less firmly and become easier to remove. Thus electron affinity and electronegativity decrease as atomic radii increase within a group.

Valence Electrons

The number of valence electrons repeats periodically across a period, if the transition elements are omitted. For example, the number of valence electrons changes from 1 to 8 in Period 2 (Li to Ne), goes back to 1 with Na, changes from 1 to 8 in Period 3 (Na to Cl), and so forth.

Within the representative elements (Groups 1, 2, 13–17), every element in a group has the same number of valence electrons. For example, every element in Group 15 has five valence electrons; every element in Group 17 has seven. This constant number of valence electrons explains the chemical similarity among elements in the same group.

PRACTICE

5.1 As you move across a period, left to right, describe what generally happens (decreases, increases, or remains the same) to

(a) the number of valence electrons

(b) the ionization energy

(c) the atomic radius.

5.2 As you move down a group, describe what generally happens to

(a) the number of valence electrons

(b) the ionization energy

(c) the atomic radius

5.3 Identify the element from the clues given below:

(a) This element has the same number of valence electrons as calcium and the same number of occupied principal energy levels as carbon.

(b) With a smaller atomic radius than phosphorus, this element has a smaller ionization energy than fluorine and is chemically similar to iodine.

(c) This element has the smallest ionization energy of any element in Period 4.

Bonding Behavior

Bonding behavior, you recall, is related to the number of valence electrons in an atom. Bonding behavior is another recurring, or periodic, trend that can be observed. Looking at Period 2 and Period 3, you will notice the following:

1. Li and Na each have one valence electron. They tend to lose this electron and form 1+ ions in ionic compounds.
2. Be and Mg each have two valence electrons. These elements tend to lose both electrons and form 2+ ions in ionic compounds.
3. B and Al have three valence electrons. Boron has a very small atomic radius, and holds its valence electrons too tightly to permit the formation of ionic compounds. Aluminum, with its larger radius, is able to lose electrons to form 3+ ions in ionic compounds.
4. C and Si have four valence electrons. These elements form covalent bonds only.
5. N and P have five valence electrons. It is difficult to gain three electrons. Hence these elements form mostly covalent bonds. In special cases, N and P may form ionic bonds, as in K_3N. In these cases, N and P become 3− ions.
6. O and S have six valence electrons. They form covalent bonds with other nonmetals, and ionic bonds with many of the metals. In forming ionic bonds, both S and O become 2− ions.

7. F and Cl have seven valence electrons. They often gain one electron and form 1− ions in ionic compounds. They can also share electrons with related elements, as is the case in the compound OF_2.
8. Ne and Ar have eight valence electrons. As far as is known, neon does not form bonds with other elements. Its neighbors Ar, Kr, Xe, and Rn do form compounds. Ar, Kr, Xe, and Rn have larger atomic radii, and their outer electrons are farther from the nucleus. Thus the electrons are less firmly held and can be involved in bond formation.

The trends outlined here for Period 2 and Period 3 repeat in the other periods, with the exception of the transition elements. Valence electrons and bonding behavior are the basis for the similarities in chemical properties of elements within groups, or families. In most cases, elements are most similar when they are directly above or below one another in the Periodic Table.

Metals and Nonmetals

Most of the elements in the Periodic Table are **metals.** The metals include all the elements of Group 1 except hydrogen, all of Group 2, all of the transition elements, and a few others: aluminum (Al), tin (Sn), and bismuth (Bi).

Metallic activity is related to the tendency to lose electrons. Hence the most active metals are in Group 1 and Group 2. Across a period, outer electrons are held more tightly as atomic radii decrease. Metallic activity therefore decreases across periods. Down a group, outer electrons are held less firmly as atomic radii increase. Metallic activity therefore increases down a group. Again, a periodic trend is evident: from strong metallic character on the left side of the Periodic Table to nonmetallic character on the right, back to strong metallic behavior on the left, and so on. As free elements, metals are good electrical conductors.

The **nonmetals** appear on the right side of the Periodic Table. All the elements in Group 17 are nonmetals. In addition, carbon (C), nitrogen (N), phosphorus (P), oxygen (O), sulfur (S), and selenium (Se) are nonmetals. One measure of the activity of a nonmetal is its ability to gain one or more electrons. In this sense, as atomic numbers increase, nonmetallic activity or character increases across a period and decreases down a group. Periodicity is evident in the recurring trend from metal to nonmetal in the periods. As free elements, nonmetals are nonconductors of electricity.

The properties of a few elements lie somewhere between those of metals and those of nonmetals. These elements, called **metalloids** or **semimetals**, are found in Group 13 through Group 16 and include boron (B), silicon (Si), germanium (Ge), arsenic (As), antimony (Sb), and tellurium (Te). When they are free elements, the metalloids are semiconductors—they conduct an electric current under some conditions but not under others.

Aluminum (Al) clearly has the properties of a metal, including conductivity, luster, and a tendency to form 3+ ions or to assume a 3+ oxidation state. However, in certain compounds, aluminum acts like a semimetal. Therefore, although aluminum is more frequently included with the metals, it may sometimes be referred to as a semimetal.

Bases and acids Metals and nonmetals form compounds called hydroxides, which can be symbolized as XOH. The X stands for a metal or a nonmetal. When the bond between X and OH is broken, the hydroxide acts as a *base*. Hydroxides of metals behave in this way. In this case, X is a metal. An example is NaOH, which forms Na^+ and OH^- ions in water. When the bond between XO and H is broken, the hydroxide acts as an *acid*. Hydroxides of nonmetals behave in this way. In this case, X is a nonmetal. An example is Cl(OH), which forms H^+ and ClO^- ions in water. Because the hydrogen is the least electronegative element in the compound, ClOH is more correctly written as HClO.

Hydroxides of metalloids can act as either acids or bases, depending on the chemical environment. They are often called **amphoteric hydroxides.** $Al(OH)_3$ reacts as a base with HCl, but reacts as an acid with the strong base, NaOH.

The Transition Elements

As you have noticed, the transition elements have been omitted from all the preceding considerations of the Periodic Table. This is because regularities in periods and groups are interrupted whenever the transition elements appear. Recall that in a transition series, the number of valence electrons is virtually constant as an inner sublevel receives additional electrons. This means that such properties as atomic radius, ionization energy, and bonding behavior change very little among the transition elements.

Some common transition elements are scandium (Sc) through zinc (Zn) in the fourth row of the Periodic Table.

The lanthanide and actinide series Elements 58–71 and 90–103 belong to the lanthanide and actinide series, respectively. The lanthanides are filling the 4*f* sublevel, and the actinides, the 5*f*. Since an *f* sublevel can hold 14 electrons, there are 14 elements in each series. The lanthanide series contains some of the rarest of all the stable elements. For this reason they are sometimes referred to as the "rare earths." The addition of 14 protons to the nucleus, across the lanthanide series, pulls the electron cloud closer to the nucleus. As a result, the transition elements in Period 6, from Hf to Hg, have smaller radii than would otherwise be expected. This "lanthanide contraction" causes many of the elements in Period 6 to be very similar in radius to those in Period 5. The actinides, elements 90–103, are chemically quite similar to the lanthanides. However, they are all radioactive, and so cannot be put to ordinary chemical use. Uranium, element 92, is the last naturally occurring element on the Periodic Table. All of the elements beyond uranium do not occur naturally, and are manmade.

Elements 104–109 are in the 6*d* transition series. Elements with still higher atomic numbers have been discovered. According to theory, these new elements should fit into the 6*d* transition series.

Period 1

Period 1 contains only two elements, hydrogen and helium. Hydrogen is usually placed above the metals of Group 1 in the Periodic Table, but it bears little or no resemblance to these metals. It is a gas at room temperature and is the least dense (the lightest) of all the elements. An atom of hydrogen, unlike an atom of a Group 1 metal, loses its *s* electron with great difficulty because it is in the first principal energy level. Hydrogen may be placed above Group 17, instead of above the Group 1 metals. But hydrogen is also very unlike the nonmetals of Group 17, except that hydrogen, fluorine, and chlorine are all gases at room temperature. Hydrogen and the Group 17 elements require a single electron to complete their outermost shells. Hydrogen does not acquire electrons readily, but the Group 17 elements do.

Helium ($1s^2$) is a representative member of the noble gas family, which is usually labeled Group 18. Helium is the least dense of the noble gases. To date, it has not formed any chemical compounds.

Periods 2 and 3

Periods 2 and 3 are the last two periods that do not contain transition elements. Each of these periods is made up of eight elements, starting with an alkali metal (lithium, Li, in Period 2; sodium, Na, in Period 3) and ending with a noble gas (neon, Ne, in Period 2; argon, Ar, in Period 3). As you go from left to right across these periods, the metallic character of the elements decreases and the nonmetallic character increases. Valence electrons increase regularly from one to eight as the $2s$, $2p$ or the $3s$, $3p$ sublevels fill with electrons. Other properties of the elements in these periods, such as trends in atomic radii and ionization energies, were described in Chapter 2.

Period 4

Period 4 has a total of 18 elements. The period begins with the active metals potassium (K) and calcium (Ca), whose valence electrons are in the $4s$ sublevel. These elements are followed by the $3d$ transition series of 10 elements, from scandium (Sc) through zinc (Zn). In these elements, the $3d$ sublevel becomes lower in energy than the $4s$ sublevel as soon as the $3d$ sublevel contains one or more electrons. Once the $3d$ sublevel is filled, the $4p$ sublevel receives electrons. The period ends with the noble gas krypton (Kr).

The $3d$ transition elements, all metals, can use both $4s$ and $3d$ electrons to form chemical bonds. In many chemical environments, the orbitals within the d sublevel have different energies. Electrons can then absorb light energy and move from one d orbital to another. When this happens, the wavelengths of light that are not absorbed are reflected. Hence most transition elements have a color in solid compounds and in water solutions. You may have seen the blue color of water solutions containing copper ions (Cu^{2+}) and the green color of solutions containing nickel ions (Ni^{2+}).

CHAPTER REVIEW

The following questions will help you check your understanding of the material presented in the chapter.

1. Which period contains elements that are all gases at STP? (1) 1 (2) 2 (3) 3 (4) 4
2. Which group contains atoms that form 1+ ions having a noble gas configuration? (1) 1 (2) 7 (3) 11 (4) 17
3. The element with an atomic number of 34 is most similar in its chemical behavior to the element with an atomic number of (1) 19 (2) 31 (3) 36 (4) 52.
4. A pure compound is blue in color. It is most likely a compound of (1) sodium (2) lithium (3) calcium (4) copper.
5. Which represents the correct order of activity for the Group 17 elements? (> means "greater than.") (1) bromine > iodine > fluorine > chlorine (2) fluorine > chlorine > bromine > iodine (3) iodine > bromine > chlorine > fluorine (4) fluorine > bromine > chlorine > iodine
6. All elements in Period 3 have (1) an atomic number of 3 (2) 3 valence electrons (3) 3 occupied principal energy levels (4) an oxidation number of +3.
7. Which group contains two metalloids? (1) 2 (2) 5 (3) 12 (4) 15
8. When the atoms of the elements of Group 18 are compared in order from top to bottom, the attractions between the atoms of each successive element (1) increase and the boiling point decreases (2) decrease and the boiling point increases (3) increase and the boiling point increases (4) decrease and the boiling point decreases.
9. The atoms of the most active nonmetals have (1) small atomic radii and high ionization energies (2) small atomic radii and low ionization energies (3) large atomic radii and low ionization energies (4) large atomic radii and high ionization energies.

10. Which of the following elements has the smallest atomic radius? (1) K (2) Ca (3) As (4) Br

11. The oxide of metal X has the formula XO. Which group in the periodic table contains metal X? (1) 1 (2) 2 (3) 13 (4) 15

12. Which period contains elements in which electrons from more than one principal energy level may be involved in bond formation? (1) 1 (2) 2 (3) 3 (4) 4

13. Which chloride is most likely to be colored in the solid state? (1) KCl (2) $NiCl_2$ (3) $AlCl_3$ (4) $CaCl_2$

14. The element whose properties are most similar to those of tellurium is (1) Be (2) S (3) O (4) Po.

15. A sulfide with the formula X_2S_3 would be formed if sulfur combined with any element in Group (1) 1 (2) 2 (3) 13 (4) 14.

16. As you go from fluorine to astatine in Group 17 the electronegativity (1) decreases and the atomic radius increases (2) decreases and the atomic radius decreases (3) increases and the atomic radius decreases (4) increases and the atomic radius increases.

17. Elements that generally exhibit multiple oxidation states and whose ions are usually colored are classified as (1) metals (2) metalloids (3) transition elements (4) nonmetals.

18. If X is the atomic number of an element in Group 12, an element with the atomic number (X + 1) will be found in Group (1) 11 (2) 2 (3) 13 (4) 3.

19. All the elements in Group 13 have the same (1) atomic radius (2) number of occupied principal energy levels (3) electronegativity (4) number of valence electrons.

20. What is the basis for the arrangement of the present Periodic Table? (1) number of neutrons in the nucleus of an atom (2) number of nucleons in the nucleus of an atom (3) atomic number of an atom (4) atomic mass of an atom

21. As you go from left to right across Period 3 of the Periodic Table, there is a decrease in (1) ionization energy (2) electronegativity (3) metallic characteristics (4) valence electrons.

22. As the elements in Group 2 are considered in order of increasing atomic number, the tendency of each successive atom to

form a positive ion generally (1) decreases (2) increases (3) remains the same.

23. Which is an example of a metalloid? (1) Fe (2) La (3) Mg (4) Si

24. Which of the following ions has the smallest ionic radius? (1) Cl^- (2) S^{2-} (3) K^+ (4) Ca^{2+}

Base your answers to questions 25 through 27 on the partial Periodic Table that follows.

	Group							
Period	***1***	***2***	***13***	***14***	***15***	***16***	***17***	***18***
2	A				E			
3		J		D		M		R
4			L				G	

25. Which two elements are least likely to react with each other and form a compound? (1) A and G (2) D and G (3) J and M (4) L and R

26. What is the probable formula for a compound formed from elements L and M? (1) L_4M (2) LM_3 (3) L_2M_3 (4) L_3M_2

27. Which element has the lowest melting point? (1) R (2) J (3) G (4) D

28. Elements with low electronegativities would most likely have (1) large atomic radii and high ionization energies (2) small atomic radii and high ionization energies (3) large atomic radii and low ionization energies (4) small atomic radii and low ionization energies.

29. How many electrons occupy the outermost principal energy level of most transition metals? (1) 1 (2) 2 (3) 3 (4) 7

CONSTRUCTED RESPONSE

1. Very little is known about the recently discovered element with the atomic number 104. Predict a probable atomic radius for this element, and explain your prediction.

2. The table of ionization energies in Appendix 4 does not list a value for astatine, element number 85 (At). Suggest a possible value for this ionization energy and explain your answer.
3. A certain element "X" forms a compound with the formula XCl_2. Because this compound is blue when dry, but pink when wet, it can be used to measure humidity. Element X is unusual, in that its atomic mass is greater than that of the elements on either side of it on the Periodic Table. What is element X? Explain how you arrived at your answer.
4. How many elements are there in Period 6? Explain how you arrived at your answer.

It's Elementary!

Born in 1778, the English chemist Sir Humphrey Davy began, at the age of 19, investigating the properties of gases, including nitrous oxide, or "laughing gas." He persuaded his friends in scientific and literary circles to inhale the gas and report its effects. His subjects included Samuel Taylor Coleridge, the poet who composed *The Rime of the Ancient Mariner.* Davy nearly died after inhaling "water gas," a mixture of hydrogen and carbon monoxide, which was sometimes used as a fuel.

Fortunately, Davy survived, and in 1800 published the results of his investigations. This established his reputation as a scientist, and he was given a post in chemisty at the Royal Institution in London. Davy investigated the relation between electricity and chemistry. He concluded that since substances with opposite charges chemically combined to form compounds, an electric current could be used to break apart compounds into their component elements. Using this process, called electrolysis, Davy isolated for the first time in history the Group 1 elements sodium and potassium, and the Group 2 elements magnesium, calcium, and barium.

In 1807, Davy presented the initial results of his work with electrolysis. He went on to lecture throughout Europe and made many more discoveries, such as proving that diamond is a form of carbon. Davy was knighted in 1812 and made a baronet in 1818. In 1815, responding to a plea from coal miners, he devised the Davy lamp, which would not cause an explosion if exposed to methane gas, a common hazard in mines. Sir Humphrey Davy died in Geneva at the age of 50.

Some Chemical Families

LOOKING AHEAD

You have seen the gradual changes in chemical properties that occur across the periods in the Periodic Table. You have observed how properties repeat regularly as you proceed from highly reactive metals through metalloids, to nonmetals, and finally to noble gases. You will now examine some of the groups in the table in greater detail, becoming familiar with the behavior of many of the more common elements.

When you have completed this chapter you should be able to:

- **Contrast** atomic radii, ionic radii, and physical states of the alkali metals and the alkaline earth metals.
- **Define** and give examples of allotropes.
- **Distinguish** between nonmetals, metalloids, and metals.
- **Describe** the chemistry of representative members of certain chemical families.
- **Account for** the regularities in Groups 1, 2, 13–17; the uniqueness of hydrogen.
- **Discuss** the origin of some common air pollutants related to elements in Group 15 and Group 16.

Group 1—Hydrogen and the Alkali Metals

Although hydrogen is very different from the alkali metals, it is often considered to be a member of Group 1. At room temperature, hydrogen is a gas while the alkali metals are solids. What these elements have in common is one valence electron in their outermost energy level.

Hydrogen

Hydrogen has the electron structure $1s^1$. Like the alkali metals, therefore, hydrogen has a single *s* electron in its outermost energy level. Unlike the alkali metals, it does not have inner energy levels. The hydrogen atom also has a small covalent radius (37 pm), which explains some of its unique properties.

The loss of the single electron liberates the free proton H^+. In aqueous solution, H^+ becomes hydrated, or aquated, forming H_3O^+, or H^+ (*aq*). Since the single *s* electron represents a half-filled orbital, an atom of hydrogen can also gain or share an electron to fill the orbital. The electronegativity of hydrogen (2.1) suggests a greater tendency to share than to gain electrons. Although hydrogen can gain an electron to form a hydride ion, as in LiH, hydrogen atoms tend to share electrons with nonmetals, forming compounds such as CH_4 and HCl. Because of these unusual characteristics, hydrogen does not fit satisfactorily into any chemical family. Largely for convenience, it is often listed with Group 1. However, it is best to think of hydrogen as neither a metal nor a nonmetal, but as a unique element, with its own singular properties.

Hydrogen is an important commercial gas. It has tremendous potential as a fuel source, although it is difficult to store because of its flammability. Considerable quantities of hydrogen are used to convert liquid oils into solid fats, a process called hydrogenation. Hydrogen is also used to provide high temperatures in oxyhydrogen torches and as a source of energy in fuel cells.

Hydrogen is prepared commercially by reacting coke (C) or natural gas (CH_4) with steam.

$$C(s) + H_2O(g) \longrightarrow CO(g) + H_2(g)$$

$$CH_4(g) + H_2O(g) \longrightarrow CO(g) + 3\ H_2(g)$$

To prepare hydrogen in the laboratory, you place a fairly active metal, such as zinc or magnesium, in a solution of dilute hydrochloric or sulfuric acid. A typical reaction would be

$$Zn(s) + 2\ HCl(aq) \longrightarrow H_2(g) + ZnCl_2(aq)$$

Because hydrogen gas is nearly insoluble in water, the gas can be collected by water displacement, as shown in Appendix 3, on page 584. To test for the presence of hydrogen, a burning splint is placed in the mouth of the test tube. The hydrogen reacts with a loud "pop" and forms water, as shown in Figure 6-1.

The Alkali Metals

The elements of the **alkali metal family** (Group 1) are lithium, sodium, potassium, rubidium, cesium and francium. (Francium is highly radioactive, and will not be included in our discussion.) Some of the properties of the alkali metals are compared in the table on page 203.

From the table, you can see that the outermost electron shell of each alkali metal has just a single electron. In each case, the loss of this electron produces the electron configuration of a noble gas. (While the Group 11 metals, Cu, Ag, and Au, also have one valence electron, they do not produce noble gas configurations

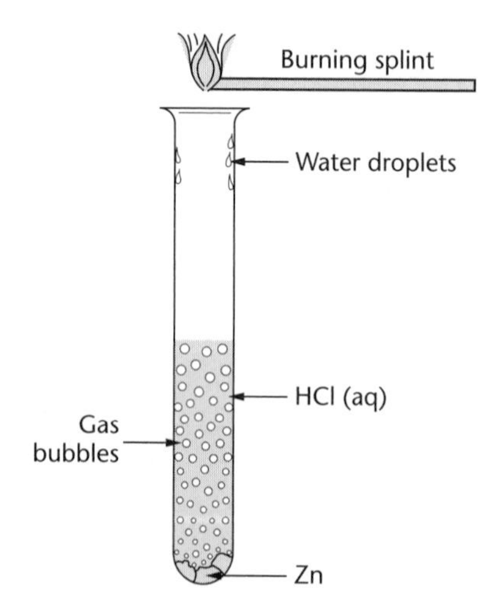

Figure 6-1 Testing for hydrogen gas

The Alkali Metals

Property	*Lithium*	*Sodium*	*Potassium*	*Rubidium*	*Cesium*
Atomic number	3	11	19	37	55
Electron structure	$1s^2$ $2s^1$ (2-1)	$1s^2$ $2s^2, 2p^6$ $3s^1$ (2-8-1)	$1s^2$ $2s^2, 2p^6$ $3s^2, 3p^6$ $4s^1$ (2-8-8-1)	$1s^2$ $2s^2, 2p^6$ $3s^2, 3p^6, 3d^{10}$ $4s^2, 4p^6$ $5s^1$ (2-8-18-8-1)	$1s^2$ $2s^2, 2p^6$ $3s^2, 3p^6, 3d^{10}$ $4s^2, 4p^6, 4d^{10}$ $5s^2, 5p^6$ $6s^1$ (2-8-18-18-8-1)
Atomic (metallic) radius (pm)	155	190.	235	248	267
Ionic radius, X^+ (pm)	60	95	133	148	169
First ionization energy (kJ/mole)	520	496	419	403	376
Electronegativity	1.0	0.9	0.8	0.8	0.8
Oxidation state	+1	+1	+1	+1	+1

*One picometer (pm) = 10^{-12} meter.

when the electron is lost.) The alkali metals each have the largest atomic radius and the lowest ionization energy in their respective periods. The ease with which these metals lose electrons makes them the most reactive of the metallic elements. Since the radii increase and the ionization energies decrease as we go down the group, the metals become progressively more active, from Li to Cs.

Sodium and Potassium Although both sodium and potassium are among the most abundant elements in Earth's crust, they were not produced, as pure metals, until 1807. Both were recognized in compounds long ago; the name "sodium" comes from "soda," the old name for baking soda, or sodium hydrogen carbonate. Yet they were not isolated from their compounds sooner, because the only practical way to separate active metals from their compounds is through the use of electricity, in a process called **electrolysis.** When an electric current is passed through molten NaCl, for example, the salt decomposes to produce sodium and chlorine:

$$2\,NaCl \longrightarrow 2\,Na + Cl_2.$$

However, scientists learned how to generate sufficient amounts of electricity only at the beginning of the nineteenth century. By using electrolysis, the British chemist Sir Humphrey Davy was able to discover sodium, potassium, calcium, magnesium, and barium—five new elements—in the space of just one year.

Sodium and potassium are both soft, silvery metals that are easily cut with a knife. Because they are so active, they are normally stored under kerosene; they would rapidly form oxides or hydroxides if exposed to air. Both metals react vigorously with water. If a small piece of sodium is dropped into water, it instantly begins to react, forming hydrogen gas and sodium hydroxide.

$$2\ \mathrm{Na}(s) + 2\mathrm{H_2O}(l) \longrightarrow 2\ \mathrm{NaOH}(aq) + \mathrm{H_2}(g)$$

Because sodium is less dense than water, the metal floats, and the hydrogen bubbles act like little propellers, moving the sodium around in the water. (See Figure 6-2.) If too large a piece of sodium is used, the heat generated by the reaction can be sufficient to ignite the hydrogen gas, resulting in a violent explosion.

Potassium, with its larger radius and smaller ionization energy, is more reactive than sodium. When potassium is placed in water, the hydrogen produced immediately bursts into flame. The flame produced by the reaction is violet in color. The flame produced by sodium is a bright yellow-orange. One of the interesting properties of the alkali metals is that their compounds impart characteristic colors to a flame. The flame colors of lithium, sodium, and potassium are red, yellow, and violet, respectively. Can you predict the color produced by rubidium? If you predicted red, you were correct.

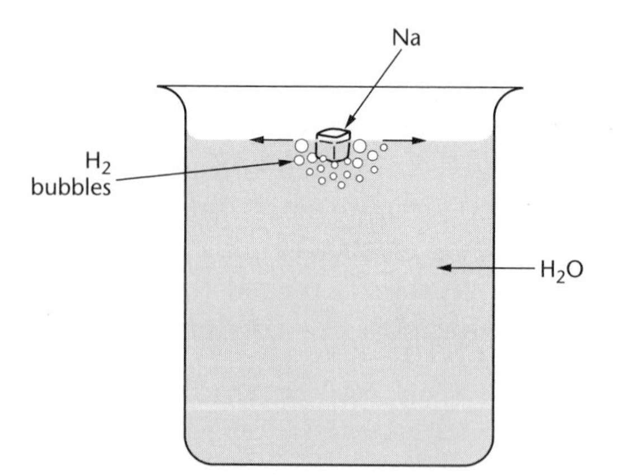

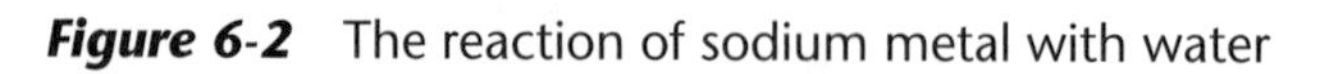
Figure 6-2 The reaction of sodium metal with water

All the alkali metals react similarly with water, producing hydrogen and a hydroxide, as shown previously for sodium. The hydroxide compounds of these metals are strong bases, also called alkalis—hence the name of the group: the alkali metals. (The general properties of bases are discussed in Chapter 12.)

PRACTICE

6.1 Write a balanced equation for the reaction that occurs when potassium is placed in water.

6.2 Write a balanced equation for the reaction that occurs when hydrogen gas burns in oxygen.

6.3 Of the following terms

(1) single replacement (2) double replacement (3) decomposition (4) combination (5) exothermic (6) endothermic

(a) Which TWO terms best describe the reaction between sodium and water?

(b) Which TWO terms best describe the reaction between hydrogen and oxygen?

6.4 The vapor of an alkali metal is often used in street lamps. Which metal is it? How do you know?

6.5 Both sodium and potassium will react with dilute hydrochloric acid to produce hydrogen, yet neither metal is ever used in the laboratory preparation of hydrogen. Why not?

Group 2—The Alkaline Earth Metals

The elements of the **alkaline earth metal family** are beryllium, magnesium, calcium, strontium, barium, and radium. Beryllium and magnesium have properties that are somewhat different from the properties of the other alkaline earth metals. Radium is radioactive.

Compare each of the alkaline earth metals with the alkali metal directly to its left in the Periodic Table. You will notice that each alkaline earth metal has an atomic number one higher than the alkali metal next to it, in Group 1. The increase in positive charge in the nucleus of the Group 2 element produces a greater attractive force on the electrons. This force tends to shrink the atom. The atomic radii of the alkaline earth metals are therefore smaller than those of the corresponding alkali metals.

Ions of the alkaline earth metals have a charge of 2+. Ions of the alkali metals have a charge of 1+. The nuclei of the alkaline earth metal ions therefore attract the remaining electrons more strongly than do the nuclei of the alkali metal ions. This means that the radii of the alkaline earth metal ions are smaller than the radii of the corresponding alkali metal ions. The following table summarizes some important properties of the alkaline earth metals.

Like their neighbors in Group 1 of the Periodic Table, the metals in Group 2 lose electrons too easily to be found uncombined in nature. With their smaller atomic radii, and higher ionization energies, the Group 2 metals are slightly less active than the Group 1 metals. The difference in chemical activity between the two groups is demonstrated by comparing the reactions of the metals with water. Recall that sodium and potassium react explosively with water to produce hydrogen and a metal hydroxide. When a strip of magnesium is placed in water, no apparent change occurs. However, when the acid-base indicator, phenolphthalein, is added to the solution, it gradually turns pink, indicating a basic solution. The reaction: $Mg + 2\,H_2O \longrightarrow Mg(OH)_2 + H_2$ does take place, although quite slowly at room temperature. Like the Group 1 metals, the Group 2 metals form bases in water; hence the name "alkaline earth" is given to the group.

How would you expect calcium to behave in water? With its larger radius and smaller ionization energy, calcium should be more active than magnesium. Calcium does react quite vigorously with water, producing hydrogen and calcium hydroxide. If sufficient calcium is used, a white precipitate will form in the water, because calcium hydroxide is only slightly soluble. The compounds of the Group 2 metals are generally less soluble than those of the Group 1 metals.

The Alkaline Earth Metals

Property	**Beryllium**	**Magnesium**	**Calcium**	**Strontium**	**Barium**	**Radium**
Atomic number	4	12	20	38	56	88
Electron structure	$1s^2$ $2s^2$ (2-2)	$1s^2$ $2s^2, 2p^6$ $3s^2$ (2-8-2)	$1s^2$ $2s^2, 2p^6$ $3s^2, 3p^6$ $4s^2$ (2-8-8-2)	$1s^2$ $2s^2, 2p^6$ $3s^2, 3p^6, 3d^{10}$ $4s^2, 4p^6$ $5s^2$ (2-8-18-8-2)	$1s^2$ $2s^2, 2p^6$ $3s^2, 3p^6, 3d^{10}$ $4s^2, 4p^6, 4d^{10}$ $5s^2, 5p^6$ $6s^2$ (2-8-18-18-8-2)	$1s^2$ $2s^2, 2p^6$ $3s^2, 3p^6, 3d^{10}$ $4s^2, 4p^6, 4d^{10}, 4f^{14}$ $5s^2, 5p^6, 5d^{10}$ $6s^2, 6p^6$ $7s^2$ (2-8-18-32-18-8-2)
Atomic (metallic) radius (pm)	112	160.	197	215	222	233
Ionic radius, X^{2+} (pm)	31	65	99	113	135	152
First ionization energy (kJ/mole)	900	736	590	549	503	503
Electronegativity	1.6	1.3	1.0	1.0	0.9	0.9
Oxidation state	+2	+2	+2	+2	+2	+2

Alkaline Earth Metal Compounds

Formula	Name	Source	Properties	Uses
$MgSO_4 \cdot 7H_2O$	Epsom salts	Epsomite (a mineral)	Forms a variety of hydrates	1. Medicinal purgative 2. Weighting cotton and silk 3. Fireproofing muslin
$CaCO_3$	Limestone or marble	Natural deposits formed by shells of sea animals	Thermal decomposition yields lime $CaCO_3(s) \rightarrow CaO(s) + CO_2(g)$	1. Source of many calcium compounds 2. Building material 3. Metallurgy of iron
CaO	Lime or quicklime	Decomposition of limestone in a kiln	Reacts with water to produce slaked lime and considerable heat $CaO + H_2O \rightarrow Ca(OH)_2 + 67\ kJ$	1. Production of slaked lime, $Ca(OH)_2$ 2. Smelting of metals 3. Drying agent
$Ca(OH)_2$	Slaked lime or limewater	Reaction of lime, CaO, with water	1. Common base 2. Reacts with halogens to form oxyhalogen compounds	1. Preparation of mortar 2. Manufacture of bleaching powder 3. Purification of sugar
$CaSO_4 \cdot 2H_2O$	Gypsum	Natural deposits	Partial dehydration yields plaster of Paris $2\ CaSO_4 \cdot 2H_2O \rightarrow (CaSO_4)_2 \cdot H_2O + 3\ H_2O$	1. Source of plaster of Paris 2. Building material
$CaCl_2$	Calcium chloride	By-product in the Solvay process for making $NaHCO_3$	Absorbs and holds water vapor (hygroscopic material)	1. Drying agent 2. Freezing mixtures
$BaSO_4$	Barite	Natural deposits and by reaction $BaO_2 + H_2SO_4 \rightarrow H_2O_2 + BaSO_4(s)$	1. Stable against heat and atmosphere 2. Absorbs X rays	1. Paint pigment and filler 2. X-ray diagnosis
$Sr(NO_3)_2$	Strontium nitrate	$SrCO_3 + 2\ HNO_3 \rightarrow Sr(NO_3)_2 + H_2O + CO_2(g)$	Produces characteristic red color in a flame	Fireworks and red flares

Calcium and Magnesium Ions and Our Water Supply

Our drinking water, especially if it comes from underground streams, is often rich in calcium and magnesium ions. Water that contains significant concentrations of either of these ions is called **hard water.** Hard water is perfectly safe to drink. However, recall that the solubilities of the compounds formed by the Group 2 metals are lower than those in Group 1. Both magnesium and calcium ions form an insoluble compound with the stearate ion, ($C_{17}H_{35}COO^-$) an ion contained in most soaps. The formation of insoluble compounds, such as calcium stearate, interferes with the washing action of the soap.

$$Ca^{2+}(aq) + 2\ C_{17}H_{35}COO^-(aq) \longrightarrow (C_{17}H_{35}COO)_2Ca(s)$$

$$Mg^{2+}(aq) + 2\ C_{17}H_{35}COO^-(aq) \longrightarrow (C_{17}H_{35}COO)_2Mg(s)$$

Since $Ca^{2+}(aq)$ and $Mg^{2+}(aq)$ cause hardness, the amount of these ions in the water must be decreased to soften it. This can be done by adding chemical agents to the water. Compounds that contain the carbonate ion (CO_3^{2-})—washing soda (Na_2CO_3), for example—may be added. The carbonate ion reacts with the metal ions and forms the insoluble compounds $CaCO_3$ or $MgCO_3$, which settle out of the water. The equations for these reactions are as follows:

$$CO_3^{2-}(aq) + Ca^{2+}(aq) \longrightarrow CaCO_3(s)$$

$$CO_3^{2-}(aq) + Mg^{2+}(aq) \longrightarrow MgCO_3(s)$$

Water may also be softened on a larger scale by the use of ion-exchange resins. Ion-exchange resins are able to substitute a desired ion for an undesirable one. In the case of hard water, the resins remove the calcium and magnesium ions from the water, replacing them with sodium ions. Sodium stearate is soluble in water, so the resulting solution is "soft" water; it will permit soap to lather normally.

Hydroxides of Calcium and Magnesium

We have seen that both the alkali metals and the alkaline earth metals form hydroxides in water; these compounds are called bases, or alkali. Less active metals tend to form weaker bases, so $Ca(OH)_2$ and $Mg(OH)_2$ are weaker than the corresponding bases of Group 1, KOH and NaOH. Sodium hydroxide, also known as lye, is an extremely toxic substance, which is lethal if swallowed, and can cause permanent skin damage if touched. Magnesium hydroxide, also known as milk of magnesia, is a mild enough base to be medically useful, both as a laxative and as an antacid. Calcium hydroxide, also called lime, is used by gardeners to "sweeten" acidic soil. It is a mild enough base not to damage the plants around it.

The sources, properties, and some additional uses of some compounds of the alkaline earth metal family of elements are summarized in the table on page 208.

Flame Tests

Like the metals of Group 1, many of the Group 2 metals form compounds that will impart a color to a flame. Calcium ions color a flame orange-red, strontium a bright intense red, and barium a yellow-green. The flame colors are often used in the laboratory to test solutions for the presence of these ions. A platinum wire with a small loop at its end is dipped into the solution to be tested, and then placed in the flame of a Bunsen burner. (See Figure 6-3.) The resulting color can be used to confirm the presence of sodium, lithium, potassium, calcium, barium, strontium, or any of several other ions in the solution.

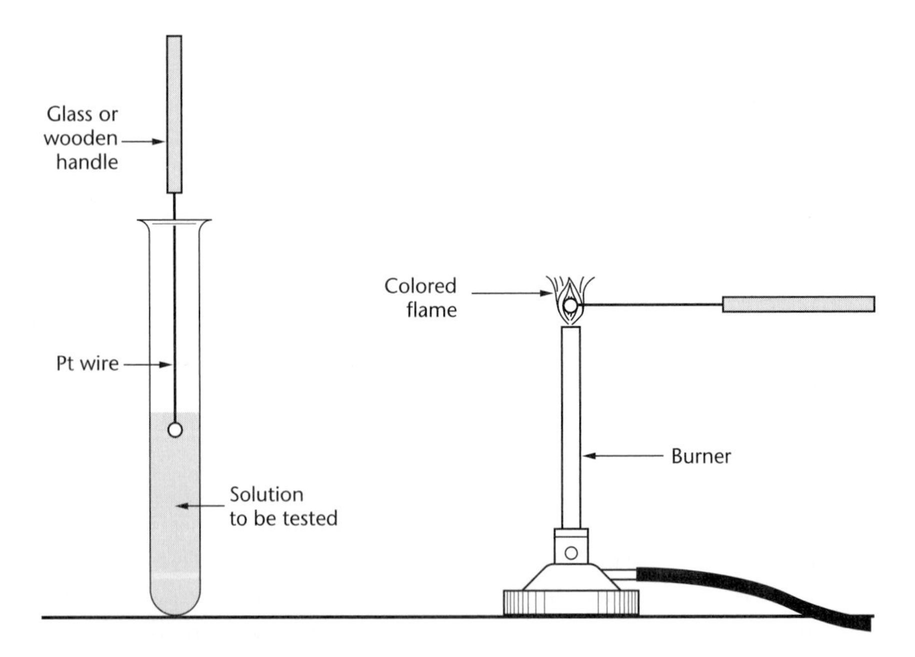

Figure 6-3 Flame test procedure

PRACTICE

6.6 What two ions are most commonly found in hard water?

6.7 Element X is an alkali metal. Element Y is an alkaline earth metal in the same period. Compare these two elements with regard to

(a) ionization energy

(b) electronegativity

(c) atomic radius

(d) size of their ions

(e) charge of their ions

(f) chemical activity

6.8 Why are alkali metals and alkaline earth metals never found uncombined in nature?

6.9 Write a balanced equation for the reaction that occurs when barium metal is placed in water. Would you expect this reaction to be faster or slower than the reaction of calcium with water? Explain your answer.

Group 13—The Aluminum Family

The elements in Group 13 include boron, aluminum, gallium, indium, and thallium. The properties of the elements in Group 13 fit into the trends that were noted for the alkali metals and the alkaline earth metals. Each element in Group 13 has two *s* electrons and one *p* electron in the outer energy level. The elements thus have a 3+ oxidation state. The heavier elements may lose only the single *p* electron. They then have an oxidation state of 1+. An example is thallium in $TlCl$.

The elements in Group 13 are less reactive than the elements in Group 1 and Group 2. The ionization energies of Group 13 atoms are higher and the tendency to form ions is weaker than in Group 1 and Group 2. Thus boron, which has the smallest atomic radius in Group 13, forms only covalent compounds, such as boron trifluoride, BF_3.

As the atoms become larger in Group 13, the tendency to lose electrons becomes stronger. The elements thus become more metallic from boron to thallium.

Aluminum is the most important element in Group 13. Aluminum is a malleable and ductile metal about one-third as dense as iron. For its weight, it has unusually high tensile strength and is used as a structural metal in aircraft, automobiles, and buildings. Aluminum is also an excellent conductor of heat and electricity and is used in many household appliances and in alloys.

Aluminum is a reactive metal. It tarnishes easily in air, forming a protective oxide coating. The oxide coating prevents further damage to the metal, so that objects made of aluminum remain structurally sound upon exposure to air and moisture. Iron, though less active than aluminum, does not form a protective coating and so may eventually rust through.

Aluminum reacts readily with acids. Cola beverages, which are often sold in aluminum cans, are very acidic and would react with the aluminum if the inside of the can were not specially coated to prevent the beverage from coming in contact with the metal.

Group 14—The Carbon Family

Group 14 includes carbon, a nonmetal; silicon and germanium, which are semimetals or metalloids; and tin and lead, which are metals. Carbon is unlike all the other elements in one way. Atoms of carbon can bond together in an almost endless variety of open, chainlike structures and closed, ringlike structures. Structures like these form compounds in organic chemistry, the subject of Chapter 14.

Carbon also forms some compounds, such as carbon dioxide and carbonates, that are not usually studied in organic chemistry.

These compounds and compounds that do not contain carbon are generally called **inorganic.**

Carbon dioxide is constantly being generated and released into the atmosphere as a result of the process of respiration in living things. Carbon dioxide is also constantly being removed from the atmosphere by green plants, which use the gas in the manufacture of their food, a process called **photosynthesis.** Carbon dioxide is also released into the atmosphere when fossil fuels, such as coal and oil, are burned.

The amount of carbon dioxide in the atmosphere has been gradually increasing. This increase is at least partly the result of the increasing use of fossil fuels as population and industrialization have grown and spread throughout the world. At the same time, plants that can remove excess carbon dioxide from the atmosphere are being destroyed. Carbon dioxide traps heat in the atmosphere, and the temperature of the Earth has been rising slightly. Scientists suspect that there is a connection between the increase in carbon dioxide levels in the atmosphere and the increase in global temperatures, but there is much disagreement about how significant that connection is. Global warming has become a serious concern of many environmentalists, who fear that dramatic climatic changes may occur in the next century if we do not decrease our use of fossil fuels. Other scientists disagree, pointing out that in the past Earth has been both much colder and much warmer than it is now, long before humans began to burn fossil fuels.

The laboratory preparation of CO_2 is given in Appendix 3.

Silicon is the second most abundant (next to oxygen) element on Earth. It is not found free in nature. It makes up a major proportion of sand, sandstone, clay, silica rock, quartz, and other common minerals. Silicon carbide, SiC, is one of the hardest substances known. It is therefore very useful as an abrasive. Carborundum is a commercially produced form of silicon carbide. Like carbon, silicon forms chainlike molecules of varying lengths. These molecules, called *silicones,* consist of atoms of Si, C, O, and H. The short-chain molecules act like oils, the medium chains act like greases, and the long chains are rubberlike. Hence silicones have a variety of interesting uses.

Germanium is a typical metalloid. It is useful as a semiconductor in electronic devices, such as transistors.

Tin and lead are metals that are widely used in the home and in industry.

Group 15—The Nitrogen Family

Group 15 includes the nonmetals nitrogen and phosphorus; the semimetals, or metalloids, arsenic and antimony; and the metal bismuth. Some of the important properties of the members of the nitrogen family are shown in the table on page 215.

Each member of the nitrogen family has five valence electrons—two *s* electrons and three *p* electrons. The ionization energies of nitrogen and phosphorus are relatively high. Therefore they do not form positively charged ions. However, antimony and bismuth can lose their three *p* valence electrons and form Sb^{3+} and Bi^{3+} ions. This illustrates the tendency of the Group 15 elements to become more metallic with increasing atomic number. The shielding effect of the inner electrons leads to an increasing tendency to lose outer electrons, as shown by decreasing ionization energy values. In combinations with the very reactive elements of the alkali metal family, nitrogen and phosphorus may form 3− ions: N^{3-} (nitride ion) and P^{3-} (phosphide ion).

With the exception of the nitrides, such as AlN, and the phosphides, such as K_3P, the members of the nitrogen family form covalent bonds. They also form polyatomic ions, such as NO_3^- and PO_4^{3-}, with oxygen. The bonding within these ions, however, is covalent, as shown in the following diagram of a nitrate ion:

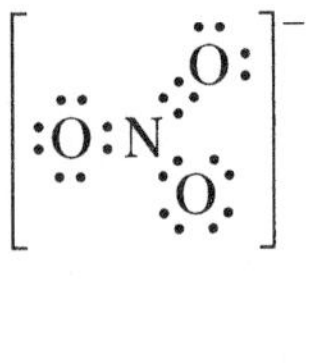

```
[        ..  ]−
        .O:
  ..   ..
 :O:N
  ..   ..
        :O:
[       ..   ]
```

Nitrogen

The nitrogen molecule has a triple covalent bond between the two N atoms. Nitrogen is therefore exceptionally stable and is difficult to convert into compounds. Yet, nitrogen is present

The Nitrogen Family

Property	Nitrogen	Phosphorus	Arsenic	Antimony	Bismuth
Atomic number	7	15	33	51	83
Electron structure	$1s^2$ $2s^2, 2p^3$ (2-5)	$1s^2$ $2s^2, 2p^6$ $3s^2, 3p^3$ (2-8-5)	$1s^2$ $2s^2, 2p^6$ $3s^2, 3p^6, 3d^{10}$ $4s^2, 4p^3$ (2-8-18-5)	$1s^2$ $2s^2, 3p^6$ $3s^2, 3p^6, 3d^{10}$ $4s^2, 4p^6, 4d^{10}$ $5s^2, 5p^3$ (2-8-18-18-5)	$1s^2$ $2s^2, 2p^6$ $3s^2, 3p^6, 3d^{10}$ $4s^2, 4p^6, 4d^{10}, 4f^{14}$ $5s^2, 5p^6, 5d^{10}$ $6s^2, 6p^3$ (2-8-18-32-18-5)
Atomic radius (pm)	92	128	139	159	170.
First ionization energy (kJ/mole)	1402	1012	944	831	703
Electronegativity	3.0	2.2	2.2	2.1	2.0
Oxidation states	All states from −3 to +5	−3, −2, +1, +3, +4, +5	−3, +3, +5	−3, +3, +5	−3, +3, +5

in proteins, which form the cells and tissues of living things, and is a valuable part of fertilizers and other important compounds.

Molecular nitrogen, N_2, must be converted into nitrogen compounds before it can be used by living things. In nature, the energy of lightning and the action of certain bacteria that live in the roots of some plants change nitrogen into usable nitrogen compounds. This process is called **nitrogen fixation.**

Some of the oxides of nitrogen are listed in the table below.

Oxide of N	Name	Oxidation Number of N
N_2O	Nitrous oxide (dinitrogen monoxide)	+1
NO	Nitric oxide (nitrogen monoxide)	+2
N_2O_3	Dinitrogen trioxide	+3
NO_2	Nitrogen dioxide	+4
N_2O_4	Dinitrogen tetroxide	+4
N_2O_5	Dinitrogen pentoxide	+5

NO_2 is a dark brown, extremely poisonous gas. It is formed when heat (such as the heat produced by the combustion of gasoline in an automobile engine) causes a reaction between atmospheric nitrogen and oxygen. Nitrogen dioxide is itself a serious air pollutant. It also is responsible for the formation of ozone, one of the harmful components of smog.

N_2O_3 dissolves in water and forms HNO_2, nitrous acid.

$$N_2O_3 + H_2O \longrightarrow 2\ HNO_2$$

N_2O_5 dissolves in water and forms nitric acid, HNO_3.

$$N_2O_5 + H_2O \longrightarrow 2\ HNO_3$$

The laboratory preparation of nitric acid is given in Appendix 3.

Ammonia One of the most important nitrogen compounds is ammonia, NH_3. Ammonia is used in the manufacture of many chemical products, including fertilizers, medicines, dyes, and explosives. Some ammonia is formed by fixation of atmospheric nitrogen by lightning. However, most of the ammonia used is made synthetically from its elements. The reaction between nitrogen and hydrogen to form ammonia can be written as

$$N_2(g) + 3\ H_2(g) \rightleftharpoons 2\ NH_3(g).$$

The air is 80 percent nitrogen, and hydrogen is readily available, so there would seem to be no problem in producing large quantities of ammonia. Unfortunately, the reaction is so slow at room temperature that it produces no observable yield of ammonia. The reaction becomes faster at higher temperatures, but it also changes direction, going predominantly *from* ammonia to hydrogen and nitrogen. In 1911, the German chemist Fritz Haber was able to produce useful quantities of ammonia by heating the gases under very high pressure, in the presence of iron and aluminum oxides. The oxides act as **catalysts.** They speed up the reaction without being permanently changed by the reaction. (Catalysts are more fully discussed in Chapter 9.) This process is called the Haber process.

In the laboratory, ammonia is generally prepared through the reaction of a strong base, such as sodium hydroxide, with an ammonium salt, such as ammonium chloride.

$$NH_4Cl + NaOH \longrightarrow NaOH + H_2O + NH_3 \uparrow$$

(An upward arrow, or (g), is often used in a chemical equation to indicate production of a gas.) A suitable apparatus for carrying out this process is illustrated in Appendix 3.

Ammonia is used to prepare nitric acid commercially by the Ostwald process. Ammonia is oxidized in the presence of a catalyst to form an oxide of nitrogen. The oxide is then converted into nitric acid.

$$(1)\quad 4\ NH_3(g) + 5\ O_2(g) \longrightarrow 4\ NO(g) + 6\ H_2O$$

$$(2)\quad 2\ NO(g) + O_2(g) \longrightarrow 2\ NO_2(g)$$

$$(3)\quad 3\ NO_2(g) + H_2O \longrightarrow 2\ HNO_3 + NO(g)$$

The NO in step 3 is recycled and reused in step 2.

Phosphorus

Under ordinary conditions, the element phosphorus exists as a P_4 molecule. Two different models of the phosphorus molecule are shown in Figure 6-4 on page 218.

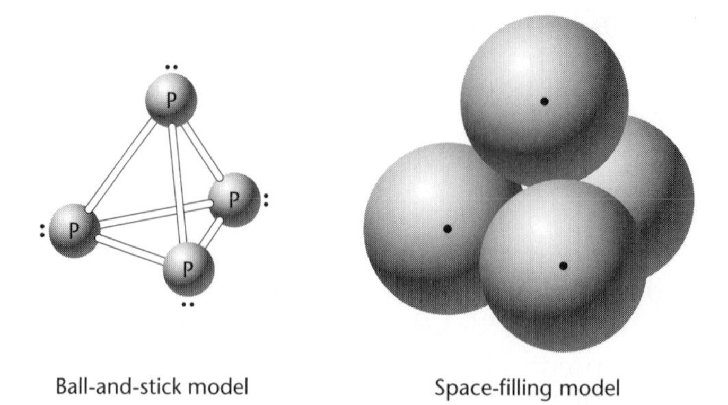

Figure 6-4 Two models of the P_4 molecule

Phosphorus is an **allotropic** element. Allotropic elements or compounds exist in two or more forms in the same physical state. Each form has its own physical and chemical properties. White phosphorus consists of individual P_4 molecules bonded by van der Waals forces. Red phosphorus consists of chains of P_4 molecules covalently bonded in a giant molecule. These structural differences are thought to account for the differences in the properties of the two forms of phosphorus.

White phosphorus forms a vapor readily (is volatile) at room temperature and catches fire so easily that it must be kept and handled under water.

$$P_4 + 5\,O_2 \longrightarrow \underset{\text{tetraphosphorus decoxide}}{P_4O_{10}}$$

Red phosphorus is much less volatile and much less reactive than the weakly bonded white phosphorus.

Phosphorus is used largely in the manufacture of phosphate fertilizers, organic phosphates, and matches. It is also used to make special bronze alloys.

Other Elements of Group 15

Arsenic is a metalloid. Small quantities of arsenic are used with lead to form an alloy that is harder than lead alone would be. Arsenates, AsO_4^{3-}, are used as insecticides. Since these compounds are poisonous, any food that has been in contact with such an insecticide spray must be washed carefully.

Antimony is also a metalloid. An alloy of antimony, lead, and tin is used to make type for printing. This alloy is unusual in that, unlike most liquids, it expands on solidifying, thus ensuring sharp impressions. Certain fabrics are treated with antimony compounds before being dyed.

Bismuth is a metal. It is used in the preparation of alloys that melt at low temperatures and are therefore useful in automatic fire sprinklers. Bismuth compounds are used as medicinals.

PRACTICE

6.10 All of the Group 13 elements except boron can readily form 3+ ions. Why do they tend to form 3+ ions? Why doesn't boron form similar ions?

6.11 In Group 14, carbon is a nonmetal, while tin and lead are metals. Explain this difference in behavior.

6.12 Account for the exceptional stability of nitrogen gas.

6.13 All of the Group 15 elements show a maximum oxidation state of +5. Why?

Group 16—The Oxygen Family

The members of Group 16 are oxygen, sulfur, selenium, tellurium, and polonium. The most important members of this group are oxygen and sulfur. These two elements and selenium are nonmetals. Tellurium and polonium are metalloids. Polonium is the first of the naturally occurring elements in the Periodic Table (with atomic number 84, it follows bismuth) that has only radioactive isotopes. Polonium will not be further discussed. The table on page 220 lists some properties of the other elements in Group 16.

The Oxygen Family

Property	***Oxygen***	***Sulfur***	***Selenium***	***Tellurium***
Atomic number	8	16	34	52
Electron structure	$1s^2$ $2s^2, 2p^4$ (2-6)	$1s^2$ $2s^2, 2p^6$ $3s^2, 3p^4$ (2-8-6)	$1s^2$ $2s^2, 2p^6$ $3s^2, 3p^6, 3d^{10}$ $4s^2, 4p^4$ (2-8-18-6)	$1s^2$ $2s^2, 2p^6$ $3s^2, 3p^6, 3d^{10}$ $4s^2, 4p^6, 4d^{10}$ $5s^2, 5p^4$ (2-8-18-18-6)
Atomic radius (pm)	65	127	140.	142
First ionization energy (kJ/mole)	1314	1000.	941	869
Electronegativity	3.5	2.6	2.6	2.1
Oxidation states	−2, −1, +1, +2	−2, +4, +6	−2, +4, +6	−2, +4, +6

All the elements in the oxygen family have six valence electrons. The size of each atom increases with increasing atomic number. This indicates that the number of occupied energy levels is increasing. As in the Group 15 elements, the shielding effect of inner electrons allows electrons to be lost more easily as the covalent radius increases. The ionization energies therefore decrease with increasing atomic number. Generally, the ionization potentials are too high to permit the formation of a 6+ ion (except in polonium and tellurium). Thus, with increasing atomic number, the nonmetallic properties of the elements of the oxygen family decrease and the metallic properties increase. These are the trends that were noted in Group 15.

When compared with H_2S, H_2Se, and H_2Te, H_2O has abnormal properties, as noted on pages 138–140. This can be explained by the presence of hydrogen bonding in water, a result of the high electronegativity value of oxygen.

Oxygen

Making up about 20 percent of the atmosphere and a part of many different compounds, oxygen is the most common element on Earth. It is essential to nearly all living things. In addition, oxygen is used in the production of iron and steel, in producing the hot flames necessary for cutting and welding metals, and as an oxidizer for liquid rocket propellants. Oxygen is obtained from natural sources, either by the fractional distillation of liquid air or by the breakdown of water by an electric current (electrolysis). The laboratory preparation of oxygen is given in Appendix 3.

Oxygen is the second most electronegative element (after fluorine). Thus, in nature, it is found combined with many other elements. It takes electrons from these other elements and becomes negatively charged, existing either as a 2− ion in ionic compounds or in the oxidation state of −2 in compounds with three elements (ternary oxygen compounds). Oxygen is thus an excellent oxidizing agent.

In the peroxide structure, oxygen has an oxidation number of −1. The best-known peroxide compound is hydrogen peroxide, H_2O_2. The electron-dot diagram for H_2O_2 is

$$\begin{array}{l} \mathrm{H\overset{x}{\cdot}\ddot{O}:} \\ \quad \mathrm{:\underset{\cdot\cdot}{\ddot{O}}\overset{x}{\cdot}H} \end{array}$$

Notice that each oxygen atom, being more electronegative than the H atom, has attracted one hydrogen electron. Each O atom therefore has an oxidation number of −1.

Oxygen exists in another gaseous molecular form, O_3, called *ozone*. Oxygen and ozone are allotropes. Ozone is prepared by passing an electric discharge through oxygen.

$$3\ O_2(g) + 286\ \text{kJ} \longrightarrow 2\ O_3(g)$$

Ozone is also formed from the air pollutant NO_2, in a series of reactions that require light (photolysis).

$$NO_2 + \text{sunlight} \longrightarrow NO + O$$

$$O + O_2 \longrightarrow O_3$$

Ozone is generally found 16 to 40 kilometers above sea level, in the region of the atmosphere called the stratosphere. It acts as a shield against the penetrating and dangerous ultraviolet radiation of the sun. Ozone in the upper atmosphere therefore protects living things on Earth.

Sulfur

Next to oxygen, sulfur is the most common member of the oxygen family. At room temperature, sulfur is a yellow, odorless solid. Like phosphorus and oxygen, sulfur exists in several different allotropic forms. One form, a commercial variety of sulfur, is called roll sulfur. If a solution of roll sulfur in carbon disulfide, CS_2, is allowed to evaporate slowly, crystals of rhombic sulfur appear. Crystals of rhombic sulfur are eight-sided. Below 96°C, rhombic sulfur remains stable. If roll sulfur is melted and kept between 96°C and 120°C, needle-shaped crystals of sulfur are formed. This allotrope is called prismatic sulfur or monoclinic sulfur.

Both rhombic and monoclinic sulfur seem to consist of units of S_8 molecules. Each unit has its own geometric shape in the crystal. A unit consists of eight atoms of sulfur joined in a puckered ring, as shown in Figure 6-5. The sulfur atoms are joined to one another in the ring by covalent bonds.

Sulfur is less reactive than oxygen. It unites with metals to form ionic compounds. It unites with nonmetals to form covalently bonded compounds. Examples of these reactions are

$$\text{(1)} \quad 2\,Na(s) + S(g) \longrightarrow 2\,Na^+ + S^{2-}$$

$$\text{(2)} \quad S_8(s) + 8\,O_2(g) \longrightarrow 8\,SO_2(g)$$

The oxidation number of sulfur changes from 0 to −2 in reaction 1. Sulfur acquires a −2 oxidation state in metal sulfides, because sulfur has a higher electronegativity than the metals. In reaction 2, the oxidation number of sulfur changes from 0 to +4. Sulfur acquires a positive oxidation state when bonding to oxygen, because sulfur has a lower electronegativity than oxygen.

Most of the sulfur mined in the world is converted into sulfur compounds, mainly sulfuric acid, H_2SO_4. Some sulfur is used to vulcanize rubber. Sulfur is also used as an ingredient in insecticides.

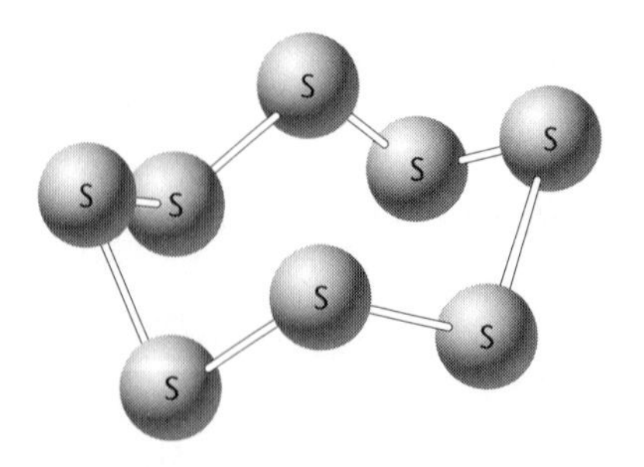

Figure 6-5 A model of the S_8 molecule

Sulfur compounds The most common sulfur compounds are hydrogen sulfide, H_2S; sulfur dioxide, SO_2; sulfuric acid, H_2SO_4; and a variety of sulfur-oxygen compounds, including sodium sulfate, Na_2SO_4; magnesium sulfate, $MgSO_4$; and sodium thiosulfate, $Na_2S_2O_3$.

Sulfur dioxide and various sulfites are used to bleach materials, such as wool and silk, that would be destroyed by chlorine bleaches. You will also find the words "contains sulfites" on most wine bottles sold in the U.S. Sulfites are added to wines to prevent deterioration that might otherwise occur as the wines age.

Sulfur dioxide is a gas with a pungent, irritating odor. The odor of burning sulfur, or of a struck match, is due to sulfur dioxide. SO_2 is a serious air pollutant. Most of the sulfur dioxide in the air comes from the burning of petroleum products and coal from which sulfur has not been sufficiently removed. Oil low in sulfur content is more expensive than "cleaner" oil, but its use decreases the dangers of sulfur dioxide to the well-being of people, animals, and plants. The laboratory preparation of sulfur dioxide is given in Appendix 3.

In the presence of moisture, such as mist in the air, SO_2 is converted to H_2SO_4. Some of this sulfuric acid then returns to Earth's surface as "acid rain." Acid rain has made some of the lakes in New York State too acidic to support life. Sulfuric acid in air also damages buildings that contain marble or limestone. The reaction between sulfuric acid and marble or limestone is

$$H_2SO_4 + CaCO_3 \longrightarrow CaSO_4 + H_2O + CO_2$$

On the other hand, sulfuric acid is probably the most important compound of sulfur and is one of the world's most widely used industrial chemicals. Sulfuric acid is used at one stage or another in many chemical processes. The major uses of sulfuric acid include

1. Production of fertilizers, such as ammonium sulfate and superphosphate of lime.
2. Removal of rust and scale from steel (pickling) before steel is plated with protective metal coatings.
3. Refining of petroleum to remove undesirable compounds from lubricating oils.

Sulfuric acid is a very corrosive liquid. It is also a powerful **dehydrating agent**—a substance that removes water or the elements that make up water from matter with which the substance is in contact. The dehydrating action of concentrated sulfuric acid is shown in the following reaction. The acid reacts with a carbohydrate (a compound of carbon, hydrogen, and oxygen), such as sugar (sucrose) or cellulose (wood, cotton).

$$\underset{\text{sucrose}}{C_{12}H_{22}O_{11}} + 11\ H_2SO_4 \longrightarrow 12\ C + 11\ H_2SO_4 \cdot H_2O$$

This reaction accounts for the charring that occurs when concentrated sulfuric acid comes into contact with carbohydrates.

When a sulfur atom replaces an oxygen atom, the name of the resulting particle has the prefix *-thio*. Hence SO_4^{2-} is the sulfate ion, and $S_2O_3^{2-}$ is the thiosulfate ion.

Compounds made up of silver and halogen (silver halides) dissolve in sodium thiosulfate and form stable complex compounds.

$$AgBr(s) + 2\ Na_2S_2O_3(aq) \longrightarrow NaBr(aq) + Na_3Ag(S_2O_3)_2(aq)$$

In photography, silver bromide, AgBr, that has not been acted on by the developer is removed from the film by this reaction. Removing the AgBr is called "fixing" the film. Sodium thiosulfate is sold commercially as "hypo" and is called a photographic fixer, or fixing agent.

Other Elements of Group 16

The remaining common elements of the oxygen family, selenium and tellurium, resemble sulfur chemically.

Selenium occurs naturally in sulfide ores. Like sulfur, it has allotropic forms. Selenium forms a number of compounds comparable to sulfur compounds, such as hydrogen selenide, H_2Se, and selenic acid, H_2SeO_4. When selenium is exposed to light, its conductivity increases considerably. Because of this property, called photoconductivity, selenium is useful in some photoelectric cells. Selenium is also used to tint glass.

Tellurium occurs in nature as a telluride of lead, silver, and gold. Tellurium forms compounds comparable to compounds of sulfur and selenium, such as H_2Te and H_2TeO_4. Tellurium is used as an additive to steel and as a coloring agent in glass. In general, however, tellurium has limited usefulness.

Group 17—The Halogens

The **halogen family**—fluorine, chlorine, bromine, iodine, and astatine—makes up Group 17. Astatine is a radioactive halogen not found in nature and is omitted from this discussion. Some of the important properties of the other halogens are listed in the table below.

The Halogens

Property	*Fluorine*	*Chlorine*	*Bromine*	*Iodine*
Atomic number	9	17	35	53
Electron structure	$1s^2$ $2s^2$, $2p^5$ (2-7)	$1s^2$ $2s^2$, $2p^6$ $3s^2$, $3p^5$ (2-8-7)	$1s^2$ $2s^2$, $2p^6$ $3s^2$, $3p^6$, $3d^{10}$ $4s^2$, $4p^5$ (2-8-18-7)	$1s^2$ $2s^2$, $2p^6$ $3s^2$, $3p^6$, $3d^{10}$ $4s^2$, $4p^6$, $4d^{10}$ $5s^2$, $5p^5$ (2-8-18-18-7)
Atomic radius (pm)	57	97	112	132
Ionic radius, X^- (pm)	136	187	195	216
First ionization energy (kJ/mole)	1681	1251	1140.	1008
Electronegativity	4.0	3.2	3.0	2.7
Oxidation states	−1	−1, +1, +3, +5, +7	−1, +1, +3, +5	−1, +1, +5, +7

All halogens have two *s* electrons and five *p* electrons in the outermost energy level. Each of the halogens can acquire the same electron configuration as the noble gas next to it (in Group 18) by one of the following methods (X represents a halogen):

1. Formation of a covalent bond between halogen atoms.

$$2\,:\ddot{\underset{\cdot\cdot}{X}}\cdot \longrightarrow :\ddot{\underset{\cdot\cdot}{X}}:\ddot{\underset{\cdot\cdot}{X}}: \quad \text{or} \quad 2\,:\ddot{\underset{\cdot\cdot}{Cl}}\cdot \longrightarrow :\ddot{\underset{\cdot\cdot}{Cl}}:\ddot{\underset{\cdot\cdot}{Cl}}:$$

2. Formation of a halide ion.

$$:\ddot{\underset{\cdot\cdot}{X}}:\ddot{\underset{\cdot\cdot}{X}}: + 2e^- \longrightarrow 2\,:\ddot{\underset{\cdot\cdot}{X}}:^- \quad \text{or} \quad :\ddot{\underset{\cdot\cdot}{F}}:\ddot{\underset{\cdot\cdot}{F}}: + 2e^- \longrightarrow 2\,:\ddot{\underset{\cdot\cdot}{F}}:^-$$

3. Formation of a covalent bond with hydrogen, carbon, or nonmetals.

$$\mathrm{H}{:}\ddot{\underset{\cdot\cdot}{\mathrm{X}}}{:} \qquad {:}\ddot{\underset{\cdot\cdot}{\mathrm{X}}}{:}\underset{{:}\ddot{\underset{\cdot\cdot}{\mathrm{X}}}{:}}{\overset{{:}\ddot{\mathrm{X}}{:}}{\mathrm{C}}}{:}\ddot{\underset{\cdot\cdot}{\mathrm{X}}}{:} \qquad {:}\ddot{\underset{\cdot\cdot}{\mathrm{X}}}{:}\underset{{:}\ddot{\underset{\cdot\cdot}{\mathrm{X}}}{:}}{\ddot{\mathrm{N}}}{:}\ddot{\underset{\cdot\cdot}{\mathrm{X}}}{:} \qquad \text{or} \qquad \mathrm{HBr},\ \mathrm{CCl_4},\ \mathrm{NI_3}$$

Of all the halogens, fluorine has the greatest attraction for electrons, according to electronegativity values. Iodine has the least. This means that F^- is an extremely stable ion. I^- is the least stable of the halide ions. Fluorine has the smallest nuclear charge of the halogens. It also has the smallest atomic radius. As a result, electrons are attracted to the valence shell of fluorine with great ease. This is the reason for the very high electronegativity value of fluorine.

Fluorine is also the most electronegative of all the known elements. This means that fluorine is the most chemically reactive nonmetal. Because of its high electronegativity, fluorine can have only a negative oxidation state (-1). The other halogens also have an oxidation state of -1. However, in combination with elements that are more electronegative, chlorine, bromine, and iodine may attain positive oxidation states as well.

The values for the ionization energies are high in the halogens. Thus it is very difficult to remove an electron from a halogen atom. In any given row of the Periodic Table, the halogen has the highest ionization energy except for that of the stable noble gas in the row. As the atomic radius of a halogen atom increases, the distance from the nucleus to the outer electrons increases. Thus, the shielding effect of the layers of electron energy levels increases, and the ionization energies therefore decrease.

Recall that intermolecular attractions increase as the molecular size of related substances increases. At ordinary conditions, fluorine and chlorine are gases, bromine is a liquid, and iodine is a solid. When heated at ordinary pressure, iodine changes directly from solid to gas. Recall that this process is called sublimation.

Elements that gain electrons are called oxidizing agents. Fluorine is the strongest oxidizing agent in Group 17. Chlorine, however, is a more common oxidizing agent. When chlorine is added to drinking water and swimming pools, it destroys harmful microorganisms by oxidizing them. The fluoride ion, F^-, is added to drinking water or toothpaste as a protection against tooth decay. Two compounds commonly used for this purpose are tin (II) fluoride (stannous fluoride), SnF_2, and sodium fluoride, NaF. Excessive amounts of these substances must be avoided, however, because excess F^- ion can cause mottled teeth and can also destroy bone.

It has already been mentioned that the halogens can assume different oxidation states. Using chlorine as a typical example, the following table summarizes the oxidation states of the halogens in representative compounds.

Oxidation State	***H Compound***	***K Compound***
−1	HCl, hydrochloric acid	KCl, potassium chloride
+1	HClO (HOCl) hypochlorous acid	KClO, potassium hypochlorite
+3	$HClO_2$ (HOClO), chlorous acid	$KClO_2$, potassium chlorite
+5	$HClO_3$ ($HOClO_2$), chloric acid	$KClO_3$, potassium chlorate
+7	$HClO_4$ ($HOClO_3$), perchloric acid	$KClO_4$, potassium perchlorate

The electron-dot diagrams below show the structures of the oxychlorine acids mentioned in the table. These acids vary in strength, with HClO the weakest and $HClO_4$ the strongest.

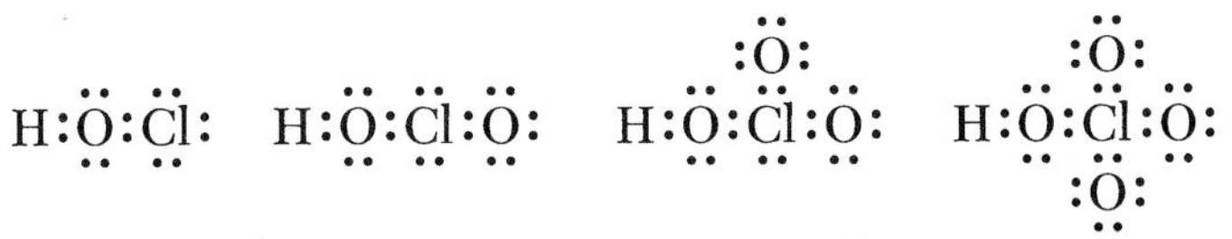

```
                                      ..          ..
                                     :O:         :O:
 .. ..       .. .. ..        .. .. ..     .. .. ..
H:O:Cl:     H:O:Cl:O:       H:O:Cl:O:    H:O:Cl:O:
 .. ..       .. .. ..        .. .. ..     .. .. ..
                                                 :O:
                                                  ..
```

The halogens are so active they do not occur uncombined in nature. Cl_2, Br_2, and I_2 are prepared by oxidizing their negative ions, Cl^-, Br^-, and I^-. F^- ions can be oxidized only by a direct electric current, that is, by electrolysis. (Why?) Cl^- ions can be oxidized by electrolysis or by a stronger oxidizing agent.

Seawater and salt wells are important commercial sources of the halogens. The laboratory preparation of the halogens is given in Appendix 3.

Chlorine is the most common halogen. However, all the elements of Group 17 are useful, as shown by the following table.

Halogen	***Some Uses***
Fluorine	Refrigerant (Freon), aerosol propellants, Teflon, antidecay additive to toothpastes and drinking water
Chlorine	Bleaches, disinfectants, insecticides, dyes, and medicines
Bromine	Tranquilizers and sedatives, antiknock additives to gasoline, manufacture of AgBr (used in photography)
Iodine	Germicides and antiseptics, iodized salt to prevent goiter

The Hydrogen Halides

The hydrogen halides HF and HCl are commonly used halogen compounds. Hydrogen fluoride is used to etch glass. The glass is first covered with a thin layer of paraffin. The design is then scratched through the paraffin so that the surface of the glass is exposed. The exposed glass is then treated with HF vapors or a solution of HF. The HF reacts with the sand (SiO_2) in the glass and forms SiF_4, which evaporates. The reaction is

$$SiO_2 + 4\ HF \longrightarrow SiF_4(g) + 2\ H_2O$$

HF in any form produces painful, slow-healing sores when it comes into contact with the skin.

Hydrochloric acid is used in the manufacture of metallic chlorides, dyes, and medicines. It is also used to pickle metals—that is, to dissolve metal oxide coatings before electroplating. The laboratory preparation of hydrochloric acid is given in Appendix 3.

The hydrogen halides are all gases at room temperature. They form acid solutions when they dissolve in water. The melting points and boiling points of the hydrogen halides are listed in the table that follows. Notice the trends in the melting points and boiling points of HCl, HBr, and HI. The increasing trends in both melting and boiling points of these compounds reflect increasing intermolecular attractions with increasing molecular size. If the trends continued for HF, the melting and boiling points of HF would be lower than those of HCl. Instead, the melting and boiling points of HF are higher than those of HCl.

The higher melting and boiling points of HF may be explained by the large difference in electronegativity between hydrogen and fluorine. This difference favors the formation of hydrogen bonds, which link HF molecules together (remember the hydrogen bonds in water). Energy is first needed to break these hydrogen bonds. Thus more energy is needed to effect the change of state.

Hydrogen Halide	***Melting Point (°C)***	***Boiling Point (°C)***
HF	−83	19.4
HCl	−112	−85
HBr	−86	−67
HI	−51	−36

Group 18—The Noble Gases

The members of the family of noble gases, Group 18, are helium, neon, argon, krypton, xenon, and radon. These gases were discovered at the close of the 19th century, although their existence was suspected at least 100 years earlier. They used to be called the inert gases because they did not form compounds. Since 1962, however, compounds of some of the inert gases have been prepared. The gases are now known as the noble gases. Some important properties of the noble gases are shown in the table on page 230.

Helium has a complete *s* sublevel. Each of the other noble gases has a complete *s* and *p* sublevel in the outermost energy level. Atoms that have eight electrons in this energy level are extremely stable.

Molecules of the noble gas elements are **monatomic** (have only one atom). Most gaseous elements are **diatomic** (have two atoms), as are H_2, O_2, and Cl_2, for example. This suggests that the noble gases have little or no tendency to form bonds. The atoms of these elements also have relatively high ionization energies. This property, too, is related to the tendency not to form bonds. The boiling points of the noble gases are all well below 0°C (273 K). Thus there cannot be strong attractive forces among the atoms. Of the elements, helium has the lowest boiling point, 4.1 K.

The Noble Gases

Property	Helium	Neon	Argon	Krypton	Xenon	Radon
Atomic number	2	10	18	36	54	86
Electron structure	$1s^2$ (2)	$1s^2$ $2s^2, 2p^6$ (2-8)	$1s^2$ $2s^2, 2p^6$ $3s^2, 3p^6$ (2-8-8)	$1s^2$ $2s^2, 2p^6$ $3s^2, 3p^6, 3d^{10}$ $4s^2, 4p^6$ (2-8-18-8)	$1s^2$ $2s^2, 2p^6$ $3s^2, 3p^6, 3d^{10}$ $4s^2, 4p^6, 4d^{10}$ $5s^2, 5p^6$ (2-8-18-18-8)	$1s^2$ $2s^2, 2p^6$ $3s^2, 3p^6, 3d^{10}$ $4s^2, 4p^6, 4d^{10}, 4f^{14}$ $5s^2, 5p^6, 5d^{10}$ $6s^2, 6p^6$ (2-8-18-32-18-8)
Melting point (°C)	−271.9 (at 26 atm)	−248.7	−189.3	−157	−111.5	−71
Boiling point (°C)	−268.9	−245.9	−185.7	−152.9	−107.1	−62
First ionization energy (kJ/mole)	2372	2081	1521	1351	1170	1037
Abundance in Earth's atmosphere (parts per million by volume)	5.2	18.2	9430.	1.1	0.09	6×10^{-14}

Compounds of radon, xenon, krypton, and argon with oxygen and fluorine have been prepared. Recall that fluorine and oxygen are the two most electronegative elements. Chemists therefore used fluorine and oxygen to try to form stable bonds with the noble gases. It is also logical to assume that the noble gases whose outermost electrons are farthest from the nucleus and thus most loosely held might form bonds with other elements. The s^2p^6 electron configuration in the valence shell is exceptionally stable, but these electrons can be "loosened up" to form bonds.

Compounds of the lighter noble gases—neon and helium—have not yet been prepared. The valence electrons of these elements are closer to the nucleus and are more tightly held than are those of the heavier noble gases. Bond formation is therefore more difficult, or perhaps is not even possible, with neon and helium.

PRACTICE

6.14 What is the most active nonmetal in Group 16?

6.15 Which element in Group 18 has the highest boiling point?

6.16 Hydrogen gas is half as dense as helium gas, and so has more "lifting power" when used in balloons and blimps. Despite this, why is helium always used in airships, rather than hydrogen?

6.17 What are the formulas of two different allotropic forms of oxygen?

CHAPTER REVIEW

The following questions will help you check your understanding of the material presented in the chapter.

1. A sodium ion, Na^+, has the same electron structure as which of the following particles? (1) Mg^{2+} (2) S^{2-} (3) Cl^- (4) Ar
2. Which of the following metals reacts explosively in cold water? (1) K (2) Mg (3) Al (4) Be

3. Which of the following is *not* correct as you go from the top to the bottom of Group 1 (1) Electronegativities decrease. (2) Ionization energies increase. (3) Oxidation states are the same. (4) Ionic radii are less than their atomic radii.
4. Which of the following is *not* correct in comparing Group 2 metals with the corresponding Group 1 metals? (1) Group 1 atoms have larger radii. (2) Group 2 ions have smaller radii. (3) Group 2 metals have higher first ionization potentials. (4) Group 2 metals have smaller electronegativity.
5. Because the nuclear charge of an alkaline earth metal is one unit greater than it is for the corresponding alkali metal, the alkaline earth metals (1) have larger atoms (2) have smaller ionic radii (3) are more active (4) are softer.
6. Element X forms a base with the formula $X(OH)_2$. Element X is most likely to be (1) an alkali metal (2) an alkaline earth metal (3) a halogen (4) a noble gas.
7. A nitrogen molecule (1) is polar (2) contains three pairs of shared electrons (3) is unstable (4) is very soluble in water.
8. Nitrogen fixation refers to the (1) decomposition of nitrogen compounds (2) formation of nitrogen atoms (3) liquefaction of nitrogen (4) conversion of atmospheric nitrogen to useful nitrogen compounds.
9. Scientists determined that the formula for the phosphorus molecule is P_4 because (1) all nonmetals have four atoms in the molecule (2) the molecular mass is found to be four times the atomic mass (3) phosphorus has four valence electrons (4) phosphorus has four allotropic forms.
10. The compound in which sulfur has the highest oxidation state is (1) H_2S (2) Na_2SO_3 (3) H_2SO_4 (4) $Na_2S_2O_3$.
11. The two allotropes of oxygen differ in (1) the number of neutrons (2) the number of valence electrons (3) the number of atoms in the molecule (4) electronegativity.
12. To complete the equation

$$MnO_2 + 2\,Br^- + \_\_\_\_ \longrightarrow Mn^{2+} + Br_2 + 2\,H_2O$$

the necessary term in the blank space is (1) 4 H_2O (2) 4 H^+ (3) 4 OH^- (4) 4 HBr.

13. Which of the following statements is *not* correct for the halogen elements? (1) Halogen elements form covalent bonds with nonmetals. (2) Halogen elements in oxyhalogen compounds usually have a positive oxidation number. (3) The electronegativities of the halogens increase with increasing atomic radius. (4) The halide ion has two *s* and six *p* electrons in its valence shell.

14. Element X is a metal that forms an oxide with the formula X_2O. Element X is in Group (1) 1 (2) 2 (3) 16 (4) 17

15. Which is an alkaline earth metal? (1) Na (2) Ca (3) Ga (4) Ta

16. Which is the atomic number of an alkali metal? (1) 10 (2) 11 (3) 12 (4) 13

17. Flame tests are most useful in identifying the elements in Groups (1) 1 and 2 (2) 2 and 13 (3) 16 and 17 (4) 1 and 18

18. Which halogen would *not* be expected to have a positive oxidation state when combined with oxygen? (1) F (2) Cl (3) Br (4) I

19. Which is the electron configuration of an alkali metal?

    (1) 2-8-1 (3) 2-8-5

    (2) 2-8-2 (4) 2-8-8

20. Which element has an ionic radius that is larger than its atomic radius? (1) Li (2) Cl (3) Mg (4) Al

21. Which is an example of a metalloid? (1) B (2) Br (3) Ba (4) Rb

22. When a fluorine atom becomes an ion, it will (1) gain an electron and decrease in size (2) gain an electron and increase in size (3) lose an electron and decrease in size (4) lose an electron and increase in size.

23. If the elements are considered from top to bottom in Group 16, the number of electrons in the outermost shell (1) decreases (2) increases (3) remains the same.

24. Which group in the periodic table contains the alkali metals? (1) 1 (2) 2 (3) 13 (4) 14

25. A nonmetal that exists in the liquid state at room temperature is (1) aluminum (2) mercury (3) hydrogen (4) bromine.

CONSTRUCTED RESPONSE

1. For each of the following, give the symbol of the element described.
   (a) An alkaline earth metal found in Period 4.
   (b) The atom in Period 4 that has the highest ionization energy.
   (c) The most metallic element found in Group 15.
   (d) An element that is stable due to the triple bonds in its molecules.
   (e) Hydrogen reacts with this element to form an acid used to etch glass.
   (f) This element reacts with hydrogen in the Haber process.
   (g) In the solid state, molecules of this element contain 8 atoms.
2. How is "acid rain" formed?
3. How might the burning of fossil fuels contribute to global warming?
4. You are given two beakers, each containing a clear, colorless, aqueous solution. You are told that one of these solutions is KBr while the other is NaBr. Explain how you could test the two solutions to determine which was which. Include the probable results of your tests.

7 Chemical Calculations

LOOKING AHEAD

Much of the information in chemistry is expressed quantitatively—through the use of numbers, formulas, and equations. In this chapter, you will once again deal with the mole concept, which was introduced in Chapter 2. You will see how the mole is used to predict the results of chemical reactions quantitatively.

When you have completed this chapter you should be able to:

- **Define** mole, gram-formula mass, molar mass, and molar volume.
- **Distinguish** between empirical formulas and molecular formulas.
- **Calculate** percent composition by mass from a given formula and determine a formula from a percent composition.
- **Convert** from one quantitative unit to another: grams to moles, moles to grams, moles to liters at STP, liters at STP to moles.
- **Solve problems** involving mass and volume relationships.

Expressing Quantity

Chemists express quantity very much the same way you do in your daily life. If you wanted to buy some apples, there are at least three different ways you could express the desired quantity. You could buy two dozen apples, you could buy 10 kilograms of apples, or you could buy one bushel basket full of apples. In the first case you are expressing the number of items, in the second the weight of the items, and in the third the volume occupied by the items. These three methods of expressing quantity answer the questions "How many?" "How heavy?" and "How big?"

Chemists measure the size of the sample, its volume, in liters (L); the mass of the sample (how heavy) in grams (g); and the number of items, usually molecules, in moles. Many chemical calculations involve nothing more than converting one of these units into another. We begin by discussing our expression of the number of particles, the mole.

The Mole Concept

Eggs are commonly counted by the dozen, one dozen being equal to 12 units. A dozen is a convenient and easily understood quantity. It is also a practical quantity for counting something the size of eggs. But it is not a very useful quantity for counting something much smaller, such as atoms or molecules. Such very small particles are more conveniently and logically counted with the help of a very large number. Thus tiny particles, like atoms, ions, molecules, and electrons, are counted by the mole, one mole being equal to 6.02×10^{23} units. A mole represents 6.02×10^{23} units of anything—atoms, bricks, or bathtubs. Recall that 6.02×10^{23} is Avogadro's number.

The magnitude represented by the number 12 is well within your understanding. But the magnitude of the number 6.02×10^{23} is probably beyond your imagination. Consider it this way: A person 64 years old has lived approximately 2×10^{9} seconds. Avogadro's number is 10^{14} (one hundred trillion) times larger than that!

How did chemists arrive at the number 6.02×10^{23}? Recall that the atomic mass unit (amu) was already defined as 1/12 the mass of carbon-12. The amu is much too small a unit for expressing the mass of ordinary objects; scientists prefer to express these masses in grams. There are 6.02×10^{23} amu in a gram. Since a mole is 6.02×10^{23} particles, whatever the mass of one particle is in amu, one mole of those particles will have that mass in grams. A sodium atom has a mass of 23 amu. A mole of sodium atoms has a mass of 23 grams. The mass of a water molecule is 18 amu, so the mass of a mole of water is 18 grams.

Gram-Molecular Mass

The mass of one mole of molecules is equal to the molecular mass expressed in grams. Since the mass of a molecule of water is 18 amu, the mass of one mole of water molecules is 18 grams. The gram-molecular mass is the same as the mass of one mole of molecules. The gram-molecular mass of water is 18 grams.

Gram-Formula Mass

You will recall that ionic compounds, such as NaCl, do not form molecules. The term *gram-molecular mass* for an ionic compound is therefore meaningless. The term *gram-formula mass* is used instead for the mass in grams of one mole of an ionic compound. For example, one mole of NaCl contains 6.02×10^{23} Na^+ ions and 6.02×10^{23} Cl^- ions in a giant lattice that has a mass of about 58 grams.

When working with moles, you often imply the type of particle being described without actually stating it. Instead of writing "1 mole of CO_2 molecules," you just write "1 mole of CO_2." Since CO_2 is the formula of a molecule, it is understood that you mean 1 mole of molecules. One mole of NaCl implies 1 mole of the formula NaCl, or 1 mole of Na^+ ions and 1 mole of Cl^- ions. Whether we are actually working with molecules or with ions, we find the mass exactly the same way. We add together the atomic masses of the component elements and express the result in grams. The mass of one mole is often called the *molar mass,* and can be expressed in the unit

$$\frac{\text{grams}}{\text{mole}}$$

The molar mass is the same as the gram-formula mass.

SAMPLE PROBLEM

PROBLEM

What is the mass of 1 mole of carbon dioxide?

SOLUTION

Carbon dioxide has the formula CO_2. The atomic mass of carbon is 12, while that of oxygen is 16. Adding the mass of the 1 carbon to that of 2 oxygens gives us the total molar mass of 44 grams. A mole of CO_2 thus has a mass of 44 grams. To find the mass of 1 mole, we express the molecular mass in grams.

Working with Moles and Grams

Since you now know how to find the mass of 1 mole of any given substance, you should also be able to find the mass of any number of moles of that substance. Recall that a mole is just a number of items, like a dozen. If you knew that 1 dozen apples weighed 2 kilograms, could you predict the weight of 4 dozen apples? Of course 4 dozen apples would weigh four times 2 kilograms, or 8 kilograms. Let's examine the relationship between moles and grams. What is the mass of 4.0 moles of CO_2? You already know that the mass of 1 mole of CO_2, the molar mass, is 44 grams. Thus 4.0 moles of CO_2 must have a mass of 4×44, or 176 grams.

The general relationship between moles and grams is often expressed as follows: moles $\times$ molar mass $=$ grams. This relationship can also be written as moles $= \dfrac{\text{grams}}{\text{molar mass}}$.

SAMPLE PROBLEMS

PROBLEM

1. What is the mass of 0.40 mole of H_2O?

SOLUTION

First, find the molar mass of water. The sum of the atomic masses in the formula is 18, so the molar mass of water is 18 grams. The formula that relates the number of moles to the mass is given on page 238—moles × molar mass = grams. Substituting the values into the formula you get 0.40 mole × 18 grams/mole = 7.2 grams.

PROBLEM

2. How many moles are there in 3.6 grams of water?

SOLUTION

Once again, you need to use the molar mass of water, which you know is 18 grams. The equation that most conveniently solves for moles is also given above,

$$\text{moles} = \frac{\text{grams}}{\text{molar mass}}$$

Substituting our values into that equation we get

$$\text{moles} = \frac{3.6 \text{ grams}}{18 \text{ grams/mole}} = 0.20 \text{ mole of water}$$

PRACTICE

7.1 Find the mass of each of the following.

(a) 0.50 mole of NaOH

(b) 4.5 moles of $CaCO_3$

(c) 0.25 mole of $Ba(NO_3)_2$

(d) 3.00 moles of sodium oxide

7.2 How many moles are contained in each of the following?

(a) 60. grams of NaOH

(b) 75 grams of $CaCO_3$

(c) 26.1 grams of $Ba(NO_3)_2$

(d) 310 grams of sodium oxide

7.3 If the mass of 0.60 mole of a certain substance is 72 grams, what is the molar mass of the substance?

Percent Composition

Suppose you wished to find the percentage of girls in a chemistry class. This particular class contains 12 boys and 18 girls. You would divide the number of girls, 18, by the total number of students in the class, 30, and then multiply your answer by 100 to get the percentage.

$$\frac{18 \text{ girls}}{30 \text{ students}} \times 100 = 60\% \text{ girls}$$

Suppose, on the other hand, you wished to find the percentage of girls in the class by mass. To do that you would need to find the mass of all the students. Let us assume that the girls' average mass is 50 kg, and the boys' average mass is 70 kg. The total mass of the girls is

$$50 \text{ kg/girl} \times 18 \text{ girls} = 900 \text{ kg}$$

The total mass of the boys is

$$70 \text{ kg/boy} \times 12 \text{ boys} = 840 \text{ kg}$$

The total mass of the students is thus

$$840 \text{ kg} + 900 \text{ kg} = 1740 \text{ kg}$$

Now you can find the percent of girls by mass. Divide the total mass of the girls, 900 kg, by the total mass of the students, 1740 kg, and multiply by 100.

$$\frac{900 \text{ kg girls}}{1740 \text{ kg students}} \times 100 = 51.7\% \text{ girls by mass}$$

To find the percent of girls by mass, you needed to know the formula for the class in terms of girls and boys and the mass of each of the items in the formula, in this case, the girls and the boys.

In chemistry, the percent composition of a compound is normally expressed by mass. Thus to find it, you need to know the formula of the compound and the mass of each element in the formula. To find the percent by mass of an element in a compound, follow these steps:

1. Find the total mass of the element in one mole of the compound. For example, in one mole of H_2O the total mass of

hydrogen is two times the atomic mass of hydrogen, or $2 \times 1.0\text{ g} = 2.0\text{ g}$.

2. Find the molar mass of the compound. For H_2O, it is 18 g.
3. Divide the answer from step 1 by the answer from step 2. In this case, to find the percent hydrogen, that would be 2.0 g/18 g.
4. Multiply by 100 to get the percent.

$$\frac{2.0\text{ g}}{18\text{ g}} \times 100 = 11\%\text{ hydrogen}$$

These steps may be summarized by the formula

$$\frac{\text{total mass of element}}{\text{total mass of compound}} \times 100 = \text{percent of element}$$

SAMPLE PROBLEMS

PROBLEM

1. Calculate the percent composition by mass of carbon dioxide (CO_2).

SOLUTION

One mole of CO_2 contains 1 mole of carbon, or 1 ~~mole~~ C × 12 g/~~mole~~ = 12 g C. It contains 2 moles of oxygen atoms, or 2 ~~moles~~ O × 16 g/~~mole~~ = 32 g O. The molar mass of CO_2 is

$$12\text{ g} + 32\text{ g} = 44\text{ g}$$

The percent carbon is

$$\frac{12\text{ }\cancel{\text{g}}}{44\text{ }\cancel{\text{g}}} \times 100 = 27\%$$

The percent oxygen is

$$\frac{32\text{ }\cancel{\text{g}}}{44\text{ }\cancel{\text{g}}} \times 100 = 73\%$$

PROBLEM

2. Calculate the percent by mass of uranium in pitchblende, which has the formula U_3O_8.

SOLUTION

First, find the total mass of uranium in a mole of pitchblende. One mole of U_3O_8 contains 3 moles of uranium,

$$3 \cancel{\text{moles}}\ \text{U} \times 238\ \text{g}/\cancel{\text{mole}} = 714\ \text{g of U}$$

The molar mass of U_3O_8 is

$$3 \cancel{\text{moles}}\ \text{U} \times 238\ \text{g}/\cancel{\text{mole}} + 8 \cancel{\text{moles}}\ \text{O} \times 16\ \text{g}/\cancel{\text{mole}} = 842\ \text{g}$$

The percent of uranium equals the mass of uranium divided by the molar mass of the compound.

$$\frac{714\ \cancel{\text{g}}}{842\ \cancel{\text{g}}} \times 100 = 84.8\%$$

PROBLEM

3. Calculate the percent by mass of water in hydrated copper sulfate, $CuSO_4 \cdot 5\ H_2O$ (The "dot" in the formula of a hydrate does not mean "times." It should be read as "bonded to" or "attached to.")

SOLUTION

You can calculate the percent water in hydrates—crystallized compounds containing water—by the same method used in the preceding sample problems. First, find the total mass of water indicated by the formula, then divide the mass of water by the total molar mass of the hydrate. Five moles of water have a mass of

$$5.0 \cancel{\text{moles}} \times 18\ \text{g}/\cancel{\text{mole}} = 90.\ \text{g}$$

The molar mass of $CuSO_4 \cdot 5\ H_2O$ (including the five waters) is

$$63.5\ \text{g} + 32\ \text{g} + 64\ \text{g} + 90.\ \text{g} = 249.5\ \text{g/mole}$$

The percent water is

$$\frac{90.\ \text{g}}{249.5\ \text{g}} \times 100\% = 36\%\ H_2O$$

PRACTICE

7.4 Calculate the percent composition to the nearest whole number of each of the following compounds.

(a) CO (b) $ZnSiO_3$ (c) H_3PO_4 (d) $Ca(NO_3)_2$

7.5 To the nearest whole number what is the percent by mass of water in $Ba(OH)_2 \cdot 8H_2O$?

Empirical Formulas

You have learned how to find the percent composition of a compound from its molecular formula. Acetic acid, which is used in vinegar, has the formula $HC_2H_3O_2$. Its percent composition is 40% C, 6.7% H, and 53.3% O. Glucose, a simple sugar, has the molecular formula $C_6H_{12}O_6$. Its percent composition is exactly the same as that of acetic acid, 40% C, 6.7% H, 53.3% O. Both of these compounds have the same percent composition, because they contain the same elements in the same ratio. A molecule of acetic acid contains 2 carbon atoms, 4 hydrogen atoms, and 2 oxygen atoms. A ratio of 2 to 4 to 2 is mathematically exactly the same as the ratio of 6 to 12 to 6 found in the glucose. In simplest terms, the ratio in both compounds is 1 to 2 to 1.

We call the formula containing the simplest whole number ratio between the elements the *empirical formula* of the substance. The empirical formula of both glucose and acetic acid is CH_2O. Compounds with the same empirical formula will always have the same percent composition. The compound formaldehyde, which has the formula HCHO, would have the same percent composition as both glucose and acetic acid.

SAMPLE PROBLEM

PROBLEM

Find the empirical formula of the following compounds:

(a) ethane, C_2H_6

(b) propane, C_3H_8

(c) butane, C_4H_{10}

(d) pentane, C_5H_{12}

SOLUTION

(b + d) The formulas for both propane and pentane are *already* empirical formulas. Ratios of 3 to 8 and 5 to 12 cannot be simplified. (a + c) The empirical formula of ethane is CH_3, since a ratio of 2 to 6 can be reduced to 1 to 3. The empirical formula of butane is similarly found to be C_2H_5.

PRACTICE

7.6 Find the empirical formula of the following compounds.
(a) H_2O_2 (b) C_7H_{12} (c) C_6H_6

Determining the Empirical Formulas

You have seen that many different compounds may have the same percent composition. Therefore, it is not possible to determine the actual molecular formula from the percent composition alone. However, it is possible to determine the empirical formula of a substance from its percent composition.

Determining an empirical formula is the reverse of determining percentage composition. If you know the percent by mass of the elements in a compound, you can determine the relative number of atoms of each element in the compound. To find the relative number of atoms (or moles of atoms) of each element in a compound, divide the percentage of each element by the atomic mass of the element. The quotients are the relative numbers of atoms (or moles of atoms) in a mole of the compound.

In this method, it is assumed that the mass of the sample of the compound is 100 grams. This assumption is made because it simplifies the arithmetic. Percentages can then be easily converted to moles or to grams. The assumption is justified because all samples of the same substance have the same percent composition.

SAMPLE PROBLEMS

PROBLEM

1. A gaseous compound of hydrogen and carbon (a hydrocarbon) has the following composition by mass: carbon, 92.3 percent; hydrogen, 7.7 percent. What is the empirical formula of the gas?

SOLUTION

Find the number of moles of each element in the compound.

$$\text{Relative number of moles of C atoms} = \frac{92.3\ \text{g}}{12\ \text{g/mole}} = 7.7\ \text{moles}$$

$$\text{Relative number of moles of H atoms} = \frac{7.7\ \text{g}}{1\ \text{g/mole}} = 7.7\ \text{moles}$$

The proportion of component atoms is $C_{7.7}H_{7.7}$, or C_1H_1. Thus the empirical formula of this compound is C_1H_1. This means that any number of moles of carbon atoms and an equal number of moles of hydrogen atoms will satisfy the stated percent composition by mass.

PROBLEM

2. The composition by mass of a compound is 72.4 percent iron and 27.6 percent oxygen. What is the empirical formula of the compound?

SOLUTION

(a) Convert the percentages given to grams (assuming a 100-gram sample), and convert the grams to moles.

$$\text{moles Fe} = \frac{72.4 \text{ g}}{55.8 \text{ g/mole}} = 1.30 \text{ moles}$$

$$\text{moles O} = \frac{27.6 \text{ g}}{16 \text{ g/mole}} = 1.73 \text{ moles}$$

The formula is $Fe_{1.30}O_{1.73}$, but these are not whole numbers of moles.

(b) If the mole ratio cannot be converted into whole numbers at sight, divide each number by the smallest number.

$$\text{Fe} \quad \frac{1.30}{1.30} = 1 \qquad \text{O} \quad \frac{1.73}{1.30} = 1.33$$

Now the same formula reads $FeO_{1.33}$.

(c) If, after step b, the mole ratio is still not in whole numbers of moles, multiply the numbers by 2, 3, 4, 5, . . . until a ratio in whole numbers results. In this example, multiply by 3.

$$Fe_{1\times3}O_{1.33\times3} = Fe_3O_{3.99}, \text{ or } Fe_3O_4$$

PROBLEM

3. A compound contains 17.5 grams of iron and 7.5 grams of oxygen. What is the empirical formula?

SOLUTION

(a) In this case, you have been given grams directly, not percentages. As always, the first step is to convert to moles.

$$\text{Fe} \quad \frac{17.5 \text{ g}}{55.8 \text{ g/mole}} = 0.31 \text{ mole}$$

$$\text{O} \quad \frac{7.5\ g}{16g/\text{mole}} = 0.47 \text{ mole}$$

(b) Dividing both numbers of moles by 0.31 gives you the formula $FeO_{1.5}$.

(c) Multiplying both numbers of moles by 2 gives you the formula Fe_2O_3.

PRACTICE

7.7 Find the empirical formula of the following compounds that contain:

(a) 75 percent C, 25 percent H

(b) 40 percent Ca, 12 percent C, 48 percent O

(c) 2.5 percent H, 57.5 percent sodium, 40 percent O

(d) 40.0 percent carbon, 6.7 percent hydrogen, and 53.3 percent oxygen

(e) 66.0 percent calcium and 34.0 percent phosphorus

(f) 8.3 percent aluminum, 32.7 percent chlorine, and 59.0 percent oxygen

7.8 Find the empirical formula for the following compounds that contain:

(a) 1.8 grams Ca and 3.2 grams Cl

(b) 8.8 grams Cs and 2.35 grams Cl

(c) 233.7 grams Al and 416 grams S

(d) 10.26 grams Ni, 4.90 grams N, and 16.8 grams O

Molecular Formulas

You have been able to use the percent composition of a compound to determine its empirical formula. As you learned, several compounds may have the same empirical formula but different molecular formulas. Glucose, $C_6H_{12}O_6$, and acetic acid, $HC_2H_3O_2$, both have the same empirical formula, CH_2O. To determine the molecular formula, the percent composition alone is not enough information. If, in addition, you know the molar mass of the substance, you can find its molecular formula.

Glucose, $C_6H_{12}O_6$, has a gram molecular mass of 180 grams. The empirical formula, CH_2O, has a mass of 30 grams. Note that the molecular mass of glucose is exactly six times the mass of the empirical formula. The molecular mass of any molecule must be an exact multiple of the mass of the empirical formula. Since the mass of glucose is six times the mass of the empirical formula, the molecular formula of glucose must be exactly six times the empirical formula. If you multiply each of the subscripts in CH_2O by six, you get $C_6H_{12}O_6$.

To find the molecular formula when the empirical formula and molecular mass are known, follow these steps:

1. Find the mass of the empirical formula.
2. Divide the molecular mass by the mass of the empirical formula.
3. Multiply each of the subscripts in the empirical formula by the answer to step 2.

SAMPLE PROBLEMS

PROBLEM

1. Oxalic acid has the empirical formula HCO_2 and a molar mass of 90 grams. Find the molecular formula of oxalic acid.

SOLUTION

The empirical formula, HCO_2, has a mass of 45 grams.

$$90\ g/45\ g = 2$$

The molecular formula is twice the empirical formula, or $H_2C_2O_4$.

PROBLEM

2. Butene is 14.3 percent hydrogen, and 85.7 percent carbon by mass. It has a molecular mass of 56 grams. Find the molecular formula of butene.

SOLUTION

First, find the empirical formula. Convert the percentages to grams, and convert grams to moles.

$$\frac{14.3 \text{ g H}}{1.01 \text{ g/mole}} = 14.2 \text{ moles H}$$

$$\frac{85.7 \text{ g C}}{12.0 \text{ g/mole}} = 7.14 \text{ moles C}$$

A mole ratio of 7.14 to 14.2 is almost 1 to 2. The empirical formula is CH_2. Now, use the empirical formula and the molecular mass to find the molecular formula. Following the three steps given on page 248, you find that the mass of the empirical formula, CH_2, is 14 g. Dividing the molecular mass, 56 g, by 14 g, gives 4. The molecular formula is 4 times the empirical formula. The molecular formula of butene is C_4H_8.

PROBLEM

3. A compound has the empirical formula C_3H_8 and a molecular mass of 44 g/mole. What is the molecular formula?

SOLUTION

The mass corresponding to the formula C_3H_8 is 44 g/mole. This is the same as the molecular mass of the compound. The empirical and molecular formulas are the same: C_3H_8.

PRACTICE

7.9 The empirical formula of benzene is CH and its molecular mass is 78 grams. What is benzene's molecular formula?

7.10 A hydrocarbon contains 85.7 percent carbon and 14.3 percent hydrogen by mass. Its molecular mass is 70 g/mole. Find its molecular formula.

7.11 The butane gas used in lighters has a molar mass of 58 g/mole. Its empirical formula is C_2H_5. Find the molecular formula of butane.

7.12 The molecular mass of aspirin is 180 g/mole. Its percent composition is 60.0 percent carbon, 4.48 percent hydrogen, and 35.5 percent oxygen. What is aspirin's molecular formula?

Mole Relationships in Chemical Reactions

Consider the reaction between aluminum metal and hydrochloric acid:

$$2\,Al + 6\,HCl \longrightarrow 2\,AlCl_3 + 3\,H_2$$

Recall that the numbers used to balance the equation, in this case 2, 6, 2, and 3, are called coefficients. The coefficients in the balanced equation tell in what proportion reactants react and in what proportion the products are formed. In this case, 2 aluminum react with 6 hydrochloric acid to produce 2 aluminum chloride and 3 hydrogen. Since it is convenient to express the number of particles in moles, we can also say that 2 moles of Al react with 6 moles of HCl to produce 2 moles of $AlCl_3$ and 3 moles of H_2. The coefficients give the mole ratio, which will enable you to predict the results of reacting any number of moles of reactant. You can also predict the number of moles of reactant needed to produce any number of moles of product.

To convert from moles of one substance to moles of any other substance in a balanced equation, multiply by the ratio of the coefficients, which we call the mole ratio. For example, you can predict the amount of HCl needed to react with 10 moles of Al. Multiply the moles of Al by the ratio moles of HCl per mole of Al. The coefficients tell us that there are 6 moles of HCl for every 2 moles of Al. Thus

$$10\,\cancel{\text{moles Al}} \times \frac{6\text{ moles HCl}}{2\,\cancel{\text{moles Al}}} = 30\text{ moles of HCl.}$$

If you want to know how many moles of Al are needed to produce 6.0 moles of H_2 in this reaction, multiply

$$6.0 \cancel{\text{moles } H_2} \times \frac{2 \text{ moles Al}}{3 \cancel{\text{moles } H_2}} = 4.0 \text{ moles of Al needed.}$$

To convert from moles of substance A to moles of substance B in a balanced equation, multiply the moles of A by the mole ratio of B to A in the balanced equation.

$$\text{moles A} \times \frac{\text{moles B}}{\text{moles A}} = \text{moles B}$$

Moles to Moles—An Alternate Method

Some students prefer an alternate method of solving moles-to-moles problems. They use the relationship

$$\frac{\text{moles}}{\text{coefficient}} = \frac{\text{moles}}{\text{coefficient}}$$

for the two substances involved. In the equation, $2\text{ Al} + 6\text{ HCl} \longrightarrow 2\text{ AlCl}_3 + 3\text{ H}_2$, let us use this method to find the amount of HCl needed to react with 10. moles of Al. We write the given number of moles directly above the substance in the balanced equation, and we write an "*x*" above the unknown substance. Our set-up would look like this:

$$\frac{10 \text{ moles}}{2\text{ Al}} + \frac{x \text{ moles}}{6\text{ HCl}} \longrightarrow 2\text{ AlCl}_3 + 3\text{ H}_2$$

We then solve the equation: $\frac{10 \text{ moles}}{2} = \frac{x \text{ moles}}{6}$. If we cross multiply, we get $2x = 60$, and $x = 30$ moles of HCl, the same answer found before.

SAMPLE PROBLEM

PROBLEM

In the reaction

$$2\text{ NO} + \text{O}_2 \longrightarrow 2\text{ NO}_2$$

how many moles of O_2 are needed to produce 3.6 moles of NO_2?

SOLUTION
Multiply the 3.6 moles of NO_2 by the mole ratio of O_2 to NO_2 in the balanced equation.

$$3.6 \text{ }\cancel{\text{moles NO}_2} \times \frac{1 \text{ mole O}_2}{2 \text{ }\cancel{\text{moles NO}_2}} = 1.8 \text{ moles of O}_2$$

This problem also could be solved using the "moles over coefficient" method.

$$\begin{array}{cccc} & \text{x moles} & & \text{3.6 moles} \\ \hline 2\text{ NO} + & O_2 & \longrightarrow & 2\text{ NO}_2 \end{array}$$

Although we do not write it, we understand that the coefficient before the oxygen is 1. Our equation becomes

$$\frac{x \text{ moles}}{1} = \frac{3.6 \text{ moles}}{2}, \text{ and } x = 1.8 \text{ moles of O}_2.$$

PRACTICE

7.13 Base your answers on the balanced equation
$N_2 + 3\ H_2 \longrightarrow 2\ NH_3$

(a) How many moles of hydrogen are needed to produce 6.0 moles of ammonia?

(b) How many moles of nitrogen are required to react with 0.60 mole of hydrogen?

(c) How many moles of ammonia are formed from 0.50 mole of nitrogen and excess hydrogen?

7.14 Base your answers on this balanced equation:

$$2\ Pb(CH_3)_4 + 15\ O_2 \longrightarrow 2\ PbO + 8\ CO_2 + 12\ H_2O$$

(a) How many moles of O_2 are needed to burn 4.6 moles of $Pb(CH_3)_4$?

(b) How many moles of CO_2 are produced by burning 5 moles of $Pb(CH_3)_4$?

(c) How many moles of water will be produced by reacting 7.5 moles of O_2?

7.15 Many people cook outdoors over burning propane gas. The combustion of propane gas produces carbon dioxide and water. If enough propane gas is burned to produce 12.0 moles of water, 9.0 moles of carbon dioxide will be produced as well. In the balanced equation for the combustion of propane, the coefficient in front of the CO_2 is 3. What is the coefficient in front of the H_2O?

Molar Gas Volume

Up to now you have studied quantitative chemistry in terms of moles and grams only. While mass is the most commonly used indication of the quantity of matter, chemists often express the quantity of a gas in liters, a unit of volume.

Recall that the volume occupied by any quantity of a gas depends only upon the temperature and pressure. This can be calculated using the expression $V = nRT/P$. In this equation, V is the volume, P the pressure, T the Kelvin temperature, n the number of moles, and R a constant. If the volume is expressed in liters and the pressure is in atmospheres, the value of R is 0.0821. You can calculate that at STP, where T is 273 K and P is 1 atm, the volume of 1 mole of a gas is 22.4 liters. As long as you are working at STP, the volume of 1 mole of any ideal gas will be 22.4 liters. Therefore, 22.4 liters is called the *molar volume* of a gas under standard conditions.

SAMPLE PROBLEM

PROBLEM

Assuming ideal gas behavior, what is the volume of 1.00 mole of propane gas at STP?

SOLUTION

You do not know the formula for propane gas? It doesn't matter! The volume of 1.00 mole of any ideal gas at STP is 22.4 L. (Propane is C_3H_8 as you will learn in Chapter 14.)

Recall that in order to find the mass of a given number of moles of a substance, you multiplied the number of moles by the molar mass. Grams = molar mass × moles (see page 238). Similarly, you can find the volume of a given number of moles of ideal gas at STP by multiplying the number of moles by the molar volume, 22.4 L/moles. Liters = 22.4 L/moles × moles.

From the number of moles, we can easily calculate the mass in grams, volume in liters (for ideal gases at STP), and total number of particles in a sample of a substance. The procedures for performing mole conversions are summarized in the next section.

Working with Moles

The following rules are useful when you work with moles:

1. To convert moles to grams, multiply the moles given by the mass of one mole.

   EXAMPLE: How many grams are there in 2.0 moles of CO_2?

$$2.0\ \cancel{\text{moles}} \times \frac{44\ \text{grams}}{\cancel{\text{mole}}} = 88\ \text{grams}$$

2. To convert grams to moles, divide the grams given by the mass of one mole.

   EXAMPLE: How many moles are represented by 88 grams of CO_2?

$$\frac{88\ \cancel{\text{grams}}}{44\ \cancel{\text{grams}}/\text{mole}} = 2.0\ \text{moles}$$

3. To convert moles to an actual number of particles, multiply the moles given by 6.02×10^{23} molecules/mole.

   EXAMPLE: How many molecules are represented by 2.0 moles of CO_2?

$$2.0\ \cancel{\text{moles}}\ \frac{(6.02 \times 10^{23}\ \text{molecules})}{\cancel{\text{mole}}} = 1.2 \times 10^{24}\ \text{molecules}$$

4. To convert a number of particles to moles, divide the number given by 6.02×10^{23} molecules/mole.

   EXAMPLE: How many moles are represented by 12.04×10^{23} CO_2 molecules?

$$\frac{12.04 \times 10^{23} \cancel{\text{molecules}}}{6.02 \times 10^{23} \cancel{\text{molecules}}/\text{mole}} = 2.00 \text{ moles}$$

5. To convert moles of a gas to liters at STP, multiply the moles given by 22.4 liters/mole.
   EXAMPLE: How many liters are represented by 2.00 moles of CO_2 at STP?

$$(2.00 \cancel{\text{moles}})(22.4 \text{ liters}/\cancel{\text{mole}}) = 44.8 \text{ liters}$$

6. To convert liters of a gas at STP to moles, divide the liters given by 22.4 liters/mole.
   EXAMPLE: How many moles are represented by 44.8 liters of CO_2 at STP?

$$\frac{44.8 \cancel{\text{liters}}}{22.4 \cancel{\text{liters}}/\text{mole}} = 2.00 \text{ moles}$$

The six processes shown above may also be carried out using the following three equations:

1. To convert between moles and grams, use the molar mass, grams/mole.

$$\text{moles} = \frac{\cancel{\text{grams}}}{\cancel{\text{grams}}/\text{mole}} \text{ or}$$

$$\cancel{\text{moles}} \times \frac{\text{grams}}{\cancel{\text{mole}}} = \text{grams}$$

2. To convert between moles and volume at STP,

$$\text{moles} = \frac{\cancel{\text{liters}}}{22.4 \cancel{\text{liters}}/\text{mole}}$$

or, $\cancel{\text{moles}} \times 22.4 \text{ liters}/\cancel{\text{mole}} = \text{liters}$.

3. To convert between moles and number of molecules,

$$\text{moles} = \frac{\cancel{\text{molecules}}}{6.02 \times 10^{23} \cancel{\text{molecules}}/\text{mole}} \text{ or,}$$

$\cancel{\text{moles}} \times 6.02 \times 10^{23} \text{ molecules}/\cancel{\text{mole}} = \text{molecules}$.

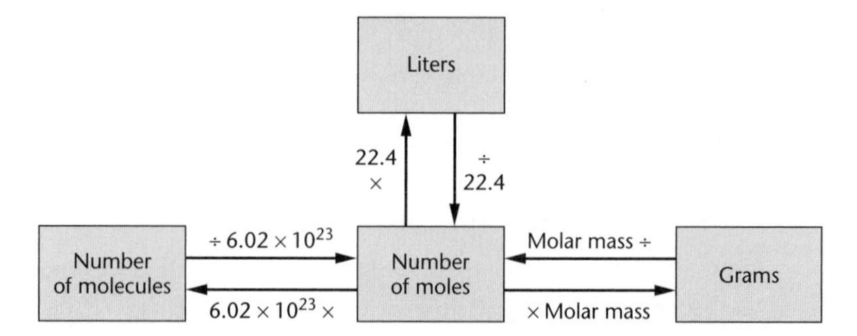

Figure 7-1 This flow chart, or concept map, may prove helpful in understanding mole conversions. Note that in each case, to go into moles, you divide. To get out of moles, you multiply. Note also, that "all roads lead to moles." Once you know the number of moles, it is easy to convert to any of the other three expressions of quantity. If you select the proper conversion, the correct answer will also show the correct units.

A concept map is often used by scientists to show the relationships among several variables. Figure 7–1 shows the relationships among moles, molecules, grams, and liters.

SAMPLE PROBLEMS

PROBLEM

1. What is the volume of 4.00 moles of CO_2 gas at STP?

SOLUTION

To convert moles of a gas at STP to liters, multiply the number of moles by 22.4 L/mole.

$$\text{moles} \times 22.4\ \text{L/mole} = \text{liters}$$

$$4.00\ \cancel{\text{moles}} \times 22.4\ \text{L}/\cancel{\text{mole}} = 89.6\ \text{L}.$$

PROBLEM

2. How many moles are there in 11.2 L of neon gas at STP?

SOLUTION

To convert liters to moles, divide by 22.4 L/mole.

$$\frac{\text{liters}}{22.4\ \text{L/mole}} = \text{moles}$$

$$\frac{11.2\ \cancel{\text{L}}}{22.4\ \cancel{\text{L}}\text{/mole}} = 0.500\ \text{mole}$$

PROBLEM

3. How many moles are there in 60. g of neon?

SOLUTION

To convert from grams to moles, divide by the molar mass. The molar mass of Ne is 20. g/mole.

$$\frac{\text{grams}}{\text{molar mass}} = \text{moles}$$

$$\frac{60.\ \cancel{\text{g}}}{20.\ \cancel{\text{g}}\text{/mole}} = 3.0\ \text{moles}$$

(Note: when converting moles to grams or grams to moles, it is not necessary to specify the temperature or pressure.)

PROBLEM

4. What is the mass of 0.20 mole of $CaCO_3$?

SOLUTION

To convert moles to grams, multiply by the molar mass.

$$\text{moles} \times \text{molar mass} = \text{grams}$$

The molar mass of $CaCO_3$ is

$$40\ \text{g} + 12\ \text{g} + (3 \times 16\ \text{g}) = 100\ \text{g}$$

$$0.20\ \cancel{\text{moles}} \times 100\ \text{g/}\cancel{\text{mole}} = 20\ \text{g}$$

PROBLEM

5. What is the volume at STP of 22.0 g of CO_2 gas?

SOLUTION

This problem is unlike the previous four, since you have not been shown a method for converting grams to liters. However, you do know how to convert grams to moles, and you know how to convert moles to liters. Therefore, you can solve the problem in two steps. You can use Figure 7.1 to find the correct path to go from grams to liters. First convert 22.0 g of CO_2 to moles, by dividing the grams by the molar mass.

$$\frac{\text{grams}}{\text{molar mass}} = \text{moles}$$

$$\frac{22.0\ \text{g}}{44.0\ \text{g/mole}} = 0.50\ \text{mole}$$

Now you can convert the moles to liters by multiplying by 22.4 L/mole.

$$0.50\ \cancel{\text{mole}} \times 22.4\ \text{L}/\cancel{\text{mole}} = 11.2\ \text{L}$$

The volume of 22 g of CO_2 at STP is 11.2 L.

PRACTICE

7.16 Change each quantity to moles.

(a) 36 g of water (b) 6.0 g of NaOH (c) 40 g of $CaCO_3$ (d) 44.8 L of CO_2 at STP (e) 11.2 L of He at STP (f) 56 L of O_2 at STP

7.17 Find the mass in grams of each of the following.

(a) 1.5 moles of H_2O (b) 0.30 mole of N_2 (c) 0.40 mole of NO_2 (d) 44.8 L of H_2 at STP (e) 56 L of Ne at STP (f) 5.6 L of O_2 at STP

7.18 Find the volume in liters at STP for each of the following ideal gases.

(a) 3.0 moles of Ar (b) 0.40 mole of CO_2 (c) 1.5 mole of Cl_2 (d) 16 g of He (e) 16 g of O_2 (f) 4.4 g of N_2O

Density Problems

Recall that density is mass per unit volume.

$$\text{density} = \frac{\text{mass}}{\text{volume}}$$

Since the volume of one mole of a gas at STP is 22.4 liters, and the mass of one mole is equal to the molar mass, you can write the formula

$$D = \frac{\text{molar mass}}{22.4 \text{ liters/mole}}$$

The density of a gas at STP in grams per liter is equal to its molar mass divided by 22.4 liters/mole. The molar mass of a gas is equal to its density at STP (in grams per liter) times 22.4 liters/mole.

$$\text{Molar mass} = D \times 22.4 \text{ liters/mole}$$

SAMPLE PROBLEM

PROBLEM

What is the density at STP of nitrogen gas, N_2?

SOLUTION

The molar mass of nitrogen, N_2, is 28.0 g. The density is equal to the molar mass divided by 22.4 L/mole.

$$\frac{28.0 \text{ grams/}\cancel{\text{mole}}}{22.4 \text{ L/}\cancel{\text{mole}}} = 1.25 \text{ grams/L}$$

PRACTICE

7.19 Find the density at STP of the following gases:

(a) CO (b) O_2 (c) Ar (d) NH_3

7.20 What is the molecular mass of a gas that has a density at STP of 1.98 grams/liter?

7.21 At STP, 4.00 liters of a certain gas has a mass of 11.4 grams. What is the molecular mass of this gas?

Quantitative Relationships in Chemical Reactions

You learned earlier in this chapter that you can convert between moles and grams as well as between moles and liters of gas at STP. If you combine these techniques with the technique of predicting moles in a balanced equation, you can predict the quantitative outcome of a chemical equation whether the quantities are expressed in moles, grams, or liters.

Mass Relationships

Using the equation

$$2\ Al + 6\ HCl \longrightarrow 2\ AlCl_3 + 3\ H_2$$

what is the maximum mass of H_2 that can be formed from 108 grams of Al? The coefficients in the balanced equation give us a mole ratio, not a gram ratio. Therefore, change the 108 grams of Al to moles. Recall that this is achieved by dividing the grams of the substance by its molar mass:

$$\frac{108\ \cancel{\text{grams}}\ \text{Al}}{27\ \cancel{\text{grams}}\ \text{Al/mole}} = 4.0\ \text{moles Al.}$$

Once you know the number of moles of Al, you can use the mole ratio in the balanced equation to find the moles of H_2.

$$4.0\ \cancel{\text{moles Al}} \times \frac{3\ \text{moles}\ H_2}{2\ \cancel{\text{moles Al}}} = 6.0\ \text{moles}\ H_2.$$

Now we can convert the 6.0 moles of H_2 to grams by multiplying by the molar mass of H_2.

$$6.0\ \cancel{\text{moles}}\ H_2 \times 2.0\ \text{grams}/\cancel{\text{mole}} = 12\ \text{grams}\ H_2.$$

Problems in which you find the mass of one substance in a reaction from the mass of another substance are called "mass-mass" problems. Note that the problem is solved in three steps. These can be summarized as follows:

Step 1. Into moles: Change the given mass to moles by dividing the mass in grams by the molar mass of the substance.

Step 2. Moles to moles: Convert from moles of the given substance to moles of the desired substance by using the mole ratio from the balanced equation.

Step 3. Out of moles: Convert the moles of substance found in Step 2 to grams by multiplying the number of moles by the molar mass of the substance.

SAMPLE PROBLEMS

PROBLEM

1. In the reaction,

$$2\ Na + 2\ H_2O \longrightarrow 2\ NaOH + H_2$$

how many grams of sodium are required to produce 6.0 grams of hydrogen gas?

SOLUTION

First, convert the 6.0 g of H_2 to moles. The molar mass of H_2 is 2.0 g/mole.

$$\frac{6.0\ \cancel{g}\ H_2}{2.0\ \cancel{g}/\text{mole}} = 3.0\ \text{moles}\ H_2$$

Next, use the mole ratio of sodium to hydrogen to find the moles of sodium.

$$3.0\ \cancel{\text{moles}\ H_2} \times \frac{2\ \text{moles Na}}{1\ \cancel{\text{mole}\ H_2}} = 6.0\ \text{moles of Na}$$

Finally, convert the 6.0 moles of Na to grams by multiplying by the molar mass of sodium.

$$6.0\ \cancel{\text{moles}}\ \text{Na} \times 23\ \text{g}/\cancel{\text{mole}} = 138\ \text{g Na}$$

PROBLEM

2. In the reaction,

$$2\ C_2H_6 + 7\ O_2 \longrightarrow 4\ CO_2 + 6\ H_2O$$

how many grams of H_2O are produced when 6.0 moles of C_2H_6 are burned?

SOLUTION

Since you were given the moles of C_2H_6 you can skip the first step, which is changing the given mass to moles. Use the mole ratio of H_2O to C_2H_6 to find the moles of water.

$$6.0 \text{ \sout{moles } } \cancel{C_2H_6} \times \frac{6 \text{ moles } H_2O}{2 \cancel{\text{moles } C_2H_6}} = 18 \text{ moles } H_2O$$

Since the question called for the answer in grams, convert the moles of H_2O to grams. The molar mass of water is 18 g/mole.

$$18 \cancel{\text{moles}}\ H_2O \times 18 \text{ g}/\cancel{\text{mole}} = 324 \text{ grams } H_2O$$

PRACTICE

7.22 $C_2H_5OH + 3\ O_2 \longrightarrow 2\ CO_2 + 3\ H_2O$

In the combustion of ethanol, shown above, how many grams of carbon dioxide are produced when 23 grams of ethanol are burned completely?

7.23 In the reaction,

$$Na_2O + H_2O \longrightarrow 2\ NaOH$$

how many moles of sodium oxide are needed to prepare 20. grams of NaOH?

7.24 In the reaction,

$$N_2 + 3\ H_2 \longrightarrow 2\ NH_3.$$

how many grams of H_2 are needed to produce 0.50 mole of NH_3?

Mass and Volume Relationships

Let us once again consider the reaction

$$2\ Al(s) + 6\ HCl(aq) \longrightarrow 2\ AlCl_3(aq) + 3\ H_2(g)$$

[This time we have indicated the state of the materials in the reaction: (*aq*) indicates that the substance is in an aqueous solution, (*s*) indicates the substance is in the solid state, (*g*) indicates that the substance is in the gaseous state, and (*l*) indicates that a substance is in the liquid state.] You have seen how to determine the number of grams of hydrogen produced by a given mass of aluminum. Suppose that instead, you wish to know the volume of hydrogen produced by a given amount of aluminum. How many liters of hydrogen may be produced at STP from 27.0 grams of aluminum? To be able to use the mole ratios in the balanced equation, you once again convert the grams of aluminum to moles.

$$\frac{27.0 \text{ \sout{grams} Al}}{27.0 \text{ \sout{grams}/mole}} = 1.00 \text{ mole Al}$$

Now, use the mole ratio to find the moles of hydrogen produced.

$$1.00 \text{ \sout{mole Al}} \times \frac{3 \text{ moles } H_2}{2 \text{ \sout{moles Al}}} = 1.50 \text{ moles } H_2.$$

Next, we convert the 1.5 moles of H_2 to liters. Recall that at STP the molar volume of a gas is 22.4 liters. To convert moles to liters, multiply the number of moles by 22.4 liters/mole

$$1.50 \text{ \sout{moles} } H_2 \times 22.4 \text{ liters/\sout{mole}} = 33.6 \text{ liters } H_2$$

As in the mass-mass problem, there were three steps to the solution: into moles, moles to moles, and out of moles.

SAMPLE PROBLEMS

PROBLEM

1. In the reaction,

$$2\,C(s) + O_2(g) \longrightarrow 2\,CO_2(g)$$

how many grams of carbon are required to react completely with 44.8 liters of oxygen at STP?

SOLUTION

First, convert the given quantity to moles. To convert liters of a gas at STP to moles, divide by 22.4 L/mole.

$$\frac{44.8 \text{ \sout{L} } O_2}{22.4 \text{ \sout{L}/mole}} = 2.0 \text{ moles } O_2$$

The second step is moles to moles. Use the mole ratio to find the moles of carbon.

$$2.0 \text{ \sout{moles O}}_2 \times \frac{2 \text{ moles C}}{1 \text{ \sout{mole O}}_2} = 4.0 \text{ moles C}$$

The third step is converting out of moles. Since the question asks for the mass in grams, multiply the number of moles of carbon by 12 grams/mole, the molar mass of carbon.

$$4.0 \text{ \sout{moles} C} \times 12 \text{ g/\sout{mole}} = 48 \text{ grams C.}$$

PROBLEM

2. In the reaction,

$$Zn(s) + 2\,HCl(aq) \longrightarrow ZnCl_2(aq) + H_2(g)$$

what is the maximum number of liters of hydrogen at STP that can be produced from 4.0 moles of HCl?

SOLUTION

Since the quantity of HCl is given in moles, you can use the mole ratio to find the moles of H_2.

$$4.0 \text{ \sout{moles HCl}} \times \frac{1 \text{ mole } H_2}{2 \text{ \sout{moles HCl}}} = 2.0 \text{ moles } H_2$$

Now you can convert the 2.0 moles of H_2 to liters at STP by multiplying by the molar volume, 22.4 L/mole.

$$2.0 \text{ \sout{moles} } H_2 \times 22.4 \text{ L/\sout{mole}} = 44.8 \text{ L}$$

PRACTICE

7.25 Hydrogen peroxide, H_2O_2, decomposes to produce water and oxygen gas. How many liters of oxygen gas are produced at STP by the decomposition of 6.8 grams of hydrogen peroxide?

7.26 $2\,Al(s) + 3\,H_2SO_4(aq) \longrightarrow Al_2(SO_4)_3(aq) + 3\,H_2(g)$

In the reaction above, how many grams of aluminum

metal would be required to produce 13.44 liters of hydrogen gas at STP?

7.27 $N_2(g) + 3\ H_2(g) \longrightarrow 2\ NH_3(g)$

In the reaction above, how many liters of H_2 at STP will react with 11.2 liters of N_2?

Volume Relationships

Consider Practice Problem 7.27. In the reaction, $N_2(g) + 3\ H_2(g) \longrightarrow 2\ NH_3(g)$ you were asked to calculate the volume of hydrogen needed to react with 11.2 liters of nitrogen. Problems in which you are asked to find the volume of one gas in a reaction from the volume of another gas in the reaction are called "volume-volume" problems.

The problem can be solved using the three-step method shown on pages 260–261. First, the 11.2 liters can be changed to moles.

$$\frac{11.2\ \text{L}}{22.4\ \text{L/mole}} = 0.500\ \text{mole}$$

Using mole ratios,

$$0.500\ \text{mol}\ N_2 \times \frac{3\ H_2}{1\ N_2} = 1.50\ \text{mole}\ H_2$$

Then, we can find the liters of hydrogen from the moles: 1.50 mole × 22.4 L/mole = 33.6 L. However, there is an easier way! Notice that in the first step you divided by 22.4, while in the last step you multiplied by 22.4. These two steps actually cancel each other—we get the same result if we omit both of them. Then, $11.2\ \text{liters}\ N_2 \times \frac{3\ H_2}{1\ N_2} = 33.6\ \text{liters}\ H_2$.

Since all ideal gases have the same molar volume under the same conditions, the volume ratio must be the same as the mole ratio. Therefore, volume-volume problems can be solved in just one step: liters of gas $A \times \frac{\text{moles}\ B}{\text{moles}\ A} =$ liters of gas B.

As in mole-to-mole problems, you may also solve volume-volume problems using the relationship $\frac{\text{liters}}{\text{coefficient}} = \frac{\text{liters}}{\text{coefficient}}$.

In the problem on page 265, we would obtain the relationship

$$\frac{11.2 \text{ liters}}{1\ N_2} = \frac{x \text{ liters}}{3\ H_2}$$

Cross multiply to get the correct answer, $x = 33.6$ liters.

Note that the one-step method of solving volume-volume problems is based on the fact that equal moles of different gases have the same volumes. However, equal moles of different gases do not have the same masses! Therefore, problems involving mass cannot be solved this way.

SAMPLE PROBLEMS

PROBLEM

1. Consider the reaction

$$C_3H_8(g) + 5\ O_2(g) \longrightarrow 3\ CO_2(g) + 4\ H_2O(g)$$

How many liters of oxygen are required to produce 45 liters of carbon dioxide?

SOLUTION

Since this is a volume-volume problem, it can be solved directly, in one step, using the mole ratio from the balanced equation.

$$45 \text{ L } CO_2 \times \frac{5 \text{ moles } O_2}{3 \text{ moles } CO_2} = 75 \text{ L } O_2$$

Or, using $\frac{\text{liters}}{\text{coefficient}} = \frac{\text{liters}}{\text{coefficient}}$ we can write

$$\frac{45 \text{ L } CO_2}{3\ CO_2} = \frac{x \text{ L } O_2}{5\ O_2};\ x = 75 \text{ L } O_2.$$

PROBLEM

2. In the same reaction, how many liters of oxygen at STP would react with 22 grams of C_3H_8?

SOLUTION

Since this problem expresses one of the quantities in grams, you cannot use the method shown in Sample Problem 1.

This problem requires a three-step solution. First, change the given mass of C_3H_8 to moles. The molar mass is 44 grams/mole.

$$\frac{22\ \cancel{g}\ C_3H_8}{44\ \cancel{g}/\text{mole}} = 0.50\ \text{mole}\ C_3H_8$$

Next, use the mole ratio to find the number of moles of O_2.

$$0.50\ \cancel{\text{mole}\ C_3H_8} \times \frac{5\ \text{moles}\ O_2}{1\ \cancel{\text{mole}\ C_3H_8}} = 2.5\ \text{moles}\ O_2$$

Finally, convert the moles of O_2 to liters at STP.

$$2.5\ \cancel{\text{moles}}\ O_2 \times 22.4\ \text{L}/\cancel{\text{mole}} = 56\ \text{L}\ O_2$$

PRACTICE

7.28 In the reaction,

$$N_2(g) + 3\ H_2(g) \longrightarrow 2\ NH_3(g)$$

how many liters of hydrogen are needed to react completely with 30 liters of nitrogen?

7.29 In the reaction,

$$S_8(s) + 12\ O_2(g) \longrightarrow 8\ SO_3(g)$$

(a) How many liters of oxygen gas are needed to produce 24 liters of SO_3?

(b) How many moles of S_8 must be burned to produce 89.6 liters of SO_3 at STP?

(c) Why CAN'T the method used in part (a) of this problem be used to find the volume of S_8 required to produce 24 liters of SO_3 at STP?

TAKING A CLOSER LOOK

Chemistry in the Real World

When chemical reactions are carried out in the laboratory, the results are not always the same as those predicted by the balanced equation. Chemists are able to adjust their calculations to the conditions in which they work. This section will familiarize you with the "real world" of chemistry.

Limiting Factors

Thus far you have predicted quantities in chemical reactions based upon the amount of just one of the reactants. For example, in the reaction

$$2\ Al + 6\ HCl \longrightarrow 2\ AlCl_3 + 3\ H_2$$

you can predict the amount of hydrogen produced from 54 grams of aluminum. The number of moles of aluminum is

$$\frac{54\ \cancel{g}\ Al}{27\ \cancel{g}/mole} = 2.0\ moles\ Al$$

The 2.0 moles of aluminum would produce 3.0 moles, or 6.0 grams, of hydrogen. This solution is based on the assumption that enough HCl is present to react with all the aluminum. Suppose, however, that you reacted 54 grams of aluminum with 145 grams of hydrochloric acid. How many grams of hydrogen would be produced?

When you are predicting the results of a reaction based on the amounts of two or more reactants, you need to know which reactant is going to be used up in the reaction. This is called the **limiting reactant.** The amount of product is determined by the amount of limiting reactant. The other reactants are said to be in excess. Some amount of excess reactants will be left unreacted when the reaction is complete.

To find the limiting reactant, you first change all quantities to moles. We already know there are 2.0 moles of aluminum. The 145 grams of HCl is 4.0 moles of HCl. Which is the limiting reactant? You have seen that 2.0 moles of Al could produce 3.0 moles of H_2. The 4.0 moles of HCl would produce 2.0 moles of H_2.

$$4.0\ \cancel{moles\ HCl} \times \frac{3\ moles\ H_2}{6\ \cancel{moles\ HCl}} = 2.0\ moles\ H_2$$

Since the HCl can produce less product, it must be the limiting reactant. The 4.0 moles of HCl can react with 1.3 moles of Al.

$$4.0 \text{ moles HCl} \times \frac{2 \text{ moles Al}}{6 \text{ moles HCl}} = 1.3 \text{ moles Al}$$

There would be 0.7 mole of Al left unreacted (2.0 moles initially −1.3 moles reacted). The amount of hydrogen produced would be 2.0 moles, or 4.0 grams. The amount of product is determined by the limiting reactant. The limiting reactant is the reactant that produces the smallest quantity of product.

Limiting reactant problems are often solved through the use of a "reaction chart." In a reaction chart, you list the initial number of moles, the change, and the final number of moles of each substance, under the balanced equation. The problem above would be laid out as follows:

	$2\,Al$	$+\ 6\ HCl \longrightarrow$	$2\,AlCl_3$	$+\ 3\,H_2$
Initial moles	2.0	4.0	0	0
Change	−1.3	−4.0	+1.3	+2.0
Final	0.7	0	1.3	2.0

Note that the limiting reactant, the HCl, is used up, so that there are no moles of HCl in the final mixture. Note also that all entries in the "change" row must be in the same ratio as the coefficients. Once we know that 4.0 moles of HCl are used up, we can calculate that there must be $4.0 \text{ mole HCl} \times \frac{2 \text{ Al}}{6 \text{ HCl}} = 1.3 \text{ mole Al}$ used up. The other changes in moles are found similarly.

SAMPLE PROBLEMS

PROBLEM

1. What is the maximum amount of water that could be formed when 4.0 grams of hydrogen gas is burned in 24 grams of oxygen?

$$2\,H_2 + O_2 \longrightarrow 2\,H_2O$$

SOLUTION

First, find the limiting factor. Change each mass to moles.

$$\frac{4.0 \text{ g } H_2}{2.0 \text{ g/mole}} = 2.0 \text{ moles } H_2$$

$$\frac{24 \text{ g } O_2}{32 \text{ g/mole}} = 0.75 \text{ mole } O_2$$

Two moles of H_2 could produce 2.0 moles of H_2O. The 0.75 mole of O_2 could produce 1.5 moles of H_2O. Therefore, the oxygen, which produces less product, is the limiting reactant. You will produce 1.5 moles of H_2O, or 1.5 moles $\times$ 18 g/mole = 27 g H_2O.

PROBLEM

2. When 4.0 grams of hydrogen is reacted with 24 grams of oxygen, as in Sample Problem 1, above, how much hydrogen is left unreacted?

SOLUTION

It has already been determined that the limiting reactant is the 24 g, or 0.75 mole of O_2. The amount of hydrogen that can react with 0.75 mole of O_2 in this reaction is

$$0.75 \cancel{\text{mole } O_2} \times \frac{2 \text{ moles } H_2}{1 \cancel{\text{mole } O_2}} = 1.5 \text{ moles } H_2.$$

Since we began with 2.0 moles of H_2 and only 1.5 moles react, there is 0.5 mole, or 1 g, of H_2 left unreacted. Note that 24 g of oxygen reacted with 4.0 g of hydrogen to produce 27 g of water (see Sample Problem 1) and left 1.0 gram of hydrogen unreacted. There was a total of 28 g of reactant before the reaction, and after the reaction there are 27 g of product and 1 g of unreacted reactant for a total of 28 g. The total mass of a reactant remains the same during any chemical reaction.

The reaction chart for the system would look like this:

	$2\ H_2$	$+\ O_2$	$\longrightarrow\ 2\ H_2O$
Initial moles	2.0	0.75	0
Change	−1.5	−0.75	+1.5
Final	0.5	0	1.5

PRACTICE

7.30 $2\ C_2H_6 + 7\ O_2 \longrightarrow 4\ CO_2 + 6\ H_2O$

Above is the reaction for the complete combustion of ethane gas.

(a) What is the maximum mass of water that can be produced when 9.0 grams of ethane and 16 grams of oxygen react until one of the reactants is completely consumed?

(b) How many moles of the excess reactant remain when the reaction is complete?

The Percent Yield

When chemical reactions are performed in the laboratory, they do not always produce the amounts of product predicted from the chemical equation. There may be other reactions taking place at the same time as the principal reaction. These are called *side reactions*. For example, the combustion of gasoline in a car engine produces carbon dioxide and water. However, the quantity of carbon dioxide produced is less than predicted from the balanced chemical equation because the reaction also produces some carbon monoxide. A badly tuned engine may produce some carbon (seen as black soot in the exhaust fumes) due to incomplete combustion of the fuel. Some reactions produce less product than expected due to equilibrium considerations. (Equilibrium will be discussed in Chapter 10.) The percent yield is a comparison of the actual amount of product, to the theoretical quantity, calculated from the balanced equation. The percent yield is found using the formula:

$$\text{percent yield} = \frac{\text{actual yield}}{\text{theoretical yield}} \times 100$$

SAMPLE PROBLEM

PROBLEM

The reaction

$$MnO_2 + 4\ HCl \longrightarrow MnCl_2 + Cl_2 + 2\ H_2O$$

is often used to prepare chlorine gas in the laboratory. When 174 grams of MnO_2 is reacted with excess HCl, 120 grams of chlorine gas is produced. What is the percent yield?

SOLUTION

First, use the balanced equation to calculate the theoretical yield.

$$\frac{174\ \cancel{g}\ MnO_2}{86.9\ \cancel{g}/\text{mole}} = 2.00\ \text{moles}\ MnO_2$$

$$2.00\ \cancel{\text{moles}\ MnO_2} \times \frac{1\ \text{mole}\ Cl_2}{1\ \cancel{\text{mole}\ MnO_2}} = 2.00\ \text{moles}\ Cl_2$$

$$2.00\ \cancel{\text{moles}}\ Cl_2 \times 70.9\ \text{g}/\cancel{\text{mole}} = 142\ \text{g}\ Cl_2$$

The theoretical amount of product is 142 g. The actual yield was 120. g. Thus the percent yield is

$$\frac{120.\ \cancel{g}}{142\ \cancel{g}} \times 100 = 84.5\%$$

PRACTICE

7.31 Oxygen is produced by the decomposition of hydrogen peroxide, through the reaction

$$2\ H_2O_2 \longrightarrow 2\ H_2O + O_2$$

Starting with 34 grams of H_2O_2, a student was able to collect 12 grams of O_2. What was the percent yield?

Using the Gas Laws

All of our calculations thus far in this chapter have been done at STP. Our laboratory work, however, is done at room temperature, which is warmer than standard temperature. You will recall that the volume of a gas depends on temperature and pressure. If you are collecting gases under conditions other than STP, you must adjust any calculations involving gas volumes. By using the gas laws (see Chapter 1), you can correct gas volumes in accordance with your working conditions.

SAMPLE PROBLEMS

PROBLEM

1. Hydrogen is often collected in the laboratory using the reaction

$$Zn(s) + 2\ HCl(aq) \longrightarrow ZnCl_2(aq) + H_2(g).$$

If at standard pressure, 0.65 gram zinc is reacted with excess hydrochloric acid, how many liters of hydrogen gas would be produced at a temperature of 20°C?

SOLUTION

First, find the volume of hydrogen that would be produced at STP. Change the grams of zinc to moles.

$$\frac{0.65\ \not{g}\ Zn}{65\ \not{g}/mole} = 0.010\ mole\ Zn$$

Since the mole ratio of H_2 to Zn is 1 to 1, 0.010 mole of H_2 must be produced. At STP, 0.010 mole $H_2 \times 22.4$ L/mole = 0.224 L of H_2. Now, apply the gas laws. Since only the temperature differs from standard, you can use Charles' Law,

$$\frac{V_1}{T_1} = \frac{V_2}{T_2}$$

V_1 is our volume at STP, 0.224 L. V_2 is the new volume. T_1 is the temperature at STP, 273 K. T_2 is our working temperature of 20°C, which must be converted to 293 K.

$$V_2 = \frac{V_1 \times T_2}{T_1}$$

$$0.240\ L = \frac{0.224\ L \times 293\ \not{K}}{273\ \not{K}}$$

PROBLEM

2. The reaction

$$2\ KClO_3 \longrightarrow 2\ KCl + 3\ O_2$$

is often used to prepare oxygen in the laboratory. If you are working at a pressure of 98.6 kilopascals and a temperature of 20°C, what is the maximum mass of $KClO_3$ needed to fill four 50-milliliter collection jars with oxygen?

SOLUTION

You need to collect 4 × 50 mL = 200 mL of oxygen. To determine the amount of $KClO_3$ required, you must first convert the quantity of oxygen to moles. Since the oxygen is not at STP, its molar volume would not be 22.4 L. You need to find the volume of oxygen at STP by using the combined gas law,

$$\frac{P_1V_1}{T_1} = \frac{P_2V_2}{T_2}$$

In this case, V_1 = 200 mL, T_1 = 293 K, P_1 = 98.6 kPa, P_2 is standard pressure, 101.3 kPa, and T_2 is standard temperature, 273 K. Using the equation above,

$$V_2 = \frac{T_2P_1V_1}{T_1P_2}$$

$$181\ \text{mL}\ O_2 = \frac{273\ \cancel{\text{K}} \times 98.6\ \cancel{\text{kPa}} \times 200\ \text{mL}}{293\ \cancel{\text{K}} \times 101.3\ \cancel{\text{kPa}}}$$

The volume at STP would be 181 mL, or 0.181 L. Now you can convert the volume to moles.

$$\frac{0.181\ \cancel{\text{L}}\ O_2}{22.4\ \cancel{\text{L}}/\text{mole}} = 0.00808\ \text{mole}\ O_2$$

You can now use the balanced equation to find the required moles of $KClO_3$.

$$0.00808\ \cancel{\text{mole}\ O_2} \times \frac{2\ \text{moles}\ KClO_3}{3\ \cancel{\text{moles}\ O_2}}$$

$$= 0.00539\ \text{mole}\ KClO_3$$

Finally, multiply 0.00539 mole $KClO_3$ by the molar mass of $KClO_3$, 122.6 g/mole.

$$0.00539\ \cancel{\text{mole}}\ KClO_3 \times 122.6\ \text{g}/\cancel{\text{mole}} = 0.661\ \text{g}\ KClO_3$$

PRACTICE

7.32 (a) How many liters of oxygen could be obtained at standard pressure, and a temperature of 25°C, from the complete decomposition of 34 grams of H_2O_2?

(b) At 25°C, how many grams of oxygen would be produced from the complete decomposition of 34 grams of H_2O_2?

Burning in Air

Most combustion reactions take place in air, rather than in pure oxygen. Since the air is roughly 20 percent oxygen, or one-fifth oxygen by volume, it takes about five times as many liters of air for complete combustion as it would pure oxygen.

SAMPLE PROBLEM

PROBLEM

How many liters of air will completely burn 1.0 liter of H_2 at STP? The reaction is $2\ H_2 + O_2 \longrightarrow 2\ H_2O$.

SOLUTION

This is a volume-volume problem, so the liters of oxygen required can be found using the mole ratio from the balanced equation.

$$1.0\ \text{L}\ H_2 \times \frac{1\ \text{mole}\ O_2}{2\ \text{moles}\ H_2} = 0.5\ \text{L}\ O_2$$

Since air is roughly one-fifth oxygen, there is five times more air required than oxygen.

$$0.5\ \cancel{\text{L}\ O_2} \times \frac{5\ \text{L air}}{1\ \cancel{\text{L}\ O_2}} = 2.5\ \text{L air}$$

PRACTICE

7.33 How many liters of air are needed to burn 11.2 liters of octane, C_8H_{18}, at STP?

Graham's Law

Thomas Graham investigated the relationship between the rate of diffusion of a gas and the mass of the gas. *Graham's law* states: At constant temperature and pressure, the rates of diffusion of gases are inversely proportional to the square root of their molecular masses (or their densities, since densities are proportional to molelecular masses). Graham's law can be expressed mathematically as

$$\frac{R_1}{R_2} = \sqrt{\frac{M_2}{M_1}}$$

In this equation, R_1 is the rate of diffusion of a gas with molecular mass M_1, and R_2 is the rate of diffusion of a second gas with molecular mass M_2.

SAMPLE PROBLEM

PROBLEM

Hydrogen gas has a molecular mass of 2. Oxygen gas has a molecular mass of 32. How do the rates of diffusion of these two gases compare?

SOLUTION

Let M_1 represent the molecular mass of hydrogen gas and R_1 the rate of diffusion of hydrogen. Let M_2 represent the molecular mass of oxygen gas and R_2 represent the rate of diffusion of oxygen. Then,

$$\frac{R_1}{R_2} = \sqrt{\frac{32}{2}} = \sqrt{16} = 4$$

The rate of diffusion of hydrogen molecules is four times the rate of diffusion of oxygen molecules.

PRACTICE

7.34 Compare the rates of diffusion of neon and krypton.

CHAPTER REVIEW

The following questions will help you check your understanding of the material presented in the chapter.

1. The number of moles in 2.16 grams of silver is (1) 2.00×10^{-2} (2) 4.59×10^{-2} (3) 2.00×10^{2} (4) 2.33×10^{2}.
2. What is the percentage by mass of oxygen in CuO? (1) 16 percent (2) 20 percent (3) 25 percent (4) 50 percent
3. The percentage by mass of hydrogen in H_3PO_4 is equal to (1) $\frac{1 \times 100}{98}$ (2) $\frac{3 \times 100}{98}$ (3) $\frac{98 \times 100}{3}$ (4) $\frac{98 \times 100}{1}$
4. What is the approximate percentage composition by mass of $CaBr_2$ (formula mass = 200)? (1) 20 percent calcium and 80 percent bromine (2) 25 percent calcium and 75 percent bromine (3) 30 percent calcium and 70 percent bromine (4) 35 percent calcium and 65 percent bromine
5. A compound contains 50 percent sulfur and 50 percent oxygen by mass. The empirical formula of this compound is (1) SO (2) SO_2 (3) SO_3 (4) SO_4.
6. Which compound contains the greatest percentage of oxygen by mass? (1) BaO (2) CaO (3) MgO (4)SrO
7. Given the reaction $N_2 + 3\ H_2 \longrightarrow 2\ NH_3$, how many grams of ammonia are produced when 1.0 mole of nitrogen reacts? (1) 8.5 (2) 17 (3) 34 (4) 68
8. The empirical formula of a compound is CH_2. The molecular formula of this compound could be (1) CH_4 (2) C_2H_2 (3) C_2H_4 (4) C_3H_3.

9. According to the balanced equation

$$Cu + 4\,HNO_3 \longrightarrow Cu(NO_3)_2 + 2\,H_2O + 2\,NO_2(g)$$

how many moles of nitric acid are necessary to react with 3.0 moles of copper? (1) 0.75 (2) 4 (3) 3.0 (4) 12

10. The number of molecules present in 76 grams of fluorine (F_2) gas is equal to

(1) $76 \times 6 \times 10^{23}$

(2) $\frac{76}{6 \times 10^{23}}$

(3) $\frac{76 \times 6 \times 10^{23}}{38}$

(4) $\frac{76 \times 6 \times 10^{23}}{19}$

11. Given the balanced equation

$$3\,PbCl_2 + Al_2(SO_4)_3 \longrightarrow 3\,PbSO_4 + 2\,AlCl_3$$

how many moles of $PbSO_4$ will be formed when 0.050 mole of $Al_2(SO_4)_3$ is consumed? (1) 0.05 (2) 0.15 (3) 0.30 (4) 0.50

12. Which of the following samples contains the greatest number of moles? (1) 40. grams of NaOH (2) 24 grams of H_2O (3) 200. g of PbS (4) 250. g of $PbSO_4$

13. How many moles of $CaCO_3$ are there in 20. grams of $CaCO_3$? (1) 0.20 (2) 5.0 (3) 20. (4) 2000

Base your answers to questions 14 and 15 on the following information.

Oxygen and hydrogen gas are produced by the electrolysis of water according to the following equation:

$$2\,H_2O(l) \xrightarrow{\text{elect.}} 2\,H_2(g) + O_2(g)$$

14. How many moles of oxygen are produced in the electrolysis of 4.0 moles of water? (1) 2.0 (2) 4.0 (3) 6.0 (4) 8.0

15. How many liters of hydrogen gas at STP are produced in the electrolysis of 90 grams of water? (1) 10.0 (2) 22.4 (3) 56.0 (4) 112

Base your answers to questions 16 through 18 on the following information.

One mole of potassium permanganate ($KMnO_4$) reacts completely with hydrochloric acid according to the reaction:

$$16\ HCl + 2\ KMnO_4 \longrightarrow 2\ KCl + 2\ MnCl_2 + 5\ Cl_2 + 8\ H_2O$$

16. How many moles of water are produced? (1) 1 (2) 2 (3) .5 (4) 4

17. How many grams of potassium chloride are produced? (1) 1 (2) 37 (3) 74 (4) 148

18. How many liters of chlorine measured at STP are produced? (1) 11.2 (2) 22.4 (3) 56.0 (4) 112

19. What is the volume of 0.500 mole of an ideal gas at STP? (1) 0.500 L (2) 11.2 L (3) 22.4 L (4) 44.8 L

20. The volume occupied by 3.01×10^{23} molecules of NO_2 gas at STP is closest to (1) 0.500 L (2) 1.00 L (3) 11.2 L (4) 22.4 L.

21. According to the reaction $N_2(g) + 3\ H_2(g) \longrightarrow 2\ NH_3(g)$, how many liters of hydrogen are required to produce exactly 3.0 liters of ammonia? (1) 1.5 (2) 2.0 (3) 4.5 (4) 6.0

22. What is the total volume, in liters, occupied by 56.0 grams of nitrogen gas at STP? (1) 11.2 (2) 22.4 (3) 33.6 (4) 44.8

23. Given the reaction $C_3H_8(g) + 5\ O_2(g) \longrightarrow 4\ H_2O(g) + 3\ CO_2(g)$, what is the total number of liters of CO_2 produced when 150 liters of O_2 react completely with C_3H_8? (1) 3.0 (2) 90 (3) 150 (4) 250

24. Eleven grams of a gas occupy 5.6 liters at STP. What is the molecular mass of this gas? (1) 11 (2) 22 (3) 44 (4) 88

25. What is the mass of 1.00 mole of a gas if 28.0 grams of this gas occupy 22.4 L at STP? (1) 1.0 g (2) 1.25 g (3) 22.4 g (4) 28.0 g

26. What is the mass in grams of 22.4 liters of O_2 gas at STP? (1) 8 (2) 16 (3) 32 (4) 64

27. Fourteen grams of a gas occupy 11.2 liters at STP. The gas may be (1) carbon monoxide (2) hydrogen sulfide (3) hydrogen chloride (4) sulfur dioxide.

28. A liter of chlorine at STP has a mass of approximately (1) 0.3 g (2) 1 g (3) 1.5 g (4) 3 g.

29. Given the reaction $H_2(g) + I_2(s) \longrightarrow 2\ HI(g)$, what is the volume of hydrogen required to produce 22.4 liters of HI at STP? (1) 1.00 L (2) 2.00 L (3) 11.2 L (4) 22.4 L

30. What is the volume occupied by 2.00 grams of helium at STP? (1) 22.4 L (2) 11.2 L (3) 4.00 L (4) 2.00 L

31. What is the volume occupied by 9.03×10^{23} molecules of an ideal gas at STP? (1) 14.9 L (2) 22.4 L (3) 33.6 L (4) 67.2 L

32. Which gas has a density of 1.34 grams/liter at STP? (1) NO_2 (2) NO (3) N_2 (4) H_2

33. If 6.02×10^{23} molecules of N_2 react according to the equation $N_2 + 3\ H_2 \longrightarrow 2\ NH_3$, the total number of molecules of NH_3 produced is (1) 1.00 (2) 2.00 (3) 6.02×10^{23} (4) 12.0×10^{23}.

34. The number of atoms in 2 grams of calcium is equal to

(1) $\dfrac{2 \times 6.02 \times 10^{23}}{40}$ (3) $\dfrac{6.02 \times 10^{23}}{2 \times 40}$

(2) $\dfrac{40 \times 6.02 \times 10^{23}}{2}$ (4) $2 \times 40 \times 6.02 \times 10^{23}$

35. What are the products of the electrolysis of one mole of water at STP? (1) 11.2 L of O_2 and 22.4 L of H_2 (2) 22.4 L of O_2 and 22.4 L of H_2 (3) 16 g of O_2 and 8 g of H_2 (4) 32 g of O_2 and 2 g of H_2

36. If the density of gas X at STP is 1.00 grams/liter, the mass of one mole of this gas is (1) 1.00 g (2) 2.00 g (3) 11.2 g (4) 22.4 g.

37. The number of atoms of hydrogen in 1.00 mole of NH_3 is equal to (1) 6.02×10^{23} (2) $2(6.02 \times 10^{23})$ (3) $3(6.02 \times 10^{23})$ (4) $4(6.02 \times 10^{23})$.

38. A 60-gram sample of $LiCl \cdot H_2O$ is heated in an open crucible until all of the water has been driven off. What is the total mass of LiCl remaining in the crucible? (1) 18 g (2) 24 g (3) 42 g (4) 60 g

CHEMISTRY CHALLENGE

The following questions will provide practice in answering SAT II-type questions.

For each question in this section, one or more of the responses given are correct. Decide which of the responses is (are) correct.

Then choose

(a) if only I is correct;
(b) if only II is correct;
(c) if only I and II are correct;
(d) if only II and III are correct;
(e) if I, II, and III are correct.

Summary: Choice	(a)	(b)	(c)	(d)	(e)
Correct statement(s):	I	II	I & II	II & III	All

1. The volume of one mole at STP is 22.4 liters for
I. water
II. neon
III. nitrogen.

2. In the reaction $N_2 + 3 H_2 \longrightarrow 2 NH_3$ the quantity of hydrogen needed to produce 34 grams of NH_3 at STP is
I. 3.0 moles
II. 67.2 L
III. 3.0 g.

3. The possible molar mass of a substance with the empirical formula CH_2 is
I. 28 g
II. 56 g
III. 140 g.

4. Two moles of nitrogen gas at STP has
I. a density of 0.625 g/L
II. a volume of 44.8 L
III. a mass of 28 g.

5. The percent composition of $C_6H_{12}O_6$ by mass is
I. 40 percent carbon
II. 50 percent hydrogen
III. 25 percent oxygen.

6. Which is (are) equivalent to 0.500 mole of Ar gas at 273 K and a pressure of 2.00 atm?
I. 5.6 L
II. 3.01×10^{23} atoms
III. 20 g

7. In the reaction

$$2 CO + O_2 \longrightarrow 2 CO_2$$

the maximum quantity of CO_2 obtainable when 4 moles of CO are reacted with 1.5 moles of O_2 at STP is

I. 2 moles

II. 132 g

III. 89.6 L.

8. The volume of 0.010 mole of He at STP is

I. 0.04 g

II. 0.224 L

III. 224 mL.

9. The quantity of gas needed to completely burn 12 grams of carbon to produce CO_2 at STP is

I. 32 g O_2.

II. 22.4 L O_2.

III. 112 L air.

10. Which molecule(s) is (are) 50% sulfur by mass?

I. SO

II. SO_2

III. S_2O_4

Solutions

LOOKING AHEAD

Many of the chemicals you use, both in the laboratory and at home, are not pure substances; they are aqueous solutions. In this chapter you will study the dissolving process and some of the types of solutions that result. You will also learn about the quantitative aspects of solutions, because the properties of solutions depend not only on their composition, but also on their concentration—the relative proportions of solute and solvent.

When you have completed this chapter, you should be able to:

- **Define** solution, solute, solvent, solubility, and saturation.
- **Describe** the effects of temperature and pressure on the solubility of various solutes.
- **Define** and **use** concentration units such as molarity, percent by mass, and parts per million.
- **Describe** the effect of a solute on the vapor pressure, boiling point, and freezing point of a liquid solvent.
- **Use** and interpret solubility curves.

What Is a Solution?

A **solution** is a homogeneous mixture consisting of a solvent and a solute. The **solvent** is what does the dissolving. The **solute** is what is dissolved. When substances of different phases, such as a solid and a liquid, are mixed to form a solution, the solvent retains its phase, while the phase of the solute changes. In a water solution of salt, for example, the solution is a liquid. Therefore water, the liquid, is the solvent, while salt, which was originally a solid, is the solute. When the substances that are mixed to form a solution start out in the same phase, the substance present in greater quantity is generally called the solvent. However, the distinction is not important.

You learned that a homogeneous mixture is uniform in composition—it has only one phase. In other words, a homogeneous mixture is a uniform system. Solutions are uniform systems. They are therefore considered to be homogeneous. If you compare samples of equal volume taken from a salt solution, the quantities of the solvent—water—and the solute—salt—are the same in each sample. The salt solution is homogeneous.

Solutions contain one distinguishable, or visible, phase. This means that you see only one phase with the naked eye or even with the help of a magnifying lens. This is called a visible, or macroscopic, view of a solution. Such a view might suggest that only one kind of matter is present in a solution, because all you see is a single liquid. If you could actually see the particles in a solution, however, you would find two kinds of matter: solute and solvent particles (molecules or ions).

Systems that contain more than one distinguishable substance are called mixtures. Consider a mixture of salt and sugar. No matter how finely ground the salt and sugar are, you can still distinguish two substances in the mixture: salt and sugar. You can distinguish the substances because boundaries separate the parts that make up the mixture. No boundaries can be detected in a solution.

Types of Solutions

There are many different types of solutions. The most common type is solids dissolved in liquids. Other types of solutions include:

1. Solutions of gases in liquids. An example is carbon dioxide gas dissolved in water (carbonated water).
2. Solutions of liquids in liquids. Alcohol dissolved in water is a solution of this type.
3. Solutions of gases in gases. Nitrogen and oxygen are dissolved in each other in air.
4. Solutions of gases in solids. An example is air dissolved in ice.
5. Solutions of solids in solids. Copper is dissolved in zinc in the alloy called brass.
6. Solutions of solids in gases or solutions of liquids in gases. Iodine vapor is a solution of iodine particles in air. Carbon tetrachloride vapor is a solution of particles of carbon tetrachloride in air. Both of these types of solutions are generally considered to be solutions of gases, or vapors, in gases.

The Nature of the Dissolving Process

When two elements react and form a compound, the properties of the compound are generally different from the properties of the elements that make up the compound. For example, sodium, a very reactive metal, combines with chlorine, a poisonous gas, and forms sodium chloride, or table salt, a common and useful part of our diet.

Solutions are different from compounds in this respect. A solution has some of the properties of the solute and some of the properties of the solvent. For example, potassium chromate, K_2CrO_4, is a yellow compound. A solution of K_2CrO_4 in water is yellow like the solute, K_2CrO_4. The solution also has many of the properties of the solvent, water. On the other hand, the boiling point, freezing point, and density of a solution are different from those of either the solute or the solvent.

Energy changes accompany the dissolving process. Heat may be given off (an exothermic change), or heat may be absorbed (an endothermic change). These energy changes indicate that there are interactions between the solvent and solute particles. We will now examine the nature of some of these interactions.

Solid in Liquid Solutions

When a solid dissolves in a liquid, the bonds that hold it together in the solid state are broken, and the resulting particles are surrounded by the liquid molecules. If the solid is ionic, for example, NaCl, the ions separate on dissolving. Molecular substances, such as sugar, break up into individual molecules, which are then surrounded by the solvent molecules. In either case, energy is required to overcome the attractions within the solid. The energy needed to separate the solid particles, whether they are ions or molecules, is called the **crystal lattice energy.**

The attractions between the solute particles and the solvent particles result in a release of energy. In aqueous solutions, this energy is called the **hydration energy.** When a solid dissolves in water, the crystal lattice energy is absorbed and the hydration energy is released. The attractions within the solid tend to prevent it from dissolving, while the attractions between the solid and the liquid tend to favor dissolving.

Ionic solids are held together by strong interionic attractions. They cannot dissolve appreciably unless there are also strong attractions between the ions and the solvent molecules. Let us examine what happens when NaCl dissolves in water. Recall that water molecules are highly polar, with negative charge on the oxygen side of the molecule, and positive charge on the hydrogen side. The water molecules can surround the Na^+ and Cl^- ions, as shown in Figure 8-1. The attractions between the water molecules and the ions are strong enough to permit the ions to separate, and form a solution. The ions shown in Figure 8-1 are said to be **hydrated**—surrounded by water molecules—and are often represented with the symbols $Na^+(aq)$ and $Cl^-(aq)$. Water is the only

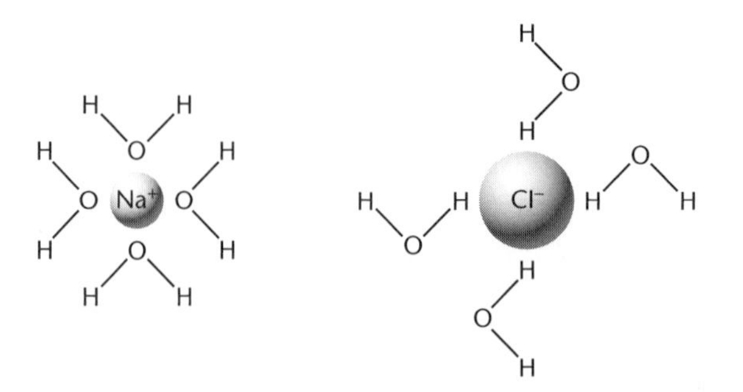

Figure 8-1 Hydrated sodium and chloride ions

commonly used solvent that is polar enough to permit ionic solids to dissolve in it.

Sugar, while not ionic, contains molecules which, like water molecules, are highly polar. The negative side of a sugar molecule is attracted by the hydrogen on the water, while the positive side is attracted by the oxygen. Like many polar substances, sugar is very soluble in water.

Solid iodine, I_2, is a nonpolar molecule. It is only slightly soluble in water, because the attractions within the water (hydrogen bonds) are much stronger than the attractions between the water and the iodine. However, iodine is extremely soluble in carbon tetrachloride, CCl_4. Carbon tetrachloride is a nonpolar molecule. The weak intermolecular attractions within the carbon tetrachloride permit other nonpolar molecules to dissolve.

Chemists often predict solubility by using a simple rule: like dissolves like. This means that nonpolar materials generally dissolve best in nonpolar solvents. Ionic and polar substances generally dissolve best in polar solvents.

Liquid in Liquid Solutions

When water and ethanol (alcohol) are mixed in any quantity they form a homogeneous mixture. Liquids that are completely soluble in one another are called *miscible* liquids. The nonpolar liquid benzene, C_6H_6, is miscible with carbon tetrachloride, CCl_4. However, when water is mixed with carbon tetrachloride, the liquids form two distinct layers (see Figure 8-2). Such liquids are said to be *immiscible.* Water is also immiscible with benzene, as shown in Figure 8-2. In each case, the top layer contains the liquid with the lower density.

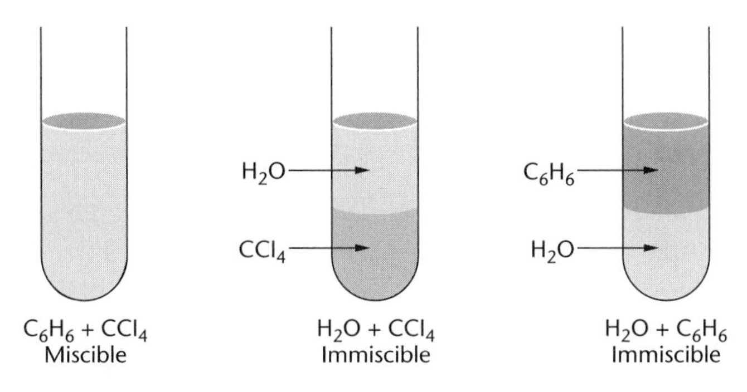

Figure 8-2 Miscible and immiscible liquids

Since water is highly polar, it tends to be miscible only with liquids that also are highly polar. It is immiscible with nonpolar liquids, such as benzene, carbon tetrachloride, and gasoline. Once again, in general, like dissolves like.

PRACTICE

8.1 When iodine crystals, I_2, are added to carbon tetrachloride, CCl_4, they dissolve readily to form a deep purple solution. When I_2 is added to water, it dissolves only slightly, producing a pale, yellow-brown mixture. If these two solutions are mixed together in a test tube and shaken, the result is shown below.

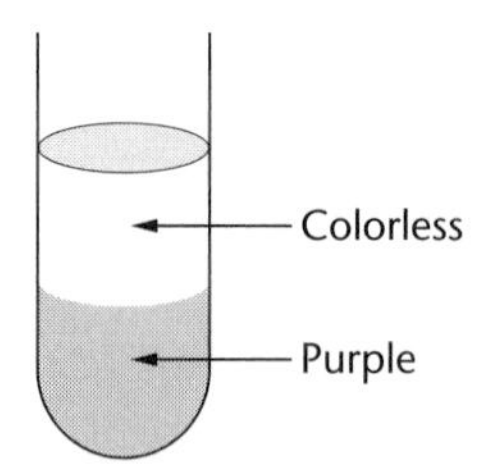

(a) Why is iodine more soluble in CCl_4 than it is in water?

(b) Why is the top layer of the mixture shown in the diagram colorless?

(c) Why are two separate layers formed in this mixture?

(d) Why is the water the top layer?

(e) If the experiment is repeated using benzene instead of CCl_4, how will the results be different?

Solutions of Gases in Liquids

Our method of predicting solubility works well for solutions of gases in liquids as well. Which should be more soluble in water, O_2 or NH_3? Since ammonia is polar, and oxygen is nonpolar, we would predict that ammonia should be more soluble in water, a polar solvent. In fact, under the same conditions ammonia is more than 10,000 times more soluble in water than is O_2.

PRACTICE

8.2 The following mixtures are prepared:

(1) sulfur and water

(2) iodine and benzene

(3) benzene and carbon tetrachloride

(4) water and gasoline

(a) Which is a mixture of immiscible liquids?

(b) Which is a mixture of miscible liquids?

(c) Which mixture contains an insoluble solid?

8.3 State whether each of the following is more soluble in (1) water or (2) carbon tetrachloride, and briefly explain your answer.

(a) HCl gas

(b) Br_2 liquid

(c) $Ni(NO_3)_2$ solid

Expressing Solubility

Sometimes you will see a compound described as "soluble," "slightly soluble," or "insoluble." Solubility expressed in these terms is sometimes too vague to be useful. It is preferable to express how much solute dissolves in a given amount of solvent at a specific temperature. This information can be determined experimentally or found in a reference table. In either case, solubility refers to the maximum amount of solute that can be dissolved in a given amount of solvent at some specific temperature. As you will soon learn, this information tells you about solubility in terms of saturated solutions.

Solubility can be expressed precisely in various ways, depending on the units that are most convenient to use. Following are some examples:

grams of solid solute per 100 grams of liquid solvent

grams of gaseous solute per 1000 mL of liquid solvent

milliliters of liquid solute per liter of liquid solvent

molarity or molality (These terms are defined on pages 299 and 304.)

parts per million

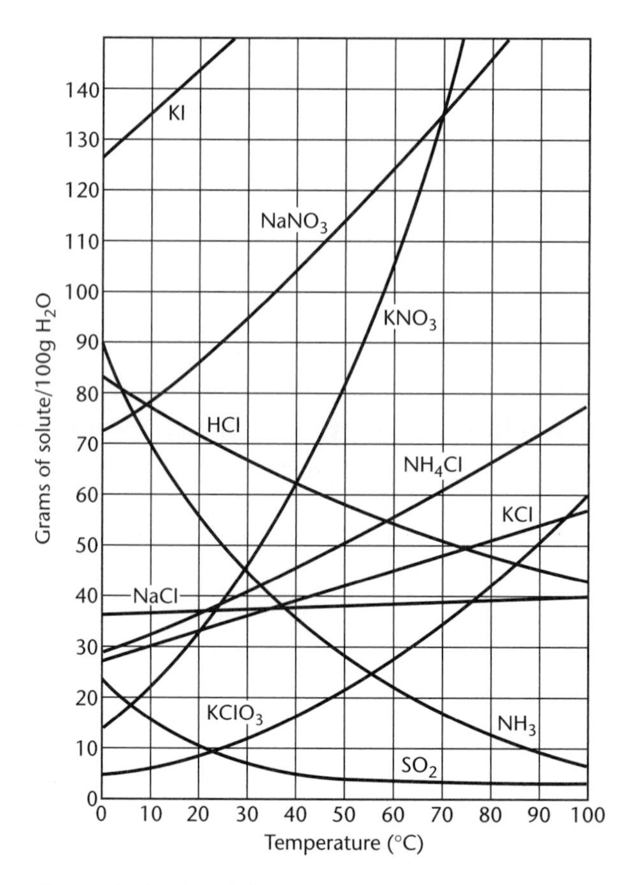

Figure 8-3 Solubility curves for some compounds

Solubility information can be shown graphically, as in Figure 8-3. Each curve shows the mass of solute that will dissolve in 100 grams of water at different temperatures.

Look at the solubility of $NaNO_3$ in 100 grams of water at 10°C. According to the graph, 79 grams of $NaNO_3$ can dissolve at this temperature. Suppose 1 additional gram of $NaNO_3$ solid were added to this solution (80 grams of $NaNO_3$ in 100 grams of water) while the temperature was kept at 10°C. What would you observe? The excess solid would settle to the bottom of the container. If this solid were separated from the solution by filtering, then dried and massed, it would have a mass of 1 gram.

At 10°C, 100 grams of water can dissolve only 79 grams of $NaNO_3$. At this temperature, the solution is **saturated.** In other words, the solution contains the maximum amount of that solute that can be dissolved at this temperature. At 40°C, an additional 25 grams of $NaNO_3$, or a total of 105 grams, are needed to saturate 100 grams of water.

If a saturated solution is cooled, the excess solute generally crystallizes. For example, at 30°C, the solubility of $NaNO_3$ is 95 g/100 g of H_2O. At 20°C, the solubility is 87 g/100 g of H_2O. If a saturated solution of $NaNO_3$ in 100 g of H_2O is cooled from 30°C to 20°C, 8 grams of solute crystallize out.

When saturated solutions of solids in water are cooled, and the solid substance becomes less soluble, crystallization normally occurs. However, under special conditions and with specific solutes (such as sodium acetate), cooling a saturated solution may not cause the excess solute to crystallize. The cooled solution then contains more solute than it normally should under the new conditions. Solutions that contain more of a given dissolved solute than will normally dissolve at a given temperature are said to be **supersaturated.** Supersaturated solutions are very unstable. The addition of one tiny solute crystal will destroy a supersaturated solution, and all of the excess solute will crystallize out.

Solutions that contain no undissolved material may be saturated, unsaturated, or supersaturated. To determine which is the case, you add one small additional solute crystal to the mixture. If the solution is unsaturated, the crystal will dissolve. If the solution is saturated, the crystal will remain undissolved. If the solution is supersaturated, additional crystals will form as the supersaturation is destroyed. The resulting solution will be saturated.

In summary, a *saturated solution* is a solution containing the maximum solute that will normally dissolve at a given temperature and pressure. An **unsaturated solution** is a solution that can dissolve more solute at a given temperature and pressure. A *supersaturated solution* is a solution containing more than the maximum solute that will normally dissolve at a given temperature and pressure.

Saturated Solutions

If you add sugar to water, at first it dissolves. However, if you continue to add sugar, the solution eventually becomes saturated, and will dissolve no more sugar at that temperature. Why does this happen? A common misconception is that the water has become full, and has no more room for additional solute. You can test this idea easily at home. Add a large quantity of sugar to a small glass of water, and stir thoroughly for a few minutes. The solution is now saturated with sugar, and has undissolved crystals sitting on the bottom. Add some unsweetened instant iced tea mix to the mixture, and stir. The tea dissolves readily. Obviously, the saturated sugar solution still has room for the tea. (Or for any other soluble material you might wish to try.) A saturated sugar solution is saturated only for sugar.

A saturated solution is actually at equilibrium. The rate of dissolving has become equal to the rate of crystallization, so no observable change occurs. As sugar is added to water, at first the rate of dissolving is much greater than the rate of crystallization, so only dissolving is observed. As more and more sugar goes into solution, the rate of crystallization increases, since sugar molecules can "find" one another more easily. Eventually, the rate of crystallization becomes equal to the rate of dissolving. For every sugar molecule that dissolves, one crystallizes, so that no change in the solution is observable. The solution has reached equilibrium and is now saturated. No additional sugar will dissolve, but any other soluble material will! If a saturated sugar solution is cooled, the rate of dissolving decreases, so that it becomes slower than the rate of crystallization. Crystals of sugar form in the solution.

PRACTICE

8.4 A single crystal of sodium thiosulfate, $Na_2S_2O_3$, is added to a clear solution of that substance. The crystal grows larger, and then a large number of additional crystals form. A considerable amount of heat is released during the crystallization process.

(a) Was the original solution saturated, unsaturated or supersaturated? How do you know?

(b) Is the dissolving of sodium thiosulfate in water endothermic or exothermic? How do you know?

Using Solubility Curves

Refer to the solubility curves on page 290. Each curve shows the quantity of solute that will saturate 100 grams of water at a given temperature. Although the chart expresses the solubility in grams per 100 grams of water, it can be used to find the solubility of a solute in any quantity of water, by setting up a simple proportion.

SAMPLE PROBLEM

PROBLEM

How many grams of $KClO_3$ are needed to saturate 250 g of water at 48°C?

SOLUTION

Using the chart, you can see that at 48°C, the solubility of $KClO_3$ in 100 g of water is 20 g. If 20 g of solute saturates 100 g of water, then how much will saturate 250 g? Set up the proportion:

$$\frac{20 \text{ g } KClO_3}{100 \text{ g } H_2O} = \frac{x \text{ g } KClO_3}{250 \text{ g } H_2O}$$

Solving the equation, you see that $x = 50$ g of $KClO_3$.

PRACTICE

8.5 At what temperature will 90 g of $NaNO_3$ exactly saturate 100 g of water?

8.6 A solution is made by adding 60 g of NH_4Cl to 100 g of water at 80°C. Is the solution saturated, unsaturated, or supersaturated?

8.7 A saturated solution of $KClO_3$ is prepared in 100 g of water at 70°C. The solution is then cooled to 30°C. How many grams of $KClO_3$ crystallize out? (The solution does not become supersaturated.)

8.8 How many grams of NaCl are needed to form a saturated solution in 25 g of water at 100°C?

8.9 (For experts) A solution was made by adding 80 g of $NaNO_3$ to 100 g of water at a temperature of 24°C. The solution was kept in an open container at that temperature, until crystals began to form.

(a) Why did crystals form?

(b) How many grams of water remained in the container when the crystals first began to form?

Factors That Influence Solubility

A ball rolls unaided down a smooth hill, losing potential energy. This is called a spontaneous change because, once the change—rolling—begins, no additional energy is needed to keep it going. The ball at the bottom of the hill has less potential energy than it had before reaching the bottom. The ball is therefore more stable at the bottom of the hill.

In a similar way, potential energy in the form of heat is released in an exothermic change. Thus exothermic changes should proceed spontaneously. For example, a dissolving process that is exothermic is like a ball rolling downhill—it should proceed spontaneously. Conversely, a dissolving process that is endothermic is like a ball trying to roll uphill—it should *not* proceed spontaneously. Experiments have shown, however, that both exothermic and endothermic processes may occur spontaneously. This fact raises a question: Is the dissolving of a solid in water, or any other change, governed by energy factors alone?

Energy and Randomness

To resolve the conflict between what we think should happen and what actually happens, let us assume that energy is not the only factor that influences solubility. The dissolving of a solid in water is, in fact, governed by another factor, in addition to the energy factor. This new factor, it turns out, is one of the most important in nature—*randomness*. Randomness is the tendency of matter to spread out, or to become as disordered as possible. What does this mean?

A scientist naming the state of a substance—solid, liquid, or gas—is describing the degree to which the particles of the substance are ordered. The opposite description, that is, the degree of disorder, or randomness, of the particles, can also be used. Substances are most ordered in the solid state, less ordered in the liquid state, and least ordered in the gaseous state. This is because the attractive forces are greatest in a solid and least in a gas. Thus the particles in a solid have the least degree of randomness, and the particles in a gas have the maximum degree of randomness.

The term **entropy**, which is symbolized by *S*, is used to describe randomness. A gas has greater entropy than has a solid.

When a solid dissolves, the attractive forces are weakened, and randomness increases. The crystal lattice breaks up and the particles become more mobile—more randomly arranged. If the solution process is endothermic, the solution cools as the process goes on. The solution has higher potential energy and lower stability than have the solute and the solvent. The dissolving process should therefore not occur spontaneously. However, during the dissolving process, randomness increases. If randomness increases enough, it outweighs the opposite effect of the energy change, and the process of solution proceeds spontaneously.

The changes in randomness and energy that occur during dissolving can be combined quantitatively, and used to accurately predict solubility. As we will see in Chapter 11, the course of most physical and chemical changes can be analyzed through calculations involving energy and entropy change.

Temperature

Look at the solubility curves in Figure 8-3 on page 290 again. Each curve describes the composition of a particular saturated solution at any specific temperature. For six of the ten solutes shown (KI, $NaNO_3$, KNO_3, NH_4Cl, KCl, and $KClO_3$), the curves rise. This indicates that the solubilities of the solids in water increase with an increase in temperature. For solid NaCl, the curve is nearly horizontal. This shows that little change in solubility takes place with increasing temperature. In other words, NaCl is just as soluble in hot water as it is in cold water. The curves for the gases NH_3, HCl, and SO_2 show a decrease in solubility with increasing temperature. This is typical of gases.

The effect of an increase in temperature on a physical or chemical change depends on whether the change is endothermic or exothermic. An increase in temperature will promote any change that is endothermic, and will inhibit any change that is exothermic. (This concept is more fully discussed in Chapter 10.) The dissolving of a solid in a liquid is usually endothermic, because more energy is absorbed in breaking the bonds within the solid, than is released when the solid interacts with the liquid solvent. The endothermic dissolving process is favored by an increase in temperature. Solids are usually most soluble at high temperatures.

Gases, however, have no crystal lattice energy. The dissolving of a gas is generally exothermic, and favored at low temperatures. The three gases, graphed in Figure 8-3 on page 290, are all most soluble at low temperatures.

Pressure

Liquids and solids are virtually incompressible. Thus an increase in pressure has little effect on the solubility of liquids and solids.

Gases, on the other hand, can easily be compressed. The solubility of a gas in water at constant temperature is proportional to the pressure of the gas. More gas will dissolve in water as the pressure of the gas is increased. In the manufacture of carbonated beverages, for example, the solubility of carbon dioxide in water is greatly increased by increasing the pressure on the gas.

The effect of pressure on the solubility of gases is demonstrated every time you open a bottle of carbonated beverage. Before you open the bottle, you observe that there are no bubbles forming in the liquid. The dissolved carbon dioxide has reached equilibrium at high pressure. When you open the bottle, you decrease the pressure, and the carbon dioxide becomes less soluble and begins to bubble out of the liquid. Eventually, the system will reach equilibrium at its new pressure and stop bubbling. The beverage has become "flat." By keeping the beverage cold and the bottle sealed, you are maintaining the conditions which maximize the solubility of a gas in a liquid—low temperature and high pressure.

PRACTICE

8.10 What is the usual effect of an increase in temperature on the solubility of solids in liquids?

8.11 What is the usual effect of an increase in temperature on the solubility of gases in liquids?

8.12 A sealed container holds carbon dioxide gas and water. What effect would decreasing the volume of the container have on the solubility of the carbon dioxide? Explain your answer.

8.13 Recently, when some miners were rescued from a deep mine, one of them developed the bends—a condition that occurs when nitrogen gas becomes less soluble in the blood, and begins to bubble out of solution. Why did the gas become less soluble when the miner was brought to the surface?

The Quantitative Composition of Solutions

Different quantities of solvent and solute can be used to make solutions. The solutions then have different concentrations—that is, the solutions have different compositions. The term *concentration* refers to the quantity of dissolved matter contained in a unit (of volume or of mass) of solvent or solution.

Several methods are used to express the concentration of a solution. The terms *dilute* and *concentrated* are used to give an approximate description of the concentration of a solution. The terms do not have precise meanings, but the following general statements about them can be made. A **dilute** solution contains a much smaller proportion of solute than of solvent. A **concentrated** solution contains a relatively high proportion of solute. If you like your tea stronger, you could dissolve a greater quantity of the solute, tea, in your cup of water. To make the tea weaker, you could use less tea, or more water. Thus the terms "strong" and "weak" used in describing tea correspond to the terms "concentrated" and "dilute" used in describing solutions. More precise methods of expressing the concentration of a solution involve the concepts of *percentage, parts per million, molarity,* and *molality.*

Percentage Concentration

The **percentage concentration** of a solution can be expressed as the parts by mass of solute per 100 parts by mass of total solution. A 10-percent salt solution by mass contains 10 grams of salt dissolved in 90 grams of water—that is, 10 grams of salt in 100 grams of solution.

The equation used to find the percent concentration by mass is

$$\% \text{ by mass} = \frac{\text{grams of solute}}{\text{grams of solution}} \times 100\%$$

Parts per Million

Solutions that are extremely dilute often have their concentrations expressed in parts per million by mass (ppm). **Parts per million** expresses concentration as parts by mass of solute per one million parts by mass of total solution. The equation used to find concentration in ppm is

$$\text{parts per million} = \frac{\text{grams of solute}}{\text{grams of solution}} \times 1{,}000{,}000$$

SAMPLE PROBLEM

PROBLEM

A sample of tap water with a mass of 200. grams is found to contain 0.040 gram of Ca^{2+} ion.

(a) What is the concentration of Ca^{2+} ion in the solution in percent by mass?

(b) What is the concentration of Ca^{2+} ion in the solution in parts per million?

SOLUTION

(a) To find the percent of Ca^{2+} by mass, use the equation

$$\% \text{ by mass} = \frac{\text{grams of solute}}{\text{grams of solution}} \times 100$$

$$\% \text{ by mass} = \frac{0.040 \text{ gram}}{200 \text{ grams}} \times 100 = 0.020\%$$

(b) To find the concentration in parts per million, use the equation

$$\text{parts per million} = \frac{\text{grams of solute}}{\text{grams of solution}} \times 1{,}000{,}000$$

$$\text{ppm} = \frac{0.040 \text{ gram}}{200 \text{ grams}} \times 1{,}000{,}000 = 200 \text{ ppm}$$

PRACTICE

8.14 Find the concentration of solute in each of the following solutions in parts per million.

(a) 0.0040 gram of sodium chloride in 1000. grams of solution

(b) 0.035 gram of sodium in 250 grams of soda.

8.15 According to the label, a certain brand of cola soda contains 27 grams of sugar per 250 grams of soda. What is the percent of sugar, by mass, in that soda?

8.16 The concentration of sodium ions in a certain brand of mineral water is 40. parts per million. If a bottle of this water contains 250. grams of the mineral water, what is the total mass of the sodium ions in the bottle?

Molarity

Chemical reactions result from the collisions of particles. Chemists therefore often need to know how many particles of solute are dissolved in a given volume of solution. Calculations that involve *molar concentration,* or **molarity**, are useful in these situations. A 1-molar (1 M) solution contains 1 mole (molar mass) of solute dissolved in 1 liter of solution. Expressed as a formula,

$$\text{molarity} = \frac{\text{moles of solute}}{\text{liters of solution}} \quad \text{or} \quad M = \frac{\text{moles}}{\text{L}}$$

If you know two of the variables in the formula, you can calculate the third.

SAMPLE PROBLEMS

PROBLEM

1. If 2.0 g of NaOH are dissolved in enough water to make 200. mL of solution, what is the molarity of the solution?

SOLUTION

Mass and volume are given; molarity is required. To calculate molarity according to the formula, you must first convert grams of solute to moles and milliliters of solution to liters.

$$\text{moles of solute} = \frac{2\ \cancel{g}}{40\ \cancel{g}/\text{mole}} = 0.050\ \text{mole}$$

$$\text{liters of solution} = \frac{200.\ \cancel{mL}}{1000.\ \cancel{mL}/\text{L}} = 0.200\ \text{L}$$

$$\text{molarity} = \frac{0.050\ \text{mole}}{0.200\ \text{L}} = 0.25\ M$$

PROBLEM

2. How many moles of NaOH are contained in 200 mL of 0.25 M NaOH?

SOLUTION

Molarity and volume are given; moles are required. After converting milliliters to liters, as in Sample Problem 1, you can solve the molarity formula.

$$M = \frac{\text{moles}}{\text{L}}$$

$$0.25\ M = \frac{x\ \text{mole}}{0.2\ \text{L}}$$

$$x = 0.05\ \text{mole}$$

PROBLEM

3. How many milliliters of 0.25 M NaOH must be taken from a stock bottle to obtain 2.0 g of NaOH?

SOLUTION
Molarity and mass are given; volume is required. After converting grams of solute to moles, you can solve for liters of solvent. Then, you can convert liters of solvent to milliliters.

$$\text{moles of NaOH} = \frac{2\ \cancel{g}}{40\ \cancel{g}/\text{mole}} = 0.05\ \text{mole}$$

$$0.25\ M = \frac{0.05\ \text{mole}}{x\ \text{L}}$$

$$x = 0.2\ \text{L}$$

$$0.2\ \cancel{L} \times \frac{1000\ \text{mL}}{\cancel{L}} = 200\ \text{mL}$$

PROBLEM

4. How do you prepare 200 mL of 0.25 M NaOH solution?

SOLUTION
Find the number of moles of solute you need, and then convert to grams.

Let x = moles of solute required.

$$0.25\ M = \frac{x\ \text{mole}}{0.2\ \text{L}}$$

$$x = 0.05\ \text{mole}$$

$$\text{Grams of solute} = 0.05\ \cancel{\text{mole}} \times 40\text{g}/\cancel{\text{mole}} = 2\ \text{g}$$

To prepare the solution, weigh out 2 grams of NaOH and dissolve in enough water to make a final volume of 200 milliliters. (If you were actually to do this, you would first dissolve the solid in a volume less than 200 mL—in, say, 150 mL. Then you would add enough water to bring the final volume to exactly 200 mL.)

PROBLEM

5. A chemist is sometimes interested in the molar concentration of a specific ion. For example, what is the concentration of Cl^- in a solution labeled 0.2 M $CaCl_2$?

SOLUTION

$CaCl_2$ dissolves in water according to the equation

$$CaCl_2\,(s) \xrightarrow{H_2O} Ca^{2+}(aq) + 2\,Cl^-(aq)$$

For each unit of $CaCl_2(s)$ that dissolves, 2 Cl^- ions are in solution. If 1 mole of $CaCl_2(s)$ were dissolved to make 1 liter of solution, the solution would contain 1 mole of Ca^{2+} ions and 2 moles of Cl^- ions.

If 0.2 mole of $CaCl_2(s)$ is dissolved to make 1 liter of solution, the concentration of Ca^{2+} is 0.2 *M* and the concentration of Cl^- is 0.4 *M*.

PRACTICE

8.17 Find the molarity of the following aqueous solutions.

(a) 2.0 moles of HCl in a volume of 500. mL

(b) 20.0 g of NaOH in a volume of 2.0 L

(c) 23 g of C_2H_5OH in a volume of 500. mL

8.18 How many grams of solute are needed to make each of the following aqueous solutions?

(a) 4.00 L of 2.00 *M* HNO_3

(b) 200. mL of 4.00 *M* glucose (Molar mass of glucose = 180 g/mole.)

8.19 What is the molarity of sodium ion, Na^+, in 50.0 mL of a solution containing 10.6 g of sodium carbonate, Na_2CO_3?

Dilution Problems

We have used the relationship,

$$M = \frac{\text{moles}}{\text{L}}, \text{ or molarity} = \frac{\text{moles solute}}{\text{liters of solution}}$$

If we are solving for moles, this equation becomes

$$\text{moles} = M \times \text{L, or moles} = \text{molarity} \times \text{liters.}$$

When we dilute a solution by adding additional solvent, we do not change the number of moles of solute. Therefore, the molarity × liters in the original solution is the same as the molarity times liters in the diluted solution. This relationship is expressed in the dilution equation

$$M_o \times V_o = M_f \times V_f$$

where M_o and V_o are the *original* molarity and volume, and M_f and V_f are the *final* molarity and volume.

SAMPLE PROBLEMS

PROBLEM

1. How much water must be added to 100 mL of a 0.2 *M* solution to obtain a 0.10 *M* solution?

SOLUTION

In this problem, the original solution is being diluted with water. You can use the dilution equation in which M_o and V_o are the original molarity and volume, and M_f and V_f are the final molarity and volume. Original volume, the original molarity and final molarity are given; final volume is required.

$$M_o \times V_o = M_f \times V_f$$

$$0.2\ M \times 100\ \text{mL} = 0.1\ M \times x\ \text{mL}$$

$$x = 200\ \text{mL}$$

The final volume of the solution must be 200 mL. However, on careful reading of the question, you find that you are asked for the amount of water to be added. Since the original volume was 100 mL, and the final volume is 200 mL, you need to add 100 mL of water.

PROBLEM

2. If 400 mL of H_2O are added to 200 mL of 0.6 *M* NaOH, what is the molarity of the new solution?

SOLUTION

Once again, you can use the dilution equation. The original volume, the original molarity, and the amount of water added are given; the final volume and the final molarity are required. The final volume is 600 mL, because 400 mL of water was added to 200 mL of solution.

$$M_o \times V_o = M_f \times V_f$$

$$0.6\ M \times 200\ \text{mL} = x\ M \times 600\ \text{mL}$$

$$x = 0.2\ M$$

PRACTICE

8.20 A chemist has 40. mL of 6.0 *M* NaOH that he wishes to dilute to a concentration of 0.50 *M*. What should be the final volume of his solution?

8.21 A chemist carefully pours 50. mL of 4.0 *M* sulfuric acid into a beaker of water. The total, final volume in the beaker is 200. mL. What is the new concentration of the acid?

8.22 A student added 50.0 mL of 2.0 *M* aqueous HCl to 450 mL of water. What is the molarity of the new solution?

8.23 (For experts) A chemist mixed 50.0 mL of 2.0 *M* HCl with 100. mL of 3.0 *M* HCl, and then added an additional 100. mL of water. What is the molarity of HCl in the resulting solution?

Molality

Calculations involving molar concentration, or molarity, are used when chemists have to know how many particles of *solute* are dissolved in a given volume of solution. Sometimes they must know how many particles of *solvent* are in a solution, or they need to know the *total number* of particles (solute *and* solvent) in a solution. In these cases, their calculations involve *molal concentration,* or **molality.**

A 1-molal solution (1 *m*) contains 1 mole of solute dissolved in 1000 grams, or 1 kg, of solvent. When the solvent is water, since 1 kg of water has a volume of 1 liter, the molarities and molalities of dilute solutions are approximately the same. In more concentrated aqueous solutions, the solute occupies a significant portion of the volume, so that 1 liter of solution contains significantly less than 1 kg of water. In such cases, a 1.0 *M* solution is more concentrated than a 1.0 *m* solution.

You can solve problems involving molality with an equation very similar to the one used for molarity problems.

$$\text{molality} = \frac{\text{moles of solute}}{\text{kilograms of solvent}} \quad \text{or} \quad m = \frac{\text{moles}}{\text{kg}}$$

SAMPLE PROBLEMS

PROBLEM

1. If 1.80 g of glucose, $C_6H_{12}O_6$, are dissolved in 50. g of H_2O, what is the molality of the solution?

SOLUTION

First, find the moles of glucose.

$$\text{moles of solute} = \frac{1.80\ \cancel{g}}{180\ \cancel{g}/\text{mole}} = 0.010\ \text{mole}$$

Then, convert grams of solvent to kilograms.

$$\text{kilograms of solvent} = \frac{50\ \cancel{g}}{1000\ \cancel{g}/\text{kg}} = 0.050\ \text{kg}$$

Now, use the formula to find the molality of the solution.

$$\text{molality} = \frac{0.010\ \text{mole}}{0.050\ \text{kg}} = 0.20\ m$$

PROBLEM

2. If 0.050 mole of sulfur is dissolved in 100. mL of benzene, C_6H_6, what is the molality of the solution? The density of benzene is 0.88 gram/milliliter.

SOLUTION

Begin by finding the kilograms of benzene.

$$\text{kilograms of solvent} = 100.\ \cancel{\text{mL}} \times 0.88\ \text{g}/\cancel{\text{mL}}$$

$$= 88\ \text{g} = 0.088\ \text{kg}$$

Now, use the formula to find the molality.

$$\text{molality} = \frac{0.050\ \text{mole}}{0.088\ \text{kg}} = 0.57\ m$$

PRACTICE

8.24 Find the molality of a solution that is made up of 34.2 g of sucrose ($C_{12}H_{22}O_{11}$) dissolved in 250 g of water.

8.25 If 0.0750 mole of iodine is dissolved in 300. mL of benzene (density = 0.880 g/mL), what is the molality of the solution?

Some Properties of Solutions

Some solutions conduct electricity, and others do not. Solutions boil at higher temperatures and freeze at lower temperatures than do the pure solvents. The number of particles in solution accounts for these properties of solutions. Let us see how.

Electrolytes and Nonelectrolytes

For a substance to conduct electricity, carriers of charge must be present. These charge carriers may be electrons, or they may be ions that are free to move about. Free electrons are present only in the lattices of metals. All other conductors of electricity must contain mobile ions.

The ability of various kinds of matter to conduct electricity can be determined with a setup such as the one shown in Figure 8-4. When the switch is closed and the electrodes are connected by something that conducts electricity, the circuit is complete

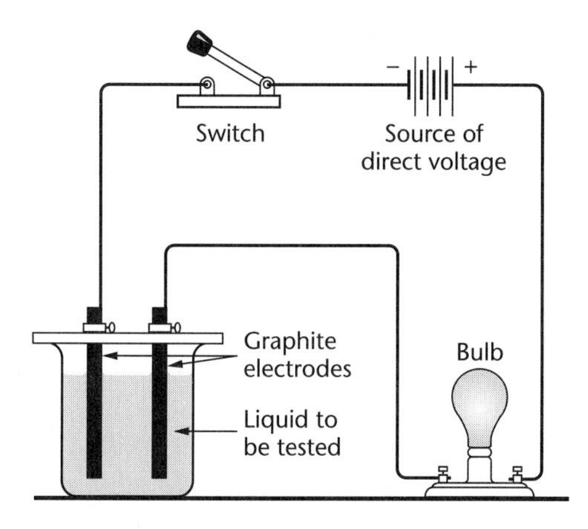

Figure 8-4 Apparatus for testing conductivity

and the bulb lights. Suppose that the electrodes are surrounded only by air. Can the air alone act as a conductor between the electrodes? No. The bulb does not light when the switch is closed. Air is a poor conductor of electricity. Now suppose that a piece of metal is placed across the electrodes. This time, when the switch is closed, the bulb glows brightly. The metal is a good conductor of electricity. Next, place the electrodes in a water solution of an ionic compound. What happens this time? The bulb lights, just as it did when the electrodes were connected by a strip of metal. The brightness of the bulb is a fairly good measure of the number of charge carriers in the solution. The solution, like the strip of metal, is a good conductor of electricity.

Compounds that conduct electricity in solution are called **electrolytes.** Electrolytes contain mobile, charged particles (ions) in the solution. Some electrolytes form large numbers of ions in solution and cause the bulb to glow brightly. Such electrolytes are called *strong electrolytes.* Other compounds, such as acetic acid (CH_3COOH) and ammonia (NH_3), are poor conductors in solution and are called *weak electrolytes.* The bulb just barely lights in these solutions. Some other compounds, such as sugar and glycerine, do not conduct electricity in solution. The bulb does not light at all in these solutions. Such compounds are called **nonelectrolytes.**

When an ionic solid dissolves, the ions separate and move about freely in the solution. (When an ionic solid melts, mobile ions also are formed as the crystal lattice breaks up.) These mobile ions are the charge carriers.

$$NaCl(s) \rightarrow Na^+(aq) + Cl^-(aq)$$

$$Cu(NO_3)_2(s) \rightarrow Cu^{2+}(aq) + 2\ NO_3^-(aq)$$

$$(NH_4)_2SO_4(s) \rightarrow 2\ NH_4^+(aq) + SO_4^{2-}(aq)$$

$$KOH(s) \rightarrow K^+(aq) + OH^-(aq)$$

$$Ca(OH)_2(s) \rightarrow Ca^{2+}(aq) + 2\ OH^-(aq)$$

Covalent compounds, such as sugar or glycerine, form hydrogen-bonded molecules when they dissolve. These molecules do not carry a charge, and the compounds are nonelectrolytes. Other covalent compounds react with molecules of the solvent and form ions. These ions are charge carriers. The following equations show the ions formed when certain covalent molecules dissolve in water:

$$HCl(g) + H_2O \rightarrow H_3O^+ + Cl^-(aq)$$

$$CO_2(g) + 2\ H_2O \rightarrow H_3O^+ + HCO_3^-(aq)$$

Boiling Points and Freezing Points

Water solutions containing a solute that does not vaporize easily (a nonvolatile solute), such as NaCl, boil at temperatures higher than the boiling point of pure water. The solutions freeze at temperatures lower than the freezing point of pure water. The elevation of the boiling point and the depression of the freezing point depend on the nature of the solute and on the concentration of the solution.

The **boiling point** is the temperature at which the vapor pressure of a liquid equals the pressure of the gas acting on the liquid. When water contains dissolved matter, the tendency of water molecules to leave the surface of the water is decreased. This is the same as saying that the vapor pressure of a solution is lower than the vapor pressure of the pure solvent. As a result, the temperature at which a solution boils is somewhat higher than the temperature at which the pure solvent boils. When the vapor pressure of a solvent has been increased enough to equal the pressure of the gas acting on it—that is, when the temperature has increased enough—the solution boils.

The **freezing point** of a liquid is the temperature at which the liquid and its solid form have the same vapor pressure. The vapor pressure of a solution is lower than the vapor pressure of pure water. It follows, then, that the freezing point of the solution will be somewhat lower than the freezing point of pure water.

Elevation of the boiling point and depression of the freezing point depend on the composition of the solution and on the concentration of the solution. For nonelectrolytes, the boiling point and the freezing point are proportional to the molal concentration of the solution.

The elevation of the boiling point per mole of solute is higher for electrolytes than it is for nonelectrolytes. For electrolytes, the elevation of the boiling point is roughly proportional to the molal concentration of the solute multiplied by the number of moles of ions formed from one mole of the solute. Similarly, the depression of the freezing point is roughly proportional to the molal concentration of the solute multiplied by the number of moles of ions formed from one mole of solute.

For approximate calculations, each mole of dissolved particles (molecules or ions) in 1000 grams of water will raise the boiling point of water 0.52°C and depress the freezing point 1.86°C. In liquid benzene, one mole of solute particles per 1000 grams of the liquid benzene raises the boiling point 2.53°C. This is almost five times the corresponding effect in pure water.

The effect of particles in solution on the freezing point is the basis for the use of antifreeze in automobile radiators. A compound such as ethylene glycol

$$\begin{array}{ccccc} & OH & & OH & \\ & | & & | & \\ H- & C & - & C & -H \\ & | & & | & \\ & H & & H & \end{array}$$

is added to the water in the radiator. The two OH groups on the molecule permit the ethylene glycol to form hydrogen bonds with the water; these strong attractions make the two liquids completely miscible. A high enough concentration of antifreeze in the water will prevent the liquid from freezing even in the coldest weather.

Compounds that form ions when dissolved in water are not used as antifreeze. The conductivity of such solutions would hasten the corrosion of the metallic parts of the radiator.

The effects of dissolved particles on the boiling and freezing points of solutions can be summarized as follows:

1. The vapor pressure of a solution is lower than that of the pure solvent.
2. The boiling-point temperature of a solution is higher than that of the pure solvent.
3. The freezing-point temperature is lower than that of the pure solvent.

Colligative Properties

Properties of solutions, such as vapor pressure, boiling point, and freezing point, are called colligative, or additive, properties. **Colligative properties** depend on the number of particles of solute in a solution, not on the chemical nature of the particles. Colligative properties, since they are additive, become more pronounced with increased concentration.

If the concentration of the solution is known, the new freezing point and boiling point of the solution can be calculated by using certain constants, which are characteristic of each solvent. These constants are called the *molal freezing point depression constant,* and the *molal boiling point elevation constant.* For water, the molal freezing point depression constant is 1.86°C/m and the molal boiling point elevation constant is 0.52°C/m. To find the change in boiling or freezing point in an aqueous solution of a nonelectrolyte, multiply the constant by the molality.

The freezing point change $\Delta t_f = 1.86°C \times m$.

The boiling point change $\Delta t_b = 0.52°C \times m$.

SAMPLE PROBLEMS

PROBLEM

1. A student dissolved 1.0 mole of sugar in 250 g of H_2O. At what temperature will this solution boil, and at what temperature will it freeze?

SOLUTION

To calculate the change in boiling or freezing point we must first know the molality of the solution.

$$m = \frac{\text{moles solute}}{\text{kg solvent}}$$

$$250 \text{ g} = 0.25 \text{ kg of solvent}$$

$$\frac{1.0 \text{ mole}}{0.25 \text{ kg}} = 4.0\ m$$

Now using the equation for freezing point depression,

$$\Delta t_f = m \times 1.86°\text{C}$$

$$\Delta t_f = 4.0 \times 1.86°\text{C} = 7.4°\text{C}.$$

The change in the freezing point is 7.4°C. Since the freezing point goes down, and the normal freezing point of water is 0°C, the new freezing point is −7.4°C. You find the new boiling point in much the same way. Recall the solution is 4.0 *m*.

$$\Delta t_b = m \times 0.52°\text{C}$$

$$4.0 \times 0.52 = 2.1°\text{C}$$

The boiling point changes by 2.1°C. Since the boiling point goes up, and the normal boiling point of water is 100°C, the new boiling point is 102.1°C.

PROBLEM

2. If 16 g of methanol, CH_3OH, are dissolved in 250 g of H_2O, at what temperature will this solution freeze?

SOLUTION

First, calculate the molality of the solution. Since molality is moles of solute per kilogram of solvent, you will need to express the quantities of solute and solvent in those units. The 16 g of CH_3OH, which has a molar mass of 32 g, is

$$\frac{16 \text{ g}}{32 \text{ g/mole}} \quad \text{or} \quad 0.50 \text{ mole.}$$

The 250 g of water is 0.25 kg of water. The molality of the solution is

$$m = \frac{0.50 \text{ mole}}{0.25 \text{ kg}} \quad \text{or} \quad 2.0\ m.$$

The change in the freezing point is 2.0 $m \times 1.86°C/m = 3.7°C$. Therefore, the freezing point of the solution is $-3.7°C$.

Solutions of Electrolytes

Ionic substances break up into their component ions when they dissolve. This process is called *dissociation*. The dissolving of NaCl, for example, can be represented by the equation

$$NaCl\ (s) \xrightarrow{H_2O} Na^+\ (aq) + Cl^-\ (aq)$$

The formation of mobile ions causes these solutions to conduct electricity. One mole of NaCl will produce one mole of sodium ions and one mole of chloride ions. In dilute solutions, these ions behave as independent particles in changing the freezing and boiling points. One mole of NaCl in water thus produces two moles of particles, since each NaCl breaks into two ions. The boiling and freezing points are changed twice as much as would be the case for a nonelectrolyte.

To calculate the freezing point of electrolytes, we introduce a new term to our equation, which becomes

$$\Delta t_f = 1.86°C \times m \times i$$

The new term, i, is the number of ions produced when the substance dissociates. For NaCl, $i = 2$. For $BaCl_2$, $i = 3$, since $BaCl_2$ dissociates to produce one Ba^{2+} ion and two Cl^- ions. For sodium sulfate, Na_2SO_4, $i = 3$. Sodium sulfate forms two sodium ions and one sulfate ion. Note that polyatomic ions, such as the sulfate, do not break up when dissolved. Boiling point problems are solved similarly, using the equation

$$\Delta t_b = 0.52°C \times m \times i$$

SAMPLE PROBLEM

PROBLEM

At what temperature will a 0.20 m solution of $CaCl_2$ freeze?

SOLUTION

$CaCl_2$ is an ionic substance; therefore, it is an electrolyte. To find the freezing point we need to consider the number of ions formed when $CaCl_2$ dissolves. For $CaCl_2$, the number of ions formed, *i*, is 3. Using the equation

$$\Delta t_f = 1.86°C \times m \times i$$

$$\Delta t_f = 1.86°C \times 0.20 \times 3 = 1.1°C$$

The freezing point of the solution is therefore −1.1°C.

PRACTICE

8.26 What is the freezing point of a solution containing 5.8 g of NaCl dissolved in 500 g of water?

8.27 Arrange the following solutions from highest freezing point to lowest:

(a) 2.0 *m* $BaCl_2$

(b) 2.0 *m* $NaNO_3$

(c) 3.0 *m* $C_6H_{12}O_6$

(d) 1.5 *m* K_2SO_4

TAKING A CLOSER LOOK

Raoult's Law

You have already seen that the vapor pressure of the solvent is decreased as solute dissolves in it. As was the case with freezing and boiling points, the vapor pressure of the solvent can be found if the concentration of the solution is known.

The vapor pressure of the solvent can be found using the relationship known as *Raoult's law.* This law is based on a simple assumption. Suppose a nonvolatile solute is added to water until it comprises one tenth of the total molecules in the solution. Then nine tenths of the molecules are water. Raoult's law assumes that if nine tenths of the molecules are water, then only nine tenths of the molecules can evaporate, and the vapor pressure is therefore reduced to nine tenths of the original vapor pressure. To use Raoult's law you need a new unit of concentration, called the *mole fraction.* The mole fraction of substance A, usually abbreviated X_A, is defined as the moles of substance A divided by the total moles of all components of the solution.

$$X_A = \frac{\text{moles } A}{\text{total moles}}$$

For example, if one mole of sugar is dissolved in 19 moles of water, the mole fraction of water is 19/20, or 0.95.

Raoult's law states that the vapor pressure of the solvent, P_A, is equal to its original pressure, P_A°, times its mole fraction. Again, if the solvent is defined as substance A, Raoult's law is

$$P_A = P_A^\circ \times X_A.$$

(If the solute is an electrolyte, the total moles include the moles of water plus the moles of each ion formed. This discussion will be confined to nonelectrolytes.)

SAMPLE PROBLEM

PROBLEM

The vapor pressure of water at 25°C is 31.7 kPa. What is the vapor pressure of a solution containing 36.0 g of glucose, $C_6H_{12}O_6$, in 86.4 g of water at 25°C?

SOLUTION

To use Raoult's law you must first find the mole fraction of solvent in the solution. Glucose has a molar mass of 180 g, so 36.0 g of glucose is

$$\frac{36.0 \text{ g}}{180 \text{ g/mole}} = 0.20 \text{ mole.}$$

Water has a molar mass of 18 g, so 86.4 g of water is

$$\frac{86.4 \text{ g}}{18 \text{ g/mole}} = 4.8 \text{ moles.}$$

The total number of moles present is

$$4.8 \text{ moles} + 0.2 \text{ mole} = 5.0 \text{ moles.}$$

Therefore the mole fraction of water,

$$X_{H_2O}, \text{ is } 4.8/5.0 = 0.96.$$

Now substituting into Raoult's law, the vapor pressure of the water in the solution is equal to the mole fraction times the original pressure, or 0.96 × 31.7 kPa. The resulting pressure is 30. kPa.

PRACTICE

8.28 Find the vapor pressure of water at 100°C in a solution containing 1.00 mole of a nonvolatile solute dissolved in 1000. g of water.

Boiling Point, Freezing Point, and Vapor Pressure

The changes in boiling point and freezing point of a solution are the direct result of the change in vapor pressure. Figure 8-5 shows the phase diagram of water, which was introduced in Chapter 1. The dashed line in the diagram shows the decreased vapor pressure, due to the presence of a solute. Note how this vapor pressure decrease results in a lower freezing point and a higher boiling point.

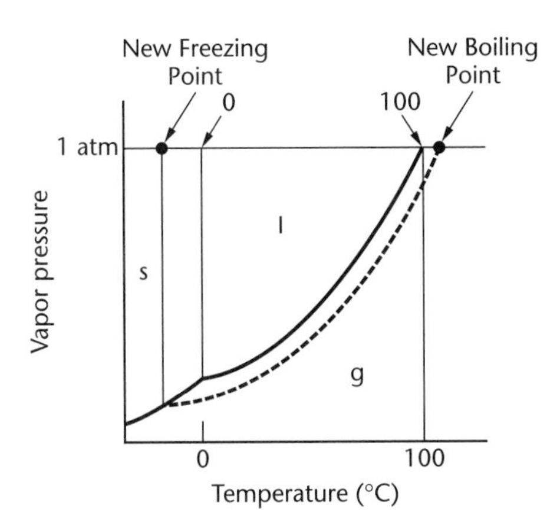

Figure 8-5 Effect of a solute on the vapor pressure, freezing point, and boiling point of water

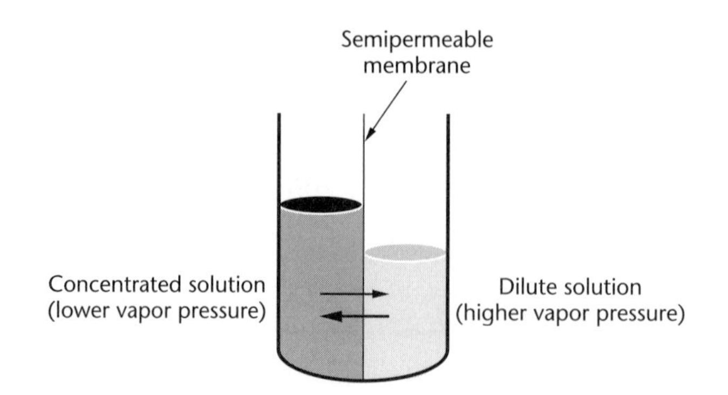

Figure 8-6 Osmosis

Osmosis

Osmosis is the process by which a liquid solvent passes through a *semipermeable membrane.* A semipermeable membrane allows the passage of solvent molecules, but not solute molecules. The membranes of animal cells are semipermeable; therefore, osmosis is an important biological process.

Osmosis may be explained on the basis of differences in vapor pressure. Suppose you have two aqueous solutions of different concentrations, separated by a semipermeable membrane. The more concentrated solution has a lower vapor pressure than the more dilute solution. Therefore water molecules will escape from the concentrated solution more slowly than they will escape the dilute solution. Since more water molecules escape from the dilute solution to the concentrated solution than vice versa, there is a net movement of water from the dilute solution to the concentrated solution. The net movement of solvent molecules in osmosis is from the region of higher vapor pressure to the region of lower vapor pressure (see Figure 8-6).

CHAPTER REVIEW

The following questions will help you check your understanding of the material presented in the chapter. All solutions are aqueous.

1. Which is a mixture? (1) $CCl_4(l)$ (2) $CaCl_2(s)$ (3) $HCl(g)$ (4) $NaCl\ (aq)$

2. A solution in which equilibrium exists between dissolved and undissolved solute must be (1) dilute (2) saturated (3) concentrated (4) unsaturated.
3. Liquid X is completely miscible with water, but immiscible with liquid Y. It is most likely that (1) molecules of X and Y are both nonpolar (2) molecules of X and Y are both polar (3) molecules of X are polar, while those of Y are nonpolar (4) molecules of Y are polar, while those of X are nonpolar.
4. According to the solubility curves on page 290, which saturated solution is most dilute at 0°C? (1) KI (2) NaCl (3) $NaNO_3$ (4) $KClO_3$
5. Find the molarity of a solution that contains 4 g of NaOH in 500 mL of solution. (Formula mass of NaOH = 40.) (1) 0.01 *M* (2) 0.2 *M* (3) 0.5 *M* (4) 2 *M*
6. NaOH is added to one beaker of distilled water, and C_2H_5OH is added to another beaker of distilled water. Both of the solutions that are formed will (1) be strong electrolytes (2) be weak electrolytes (3) have a lower boiling point than pure water (4) have a lower freezing point than pure water.
7. How many moles of KNO_3 are required to make 0.50 L of a 2.0 *M* solution of KNO_3? (1) 0.50 (2) 1.0 (3) 2.0 (4) 4.0
8. According to the solubility curves on page 290, a solution containing 100 g of KNO_3 per 100 g of H_2O at 50°C is considered to be (1) dilute and unsaturated (2) dilute and saturated (3) concentrated and unsaturated (4) concentrated and supersaturated.
9. A sample of water with a mass of 2000. grams is found to contain 0.100 gram of lead. What is the concentration of the lead in parts per million? (1) 5 ppm (2) 50 ppm (3) 200 ppm (4) 500 ppm
10. In 500.0 mL of solution, there are 10.0 g of NaOH. The molarity of this solution is (1) 1.0 *M* (2) 0.50 *M* (3) 0.25 *M* (4) 0.10 *M*.
11. According to the solubility curves on page 290, as the temperature increases from 30°C to 40°C, the solubility of potassium nitrate in 100 g of water increases by approximately (1) 5 g (2) 10 g (3) 17 g (4) 25 g.
12. At 23°C, 100 mL of a saturated solution of NaCl are in equilibrium with 1 g of solid NaCl. The concentration of the solu-

tion will be changed most when (1) 80 mL of water are added to the solution (2) additional solid NaCl is added to the solution (3) the pressure on the solution is increased (4) the temperature of the solution is decreased 5°C.

13. How many moles of $AgNO_3$ are dissolved in 10 mL of a 1 *M* $AgNO_3$ solution? (1) 1 (2) 0.1 (3) 0.01 (4) 0.001

14. How many moles of $AgNO_3$ are in 500.0 mL of a 5-molar solution of $AgNO_3$? (1) 2.5 (2) 5.0 (3) 10.0 (4) 170.0

15. What is the molarity of a solution containing 20.0 g of NaOH in 0.50 L of solution? (1) 0.50 (2) 1.0 (3) 2.0 (4) 10.

16. A sample of water is found to contain 0.0200 gram of calcium ions. If the concentration of the calcium ions in the solution is 500 ppm, what is the mass of the total sample? (1) 10. grams (2) 20. grams (3) 40. grams (4) 400. grams

17. One hundred grams of water at 10°C contain 60 g of $NaNO_3$. According to the solubility curves on page 290, how many additional grams of $NaNO_3$ must be added to form a saturated solution at 10°C? (1) 19 (2) 39 (3) 60 (4) 79

18. Which gas is most soluble in water at STP? (1) O_2 (2) CO_2 (3) NH_3 (4) N_2

19. What is the total number of moles of $CaCl_2$ needed to make 500 mL of 4 *M* $CaCl_2$? (1) 1 (2) 2 (3) 4 (4) 8

20. Two liters of a solution of sulfuric acid contain 98 g of H_2SO_4. The molarity of this solution is (1) 0.50 (2) 1.0 (3) 1.5 (4) 2.0.

21. A solution with a total mass of 1.0 kilogram contains 10. grams of HCl. The concentration of HCl in the solution could correctly be expressed as (1) 1.0% (2) 100 ppm (3) 10 molar (4) 10%.

22. The solution with the lowest freezing point will be produced when 1.0 g of $C_6H_{12}O_6$ is dissolved in (1) 18 g of H_2O (2) 100 g of H_2O (3) 180 g of H_2O (4) 1000 g of H_2O.

23. Which expression represents the molarity (*M*) of a solution?

(1) $\frac{\text{moles of solvent}}{\text{1 kg of solution}}$ (3) $\frac{\text{moles of solvent}}{\text{1 L of solution}}$

(2) $\frac{\text{moles of solute}}{\text{1 kg of solution}}$ (4) $\frac{\text{moles of solute}}{\text{1 L of solution}}$

Base your answers to questions 24 through 27 on the diagram, which represents the solubility curve of salt X. The four points on the diagram represent four solutions of salt X.

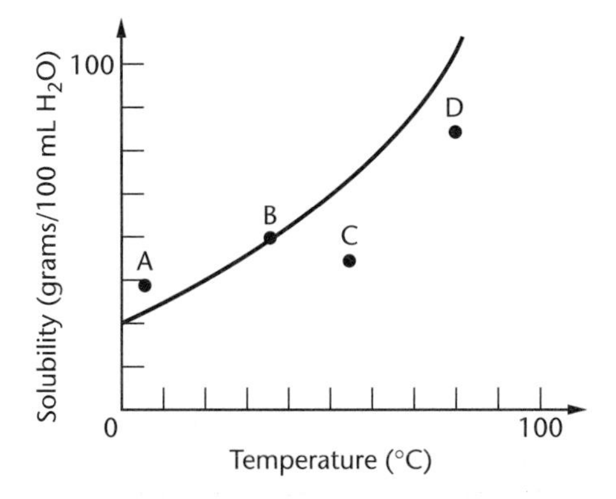

24. Which point represents a supersaturated solution of salt X? (1) A (2) B (3) C (4) D

25. Which point represents the most concentrated solution of salt X? (1) A (2) B (3) C (4) D

26. Which point represents a saturated solution of salt X? (1) A (2) B (3) C (4) D

27. Which points represent unsaturated solutions of salt X? (1) A, B (2) B, C (3) C, D (4) A, C

28. A 1-molal solution of $MgCl_2$ has a higher boiling point than a 1-molal solution of (1) $FeCl_3$ (2) $CaCl_2$ (3) $BaCl_2$ (4) NaCl.

29. Which 1-molal solution will have the highest boiling point? (1) KNO_3 (2) $Mg(NO_3)_2$ (3) $Al(NO_3)_3$ (4) NH_4NO_3

30. The freezing point of a kilogram of water will be lowered most by adding one mole of (1) $CaCO_3$ (2) $CaSO_4$ (3) $CaSO_3$ (4) $CaCl_2$.

Dry Cleaning: How Dry Is It?

A comedian once said about dry cleaners and the clothes they work on: "I know they're getting them wet!" Flicking his finger at an imaginary speck of food on his tie, he noted, "That's dry cleaning!"

All joking aside, there is some truth in those remarks. Dry cleaners use a variety of liquid solvents to clean clothing that, for one reason or another, should not be washed with water and detergent. A shrewd dry cleaner has to know a bit of chemistry to decide which solvent to use for a particular type of stain. Although water is often called "the universal solvent" because of its ability to dissolve many substances, water is a polar substance—its molecules have a slightly positively charged end and a slightly negatively charged end, similar to a bar magnet. Because of this, water is good at dissolving polar substances and ionic compounds, such as salts, which are composed of oppositely charged ions.

However, water is not good at dissolving nonpolar substances. The positive and negative charges on these molecules are more or less evenly distributed. Nonpolar substances, such as lipids (fats and oils), are not water-soluble, and they cause some of the most stubborn stains on clothing. Following the axiom "like dissolves like," dry cleaners rely on a number of nonpolar solvents to clean clothing soiled by nonpolar substances such as butter, wax, and grease.

Carbon tetrachloride, CCl_4, is a nonpolar solvent that came into widespread use early in the twentieth century. However, by the 1950s, doctors realized that exposure to CCl_4 was causing serious illnesses among dry cleaners, such as liver disease and central nervous system damage. By the 1960s, the solvent called perchloroethylene, which was thought to be safer, largely replaced CCl_4. Perchloroethylene, or "perc" is a chlorinated hydrocarbon compound derived from petroleum. However, it was soon found that perc also has dangers. It is toxic in high doses and chronic exposure to even small amounts is suspected to cause cancer. When released into the environment, perc can contaminate drinking-water supplies.

Nevertheless, perchloroethylene is still the solvent of choice among dry cleaners. Advances in dry-cleaning machinery have reduced (but not eliminated) the exposure of workers to perc and the amount of perc being released into the environment. Meanwhile, researchers are trying to develop safer and more environmentally friendly cleaning solvents. One interesting possibility is using liquid carbon dioxide as a dry cleaning solvent.

Rates of Reactions: Kinetics

LOOKING AHEAD

You probably know that given enough time, milk will turn sour. What factors determine how long a container of milk will remain drinkable? You probably have also observed that the rate at which the milk turns sour is related to its temperature. Temperature is one of several factors that affect the rate of a chemical reaction. Kinetics is the branch of chemistry that studies rates of reaction. In this chapter you will examine how reaction rates are determined and how they can be changed.

When you have completed this chapter you should be able to:

- **Define** reaction rate, energy of activation, and mechanism.
- **Explain** the effect of changes in temperature, concentration, and pressure on reaction rate.
- **Interpret** the energy profile of a reaction.
- **Relate** reaction rate to the surface area of solids.
- **Predict** changes in reaction rate on the basis of collision theory.

Factors That Determine Reaction Rates

$$H_2(g) + I_2(g) \rightarrow 2\ HI(g)$$

This equation tells you that 2 moles of hydrogen iodide gas are formed by the reaction between 1 mole of hydrogen gas and 1 mole of iodine gas. But the equation does not tell you anything about the rate at which this reaction occurs. It gives you no information about the **rate of reaction**: how fast the reactants are consumed and the products are formed.

Consider what must happen for the reaction to take place. H—H and I—I bonds must be broken before H—I bonds can be formed. This bond-breaking and bond-forming cannot take place unless the particles involved come into contact with one another—unless they collide. But collisions of particles are not enough. The collisions must have a minimum energy, or they will not produce a chemical change. And the collisions must occur at a favorable angle, or, again, they will not produce a chemical change. As you know, a head-on collision between two cars produces more change than does a sideswipe collision. In like manner, some angles of collision between particles are more effective (they produce more change) than are other angles of collision.

The rate of a chemical reaction, then, depends on several factors. It depends on the nature of the bonds that must be broken in the reactants before new bonds can be formed in the products. It depends on the rate of collisions between reacting particles and on the effectiveness of the collisions.

The rate of reaction is measured experimentally by observing some change taking place during the reaction that indicates a reactant is being consumed or a product is forming. Such a change could be a difference in color intensity, a change in pressure (if gases are involved), or a change in concentration of a reactant or a product. The ratio of the change with respect to time is the rate of the reaction.

Units appropriate to the change being measured are used. For example, a reaction rate based on a change of concentration is typically expressed as moles per liter per minute or moles per liter per second.

Now you will consider in more detail the factors that determine rates of reaction.

Energy of the Collision

For a chemical change to take place, particles must collide with a certain minimum energy. This minimum energy is called the energy of activation. **Activation energy** is the energy needed to weaken or break bonds before new bonds can be formed. Thus the energy of activation depends on the nature of the reacting particles and is different for different reactions.

Let us examine one reaction as an example.

$NO_2\,(g)$	+	$CO\,(g)$	→	$NO\,(g)$	+	$CO_2\,(g)$
nitrogen dioxide		carbon monoxide		nitric oxide		carbon dioxide

An energy profile of this reaction is shown in Figure 9-1. In this energy profile, the vertical axis indicates energy and the horizontal axis indicates the direction of the reaction. The horizontal axis is called the reaction coordinate. Notice that the reactants, NO_2 + CO, have more energy than the products, NO + CO_2, have. The reactants are higher on the energy axis, at point A, than are the products, at point B. From this fact, you know that the reaction between NO_2 and CO is exothermic. The energy profile also shows that the path from A to B is not direct. The colliding NO_2 molecules and CO molecules must acquire sufficient energy before the products, NO and CO_2, can be formed. This quantity of energy, represented by arrow C in the energy profile, is the minimum energy required for an effective collision. In

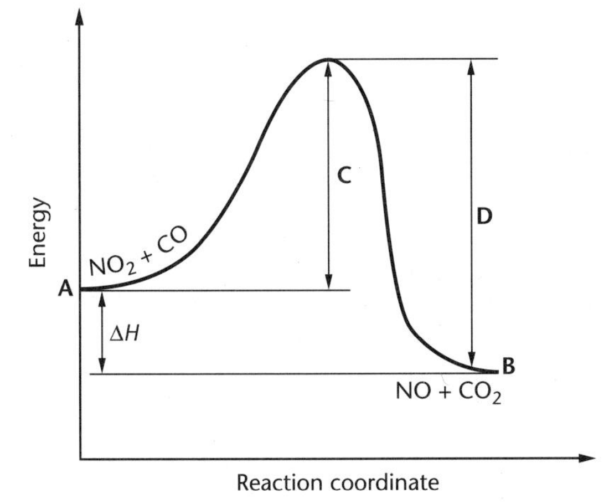

Figure 9-1 Energy profile: NO_2 + CO

other words, arrow C represents the energy of activation. In all reactions, the energy of activation can be pictured as a potential energy barrier that the reacting particles must overcome before they can change into products.

The ΔH, the **heat of reaction**, is independent of the path by which the reaction proceeds from A to B. ΔH is equal to the difference in energy between B and A, always found by subtracting the energy of the reactant from the energy of the product. ($\Delta H = H_{product} - H_{reactant}$) Since in this case, the products have less energy than the reactants, ΔH will be negative; it results from subtracting a larger quantity from a smaller quantity. ΔH is always negative in an exothermic reaction.

Arrow D on our energy profile, connects the energy at the top of the curve with the energy of the products. It represents the activation of the **reverse** reaction, which is an endothermic reaction.

For some reactions, the energy of activation (potential energy barrier) is relatively small. An energy profile of such a reaction is shown in the left part of Figure 9-2. For other reactions, the energy of activation is relatively large, as shown by the energy profile in the right part of Figure 9-2.

As you know, many reactions may proceed in either direction. That is, a reaction may proceed from A + B to form AB or from AB to re-form A + B.

$$A + B \rightarrow AB$$

$$AB \rightarrow A + B$$

The forward and the reverse reactions both have a specific activation energy requirement. As you saw in Figure 9-1, the activation

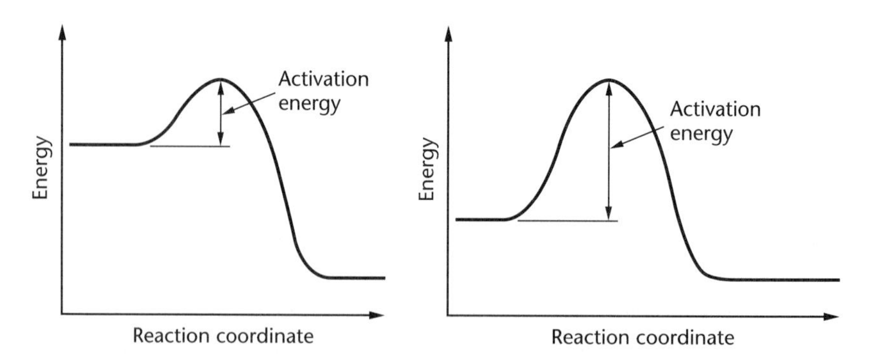

Figure 9-2 Energy profiles: different energies of activation

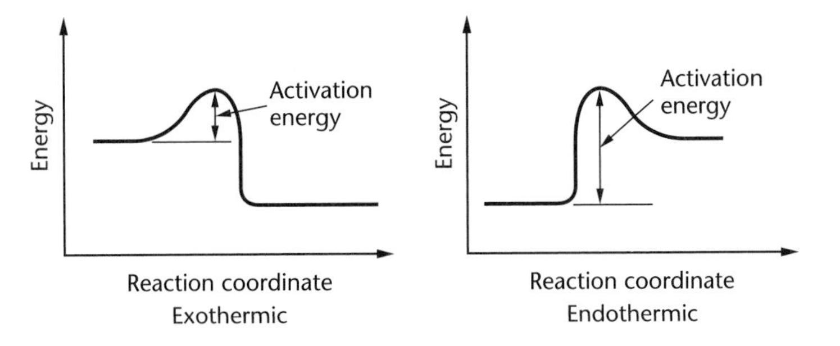

Figure 9-3 Energy profiles: exothermic and endothermic reactions

energy of the exothermic reaction (arrow C) was smaller than the activation energy of the reverse, endothermic reaction (arrow D). The same relationship exists in all reversible reactions; the endothermic reaction has a greater activation energy than the exothermic reaction. In Figure 9-3 an exothermic, and the reverse, endothermic reaction are plotted side by side. The activation energy is greater in the endothermic reaction.

When particles collide with enough energy (and at a favorable angle, remember), their bonds are weakened and rearranged. It is believed that the reactant particles become partially bonded, very briefly forming a structure called an activated complex. The **activated complex** is a temporary, high-energy, transitional (in-between) structure, somewhere between reactants and products. It exists for only the short time that the colliding particles are in contact.

In an energy profile, such as the one in Figure 9-1 on page 323, the activated complex would be at the top of the crest that forms the potential energy barrier. If the collision energy is large enough, the activated complex changes into the products of the reaction. Some part of the complex, however, may return to the unchanged reactant form. If the collision energy is not large enough, the activated complex does not form. Instead, the reactants repel one another and remain unchanged. It is very much like trying to roll a ball over a hill. With a strong push (high energy), the ball rises and gets over the hill. If the push is not strong, the ball rises a bit and then falls back to its original position.

Angle of Collision

Two of many possible angles of collision between H_2 and I_2 molecules are shown in Figure 9-4. The angle of collision A is

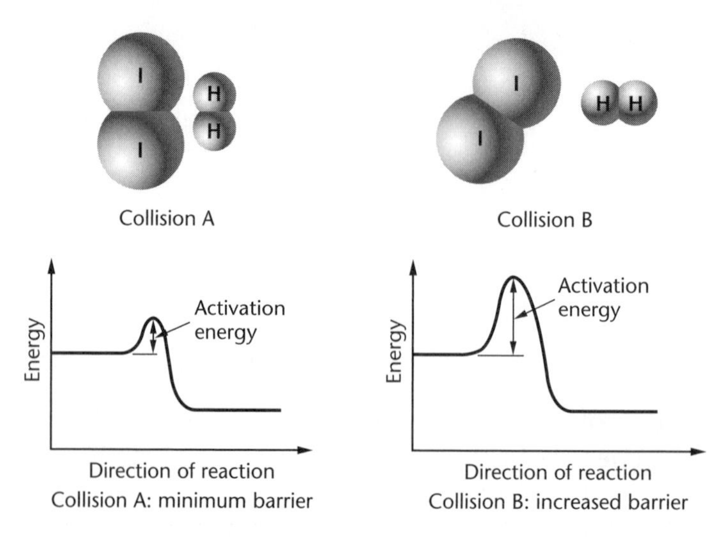

Figure 9-4 Angles of collision

much more favorable for bond-breaking than is the angle of collision B. With a favorable angle of collision, the complete reaction can occur, provided the collision has enough energy. If the angle of collision is unfavorable, the potential barrier is raised, as shown in Figure 9-4. Then more than minimum energy is needed for the reaction to take place.

The activation energies given in tables or shown in diagrams are usually minimum energies—that is, the energies required for reactions to proceed when the angles of collision are favorable.

PRACTICE

9.1 Name two factors that determine whether a given collision between molecules results in a chemical reaction.

9.2 For a certain reaction, the activation energy of the forward reaction is greater than that in the reverse reaction. What is the sign of ΔH for the forward reaction? How do you know?

9.3 Which of the following units is most appropriate for measuring the rate of a reaction?

(1) meters/sec
(2) moles per liter per second
(3) grams per liter
(4) percent by mass

Mechanism of Reaction

It is not likely that a reaction takes place in a single step—that is, that the products are formed immediately from the collision of the reactant particles. Experiments involving rates of reaction have provided evidence that many reactions take place in a number of small steps. Each step consists of collisions between two particles, and the products of one step become the reactants of the next step. The net chemical change of the reaction is the sum of all these steps. The sequence of steps involved in a reaction is called the **mechanism of reaction.**

Suppose a reaction takes place in six steps. Each of the steps must have its own rate. The overall reaction then cannot take place faster than the slowest of these steps. Let's say that steps 1, 2, 4, 5, and 6 take place very rapidly—only a few seconds are required for each step. Step 3, however, takes ten minutes to complete. The overall reaction obviously cannot take place in less than ten minutes. Step 3 is the rate-determining step. The slowest of the steps in a reaction mechanism is called the **rate-determining step.**

Changing Reaction Rates

Rates of reaction can be changed in various ways, depending on the nature of the reactants. The rate of a reaction is also influenced by the concentration of the reactants and by the temperature at which a reaction takes place. Some reactions are speeded up by the addition of a catalyst. How each of these factors works will now be discussed.

Nature of the Reactants

In general, reactions in which chemical bonds are broken are slower than reactions in which particles are rearranged without bond-breaking. The following reaction is an example of a reaction in which bond-breaking is required:

$$H_2\ (g) + Cl_2\ (g) \rightarrow 2\ HCl\ (g)$$

This reaction is quite slow. H—H and Cl—Cl bonds must be broken before the atoms can be rearranged into HCl molecules. However, very few collisions result in breaking either kind of bond. The reaction can be speeded up by exposure to light of the proper frequency. The light energy helps to break the Cl—Cl bonds in the Cl_2 molecules.

The next reaction, in contrast, takes place instantly upon mixing of the solutions. Ions are already present and need only to be rearranged for the reaction to occur.

$$AgNO_3\ (aq) + NaCl\ (aq) \rightarrow AgCl\ (s) + NaNO_3\ (aq)$$

Since the reaction is essentially ionic, it can also be written

$$Ag^+\ (aq) + NO_3^-\ (aq) + Na^+\ (aq) + Cl^-\ (aq) \rightarrow AgCl\ (s) + Na^+\ (aq) + NO_3^-\ (aq)$$

Reactions in which relatively strong bonds must be broken are generally slower than reactions in which relatively weak bonds must be broken. For example, when $KClO_3$, an ionic solid, is heated, solid KCl and gaseous O_2 are formed. The bond between K^+ and ClO_3^- is a strong ionic bond. The three chlorine and oxygen bonds

```
 O
 |
 Cl—O
 |
 O
```

are relatively weak. Heating $KClO_3$ breaks the weaker bonds, and O_2 gas forms. Prolonged heating at very high temperatures is needed to break the very stable K—Cl bond. Thus,

$$2\ KClO_3 \rightarrow 2\ KCl + 3\ O_2$$

Concentration of the Reactants

Reactant particles are colliding constantly, but only a few of the collisions result in chemical change. When the concentration of reactants is increased, the number of reactant particles in a given volume increases. There are then more collisions between particles, and the likelihood of a large number of effective collisions increases.

At constant temperature, the concentration of a gas is proportional to its pressure. Thus, if the reactants are gases, their concentration can be increased by increasing the pressure. Increasing the pressure of a gaseous system increases the frequency of collisions between particles.

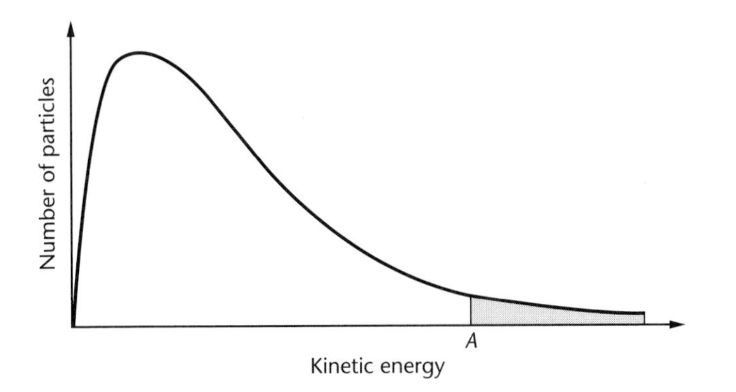

Figure 9-5 Distribution of kinetic energies at a given temperature

Temperature

The temperature of a system is a measure of the average kinetic energy of the particles in the system. You will recall that the particles in a system do not all have the same kinetic energy at the same temperature. The distribution of kinetic energies for a system at a given temperature is shown by the curve in Figure 9-5. The curve shows that relatively few particles have either very low or very high kinetic energy. Most of the particles have moderate kinetic energy.

Suppose that the energy of activation is at point A on the energy axis. The total number of particles that have at least this amount of energy is represented by the shaded area under the curve. As you can see, only a small fraction of all the particles in the system have enough energy for an effective collision.

Now suppose that the temperature of the system is increased. In Figure 9-6 on page 330, curve T_1 represents the distribution of energies at temperature T_1. Curve T_2 represents the distribution of energies at the higher temperature T_2. As in Figure 9-5, the energy of activation is at point A. Compare the areas under the curve from point A to the right. How does the number of particles that have enough energy for effective collisions at temperature T_2 (these particles are in the hatched and shaded areas) compare with the number of particles with this amount of energy at temperature T_1 (in the shaded area only)?

As the temperature of a system increases, more particles obtain sufficient energy for effective collisions. Since an increase in temperature increases the velocity of molecules, there will also be an

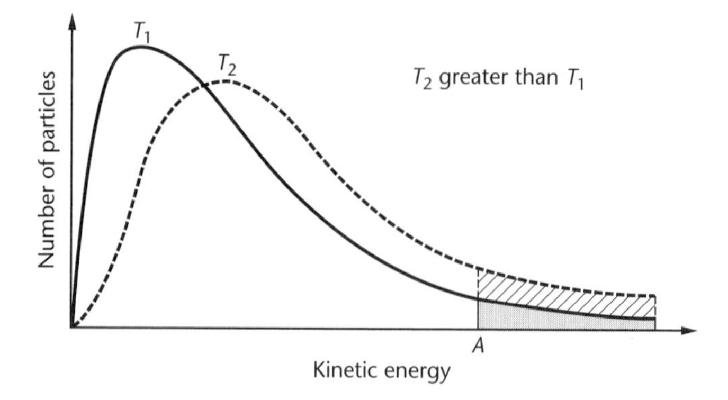

Figure 9-6 Distribution of kinetic energies at different temperatures

increase in collision frequency. Faster moving particles will collide with each other more often. However, the major effect of increased temperature is an increase in collision efficiency. Since faster-moving particles collide with more force, a higher percentage of the collisions will attain the necessary activation energy. The shape of the curve shows that a *small* increase in temperature produces a *large* increase in the total number of particles with the necessary activation energy. The precise effect of a temperature change on the reaction rate varies greatly from one reaction to another. The greater the activation energy, the greater the effect of temperature change on the rate. For many reactions, a temperature increase of only 10°C can more than double the reaction rate.

Catalysts

The rusting of iron is a chemical change. The rate of reaction is generally slow and depends on the environment. The environment must contain oxygen and moisture for the reaction to occur. However, if the environment also contains carbon dioxide, the rate of rusting is considerably faster. Yet, CO_2 is not essential to the reaction. In the presence of moisture, Fe and O_2 will react and form Fe_2O_3, whether or not CO_2 is also present. The CO_2 acts as a catalyst. A **catalyst** is a substance that changes the rate of a reaction without being permanently changed itself.

Catalysts affect the reaction rate through several different mechanisms. In general, chemists agree that catalysts change the activation energy requirements of reactions. Thus a catalyst speeds up a reaction by decreasing the activation energy of the reaction.

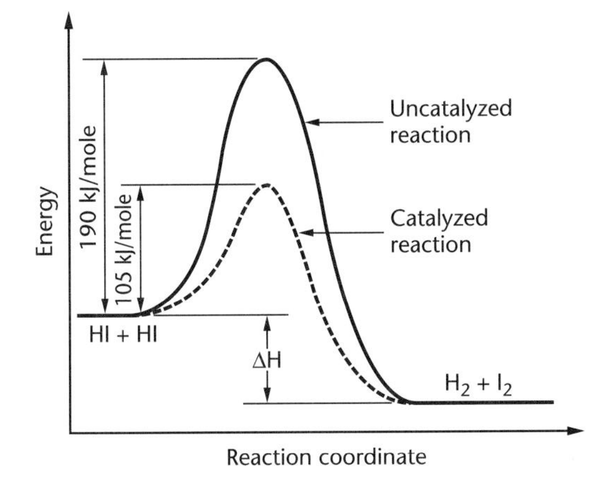

Figure 9-7 Energy profile: catalyzed reaction

Let's take an example. Hydrogen iodide gas breaks down into hydrogen gas and iodine gas. The activation energy ($\Delta H_{act.}$) of the reaction is about 190 kilojoules per mole.

$$2\ HI\ (g) \rightarrow H_2\ (g) + I_2\ (g) \qquad \Delta H_{act.} = 190\ \text{kJ/mole}$$

The metal platinum acts as a catalyst in this reaction. In the presence of platinum, the activation energy requirement of the reaction drops to 105 kJ per mole. The energy profiles of the catalyzed and the uncatalyzed reactions are shown in Figure 9-7. Note that the catalyst changes the activation energy, making the activation energy lower in both the forward and reverse reaction. However, the catalyst does *not* change the energy of the products or reactants, and therefore has no effect on the value of ΔH for the reaction.

How do catalysts work? One explanation that applies to many catalysts is that they work by changing the reaction mechanism, or pathway. Suppose the reaction A + B → AB occurs at a very slow rate. When catalyst C is present, the rate is speeded up because an intermediate product, AC, which offers a new reaction path, is formed. This new path requires lower activation energy. The AC reacts with B to re-form catalyst C.

Uncatalyzed Reaction	***Catalyzed Reaction***
1 Step	2 Steps
A + B → AB (slow)	Step 1: A + C → AC (fast)
	Step 2: AC + B → AB + C (fast)

Note that the catalyst, C, enters the catalyzed reaction for a moment, in step 1. However, it is liberated unchanged in step 2. The catalyst is not changed by the net reaction.

Some reactions are catalyzed by finely divided solids that attract the reacting materials to their surfaces. Particles that are held and concentrated on the surface of a catalyst are said to be *adsorbed.* If a catalyst is finely divided, it will have a very large surface area and will be able to adsorb very large numbers of reactant particles. A finely divided metallic catalyst may adsorb several hundred times its own volume of gaseous molecules.

By decreasing the activation energy, a catalyst permits lower energy collisions to be effective. Thus a catalyst increases collision efficiency. The rate of a chemical reaction can be changed either by changing collision frequency, collision efficiency, or both. An increase in concentration, as we have seen, increases collision frequency. An increase in temperature increases both frequency and efficiency of collisions.

Surface Area

A log may burn in a fireplace for several hours. If the same log is ground down to sawdust and thrown into a fire, it will be consumed almost instantaneously. By decreasing the particle size, you increase the number of particles exposed to the oxygen needed for combustion. The sawdust particles have a much larger exposed surface than had the log. The rate of a reaction involving solids is increased by decreasing the size of the solid particles, which increases the exposed surface area. An increase in surface area increases the frequency of collisions, and so increases the reaction rate.

In summary, any change that increases either collision efficiency or collision frequency will increase the rate of a reaction.

PRACTICE

9.4 For each pair of conditions, indicate which would result in a faster reaction and why.

(a) zinc strips in hydrochloric acid or zinc powder in hydrochloric acid

(b) magnesium ribbon in concentrated acid or magnesium ribbon in dilute acid

(c) cold milk getting sour or warm milk getting sour

(d) one mole of iron and one mole of oxygen in a 1.00-liter container or one mole of iron and one mole of oxygen in a 10.0-liter container

---

Potential Energy Diagrams

We have frequently used energy profiles to help us understand the energy changes that occur during a chemical reaction. In these diagrams, the energy is sometimes more specifically identified as potential energy. The energy profile is then frequently called a potential energy diagram. Let us take a close look at the information contained in these diagrams.

Figure 9-8 shows the potential energy curve for an exothermic reaction. You know that it is exothermic, because the potential energy of the products is less than that of the reactants. The six arrows drawn on the curve each represent a specific amount of energy.

Arrow 1 represents the potential energy of the reactant. Any arrow that goes down to the bottom of the graph corresponds to the potential energy of a component in the system.

Arrow 2 corresponds to the difference in potential energy between the starting material (the reactant) and the highest point on the curve (the activated complex). The energy required for the reactant to form the activated complex is the activation energy. Arrow 2 shows the activation energy of the reaction.

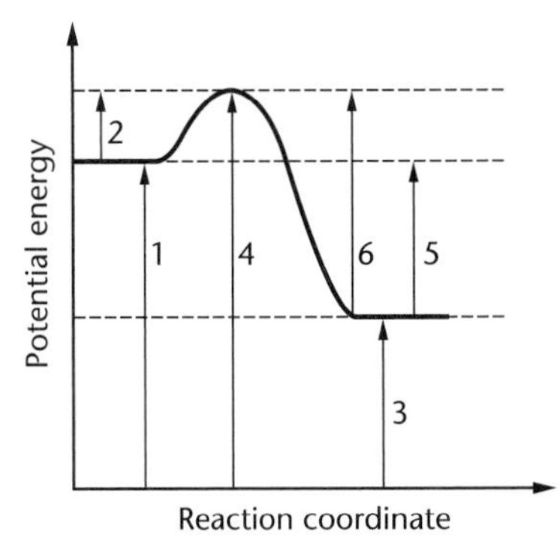

Figure 9-8 Potential energy diagram: exothermic reaction

Arrow 3 shows the potential energy of the product.

Arrow 4 shows the potential energy of the activated complex.

Arrow 5 covers the interval between the potential energy of the product and the potential energy of the reactant. You will recall that this interval corresponds to ΔH, the heat of reaction. The heat of reaction is equal to the potential energy of the product minus the potential energy of the reactant. In this case, that is equivalent to arrow 3 minus arrow 1. Since the potential energy shown by arrow 1 is greater than that shown by arrow 3, the heat of reaction is negative. Recall that ΔH is always negative for an exothermic reaction.

Arrow 6 shows the energy difference between the product and the activated complex. This corresponds to the activation energy of the reverse reaction.

Now look at Figure 9-9, which shows a potential energy diagram for an endothermic reaction. In this case, the potential energy of the product is greater than that of the reactant. The numbered arrows on this diagram correspond to the numbered arrows on Figure 9-8, described above. In this case, however, arrow 3, the potential energy of the product, is greater than arrow 1, the potential energy of the reactant. Therefore, if arrow 1 is subtracted from arrow 3, the result, ΔH, is positive. For an endothermic reaction, ΔH is always positive.

If the forward reaction is exothermic, then the reverse reaction must be endothermic, and vice versa. The heat of reaction has exactly the same absolute value in either direction, but is negative in the exothermic direction and positive in the endothermic direction. The activation energy is always greater in the endothermic direction than it is in the exothermic direction.

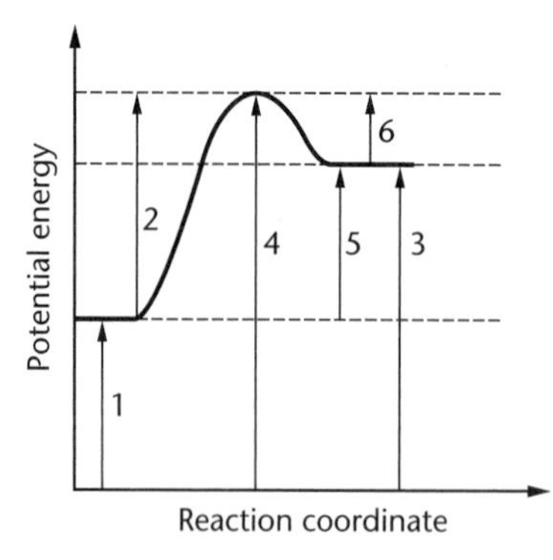

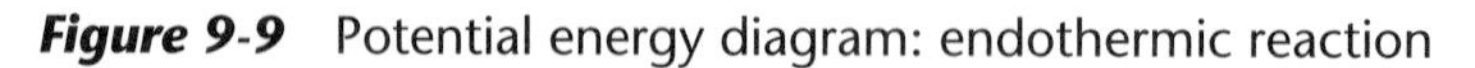

Figure 9-9 Potential energy diagram: endothermic reaction

PRACTICE

Base your answers to questions 9.5–9.9 on Figures 9-8 and 9-9.

9.5 If a catalyst is added to either of these reactions, three of the six arrows would change. Identify the three arrows that would change.

9.6 In Figure 9-8, suppose the activation energy of the reaction is 15 kJ and the heat of the reaction is −45 kJ. What is the activation energy of the reverse reaction?

9.7 In Figure 9-9, if the activation energy is 50 kJ, and ΔH is 35 kJ, what is the activation energy of the reverse reaction?

9.8 In Figure 9-9, if ΔH for the forward reaction is 35 kJ what is the value of ΔH for the reverse reaction?

9.9 Figure 9-7 on page 331 shows the effect of a catalyst on a reaction. You can see that the catalyst lowers the energy of the activated complex. What effect does a catalyst have on the rate of the reverse reaction? Why?

CHAPTER REVIEW

The following questions will help you check your understanding of the material presented in the chapter.

1. In order for nitrogen and oxygen to react according to the equation

$$N_2\ (g) + O_2\ (g) \rightarrow 2\ NO\ (g)$$

(1) nitrogen and oxygen particles must first collide (2) the gases must be liquefied (3) a catalyst must be used (4) the product must be dissolved in water.

2. When the energy of activation of the reverse reaction is greater than the energy of activation of the forward reaction (1) the reaction stops (2) the forward reaction is exothermic (3) the product is unstable (4) no collisions take place.

3. In a reaction, the activated complex (1) has more energy than the reactants or the products (2) acts as a catalyst (3) always forms products (4) is a stable compound.

4. If the energy of activation for a reaction is high without a catalyst but is lower with a catalyst, the addition of a catalyst causes the heat of the reaction to (1) decrease (2) increase (3) remain the same.
5. In a chemical reaction, if the amount of substance that changes per unit time increases, the rate of the reaction (1) decreases (2) increases (3) remains the same.
6. When the energy of activation of a system increases, the height of the potential energy barrier (1) decreases (2) increases (3) remains the same.
7. When the strength of the bonds in the reacting particles increases, the rate of reaction of the particles (1) decreases (2) increases (3) remains the same.
8. When the concentration of reacting particles in a chemical system increases, the number of collisions per second (1) decreases (2) increases (3) remains the same.
9. If a reaction releases energy without the use of a catalyst, the amount of energy released with the addition of a catalyst (1) decreases (2) increases (3) remains the same.
10. The compound whose molecules have the highest average kinetic energy is (1) NO (*g*) at 25°C (2) N_2O (*g*) at 15°C (3) NO_2 (*g*) at 30°C (4) N_2O_3 (*g*) at 20°C.
11. As the average kinetic energy of the molecules of a sample increases, the temperature of the sample (1) decreases (2) increases (3) remains the same.
12. According to the potential energy diagram, what is the reaction A + B → C? (1) endothermic and ΔH is positive (2) endothermic and ΔH is negative (3) exothermic and ΔH is positive (4) exothermic and ΔH is negative

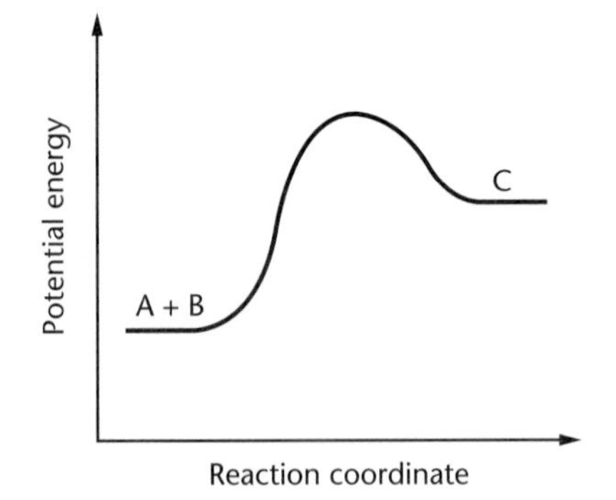

For each statement in questions 13 through 15, write the number of the potential energy diagram that is described by that statement.

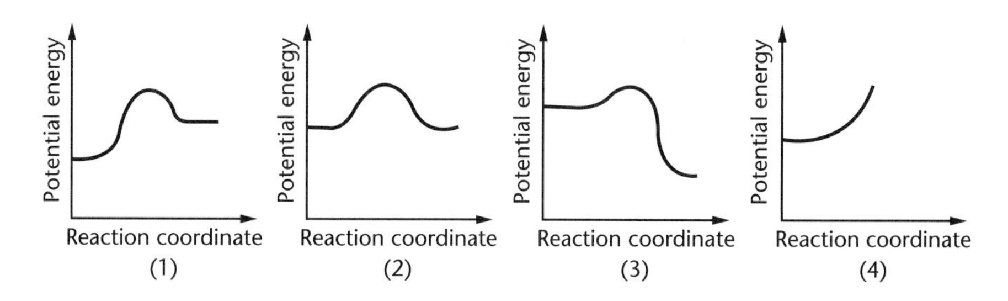

13. The potential energy of the products is less than the potential energy of the reactants.
14. There is insufficient activation energy for the reaction to occur.
15. The activation energy is greater in the forward direction than in the reverse direction.
16. If the concentration of one of the reactants in a chemical reaction is increased, the rate of the reaction (1) decreases (2) increases (3) remains the same.

Base your answers to questions 17 through 19 on the following information.

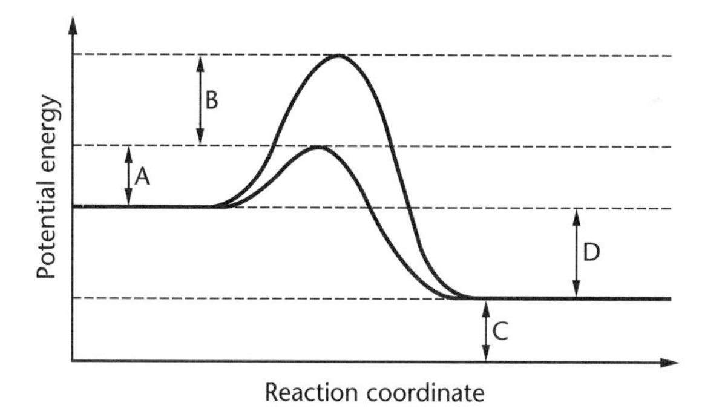

17. Which represents the energy of activation for the uncatalyzed forward reaction? (1) A + B (2) B (3) C + D (4) D
18. Compared with the potential energy of the reactants, the potential energy of the products is (1) less (2) the same (3) greater (4) impossible to determine.
19. Which represents the heat of reaction for the catalyzed reverse reaction? (1) A (2) B (3) C (4) D

20. Given the potential energy diagram:

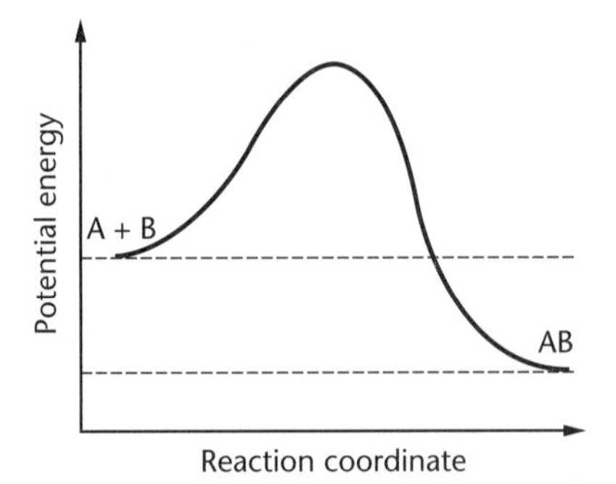

With reference to energy, the reaction A + B → AB can best be described as (1) endothermic, having a $+\Delta H$ (2) endothermic, having a $-\Delta H$ (3) exothermic, having a $+\Delta H$ (4) exothermic, having a $-\Delta H$.

21. Which is changed when a catalyst is added to a chemical reaction? (1) the potential energy of the reactants (2) the potential energy of the products (3) the activation energy (4) the heat of reaction

22. As the rate of a given reaction increases due to an increase in the concentration of the reactants, the activation energy for that reaction (1) decreases (2) increases (3) remains the same.

23. Which statement best explains why an increase in temperature usually increases the rate of a chemical reaction? (1) The activation energy of the reaction is decreased. (2) The ΔH of the reaction is increased. (3) The free energy of the reaction is decreased. (4) The effectiveness of collisions between particles is increased.

24. When a catalyst lowers the activation energy, the rate of a reaction (1) decreases (2) increases (3) remains the same.

25. A certain exothermic reaction produces 40 kJ of heat. If the activation energy of the forward reaction is 15 kJ, what is the activation energy of the reverse reaction? (1) 55 kJ (2) 25 kJ (3) 40 kJ (4) 15 kJ

26. If the activation energy for the forward reaction is 25 kJ, and the activation energy of the reverse reaction is 80 kJ, what is the value of ΔH for the reaction? (1) +55 kJ (2) −55 kJ (3) +105 kJ (4) −105 kJ

CONSTRUCTED RESPONSE

1. For the reaction $N_2 + 3\ H_2 \rightarrow 2\ NH_3$, ΔH is −92 kJ.
   (a) Draw the potential energy diagram for the reaction.
   (b) On your potential energy diagram, indicate with a dotted line how the diagram would change if a catalyst were added.
   (c) N_2 has very strong, triple bonds. What effect does this fact have on the activation energy of the reaction?
   (d) What is the value of ΔH for the reaction $2\ NH_3 \rightarrow N_2 + 3\ H_2$?
2. Solutions of hydrogen peroxide in water decompose to produce oxygen gas and water. What are three things you could do to increase the rate of formation of oxygen?

Approaching Kinetics by Halves

The rate of a reaction can be described in terms of the change in concentration of a reactant per unit of time. In the reaction of zinc with hydrochloric acid, you might express the rate of change in the concentration of hydrochloric acid (HCl) in terms of molarity per second ($2\ HCl + Zn \rightarrow ZnCl_2 + H_2$). Suppose that a 2.00 molar (M) solution of HCl reacted with zinc at the rate of 0.100 molar per second. What would the concentration of HCl be after 10.0 seconds? Seems simple enough—0.100 molar per second multiplied by10.0 seconds is 1.00 M. If the reaction produced a change of 1.00 M, and you began with 2.00 M HCl, the solution should now be 1.00 M.

There is a problem with this simple answer, however. Recall that the concentration of reactant affects the reaction rate. The HCl was reacting at 0.100 molar per second when its concentration was 2.00 M. However, as the reaction proceeds, the concentration of HCl decreases. Therefore, the rate decreases steadily.

Predicting the change in concentration resulting from a constantly changing rate of reaction is not a simple problem!

In many cases, the rate of reaction varies directly with the concentration of a reactant. This means that if the concentration is halved, the rate of reaction is halved. Such reactions are said to be "first-order." Suppose that it took 90.0 seconds for the concentration of HCl to change from 2.00 M to 1.00 M. What would the concentration be 90.0 seconds after that? Assuming that the reaction is first-order, the concentration would be 0.50 M. Let's see why.

In the first 90.0 seconds, the change in concentration was 1.00 M. At this time, since the concentration is one-half the original concentration, the rate is one half the original rate. The average rate of the reaction, as the concentration of HCl goes from 1.00 M to 0.50 M, is exactly half of what the rate was when going from 2.00 M to 1.00 M. In the same amount of time, but at half the rate, the concentration changes half as much, or 0.50 M. Therefore, the amount of time needed to go from 2.00 M to 1.00 M must be exactly the same as the amount of time to go from 1.00 M to 0.50 M. First-order reactions have a varying rate, but a constant half-life. This means that the amount of time required to halve the concentration of a reactant is constant at constant temperature. If the HCl reacts with a half-life of 90.0 seconds, it is possible to predict its concentration at any time. In the same reaction, how long would it take for a 4.00M solution of HCl to reach 0.50 M? In the first 90.0 seconds the concentration would change from 4.00 M to 2.00 M. In the next 90.0 seconds the concentration would halve again going from 2.00 M to 1.00 M. In the next 90.0 seconds the concentration would change from 1.00 M to 0.50 M. It would take 3×90.0 seconds, or 270 seconds.

The concept of constant half-life is particularly important in the study of nuclear chemistry (Chapter 15). Radioactive decay always shows first-order kinetics.

Equilibrium

LOOKING AHEAD

Why does the water in a sealed container seem to stop evaporating? Why does the solute in a saturated solution appear to stop dissolving? You should know that in both cases, the systems are at equilibrium. The equilibrium between a liquid and its vapor was discussed in Chapter 1, and the equilibrium between dissolved and undissolved solute was discussed in Chapter 8. In this chapter, you will apply the concept of equilibrium to chemical reactions, as well as to physical changes.

When you have completed this chapter you should be able to:

- **Explain** how a chemical reaction reaches equilibrium.
- **Predict** how reactions at equilibrium will shift in response to a change in conditions (Le Chatelier's Principle).
- **Make predictions** about chemical systems based on K_{EQ}, the equilibrium constant.
- **Write** the equilibrium expression for a given chemical reaction.
- **Use** equilibrium constants to predict the solubilities of ionic substances in water.
- **Use** solubility guidelines to predict precipitation reactions.

Kinetics and Chemical Equilibrium

Let us begin the study of chemical equilibrium by applying our knowledge of kinetics to a simple chemical reaction, the reaction of hydrogen gas with iodine vapor to produce gaseous hydrogen iodide. When H_2 and I_2 molecules are mixed, they collide with one another. A certain fraction of the collisions will be favorable for chemical change, and HI gas molecules will form.

$$H_2(g) + I_2(g) \rightarrow 2\ HI(g)$$

Slowly, more and more HI molecules form as a result of favorable collisions. If the reaction takes place in a closed vessel, the product HI molecules will also collide with one another. As the concentration of HI increases, the frequency of HI—HI collisions will increase. Some of the HI—HI collisions will have enough energy and will occur at the proper orientation to change back to H_2 and I_2. This reaction is the reverse of the initial one. Soon, the rate at which H_2 and I_2 molecules produce HI and the rate at which HI molecules collide and produce H_2 and I_2 become equal. The system is then in a state of equilibrium. The reactions are shown graphically in Figure 10-1.

When the system is at equilibrium, H_2 and I_2 molecules still collide and form HI, and HI molecules still collide and form H_2 and I_2. However, the rates of the opposing reactions are equal.

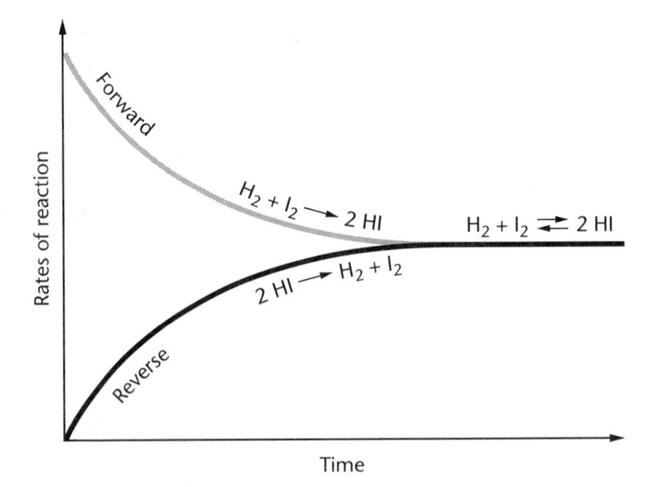

Figure 10-1 Reaction rates for the reaction of hydrogen with iodine

There is therefore no net change in the concentrations of H_2, I_2, and HI.

$$H_2 + I_2 \underset{r_2}{\overset{r_1}{\rightleftarrows}} 2\ HI$$

$$r_1 = \text{forward rate}$$

$$r_2 = \text{reverse rate}$$

$$r_1 = r_2$$

When a reaction reaches equilibrium, both the forward and the reverse reactions continue to take place. However, there is no further change in the concentrations of the reactants and products. In some reactions, there may be more products than reactants at equilibrium. In other reactions, there may be more reactants than products. Equal concentrations of reactants and products are *not* necessarily formed when equilibrium is reached.

Physical Equilibrium

When two opposing processes go on at the same time and at the same rate, **equilibrium** exists. You have just examined an example of chemical equilibrium. Equilibrium can exist between opposing physical processes too. For example, *phase equilibrium* can be reached when a liquid becomes a solid (freezes) and the solid melts and re-forms the liquid. Phase equilibrium can also exist between a liquid and its vapor (gas).

Let us look at one example of phase equilibrium involving a liquid and its vapor.

Phase Equilibrium If you were to observe, over a period of time, a stoppered flask half-filled with water, the level of liquid in the flask would appear to remain the same. How could you account for this observation? After all, a liquid continually evaporates at some fixed rate that depends on the temperature.

It can be shown experimentally that as the evaporation of a liquid occurs, an opposite change also occurs. In this opposite change, molecules of vapor collide at the surface of the liquid and re-form molecules of liquid. In other words, condensation—the change of state from a vapor, or gas, to a liquid—occurs at the same time that evaporation occurs.

When a liquid is placed in a sealed flask, only evaporation occurs at first. Soon, however, molecules of vapor condense. In time, the rates of these opposing changes—evaporation and condensation—become equal. At this point, an equilibrium exists. No visible changes seem to be occurring in the flask. However, at the molecular level, a balance exists between evaporation and condensation. Both processes continue to occur at the same rate.

Solution Equilibrium

Solubility was expressed earlier in terms of a saturated solution: Solubility refers to the maximum amount of solute that can be dissolved in a given amount of solvent at a specific temperature and pressure. Solubility can also be expressed in terms of equilibrium: In a saturated solution of a solid, the excess, undissolved solid is in equilibrium with the solute ions. Both ways of expressing solubility are useful.

Let us consider two kinds of solutions at equilibrium—gases in liquids and solids in liquids.

Gases in Liquids Seltzer, or soda water, is a solution of CO_2 in water, under high pressure. In a closed bottle of seltzer (see Figure 10-2), some of the CO_2 molecules constantly escape from the solution into the space above the liquid. At the same time, CO_2 molecules go from the space above the liquid into the liquid. As long as the bottle remains sealed at a constant temperature, the

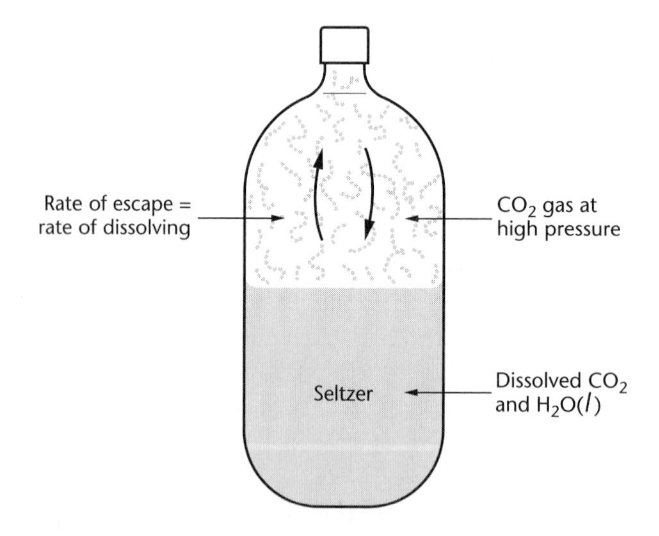

Figure 10-2 Equilibrium in a sealed seltzer bottle

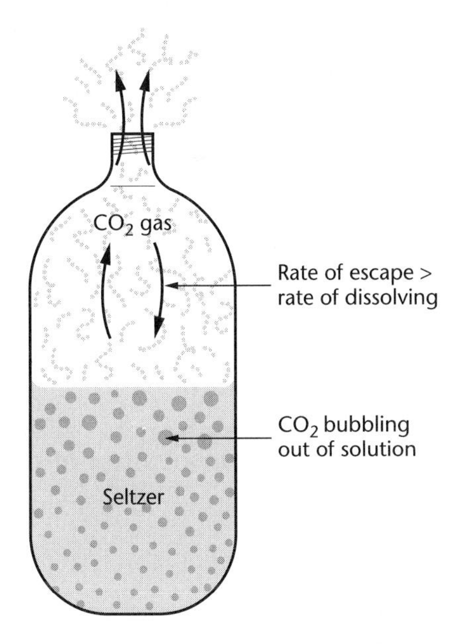

Figure 10-3 Escape of CO_2 from an open seltzer bottle

rate at which CO_2 molecules leave the solution will be the same as the rate at which CO_2 molecules enter the solution. The system remains at equilibrium, and no gas bubbles are formed. The concentration of CO_2 in the solution remains constant.

When the bottle is opened, some of the CO_2 molecules escape from the bottle into the air, decreasing the pressure and concentration of the gas above the liquid. The rate at which the CO_2 molecules enter the liquid is now decreased. (See Figure 10-3.) The system is no longer at equilibrium; CO_2 gas now escapes the liquid faster than it dissolves, forming the bubbles we enjoy when we drink carbonated beverages.

We can represent this equilibrium with the equation

$$CO_2\ (g) \rightleftharpoons CO_2\ (aq)$$

An increase in pressure, as provided by the bottling plant, drives this equilibrium to the right. A decrease in pressure, provided when you open the bottle, drives the equilibrium to the left.

Solids in Liquids Now consider a saturated solution of $NaNO_3$ in water at a fixed temperature. Excess $NaNO_3$ is a solid at the bottom of the container. Equilibrium exists between the dissolved

$NaNO_3$ and the solid $NaNO_3$. The solid $NaNO_3$ continues to dissolve, releasing Na^+ and NO_3^- ions into the water. At the same time and rate, Na^+ and NO_3^- ions return from solution to the solid (they crystallize). These changes are shown by the equation:

$$NaNO_3\,(s) \underset{\leftarrow \text{crystallizing}}{\overset{\text{dissolving}\rightarrow}{\rightleftarrows}} Na^+\,(aq) + NO_3^-\,(aq)$$

The Composition of an Equilibrium Mixture

From the examples of chemical equilibrium you have considered, you can conclude that at a given temperature and pressure the concentrations of reactants and products in any equilibrium mixture are fixed. What determines the numerical values of these concentrations?

To answer this question, let us begin with the generalized reaction

$$2\text{ A} + 3\text{ B} \rightleftarrows \text{C} + \text{D}$$

Assume that 2 moles of A is allowed to react with 1 mole of B in a one liter container. Suppose that when equilibrium is reached, measurements show that 0.2 mole of C has formed. You can now calculate the quantities of all of the other components in the mixture at equilibrium. If 0.2 mole of C is formed, the molar relationships show that 0.2 mole of D is also formed. Twice as much A, or 0.4 mole, must have been used in the reaction, and three times as much B, or 0.6 mole, must have been used. The composition of the mixture at equilibrium is as follows:

A = 2 moles − 0.4 mole = 1.6 moles

B = 1 mole − 0.6 mole = 0.4 mole

C = 0.2 mole (given)

D = 0.2 mole (from the molar relationships in the equation)

The composition of an equilibrium mixture can be found conveniently by setting up a reaction chart, as follows.

Reaction:	2A	+	3B	$\rightleftarrows$	C	+	D
Initial:	2 moles		1 mole		0		0
Change:	−0.4 mole		−0.6 mole		+0.2 mole		+0.2 mole
Equilibrium:	1.6 mole		0.4 mole		0.2 mole		0.2 mole

When using a reaction chart of this type, be sure that all entries in the "Change" row are in exactly the same ratio as the coefficients in the balanced equation. Also note that the signs must be opposite on opposite sides of the yields sign. If reactant is decreasing, the product must be increasing, and vice versa.

PRACTICE

10.1 Use a reaction chart to solve the following problem. For a generalized reaction with the equation

$$A + 2\,B \rightleftarrows C + 3\,D$$

When 1 mole of A is mixed with 1 mole of B in a 1-liter container, and the reaction is allowed to reach equilibrium, only 0.6 mole of A remains. Find the number of moles of B, C, and D in the container at equilibrium.

10.2 When a seltzer bottle is opened, as shown in Figure 10-3 on page 345, the mixture begins to fizz as the carbon dioxide comes out of solution. What happens when the bottle is resealed? Explain your answer.

Factors That Affect Chemical Equilibrium

Once equilibrium has been reached, it can be upset only by a factor that affects the forward and reverse reactions differently. The rate of a reaction is determined, you will recall, by the presence of a catalyst, by concentration, and by temperature. Do any of these factors affect equilibrium?

A catalyst does not. A catalyst has the same effect on both the forward and the reverse reactions. A catalyst permits a reaction to reach equilibrium in a shorter time. But once equilibrium has been reached, a catalyst has no effect. However, changes in the concentrations of the reactants, changes in pressure when gases are involved, and changes in temperature affect the rates of the forward and reverse reactions differently. The composition of the equilibrium mixture adjusts to accommodate these changes.

Le Chatelier's Principle

The principle that governs the effect of such changes on a system in equilibrium was formulated in 1887 by the French chemist Henry-Louis Le Chatelier. This principle was mentioned in Chapter 8. **Le Chatelier's Principle** can be stated as follows: If a system in equilibrium is subjected to a stress (a change in concentration, pressure of a gas, or temperature), the system adjusts to partially relieve the stress and restore equilibrium. This is the same as saying that a change in the concentrations of the reactants (or pressures, if the reactants are gaseous) or a change in the temperature of the system favors either the forward or the reverse reaction. The change produces a new equilibrium mixture.

The factors that can place a stress on a system in equilibrium will now be considered in more detail.

Effect of Changes in Concentration

As the concentrations of reactants increase, the likelihood of effective collisions between molecules increases. Thus the rate of reaction increases. If a system is at equilibrium and the concentration of one or more reactants increases, the equilibrium shifts toward the right. In other words, the equilibrium shifts to form a larger concentration of products. The shift relieves the stress on the system, in accordance with Le Chatelier's Principle. The new equilibrium mixture now has more products than would have been present had the system remained at its original equilibrium.

The opposite is also true. If a system is at equilibrium and concentration of products increases, the equilibrium shifts toward the left. The stress is offset by the formation of more reactants. In this case, the new equilibrium mixture contains more reactants than would have been present if the original equilibrium had been maintained.

Equilibrium systems are of importance in many practical situations. The commercial preparation of hydrogen chloride gas from NaCl is an example. The reaction takes place at a temperature below 500°C. The equation for the reaction is

$$NaCl + H_2SO_4 \rightleftarrows NaHSO_4 + HCl\ (g)$$

A high concentration of H_2SO_4 drives the reaction to the right. Most of the NaCl is used up. In addition, because the product, HCl, is a gas, it is continuously being removed. This, too, acts as a stress on the system (decrease in concentration of the product). To relieve this stress, the equilibrium shifts to form more HCl (more product). As the result of these shifts in equilibrium, a maximum amount of product can be obtained from a given amount of NaCl.

SAMPLE PROBLEM

PROBLEM

Suppose the following system is at equilibrium:

$$SO_3\ (g) + NO\ (g) \rightleftarrows SO_2\ (g) + NO_2\ (g)$$

If additional NO is added to the system, what happens to the concentrations of SO_3, SO_2, and NO_2?

SOLUTION

The addition of reactant causes the system to shift toward the right, forming more product. Therefore the concentrations of SO_2 and NO_2, the products, increase. When the reaction shifts to the right, the concentrations of reactants must decrease. Thus the concentration of SO_3 decreases.

PRACTICE

10.3 In the equilibrium system

$$2\ SO_2\ (g) + O_2\ (g) \rightleftarrows 2\ SO_3\ (g)$$

predict what would happen to the concentration of SO_2 gas as a result of the following changes:

(a) SO_3 is added to the system.

(b) Some O_2 is removed from the system.

The Equilibrium Constant

Consider the reaction

$$H_2\ (g) + I_2\ (g) \rightleftarrows 2\ HI\ (g)$$

When the reaction reaches equilibrium, the rate of the forward reaction must equal the rate of the reverse reaction. The rate of the forward reaction depends on the concentration of the reactants, at the same time, the rate of the reverse reaction depends on the concentration of the products. For the system to be at equilibrium, there must be a special relationship between the concentrations of products and reactants. A ratio called the *reaction quotient, Q*, is used to compare the concentrations of products and reactants. The numerator contains the concentrations of the products, and the denominator the concentrations of reactants. The square brackets ([ ]) around the symbols for the products and reactants indicate "concentration of . . . ." For the previous reaction,

$$Q = \frac{[HI]^2}{[H_2][I_2]}$$

The concentrations of the components are multiplied in this expression. Since there are two moles of HI in the product, the numerator could be written as [HI][HI], which is $[HI]^2$. All coefficients in the balanced equation become exponents in the reaction quotient.

For the system to be at equilibrium at a given temperature, the reaction quotient must have a specific value, called the equilibrium constant, K_{EQ}. At equilibrium, then,

$$\frac{[HI]^2}{[H_2][I_2]} = K_{EQ}$$

For this reaction at 298 K, the value of K_{EQ} is 600. This means, for this system to be at equilibrium the numerator must be 600 times greater than the denominator. The value of an equilibrium constant changes when the temperature changes. It is not affected, however, by changes in pressure or concentration.

The ratio of product concentration to reactant concentration at equilibrium, when properly expressed, is equal to the **equilibrium constant.** The resulting equation, such as that shown above for the formation of HI, is called an equilibrium expression. To write equilibrium expressions correctly, remember the following rules:

1. Brackets ([ ]) are used to indicate "concentration of . . . ." $[H_2]$ means "concentration of hydrogen."
2. The products go in the numerator, the reactants in the denominator. Concentrations are multiplied.
3. Coefficients in the balanced equation become exponents in the equilibrium expression.
4. Solids are omitted from equilibrium expressions.

For reactions in solution, the usual unit of concentration in an equilibrium expression is molarity (moles per liter). For reactions involving gases, however, the concentration of a gas is directly proportional to its pressure. In these systems, a pressure unit, the atmosphere, is often used instead of the concentration unit, molarity. When an equilibrium constant is computed using pressure units, it is called a K_P. When concentration units are used, the equilibrium constant is the K_C.

SAMPLE PROBLEM

PROBLEM

Write the equilibrium expression for the reaction

$$N_2\ (g) + 3\ Cl_2\ (g) \rightleftarrows 2\ NCl_3\ (g)$$

SOLUTION

Place the formula of the product over the formulas of the reactants, use brackets, remember to multiply, and change all coefficients to exponents.

$$K_{EQ} = \frac{[NCl_3]^2}{[N_2][Cl_2]^3}$$

PRACTICE

10.4 Write correct equilibrium expressions for each of the following chemical reactions:

(a) $N_2\ (g) + 3\ H_2\ (g) \rightleftarrows 2\ NH_3\ (g)$

(b) $F_2\ (g) + 2\ HCl\ (g) \rightleftarrows 2\ HF\ (g) + Cl_2\ (g)$

(c) $C\ (s) + H_2O\ (g) \rightleftarrows CO\ (g) + H_2\ (g)$

You have already seen how a system at equilibrium shifts as a result of changes in concentration. Let us examine these changes again and consider their effect on the equilibrium expression. Recall that at equilibrium, the reaction quotient, Q, must equal the equilibrium constant, K_{EQ}. For the system

$$2\ SO_2\ (g) + O_2\ (g) \rightleftarrows 2\ SO_3\ (g)$$

at equilibrium,

$$Q = K_{EQ} = \frac{[SO_3]^2}{[SO_2]^2[O_2]}.$$

Suppose we add some O_2 to the system. The system is no longer at equilibrium. The K_{EQ} does not change, as long as the temperature is constant. However, since we have increased one of the quantities in the denominator of the reaction quotient, Q is now less than K_{EQ}. To get Q equal to K_{EQ}, the numerator must increase, and the denominator must decrease. To get back to equilibrium, the reaction shifts to the right, increasing the concentration of product and decreasing the concentration of reactant. Just as predicted, an increase in the concentration of a reactant drives the reaction to the right. The equilibrium constant remains the same.

Effect of Changes in Pressure

Solids and liquids are virtually incompressible. Thus equilibrium systems involving solids and liquids do not react appreciably to changes in pressure. But the volume of a gas at constant temperature is determined by its pressure. Thus an increase in the pressure on a gas decreases the volume, and the concentration of the gas increases. Pressure therefore can have an effect on equilibrium in systems involving gases. Gaseous equilibrium systems are affected by pressure changes only under certain conditions, however. Let's see why.

According to Avogadro, equal volumes of gases at the same temperature and pressure contain the same number of molecules (or moles). The volume of a gas therefore is proportional to the number of molecules (or moles).

The following is an equilibrium reaction involving gases:

$$H_2\ (g) + Br_2\ (g) \rightleftharpoons 2\ HBr\ (g)$$

The equation shows that 1 mole of H_2 gas reacts with 1 mole of Br_2 gas to form 2 moles of HBr gas. Since the volumes of the gases are proportional to the number of moles, the equation also states that 1 volume of H_2 reacts with 1 volume of Br_2 to form 2 volumes of HBr. The total volume of the reactants equals the volume of the product.

If the pressure on this system is increased, the volume of the system will decrease. However, since both the products and the reactants have the same total volume, no shift in the reaction occurs. A change in pressure has no effect on equilibria in which there are equal volumes of products and reactants. This can be demonstrated by using the equilibrium expression for the reaction.

$$K = \frac{[HBr]^2}{[H_2][Br_2]}$$

At constant temperature you can increase the pressure of a system by changing its volume. Suppose you halve the volume. The pressures and concentrations of all of the gases would then double. How would that affect the reaction quotient? The [HBr] would double. Since [HBr] is squared in our expression, the value of the numerator would now be four times greater. The $[H_2]$ and $[Br_2]$ would each double. Multiplying their new concentrations would result in a denominator four times larger. Since the numerator and denominator have both been multiplied by four, the value of the quotient has not changed. The system is still at equilibrium, and does not shift in either direction.

The situation is different in the equilibrium system shown by the next equation.

$$PBr_3\ (g) + Br_2\ (g) \rightleftharpoons PBr_5\ (g)$$

In this case, 2 moles of gas (1 mole of PBr_3 and 1 mole of Br_2) react and form 1 mole of gas, PBr_5. Or, 2 volumes of reactants form 1 volume of product. An increase in pressure will not have the same impact on the reactants as on the product. Increasing the pressure will drive the reaction toward the side with the smaller volume, in this case, the product. A decrease in pressure will drive a reaction toward the side with the larger volume, in this case, the reactant. Let us see how an increase in pressure would affect the reaction quotient,

$$Q = \frac{[PBr_5]}{[PBr_3][Br_2]}$$

To increase the pressure of the equilibrium system at constant temperature you would need to decrease the volume. Suppose once again, you halve the volume. All of the pressures and concentrations would double. $[PBr_5]$ doubles, making the numerator of the quotient two times larger. Both $[PBr_3]$ and $[Br_2]$ double, which makes the denominator four times larger. Since the denominator has increased more than the numerator, the value of the reaction quotient has decreased and is now less than the equilibrium constant. The system is no longer at equilibrium. To get back to equilibrium the reaction must shift to the right, toward the products. When Q again equals K_{EQ}, the numerator (product) has increased and the denominator (reactants) decreased.

The effect of a change in pressure at constant temperature (which means a change in volume) on an equilibrium involving gases can be summarized as follows:

1. An increase in pressure favors the direction that produces a smaller number of moles of gas (smaller volume).
2. A decrease in pressure favors the direction that produces a larger number of moles of gas (larger volume).
3. A change in pressure has no effect on the equilibrium if the number of moles (or volumes) of gaseous products equals the number of moles (or volumes) of gaseous reactants.
4. A change in pressure does not change the value of the equilibrium constant.

PRACTICE

10.5 For each reaction, predict how the following changes would affect the amount of Cl_2 gas at equilibrium.

(a) $CO\ (g) + Cl_2\ (g) \rightleftharpoons COCl_2\ (g)$, an increase in pressure

(b) $MnO_2\ (s) + 4\ HCl\ (aq) \rightleftharpoons 2\ H_2O\ (l) + MnCl_2\ (aq) + Cl_2\ (g)$, an increase in pressure

(c) $H_2\ (g) + Cl_2\ (g) \rightleftharpoons 2\ HCl\ (g)$, an increase in pressure

(d) $SCl_4\ (g) \rightleftharpoons SCl_2\ (g) + Cl_2\ (g)$, an increase in the volume of the container

10.6 (For experts) A teacher asked a class to predict what would happen if the volume of the container were decreased in the equilibrium system $SO_3\ (g) + NO\ (g) \rightleftharpoons SO_2\ (g) + NO_2\ (g)$. One student answered that the concentrations of

SO_2 and NO_2 would increase. A second said that the amounts of SO_2 and of NO_2 would remain the same. Why are both students correct?

Effect of Changes in Temperature

The effects of a change in temperature on an equilibrium system can also be predicted with the use of Le Chatelier's Principle. First the direction in which the reaction liberates or absorbs heat must be determined. For example, in the generalized equation

$$A\ (g) + B\ (g) \rightleftharpoons C\ (g) + \text{heat}$$

the forward reaction is exothermic. The reverse reaction is endothermic.

According to Le Chatelier, a change in temperature, which is a stress, is offset in the direction that partially relieves the change, that is, overcomes the temperature change. Thus, if a system comes to equilibrium with a net release of heat (is exothermic), an increase in temperature favors the endothermic reaction. An example of such an equilibrium system is

$$H_2\ (g) + Br_2\ (g) \rightleftharpoons 2\ HBr_2\ (g) + \text{heat}$$

An increase in temperature drives this system to the left, which is the endothermic reaction. The concentrations of the reactants in the equilibrium mixture are increased. The concentration of the product is decreased.

If a system comes to equilibrium with a net absorption of heat (is endothermic), an increase in temperature again favors the endothermic reaction. Thus, increasing the temperature in the equilibrium system

$$N_2\ (g) + O_2\ (g) + \text{heat} \rightleftharpoons 2\ NO\ (g)$$

drives the system to the right, because this is the endothermic reaction requiring the absorption of heat. The concentration of the product in the equilibrium mixture is increased.

The two other stresses that have been considered, change in pressure and change in concentration, had no effect on the value of the equilibrium constant. A change in temperature, however, does change the value of the K_{EQ}. This can be explained by considering the effect of temperature change on the rates of the forward and the reverse reactions. You will recall that an increase in temperature always increases the rate of a chemical reaction. How much the rate increases depends upon the activation energy of the reaction. The greater the activation energy, the greater the effect of temperature change on the reaction rate. Figure 10-4 illustrates the energies of activation in an exothermic reaction.

Notice that the activation energy of the reverse reaction is greater than that of the forward reaction. An increase in temperature will increase the rates of both the forward and the reverse reactions. Since the reverse reaction has a higher activation energy than the forward reaction, its rate will show a larger increase than the rate of the forward reaction. With the rate of the reverse reaction now greater than the rate of the forward reaction, the system shifts to the left, producing more reactants and less product when equilibrium is reestablished. The value of the equilibrium constant has decreased, since there is less product and more reactant in the new equilibrium mixture. An increase in temperature decreases the value of the equilibrium constant for an exothermic reaction.

If the temperature is decreased, the rates of both the forward and the reverse reactions decrease. However, since the reverse reaction has the greater activation energy, its rate decreases more than the rate of the forward reaction. The reverse reaction is now slower than the forward reaction, so the system shifts to the right, producing more product and less reactant. The value of the equilibrium constant increases.

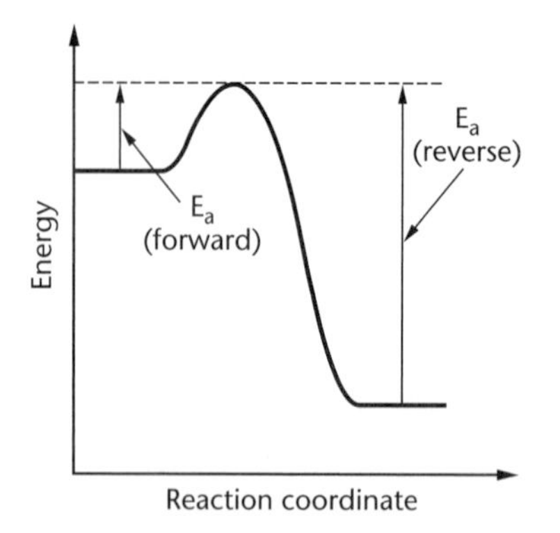

Figure 10-4 Energies of activation, E_a, for an exothermic reaction

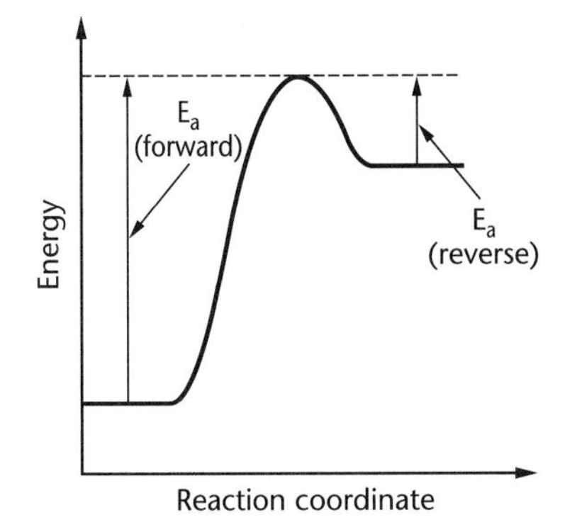

Figure 10-5 Energies of activation, E_a, for an endothermic reaction

Figure 10-5 illustrates the energies of activation in an endothermic reaction. This time the activation energy is greater for the forward reaction than for the reverse reaction. The rate of the forward reaction will be affected more by temperature change than the rate of the reverse reaction. An increase in temperature drives the reaction in the forward direction, increasing the value of the equilibrium constant. A decrease in temperature drives the reaction in the reverse direction, decreasing the value of the equilibrium constant.

The endothermic reaction, whether it is forward or reverse, always has a greater activation energy than the exothermic reaction. Increases in temperature always favor the endothermic reaction and increase the value of its equilibrium constant. Decreases in temperature always favor the exothermic reaction and increase the value of its equilibrium constant.

SAMPLE PROBLEM

PROBLEM

The K_{EQ} for the reaction at 298 K

$$N_2\ (g) + 3\ H_2\ (g) \rightleftarrows 2\ NH_3\ (g)$$

is 6.7×10^5. The ΔH for this reaction at 298 K is -92 kJ. If the temperature is raised to 373 K, how would this affect

(a) the concentration of N_2 at equilibrium,

(b) the concentration of H_2 at equilibrium,

(c) the concentration of NH_3 at equilibrium,

(d) the value of the equilibrium constant, K_{EQ}.

SOLUTION

You can tell by the sign of ΔH that the forward reaction is exothermic. An increase in temperature favors the endothermic reaction, which in this case is the reverse reaction. The reaction shifts to the left. The concentrations of the reactants, H_2 and N_2, both increase. The concentration of the product, NH_3, decreases. The value of the equilibrium constant decreases, because there is less product and more reactant under the new conditions. (In fact, small temperature changes can cause very large changes in the value of the K_{EQ}. In this case, K_{EQ} is 6.7×10^5 at 298 K. At 373 K, K_{EQ} is 3.8×10^2, which is about 2000 times lower.)

PRACTICE

10.7 For the reaction at equilibrium,

$$C\ (s) + O_2\ (g) \rightleftarrows CO_2\ (g) + 393\ \text{kJ}$$

predict the effect of an increase in temperature on

(a) the concentration of O_2,

(b) the concentration of CO_2,

(c) the value of the equilibrium constant.

10.8 For the equilibrium system,

$$\frac{1}{2}\ N_2\ (g) + \frac{1}{2}\ O_2\ (g) \rightleftarrows NO\ (g), \qquad \Delta H = +90.2\ \text{kJ}$$

predict the effect of a decrease in temperature on

(a) the concentration of O_2,

(b) the concentration of NO,

(c) the value of the equilibrium constant.

Summary: Gaseous Equilibrium Systems

The table that follows summarizes the effects of changes of concentration, pressure, and temperature in gaseous equilibrium systems.

Stress	*Direction of Shift of Equilibrium*	*Change in the Value of K_{EQ}*
$A(g) + B(g) \rightleftharpoons C(g) + \text{heat}$		
Increased concentration of A or B	$\longrightarrow$	No change
Increased concentration of C	$\longleftarrow$	No change
Increased pressure	$\longrightarrow$	No change
Decreased pressure	$\longleftarrow$	No change
Increased temperature	$\longleftarrow$	Decreases
Decreased temperature	$\longrightarrow$	Increases
$2\,A(g) + B(g) + \text{heat} \rightleftharpoons 2\,C(g) + 2\,D(g)$		
Increased concentration of A or B	$\longrightarrow$	No change
Increased concentration of C or D	$\longleftarrow$	No change
Increased pressure	$\longleftarrow$	No change
Decreased pressure	$\longrightarrow$	No change
Increased temperature	$\longrightarrow$	Increases
Decreased temperature	$\longleftarrow$	Decreases

Le Chatelier's Principle and the Haber Process

Ammonia, NH_3, is a very important raw material used in the preparation of nitrates and ammonium compounds. Considerable amounts of ammonia are prepared commercially by the Haber process. In this process, several effects predicted by Le Chatelier's Principle must be considered to obtain the highest possible yield.

The equation for the reaction is

$$N_2(g) + 3\,H_2(g) \rightleftharpoons 2\,NH_3(g) + 92\text{ kJ}$$

Notice that all the substances in the reaction are gases at equilibrium. If the pressure of the system is increased, the volume decreases and the concentrations of both reactants and product will tend to increase. But the number of moles (volume) of the product is only half the total number of moles (volume) of the reactants. For this reason, an increase in pressure is offset by the formation of more product—the equilibrium shifts in the direction of the smaller volume (product). That is, increasing the pressure increases the yield of NH_3.

The equilibrium constant of this reaction at 298 K is 6.7×10^5, or 670,000. This very large number indicates that at equilibrium, nearly all of the reactant has been converted to product.

However, if you release some hydrogen gas into the air, which is 80% nitrogen, you do not observe the formation of any ammonia. Although the reaction has a very high equilibrium constant, it is so slow at room temperature that no observable ammonia is formed. The reaction can be made to go faster by increasing the temperature. However, the formation of ammonia is exothermic. An increase in temperature will favor the reverse, endothermic reaction, and decrease the equilibrium concentration of ammonia. A temperature must be chosen at which the reaction proceeds relatively quickly, while the equilibrium constant is still high enough to produce a reasonable yield of ammonia.

All of these factors are taken into account in the Haber process. A catalyst is used so that the system reaches equilibrium rapidly. Pressures of about 400 atm are used to increase the yield of NH_3. The temperature is adjusted to around 720 K, high enough to produce ammonia rapidly without too large a decrease in the equilibrium constant of the reaction.

By adjusting such factors as concentration, pressure and temperature, you can drive a chemical reaction in the desired direction. You can also predict the effects of each of these changes on the components of a given system.

SAMPLE PROBLEMS

PROBLEM

1. Consider the reaction

$$2\ CO\ (g) + O_2\ (g) \rightleftharpoons 2\ CO_2\ (g) + 565.1\ kJ$$

To increase the quantity of CO_2 at equilibrium, what must be done to (a) the temperature, (b) the pressure, (c) the concentration of CO?

SOLUTION

(a) The reaction as written is exothermic. Exothermic reactions are favored by a decrease in temperature. Decreasing the temperature would drive the reaction to the right, producing more CO_2.

(b) There are 3 moles of gaseous reactants (2 CO + 1 O_2). There are 2 moles of gaseous product. The product has a smaller volume than the reactant. The side with the smaller volume is favored by an increase in pressure, so an increase in pressure will drive this reaction to the right, producing more CO_2.

(c) Increasing the concentration of a reactant will drive the reaction to the right, toward product. An increase in the concentration of CO will result in an increase in the quantity of CO_2.

PROBLEM

2. For the generalized reaction,

$$A(g) + B(g) + \text{heat} \rightleftharpoons 2\,C(g) + 3\,D(g)$$

predict the effect of the following on the equilibrium quantity of substance C:

(a) increase in temperature, (b) addition of D,

(c) decrease in pressure, (d) addition of a catalyst.

SOLUTION

(a) The reaction as written is endothermic. An increase in temperature favors endothermic reactions, so this reaction will shift to the right, producing more C.

(b) Addition of product drives a reaction to the left, toward reactants. Shifting the reaction to the left results in the production of less C.

(c) A decrease in pressure drives a reaction in the direction that has the larger volume. In this case, the reactants provide 2 moles of gas, while the products provide 5 moles of gas. A decrease in pressure would drive this reaction to the right, resulting in an increase in the quantity of C.

(d) A catalyst gets the reaction to equilibrium faster, but does not change any of the equilibrium amounts. The quantity of C would remain the same.

PRACTICE

10.9 The following system is at equilibrium:

$$2\ NO\ (g) + O_2\ (g) \rightleftharpoons 2\ NO_2\ (g) + 113\ kJ$$

Predict the effect of the following on the equilibrium quantity of oxygen.

(a) removing some NO, (b) increasing the pressure,
(c) decreasing the temperature, (d) adding some NO_2

10.10 For the reaction

$$2\ SO_3\ (g) \rightleftharpoons 2\ SO_2\ (g) + O_2\ (g), \qquad \Delta H = +197\ kJ$$

Describe four actions you could take to increase the quantity of oxygen present at equilibrium.

Solutions and Equilibrium

In Chapter 8, you learned how ionic substances dissociate into hydrated ions when dissolved in water. Substances that produce such ions in water are called electrolytes, because the mobile ions in solution permit the solution to conduct electricity. In Chapter 12, you will learn that ionic substances are not the only electrolytes; however, the discussion of solution equilibria will be limited to ionic solids in water.

Soluble ionic solids are strong electrolytes. In solution, these substances form ions as shown in the following general equation:

$$M^+Z^-\ (s) \xrightarrow{H_2O} M^+\ (aq) + Z^-\ (aq)$$

If MZ is a soluble ionic substance, this reaction proceeds almost completely to the right. The dissolving process separates the ions, so that only the tiniest fraction of M^+ and Z^- ions remain attached to each other. However, if MZ is only slightly soluble, it readily forms a saturated solution, and equilibrium is established: $M^+Z^-\ (s) \rightleftharpoons M^+\ (aq) + Z^-\ (aq)$. We can write the equilibrium expression for this process, $K = [M^+][Z^-]$. Since solids are omitted from equilibrium expressions, this equilibrium expression has no denominator; the constant, K, is equal to the *product* of concentrations of the two ions. This equilibrium constant is called the K_{SP}, or **solubility product constant.** Like any other equilibrium constant, the value of the K_{SP} will change only if the temperature is changed. Since the dissolving of a solid is generally endothermic, an increase in temperature will generally increase the K_{SP}. In general, the lower the K_{SP}, the lower the solubility of the salt.

Let us write the expressions for the K_{SP} for two salts. AgCl has a K_{SP} of 1.8×10^{-10}. This means that in a saturated solution of AgCl at 298 K, the product, $[Ag^+][Cl^-] = 1.8 \times 10^{-10}$. PbI_2 has a K_{SP} of 7.1×10^{-9}. In a saturated solution of lead (II) iodide, the product $[Pb^{2+}][I^-]^2 = 7.1 \times 10^{-9}$. Since lead iodide has a higher K_{SP} than silver chloride, you would expect lead iodide to be the more soluble of the two. Notice that the $[I^-]$ term is squared in the K_{SP} expression. That is because the equation for the dissolving of PbI_2 is

$$PbI_2 \rightarrow Pb^{2+}\,(aq) + 2I^-\,(aq)$$

As in all equilibrium expressions, the coefficients in the balanced equation become exponents. When writing K_{SP} expressions be sure to use exponents to express the number of ions of each type that are formed per mole of solute.

SAMPLE PROBLEM

PROBLEM

Write the K_{SP} expression for silver sulfate, Ag_2SO_4.

SOLUTION

The ions formed when silver sulfate dissolves are two silver ions, and one sulfate ion.

$$Ag_2SO_4 \rightleftharpoons 2\,Ag^+ + SO_4{}^{2-}$$

The expression is $K_{SP} = [Ag^+]^2[SO_4{}^{2-}]$.

PRACTICE

10.11 Write the expression for the K_{SP} for each of the following slightly soluble ionic solids.

(a) $BaSO_4$ (b) Li_2CO_3 (c) $Fe(OH)_3$

10.12 Arrange the following salts from most soluble to least soluble.

AgBr, $K_{SP} = 5.0 \times 10^{-13}$; AgI, $K_{SP} = 8.3 \times 10^{-17}$;
$ZnCO_3$, $K_{SP} = 1.4 \times 10^{-11}$; $CaSO_4$, $K_{SP} = 9.1 \times 10^{-6}$

The Common Ion Effect

The K_{SP} of AgCl, 1.8×10^{-10}, is a low number, so we would expect silver chloride to be only very slightly soluble in water. How would the solubility of AgCl be affected if we tried to dissolve it not in pure water, but in salt water? The solubility of AgCl is determined by the equilibrium AgCl (*s*) $\rightleftarrows$ Ag^+ (*aq*) + Cl^- (*aq*). Salt water contains Na^+ and Cl^- ions. The sodium ions do not significantly affect the silver chloride. The effect of the chloride ions, however, can be predicted using Le Chatelier's Principle. Increasing the concentration of Cl^- ions drives the equilibrium to the left, forming more undissolved AgCl, resulting in fewer silver ions in solution. We refer to the chloride as a common ion because it is common to both of our solutes. The presence of a common ion decreases the solubility of a given solute. The **common ion effect** describes the decrease in solubility of a solute caused by the presence in solution of another compound with which it has an ion in common.

PRACTICE

10.13 Which of the following substances, in solution, would decrease the solubility of $BaSO_4$? Why?

(a) $BaCl_2$ (b) K_2SO_4 (c) Na_2S (d) $Ca(NO_3)_2$

Reactions That Go to Completion

In Chapter 7, you were shown how to solve problems based on balanced chemical equations, such as this:

$$H_2\,(g) + Cl_2\,(g) \rightarrow 2\,HCl\,(g)$$

How many moles of hydrogen chloride are formed when 2.0 moles of hydrogen gas react with excess chlorine? Since the mole ratio of HCl to H_2 is 2 : 1, you would expect to form 4.0 moles of HCl. In solving this problem you assume that all of the H_2 is converted to HCl. At the end of the reaction, there are 4.0 moles of HCl, and no moles of H_2 remaining. A reaction that proceeds until one of the reactants is completely used up is said to *go to completion.* Under what conditions will reactions go to completion?

A reaction will go to completion if it has a large equilibrium constant. The equilibrium constant for the formation of HCl is 2.5×10^{33}. For the reaction quotient to reach a number this large, one or both of the reactants must approach a concentration of 0 *M*.

Another reaction that goes to completion is

$$2\,KClO_3\,(s) \rightarrow 2\,KCl\,(s) + 3\,O_2\,(g).$$

(This reaction is very slow at room temperature, and is generally heated and catalyzed.) It goes to completion because the oxygen gas leaves the system. When a reaction produces a gas that leaves the system, the reaction goes to completion. Such a reaction never reaches equilibrium because the product is removed as fast as it is formed so the reverse reaction cannot occur.

Reactions between ions in aqueous solution are said to go to completion when they form precipitates (insoluble solids), gases of low solubility, or weak electrolytes. Water itself is a weak electrolyte, so reactions that form water generally go to completion.

Predicting Precipitation

When clear aqueous solutions of $AgNO_3$ and NaCl are mixed, a white precipitate immediately forms. Because you know that AgCl has a very low K_{SP}, and is therefore nearly insoluble, you might have been able to predict that the precipitate is AgCl. In order to predict the formation of a precipitate, it is necessary to know which solids are insoluble. You can use a table, such as the "Solubility Guidelines," Table F in Appendix 4 to predict precipitation.

Let us use this table to analyze the reaction between $AgNO_3$ and NaCl. You may recall, from Chapter 4, that this is a double replacement reaction.

$$AgNO_3 + NaCl \rightarrow NaNO_3 + AgCl.$$

The table tells us that all Group 1 ions, such as sodium, form soluble compounds. Therefore, $NaNO_3$ is soluble. It also tells us that the halides (Cl^-, Br^-, I^-) are soluble except when combined with Ag^+, Pb^{2+}, and Hg_2^{2+}. We conclude from this that AgCl is insoluble. (Note that while the K_{SP} of AgCl is a low number, it is not zero. Therefore, a tiny amount of AgCl is in solution. Those compounds that are listed as insoluble should more precisely be considered nearly insoluble)

In solution, these ionic substances are dissociated into aqueous ions; we can write the *ionic* equation $Ag^+ (aq) + NO_3^- (aq) + Na^+ (aq) + Cl^- (aq) \rightarrow AgCl (s) + Na^+ (aq) + NO_3^- (aq)$ The soluble ionic solids are written as ions, while the insoluble AgCl is shown as a solid. Note that the Na^+ and NO_3^- ions appear on both sides of the equation. These ions do not participate in this reaction; they are called **spectator ions.** We can rewrite the precipitation reaction with the spectator ions omitted:

$$Ag^+ (aq) + Cl^- (aq) \rightarrow AgCl (s)$$

This last method of expressing the reaction is called the **net ionic equation.** The following Sample Problems illustrate the use of the Solubility Guidelines.

SAMPLE PROBLEMS

PROBLEM

1. What precipitate will form when sodium carbonate and calcium chloride solutions are mixed?

SOLUTION

Write the chemical equation for the reaction.

$$Na_2CO_3 + CaCl_2 \rightarrow CaCO_3 + 2\ NaCl$$

Use the Solubility Guidelines, Table F, to check the solubilities of the products. The first line in the list of soluble compounds, on the left side of the table, indicates that Group 1 ions, such as sodium, always form soluble compounds. Therefore, NaCl is soluble. The first line on the list of insoluble compounds, on the right side of the table, indicates that carbonates are insoluble except when combined with Group 1 ions or with ammonium ion. Calcium is a Group 2 metal. You can conclude that calcium carbonate is insoluble. The precipitate is $CaCO_3$.

PROBLEM

2. Which of the following solutions will form a precipitate when added to a solution of $BaCl_2$? (a) $Pb(NO_3)_2$ (b) Na_2SO_4 (c) CaI_2

SOLUTION

(a) The table indicates that Cl^- forms soluble compounds except when combined with Ag^+, Pb^{2+}, and Hg_2^{2+}. Therefore a precipitate of $PbCl_2$ will form.

(b) The table lists next to "sulfates" those ions that form insoluble sulfates. The list includes Ba^{2+}. Therefore $BaSO_4$ will precipitate.

(c) Calcium is not listed as an ion that forms insoluble chlorides. Barium ion does not form insoluble iodides. Since both $CaCl_2$ and BaI_2 are soluble, no precipitate forms.

PRACTICE

10.14 Use Table F on page 592 to identify the precipitate(s) formed, if any, when the following aqueous solutions are mixed:

(a) $AgNO_3$ with NaCl

(b) $CuSO_4$ with $Ba(OH)_2$

(c) $NaNO_3$ with KI

10.15 Which of the reactions in question 10.14 (above) would not go to completion?

10.16 Predict whether each of the following substances is soluble or insoluble in water.

(a) $CaBr_2$ (b) $Fe(OH)_3$ (c) $KClO_3$ (d) $PbSO_4$

10.17 (For experts) Give the formula of a soluble substance, which, when added to a solution of $Ba(OH)_2$, will form 2 precipitates. (There are many possible correct answers)

TAKING A CLOSER LOOK

Quantitative Aspects of Equilibrium

So far, equilibrium has been considered in a qualitative way. The principles that govern equilibrium were used to predict how different conditions affect reacting substances and their products. You have also seen that the equilibrium condition is expressed using a number called the equilibrium constant, K_{EQ}. The equilibrium constant allows a quantitative approach to equilibria: to predict the exact concentrations of products and reactants in an equilibrium system.

Use of the Equilibrium Constant

The numerical value of K_{EQ} indicates whether the products or the reactants are favored when a reaction reaches equilibrium. Recall that the products of a reaction make up the numerator of the K_{EQ} fraction. The reactants make up the denominator. If the value of K_{EQ} is greater than one, therefore, the concentrations of the products have to be greater than the concentrations of the reactants. Furthermore, the larger the value of K_{EQ}, the more the products are favored at equilibrium. On the other hand, if the value of K_{EQ} is less than one, the reactants are favored at equilibrium. And, the smaller the value of K_{EQ}, the smaller are the concentrations of the products at equilibrium.

Let us see how the K_{EQ} of two reactions can be put to use.

$$2\ HCl\ (g) \rightleftharpoons H_2\ (g) + Cl_2\ (g) \qquad K_{EQ} = 4 \times 10^{-34} \text{ at } 298\ K$$

When this reaction is at equilibrium, the concentration of HCl molecules is much greater than the concentrations of hydrogen and chlorine molecules. A low value of the K_{EQ} indicates that the system favors the formation of reactants, and the reaction tends to go from right to left in the chemical equation. This can be indicated in the chemical equation by varying the lengths of the arrows:

$$2\ HCl\ (g) \rightleftarrows H_2\ (g) + Cl_2\ (g)$$

In the second example,

$$H_2\ (g) + I_2\ (g) \rightleftarrows 2\ HI\ (g) \qquad K_{EQ} = 600$$

the concentration of HI molecules at equilibrium is greater than the concentration of H_2 and I_2 molecules.

One reaction can also be compared with another. For example, a reaction in which $K_{EQ} = 10^{-6}$ favors reactants more at equilibrium than does a reaction in which $K_{EQ} = 10^{-2}$. A reaction in which $K_{EQ} = 10^{6}$ favors products more than does a reaction in which $K_{EQ} = 10^{2}$.

K_{EQ} can be determined from an experimental study of an equilibrium system. Occasionally, observable color changes enable measurements of equilibrium concentrations to be made. A device called a spectrometer can be used to determine the concentration of a colored solute. The **spectrometer** reads the amount of light of a given frequency that is absorbed by a solute. The reading, called the "absorbance" is directly proportional to the concentration of the colored material. For example, solutions containing the Fe^{3+} ion react with those containing the SCN^{-} ion (thiocyanate) to form the $FeSCN^{2+}$ ion, which has a deep red color. A spectrometer can be used to determine the concentration of $FeSCN^{2+}$ in an equilibrium mixture. Assume the initial concentration of Fe^{3+} *(aq)* is 4.0×10^{-2} mole/liter and that of SCN^{-} *(aq)* is $1.0\ 10^{-3}$ mole/liter. The equilibrium concentration of $FeSCN^{2+}$ *(aq)* is measured and found to be 9.2×10^{-4} mole/liter.

The equation that represents the equilibrium is

$$Fe^{3+}\ (aq) + SCN^{-}\ (aq) \rightleftarrows \underset{\text{thiocyanoiron(III)}}{FeSCN^{2+}(aq)}$$

The remaining equilibrium concentrations may be determined by using a reaction chart, as shown earlier in this chapter.

The equilibrium concentration of $FeSCN^{2+}$ is 9.2×10^{-4} *M*, and its change in concentration is $+9.2 \times 10^{-4}$ *M*. The coefficients in the balanced equation are all one, so the changes in concentration are the same for all of the substances. The concentrations of the reactants must each have decreased by 9.2×10^{-4} *M*.

Reaction:	Fe^{3+} (*aq*)	+	SCN^- (*aq*)	$\rightleftharpoons$	$FeSCN^{2+}$ (*aq*)
Initial conc.:	4.0×10^{-2} *M*		1.0×10^{-3}*M*		0 *M*
Change:	-9.2×10^{-4} *M*		-9.2×10^{-4}*M*		$+9.2 \times 10^{-4}$ *M*
Equilibrium:	3.9×10^{-2} *M*		8.0×10^{-5}*M*		9.2×10^{-4} *M*

Substituting our values for the equilibrium concentrations into the equilibrium expression, you get

$$K_{EQ} = \frac{[FeSCN^{2+}\ (aq)]}{[Fe^{3+}\ (aq)][SCN^-\ (aq)]}$$

$$= \frac{9.2 \times 10^{-4}}{3.9 \times 10^{-2} \times 8.0 \times 10^{-5}} = 290 \text{ (approximately)}$$

If K_{EQ} is known, the equilibrium concentrations of the products can be calculated from the initial concentrations of the reactants.

SAMPLE PROBLEMS

PROBLEM

1. For the reaction $H_2\ (g) + I_2\ (g) \rightleftharpoons 2\ HI\ (g)$, $K_{EQ} = 600$. What is the equilibrium concentration of HI if the $[H_2]$ is 2.0 molar, and the $[I_2]$ is 3.0 molar at equilibrium?

SOLUTION

The equilibrium expression is $\frac{[HI]^2}{[H_2][I_2]} = 600.$

If we let x = the [HI] at equilibrium, this equation becomes $\frac{x^2}{2.0 \times 3.0} = 600$. Then $x^2 = 3600$, and x is the square root of 3600, which is 60. molar.

PROBLEM

2. In the same reaction, what is the concentration of HI at equilibrium, if the initial concentrations of hydrogen and iodine gases (before any reaction occurs) are both 1.00 *M*?

SOLUTION

This is different from the previous problem, because we were not given any of the concentrations at equilibrium; we were given only the initial concentrations. We will set up a reaction chart, and let x = the molarity of H_2 and I_2 that react to form HI.

Reaction:	H_2 (*g*)	+	I_2 (*g*)	$\rightleftarrows$	2 HI (*g*)
Initial conc.:	1.00 *M*		1.00 *M*		0
Change:	$-x$		$-x$		$+2x$
Equilibrium:	$1.00 - x$		$1.00 - x$		$2x$

Putting the equilibrium values into the equilibrium expression gives us

$$\frac{[2x]^2}{[1 - x][1 - x]} = 600$$

These equations are often difficult to solve, and may require use of the quadratic formula, which you may not have learned as yet in your mathematics class. In this case, though, you can take the square root of both sides of the equation, to get $\frac{2x}{1 - x} = 24.5$.

This can be solved with simple algebra, and $x = 0.924$ *M*. Since we defined the [HI] as $2x$, its concentration at equilibrium is 1.85 *M*.

PRACTICE

10.18 In the reaction N_2 (*g*) + 3 H_2 (*g*) $\rightleftarrows$ 2 NH_3 (*g*), at a certain high temperature, the equilibrium concentrations are found to be $[N_2] = 2.00\ M$, $[H_2] = 3.00\ M$, and $[NH_3] = 9.00\ M$. What is the K_{EQ} for the reaction at that temperature?

10.19 Given that $K_{EQ} = 4.0 \times 10^{-34}$ for the reaction $2\ HCl\ (g) \rightleftarrows H_2\ (g) + Cl_2\ (g)$. What is the concentration of HCl in an equilibrium mixture in which both the H_2 and the Cl_2 concentrations are 1.0×10^{-17} molar?

Quantitative Applications of K_{SP}

The solubility of a salt in moles per liter can be found from the K_{SP}, and vice versa. The solubility of AgCl is equal to the square root of its K_{SP}. $[Ag^+][Cl^-] = 1.8 \times 10^{-10}$. Since the $[Ag^+] = [Cl^-]$ in a solution of AgCl, our equation became $x^2 = 1.8 \times 10^{-10}$, where x is the solubility in moles per liter. As long as the salt forms only 1 mole of positive ion, and 1 mole of negative ion per mole of salt, the K_{SP} will equal the solubility squared. However, consider the salt PbI_2. The K_{SP} is 7.1×10^{-9}. What is the solubility of PbI_2? The expression for the solubility product constant is $K_{SP} = 7.1 \times 10^{-9} = [Pb^{2+}][I^-]^2$. Let x = the solubility of PbI_2 in moles per liter. From x moles of PbI_2, x moles of Pb^{2+} form, but twice as many, or $2x$ moles, of I^- form. The equation becomes

$$7.1 \times 10^{-9} = [x][2x]^2$$

$$4x^3 = 7.1 \times 10^{-9}$$

$$x = 1.2 \times 10^{-3}\ M$$

SAMPLE PROBLEM

PROBLEM

The solubility of $PbCl_2$ in water at 298 K is 1.59×10^{-2} moles/L. What is the K_{SP} of $PbCl_2$?

SOLUTION

$K_{SP} = [Pb^{2+}][Cl^-]^2$

Each mole of $PbCl_2$ that dissolves forms 1 mole of lead ions and 2 moles of chloride ions. Since the solubility is 1.59

$\times 10^{-2}$ *M*, $[Pb^{2+}] = 1.59 \times 10^{-2}$ *M*, and $[Cl^-] = 1.59 \times 10^{-2} \times 2$, or 3.18×10^{-2} *M*.

$$K_{SP} = [1.59 \times 10^{-2}][3.18 \times 10^{-2}]^2$$

$$K_{SP} = 1.6 \times 10^{-5}$$

PRACTICE

10.20 Find the K_{SP} of the following salts from their molar solubility:

(a) $PbCO_3 = 2.72 \times 10^{-7}$ *M*

(b) $Mn(OH)_2 = 3.63 \times 10^{-5}$ *M*

10.21 The K_{SP} for AgCl is 1×10^{-10} and for Ag_2CrO_4 the K_{SP} is 1.1×10^{-12}. Use these values to determine which salt has a greater solubility in moles per liter.

K_{SP} and the Common Ion Effect

Le Chatelier's Principle can be used to predict that the presence of a common ion will decrease the solubility of a salt. (See page 364.) You can now apply the K_{SP} to see the quantitative effect of a common ion. The solubility of AgCl in water is equal to the square root of its K_{SP}. The K_{SP} is 1.8×10^{-10}, so the solubility is 1.3×10^{-5} molar. Now suppose we added KCl to the solution, so that the concentration of Cl^- becomes 1.0 molar. Using the K_{SP} expression $[Ag^+][Cl^-] = 1.8 \times 10^{-10}$, we can see that if the Cl^- concentration is 1.0 molar, the maximum silver concentration can only be 1.8×10^{-10} molar. The solubility of AgCl in a 1.0 molar KCl solution is only 1.8×10^{-10} molar; that is about 100,000 times smaller than its solubility in water. The presence of a common ion greatly decreases the solubility of a salt.

CHAPTER REVIEW

The following questions will help you check your understanding of the material presented in the chapter.

1. In the equilibrium reaction A (*g*) + 2 B (*g*) + heat $\rightleftarrows$ AB_2 (*g*) the rate of the forward reaction will increase if there is (1) an increase in pressure (2) an increase in the volume of the reaction vessel (3) a decrease in temperature (4) a decrease in the concentration of A (*g*).
2. When a catalyst is added to a reaction at equilibrium, the rate of the forward reaction (1) decreases and the rate of the reverse reaction decreases (2) decreases and the rate of the reverse reaction increases (3) increases and the rate of the reverse reaction decreases (4) increases and the rate of the reverse reaction increases.
3. When a catalyst is added to a system at equilibrium, there is a decrease in the activation energy of (1) the forward reaction only (2) the reverse reaction only (3) both forward and reverse reactions (4) neither forward nor reverse reaction.
4. A chemical reaction has reached equilibrium when the (1) reverse reaction begins (2) forward reaction ceases (3) concentrations of the reactants and products become equal (4) concentrations of the reactants and products become constant.
5. For the reaction 2 SO_2 (*g*) + O_2 (*g*) $\rightleftarrows$ 2 SO_3 (*g*) + 196 kJ, which conditions favor the production of SO_3? (1) high temperature and high pressure (2) high temperature and low pressure (3) low temperature and high pressure (4) low temperature and low pressure
6. For a given system at equilibrium, lowering the temperature will always (1) increase the rate of reaction (2) increase the concentration of products (3) favor the exothermic reaction (4) favor the endothermic reaction.
7. Equilibrium is reached in all reversible chemical reactions when the (1) forward reaction stops (2) reverse reaction stops (3) concentrations of the reactants and the products become equal (4) rates of the opposing reactions become equal.
8. Given the equation AgCl (*s*) $\rightleftarrows$ Ag^+ (*aq*) + Cl^- (*aq*). As NaCl (*s*) dissolves in the solution, temperature remaining constant, the Ag^+ (*aq*) concentration will (1) decrease as the amount of AgCl (*s*) decreases (2) decrease as the amount of AgCl (*s*) increases (3) increase as the amount of AgCl (*s*) decreases (4) increase as the amount of AgCl (*s*) increases.

9. Given the reaction A (g) + B (g) $\rightleftarrows$ AB (g). As the pressure increases at a constant temperature, the rate of the forward reaction (1) decreases (2) increases (3) remains the same.

10. As the atmospheric pressure decreases, the temperature at which water will boil in an open container (1) decreases (2) increases (3) remains the same.

Base your answers to questions 11 through 13 on the following system at equilibrium:

$$2\ Cl_2\ (g) + 2\ H_2O\ (g) \rightleftarrows 4\ HCl\ (g) + O_2\ (g)$$

$$\Delta H = +113\ \text{kJ}$$

11. If the temperature of the system is increased at a constant pressure, the rate of the forward reaction will (1) decrease (2) increase (3) remain the same.

12. If O_2 is added to the system at a constant pressure and temperature, the number of moles of HCl will (1) decrease (2) increase (3) remain the same.

13. If the pressure on the system is increased at a constant temperature, the value of the equilibrium constant for the reaction will (1) decrease (2) increase (3) remain the same.

14. Given the equilibrium reaction H_2 (g) + Cl_2 (g) $\rightleftarrows$ 2 HCl (g). At constant temperature, which change will occur if the pressure on this system is increased? (1) The concentration of the reactants will decrease. (2) The rate of the reaction will decrease. (3) The activation energy will increase. (4) The frequency of collisions in the system will increase.

15. Given the equilibrium reaction A + B $\rightleftarrows$ C + D + heat. What change in the reaction system will change the value of the equilibrium constant? (1) an increase in the concentration of A and B (2) an increase in the concentration of C and D (3) an increase in temperature (4) an increase in pressure

16. The system $H_3PO_4 + 3\ H_2O \rightleftarrows 3\ H_3O^+ + PO_4^{-3}$ is at equilibrium. If Na_3PO_4 (s) is added, there will be a decrease in the concentration of (1) Na^+ (2) PO_4^{-3} (3) H_3O^+ (4) H_2O.

17. In the reaction $HC_2H_3O_2 + H_2O \rightleftarrows H_3O^+ + C_2H_3O_2^-$, the addition of solid sodium acetate ($Na^+C_2H_3O_2^-$) results in a decrease in the concentration of (1) $C_2H_3O_2^-$ (2) H_3O^+ (3) Na^+ (4) $HC_2H_3O_2$.

Questions 18–20 are based upon the general reaction

$$2\,A\,(g) + B\,(g) = 2\,C\,(g) + 2\,D\,(g)$$

$$\Delta H^\circ = +209 \text{ kJ}$$

18. If the temperature of the equilibrium system were increased, the quantity of (1) A would increase, and B would increase (2) A would decrease, and C would increase (3) C would increase, and D would decrease (4) B would increase, and D would decrease.

19. If additional C were added to the system, the quantity of (1) A would increase, and D would decrease (2) A would increase, and B would decrease (3) A would decrease, and D would increase (4) A would decrease, and B would increase.

20. If the volume of the container were decreased, the quantity of (1) C would increase, and D would increase (2) C would increase, and A would decrease (3) A would increase, and B would decrease (4) B would increase, and C would decrease.

Base your answers to questions 21 through 23 on the following information and equation:

$$A\,(g) + B\,(g) \rightleftharpoons AB\,(g) + \text{heat}$$

equilibrium constant (K) at 25°C = 0.50

21. If a catalyst is added to the system at equilibrium with temperature and pressure remaining constant, the concentration of A will (1) decrease (2) increase (3) remain the same.

22. When chemical equilibrium is reached, the concentration of AB compared with the concentration of A times the concentration of B is (1) less (2) greater (3) the same.

23. As the concentration of B is increased at constant temperature, the value of K (1) decreases (2) increases (3) remains the same.

24. Consider the system $AgCl\,(s) \rightleftharpoons Ag^{+}\,(aq) + Cl^{-}\,(aq)$ at equilibrium. As chloride ions are added to this system and the temperature is kept constant, the value of the equilibrium constant (1) decreases (2) increases (3) remains the same.

25. Consider the system $N_2\,(g) + 3\,H_2\,(g) \rightleftharpoons 2\,NH_3\,(g)$ at a constant temperature. An increase in pressure on this system will (1) shift the equilibrium to the left (2) shift the equilibrium to the right (3) have no effect on the equilibrium (4) change the value of the equilibrium constant.

26. When at equilibrium, which reaction will shift to the right if the pressure is increased and the temperature is kept constant?

(1) $2\ H_2\ (g) + O_2\ (g) \rightleftarrows 2\ H_2O\ (g)$
(2) $2\ SO_3\ (g) \rightleftarrows 2\ SO_2\ (g) + O_2\ (g)$
(3) $2\ NO\ (g) \rightleftarrows N_2\ (g) + O_2\ (g)$
(4) $2\ CO_2\ (g) \rightleftarrows 2\ CO\ (g) + O_2\ (g)$

27. Given the equilibrium expression $K = [A][B]/[C][D]$. Which pair represents the reactants of the forward reaction? (1) A and B (2) B and D (3) C and D (4) A and C

28. Which is the equilibrium constant of a chemical reaction that goes most nearly to completion? (1) 2×10^{-15} (2) 2×10^{-1} (3) 2×10^{1} (4) 2×10^{15}

29. Which expression represents the solubility product constant, K_{SP}, of AgCl (*s*)?

(1) $K_{SP} = [Ag^+][Cl^-]$ (3) $K_{SP} = \frac{[Ag^+]}{[Cl^-]}$

(2) $K_{SP} = [Ag^+] + [Cl^-]$ (4) $K_{SP} = \frac{[Cl^-]}{[Ag^+]}$

30. Which equilibrium system will contain the largest concentration of products at 25°C?

(1) $AgI\ (s) \rightleftarrows Ag^+\ (aq) + I^-\ (aq)$ $K_{EQ} = 8.5 \times 10^{-17}$
(2) $CH_3COOH\ (aq) \rightleftarrows H^+\ (aq) + CH_3COO^-\ (aq)$
$K_{EQ} = 1.8 \times 10^{-5}$
(3) $Pb^+\ (aq) + 2\ Cl^-\ (aq) \rightleftarrows PbCl_2\ (s)$ $K_{EQ} = 6.3 \times 10^{4}$
(4) $Cu(s) + 2\ Ag^+\ (aq) \rightleftarrows Cu^{2+}\ (aq) + 2\ Ag\ (s)$
$K_{EQ} = 2.0 \times 10^{15}$

31. Which saturated solution has the highest S^{2-} ion concentration?

(1) CdS (K_{SP} at 18°C = 3.6×10^{-29})
(2) CoS (K_{SP} at 18°C = 3.0×10^{-26})
(3) PbS (K_{SP} at 18°C = 3.4×10^{-28})
(4) FeS (K_{SP} at 18°C = 3.7×10^{-19})

For each of questions 32 through 33, select the number of the expression, chosen from the following list, that best answers that question.

EXPRESSIONS

(1) $[A^+][B^-]$ (2) $\frac{[A^+]}{[B^+]}$ (3) $\frac{[A^+][B^-]}{[AB]}$ (4) $\frac{[A^+]}{[B^-]}$

32. The equation for the reaction of a metal replacing a metallic ion B^+ from aqueous solution is

$$A\ (s) + B^+\ (aq) \rightleftarrows A^+\ (aq) + B\ (s)$$

Which expression is equal to its equilibrium constant, K_{EQ}?

33. Given the dissociation equation of a saturated solution of a slightly soluble salt:

$$AB\ (s) \rightleftarrows A^+\ (aq) + B^-\ (aq)$$

Which expression is equal to its solubility product constant, K_{SP}?

34. Which sulfide forms the most concentrated saturated solution at 25°C?

(1) CdS ($K_{SP} = 3.6 \times 10^{-29}$)
(2) CoS ($K_{SP} = 3.0 \times 10^{-26}$)
(3) FeS ($K_{SP} = 1.3 \times 10^{-17}$)
(4) HgS ($K_{SP} = 2.4 \times 10^{-52}$)

35. Given the system $2\ A\ (g) + B\ (g) \rightleftarrows 3\ C\ (g)$. The equilibrium expression for this system is

(1) $K = \frac{[A]^2[B]}{[C]^3}$
(2) $K = \frac{[2\ A]^2[B]}{[3\ C]^3}$
(3) $K = \frac{[C]^3}{[A]^2[B]}$
(4) $K = \frac{[3\ C]^3}{[2\ A]^2[B]}$

36. Based on the information in Table F, which salt will react with a solution of NaBr to form a precipitate?

(1) KCl (2) $Cu(NO_3)_2$ (3) $AgNO_3$ (4) $MgSO_4$

37. Which silver salt is most soluble in water?

(1) silver chloride—K_{SP} at 25°C = 1.1×10^{-10}
(2) silver bromide—K_{SP} at 25°C = 7.7×10^{-13}
(3) silver iodide—K_{SP} at 25°C = 1.5×10^{-16}
(4) silver sulfide—K_{SP} at 25°C = 1.0×10^{-50}

38. Carbon dioxide is most soluble in water under conditions of (1) high pressure and high temperature (2) high pressure and low temperature (3) low pressure and low temperature (4) low pressure and high temperature.

39. Which reaction is represented by the equilibrium expression

$$K = \frac{[C]^2}{[A]^3[B]}?$$

(1) $2\ C\ (g) \rightleftarrows 3\ A\ (g) + B\ (g)$
(2) $2\ C\ (g) \rightleftarrows 3\ AB\ (g)$
(3) $3\ A\ (g) + B\ (g) \rightleftarrows 2\ C\ (g)$
(4) $3\ AB\ (g) \rightleftarrows 2\ C\ (g)$

CHEMISTRY CHALLENGE

The following questions will provide practice in answering SAT II-type questions.

Each question in this section consists of a statement and a reason. Select

(a) if both the statement and reason are true, and the reason is a correct explanation of the statement;
(b) if both the statement and reason are true, but the reason is not a correct explanation of the statement;
(c) if the statement is true, but the reason is false;
(d) if the statement is false, but the reason is true;
(e) if both the statement and reason are false.

Statement	*Reason*
1. An increase in temperature generally increases the K_{SP} of a salt.	The dissolving of a salt is generally endothermic.
2. The addition of NaCl to a solution of AgCl decreases the solubility of the AgCl.	NaCl forms hydrated Na^+ ions upon dissolving in water.

Statement	*Reason*
3. In the reaction $N_2\ (g) + 3\ H_2\ (g) = 2\ NH_3\ (g)$ an increase in pressure favors the production of the product.	An increase in pressure causes an increase in the value of the equilibrium constant of the reaction.
4. For the exothermic reaction, $H_2 + I_2 = 2HI$ an increase in temperature favors the production of HI.	Exothermic reactions are favored by increases in temperature.
5. For the gaseous reaction $A + B \rightleftharpoons C + D$ at equilibrium, the removal of substance C will lead to an increase in the amount of D and a decrease in the amount of A.	The removal of a product causes the reaction to shift to the right.
6. In the reaction $2\ NO_2\ (g) \rightleftharpoons N_2O_4\ (g) + \text{heat}$, an increase in temperature will decrease $[NO_2]$.	An increase in temperature will drive a reaction in the direction in which heat is absorbed.
7. Adding a common ion decreases the K_{SP} of a salt.	The presence of a common ion decreases the solubility of the salt.
8. When the reaction: $2\ NO_2\ (g) \rightleftharpoons N_2O_4\ (g)$ is at equilibrium, the concentration of NO_2 must equal the concentration of N_2O_4.	At equilibrium, the concentration of the products must equal the concentration of the reactants.

"They're Better Than Day Rates"

If you are asked, What do you know about nitrates? you might give as a comical answer the title of this feature. Or you could respond that nitrates are among the most important products of the chemical industry. They are used in fertilizers and in gunpowder. Modern agriculture depends on an adequate supply of nitrates; without them there would not be enough food for Earth's growing population.

Nitrates can be produced from ammonia, NH_3, but until 1912, scientists had no method of producing adequate quantities of ammonia. The reaction

$$N_2\,(g) + 3\,H_2\,(g) \rightarrow 2\,NH_3\,(g)$$

has an equilibrium constant of 7×10^5 at room temperature, so you would expect to produce large amounts of ammonia just by mixing nitrogen gas and hydrogen gas. However, the reaction is so slow at room temperature that almost no ammonia is produced. When a reaction has a large equilibrium constant, but occurs very slowly, it is said to be thermodynamically feasible, but kinetically unfeasible. If you raise the temperature, the reaction becomes faster, but the reaction is exothermic. Therefore, an increase in temperature also shifts the equilibrium to the left favoring the reactants. At 1000 K, the equilibrium constant is only 3×10^{-6}. The reaction is no longer thermodynamically feasible.

In 1912, the German chemist Fritz Haber found that by using iron as a catalyst and maintaining a pressure of 200 atmospheres he could produce significant yields of ammonia at temperatures of about 700 K. Haber's method of producing ammonia, now known as the Haber process, enabled Germany to produce enough nitrates to maintain a sufficient supply of gunpowder throughout World War I, which began just two years after the process was developed.

Let's not be too hard on Fritz Haber, even though his work probably led to thousands of deaths on both sides during the war. The Haber process is used today to produce more than 110 million tons of ammonia. The nitrates produced from that ammonia may have saved more lives, through increasing food supplies, than were lost through the use of the process during the war. In any case, although Haber was a Nobel Prize winner and a great German patriot, he was forced to leave his homeland in 1933. Despite his conversion to Christianity, Haber was still a Jew to the German Nazis.

The Forces That Drive Reactions

LOOKING AHEAD

What happens when a piece of copper is placed in a solution of zinc chloride? We can write a balanced equation: $Cu + ZnCl_2 \rightarrow CuCl_2 + Zn$. However, when you actually place a piece of copper in a zinc chloride solution, nothing happens at all! Chemists say that such a reaction is not spontaneous. Under a given set of conditions and without the addition of energy (usually in the form of electricity), any change that does occur is called a spontaneous change. In this chapter you will examine those factors that control the direction of spontaneous change.

When you have completed this chapter you should be able to:

- **Define** heat of formation.
- **Relate** ΔH to changes in potential energy.
- **Find** ΔH for a given reaction, using heats of reaction tables.
- **Define** entropy.
- **Predict** the sign of the entropy change, ΔS, for a given physical or chemical change.
- **Use** ΔH and ΔS to predict the direction of a spontaneous change.
- **Determine** the sign of ΔG, the free energy, for a given physical or chemical change.

Chemical Energy

All matter has a certain amount of energy in each of the states in which it exists. For example, all matter has *kinetic energy.* Kinetic energy is responsible for the motions of molecules. In gases, the molecules exhibit three types of motion (Figure 11-1). The molecules move randomly, in straight lines. This is called translational motion. The molecules also spin or turn end over end. This is called rotational motion. And the atoms of the molecules can move back and forth. This is called vibrational motion. The molecules of liquids exhibit mostly rotational and vibrational motion. The molecules of solids exhibit only vibrational motion.

All matter also has *potential energy.* Potential energy is the stored energy of chemical bonds.

The changes that occur in matter—changes of state, chemical changes, and dissolving—all involve energy. In exothermic changes, potential energy is converted into kinetic energy, while in endothermic changes, kinetic energy is converted into potential energy. The **law of conservation of energy** states that in any chemical or physical process, energy is neither created nor destroyed. For example, when you apply a cold pack to your sore shoulder, heat flows from your shoulder to the cold pack. The amount of heat gained by the cold pack is equal to the amount of

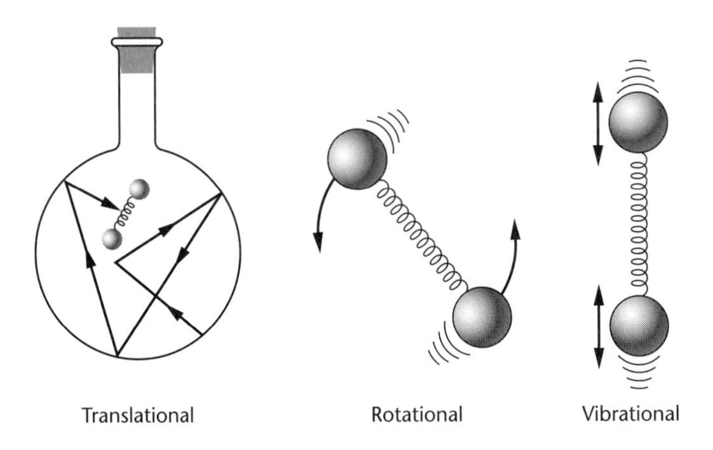

Figure 11-1 Molecular motions (Each ball represents an atom of a diatomic molecule.)

heat lost by your shoulder. In an exothermic reaction, the amount of heat released in the reaction is equal to the decrease in the potential energy of the system.

Chemical changes generally involve greater amounts of energy than do physical changes. For example, consider the following changes:

CHANGE OF STATE AT 100°C

$H_2O\ (l) \rightarrow H_2O\ (g)$ absorbs 40.7 kJ/mole

CHEMICAL CHANGE

$H_2\ (g) + \frac{1}{2}\ O_2\ (g) \rightarrow H_2O\ (l)$ releases 286 kJ/mole

DISSOLVING PROCESS

$HCl\ (g) + H_2O\ (l) \rightarrow HCl\ (aq)$ releases 75.1 kJ/mole

Whether energy is absorbed (endothermic change) or released (exothermic change) depends primarily on the potential energy of the reactants as compared with the potential energy of the products. Endothermic and exothermic changes will be discussed in the next section.

Chemical bond energy makes up the largest part of the heat content of a molecule. **Heat content** and **enthalpy** are terms used to describe the energy stored in a sample of matter. Each is expressed in joules or kilojoules and is symbolized by the letter H.

Potential energy, stability, and enthalpy are closely related to one another. Matter tends to reach a condition of lowest potential energy. This is also the condition of greatest stability. Recall the example of the ball rolling downhill until it arrives at the lowest point possible. This point is the position of minimum potential energy and maximum stability. Once started, the ball continues to roll unaided. Its action is therefore said to be spontaneous.

The enthalpy of a substance is so closely related to the potential energy of its particles that, for all practical purposes, enthalpy and potential energy can be used interchangeably. For a chemical system, the natural drive toward maximum stability is the drive toward minimum enthalpy.

Enthalpy

In the next few pages, you will be concerned with the enthalpy of various changes that take place in matter. Strictly speaking, the change in enthalpy, ΔH, is equal to the heat, q, only when the pressure on the system is constant. However, even when the pressure is not constant, the differences are very small, and of no concern to you at this time. Assume that the enthalpy change and the heat of reaction are the same.

Enthalpy Change: Endothermic

Consider, first, the general equation

$$\text{A} + \text{B} + \text{heat} \rightarrow \text{AB}$$

An equation that includes a heat term (+ heat) is called a *thermochemical equation.* The preceding thermochemical equation shows that energy must be added to change A + B into AB. The equation is for an endothermic reaction, one in which heat must be supplied to convert the reactants to the products. In an equation for an endothermic reaction, heat is shown as a term on the same side as the reactants.

In the change of A + B into AB, the enthalpy (H) of the product is higher than the enthalpy of the reactants. The change of enthalpy can be shown in a diagram, as in Figure 11-2.

Compare Figure 11-2 to the diagram of an endothermic reaction in Chapter 9, on page 334. Figure 11-2 does not show the path from reactant to product; it omits the activated complex. The value of ΔH depends only on the relative enthalpies of the products and reactants; it is independent of path. Variables such as ΔH, that depend only on the initial and final conditions, are called **state functions.**

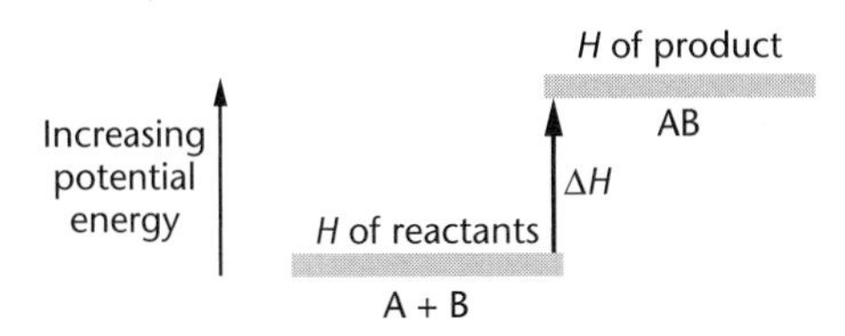

Figure 11-2 Energy diagram: endothermic reaction

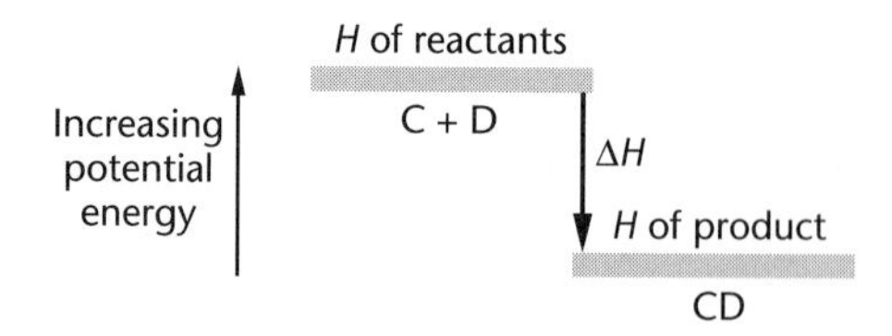

Figure 11-3 Energy diagram: exothermic reaction

The change in potential energy is symbolized in the diagram by ΔH. Since $H_{product}$ is higher than $H_{reactants}$, the change in enthalpy (ΔH) is positive. The reaction involves an increase in enthalpy, a direction opposite to that which nature favors. Just as a ball cannot roll uphill unaided, this reaction cannot proceed spontaneously, unless there is some other factor to push it.

Enthalpy Change: Exothermic

Now consider the general reaction

$$C + D \rightarrow CD + \text{heat}$$

Heat energy is included as a term on the product side in this thermochemical equation. This is an exothermic reaction, one in which heat is released. $H_{product}$ is lower than $H_{reactants}$. See Figure 11-3.

Since the enthalpy of the product is lower than the enthalpy of the reactants, ΔH is negative. The difference in the enthalpies represents energy that is released by the reaction as it goes from reactants to product. The reaction follows the direction that nature favors—a decrease in enthalpy. This suggests that this reaction can occur spontaneously, as a ball can roll downhill.

The following table summarizes the differences between endothermic and exothermic reactions.

Endothermic Reaction	***Exothermic Reaction***
H of products is higher than H of reactants	H of products is lower than H of reactants
Potential energy, enthalpy, increases	Potential energy, or enthalpy, decreases
Heat is a reactant term in the thermochemical equation	Heat is a product term in the thermochemical equation
The sign of ΔH is positive	The sign of ΔH is negative
Nature does not favor this process; stability decreases	Nature favors this process; stability increases

Heats of Reaction

For a chemical reaction, ΔH is called the heat of reaction. It tells you the amount of heat produced or absorbed when the reaction proceeds as shown. Consider the combustion of carbon monoxide.

$$2\ CO\ (g) + O_2\ (g) \rightarrow 2\ CO_2\ (g)$$

When two moles of carbon monoxide react as shown, 566.0 kJ of heat are released. Since the reaction is exothermic, the sign of ΔH is negative, so that ΔH for the reaction is shown as -566.0 kJ. You can also indicate the heat of reaction by including the heat as part of the equation. If the reaction is exothermic, then the heat is a product and appears on the right side of the equation.

$$2\ CO\ (g) + O_2\ (g) \rightarrow 2\ CO_2\ (g) + 566.0\ kJ$$

Do not be misled by the plus (+) sign in front of the 566.0 kJ. When heat is written on the right side of the equation, the sign of ΔH is negative.

Some heats of reactions are listed in Table I in Appendix 4.

You can find the above reaction on the table, where its heat of reaction is listed as -566.0 kJ. In using Table I, bear in mind that the heats shown are for the number of moles in the reaction as written. For example, the chart lists the reaction $2\ C\ (s) + 2\ H_2\ (g) \rightarrow C_2H_4\ (g)$ with a ΔH of $+52.4$ kJ. Other charts may list the same reaction this way: $C(s) + H_2\ (g) \rightarrow \frac{1}{2}\ C_2H_4\ (g)$. In this case, the ΔH would be listed as 26.2 kJ. Since the second equation shows half the number of moles, the heat absorbed must be half as much.

SAMPLE PROBLEMS

PROBLEM

1. Using the heats of reaction in Table I of Appendix 4, how much heat is produced when 2.0 moles of methane, CH_4, are burned in excess oxygen?

SOLUTION

The table gives the value of ΔH for the reaction

$$CH_4\ (g) + 2\ O_2\ (g) \rightarrow CO_2\ (g) + 2\ H_2O\ (l)$$

as -890.4 kJ. This equation shows one mole of CH_4. Therefore, burning two moles of CH_4 would produce twice as much heat, or 1781 kJ.

Notice that our answer is expressed as a positive number. This is because the question asked "How much heat is produced?" If the question had read "What is the heat of reaction for the combustion of 2.0 moles of methane?" the answer would have been -1781 kJ.

PROBLEM

2. Using the heats of reaction table, find ΔH for the combustion of 2.00 grams of hydrogen, to form liquid water.

SOLUTION

The table gives the value of ΔH for the reaction

$$2\ H_2\ (g) + O_2\ (g) \rightarrow 2\ H_2O\ (l) \text{ as } -571.6 \text{ kJ}.$$

To use this table, we need to express our quantities in moles. Recall that the number of moles equals the number of grams divided by the molar mass.

The molar mass of H_2 is 2.00 g, so the number of moles is

$$\frac{2.00\ g}{2.00\ g} = 1.00 \text{ mole.}$$

The table tells us that ΔH for the reaction is -571.6 kJ for 2 moles of H_2. Therefore, ΔH for the combustion of 1.00 mole of H_2 is half as much, or -286 kJ.

PRACTICE

11.1 How much heat is absorbed in the formation of 1.00 mole of HI (g) from H_2 (g) and I_2 (g)?

11.2 How much heat is released when 80.00 grams of NaOH (molar mass = 40 g) dissolve in water?

11.3 Of the changes shown in Table I, which one indicates the smallest difference in potential energy between the product and the reactant?

11.4 According to Table I, how many grams of glucose, $C_6H_{12}O_6$ (molar mass = 180 g) must be burned to produce 280.4 kJ?

Enthalpy of Solution

The dissolving process generally involves the breaking of bonds (with absorption of energy) and the formation of new bonds (with the release of energy). For example, when ionic solids dissolve in water, energy must be absorbed to separate the positive and negative ions held together by ionic bonds in the crystal lattice. Energy is released when the separated positive and negative ions react with the water molecules and form hydrated ions.

The overall process is exothermic or endothermic depending on which is greater—the energy absorbed or the energy released. Some typical enthalpies of solution follow:

$$(1)\ HCl\ (g) \rightarrow H^+\ (aq) + Cl^-\ (aq) \qquad \Delta H = -75.1\ kJ$$

$$(2)\ NaCl\ (s) \rightarrow Na^+\ (aq) + Cl^-\ (aq) \qquad \Delta H = +4.26\ kJ$$

In reaction 1, the energy given off by the hydration of the hydrogen ions and chloride ions must be greater than the energy needed to break the covalent H—Cl bond. In reaction 2, the energy needed to break down the crystal lattice of sodium chloride to separate the ions must be greater than the energy released when the sodium and chloride ions become hydrated.

The dissolving of NaCl in water is an endothermic process, and the enthalpy change is positive. The change should not occur spontaneously. However, you know that salt does dissolve in water spontaneously. There is an apparent conflict between fact and theory. You were concerned with this conflict in Chapter 8. Let us now examine the second fundamental factor that influences spontaneity in chemical reactions. Then the question about the dissolving of salt can be answered.

Entropy

Matter tends to attain minimum potential energy. Another fundamental property of matter is that it tends to attain a state of maximum randomness. This is one of the universal laws in nature. Matter tends to attain maximum entropy—that is, to assume the least organized, or most random, state possible.

Imagine a box with a vertical partition, such as is shown in Figure 11-4. One compartment contains gas A. The other contains gas B. There is some degree of order, or organization, in this system, because each gas is present in its own compartment.

What will happen if you remove the partition? There will be a spontaneous change. Molecules of gas A will soon be found where there were only B molecules. B molecules will be found where there were only A molecules. Less order and more randomness will now be present in the system.

The opposite change, with A molecules returning to one compartment and B molecules to the other, has a low probability of taking place spontaneously. This would be a change toward more order and less randomness—in the direction opposite to the one that matter tends to take.

Since the gaseous state is the least ordered (most random) and the state that nature favors, why is not all matter gaseous? Bonding forces—forces that lead to greater order and to lower energy—tend to counteract the natural drive toward randomness.

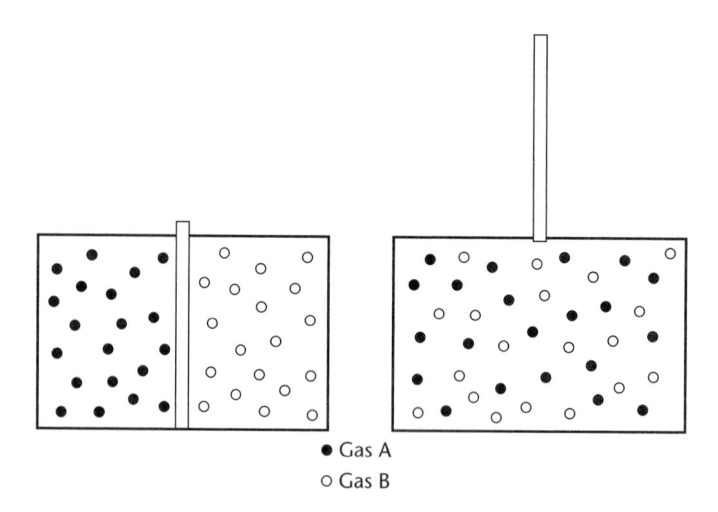

Figure 11-4 Drive toward maximum randomness

Measuring Entropy

The symbol for entropy is S, and so the symbol for the change in entropy is ΔS. Entropy is easiest to measure for a system at equilibrium. At equilibrium $\Delta S = \frac{\Delta H}{T}$.

For example, at 100°C, or 373 K, we know that water is at its boiling point, where the liquid and the vapor are at equilibrium. ΔH for the boiling of water is 40.7 kJ/mole. Therefore, ΔS for the change $H_2O\ (l) \rightarrow H_2O\ (g)$ is $\frac{40.7\text{kJ/mole}}{373\text{ K}} = 0.109$ kJ/mole-K, or 109 J/mole-K. Notice that the unit of entropy is joule per mole-Kelvin.

The entropy change in this case was positive. Water vapor is in a more random state than is liquid water. While measuring the exact change in entropy for a system can be difficult, predicting the sign of the change is often very easy. Gases have the greatest entropy, while solids have the lowest. In a reaction involving gases only, the greater entropy is found on the side of the reaction containing the greater number of moles of gas.

SAMPLE PROBLEMS

PROBLEM

1. Predict the entropy change for the reaction

$$2\ H_2\ (g) + O_2\ (g) \rightarrow 2\ H_2O\ (l)$$

SOLUTION

Since the reactants are gases, while the product is a liquid, the reactants should have greater entropy than the products. The entropy decreases, and the sign of ΔS is $-$.

(The actual value of ΔS for this change at 298 K is -327 J/K)

PROBLEM

2. Predict the entropy change for the reaction

$$N_2\ (g) + 3\ H_2\ (g) \rightarrow 2\ NH_3\ (g)$$

SOLUTION

In this case, all of the substances in the reaction are gases. However, there are four moles of gas in the reactant, and only two moles of gas in the product. The entropy decreases, and ΔS should be negative. (The actual value is -200 J/K)

PRACTICE

11.5 For each of the following changes, predict the entropy change, and explain your answer.

(a) $CO_2\ (s) \rightarrow CO_2\ (g)$

(b) $2\ CO\ (g) + O_2\ (g) \rightarrow 2\ CO_2\ (g)$

(c) $Zn\ (s) + 2\ HCl\ (aq) \rightarrow ZnCl_2\ (aq) + H_2\ (g)$

(d) $NaCl\ (s) \rightarrow NaCl\ (aq)$

Other Factors That Determine Entropy

Thus far, you have used the state of matter to predict the sign of an entropy change. Even in changes that involve neither changes in state, nor changes in the number of moles of gas, there are entropy changes. For example, suppose you increased the volume of the container in a gaseous system. The gas molecules would now have more space available to them, and so would be distributed more randomly. The entropy would increase. Other factors that increase entropy are increased temperature, and decreased pressure. The following table summarizes some of the factors that influence the entropy of the system.

Factors That Influence Entropy

Low Entropy	High Entropy
Solids	Gases
Undissolved solid	Aqueous solution
Low temperature	High temperature
High pressure	Low pressure
Small, simple molecules	Large, complex molecules

Standard Molar Entropies

Because entropy is dependent on pressure and concentration, scientists compare the entropies of different substances under one defined set of conditions. When the entropy is measured for 1 mole, at 298 K, and 1 atmosphere for gases, and at 1 molar for solutes, the resulting value is called the **standard molar entropy**, and has the symbol S^0. Notice how the standard molar entropies in the table below illustrate some of the general trends in the table on page 392.

Standard Molar Entropies

Substance	*Standard molar entropy, S°, in J/mole-K*
O_2 (*g*)	205.0
O_3 (*g*)	237.6
Br_2 (*l*)	152.3
Br_2 (*g*)	245.3
Br (*g*)	174.9

PRACTICE

Base your answers on the table of Standard Molar Entropies, above.

11.6 Why is the entropy value for O_3(*g*) greater than that of O_2(*g*)?

11.7 Explain the increases in entropy from Br_2 (*l*), to Br (*g*) to Br_2 (*g*).

11.8 (For experts) Find the standard entropy change, $\Delta S°$, for the reaction $2\ O_3\ (g) \rightarrow 3\ O_2\ (g)$.

Does your answer have the sign you expected? Explain.

Spontaneous Reactions

You have now examined two factors that influence whether a reaction is, or is not, spontaneous. These are the enthalpy change, ΔH, and the entropy change, ΔS. Change is favored when there is a decrease in potential energy and the sign of ΔH

is negative. Such changes are exothermic. In addition, change is favored when there is an increase in the entropy and the sign of ΔS is positive. Such changes increase the randomness of the system.

Consider the combustion of ethane, C_2H_6.

$$2\ C_2H_6\ (g) + 7\ O_2\ (g) \rightarrow 4\ CO_2\ (g) + 6\ H_2O\ (g) \qquad \Delta H = -2850\ \text{kJ}$$

Since there are 9 moles of gas in the reactants, and 10 moles of gas in the products, we would predict that ΔS should be positive. In fact, ΔS for this reaction is +92.6 J/K. In this reaction, both the enthalpy change and the entropy change are in the directions favorable for a spontaneous change. In fact, this reaction is spontaneous. Any reaction in which ΔH is negative and ΔS is positive must be spontaneous. Spontaneous change is most favored by a decrease in potential energy and an increase in entropy.

Now consider the reverse reaction:

$$4\ CO_2\ (g) + 6\ H_2O\ (g) \rightarrow 2\ C_2H_6\ (g) + 7\ O_2\ (g) \quad \Delta H = +2850\ \text{kJ}$$

In this case, ΔH is positive and ΔS is negative. We would predict that such a reaction will not be spontaneous. When there is an increase in potential energy, and a decrease in entropy, both factors are unfavorable. Such changes are never spontaneous.

There are two other possibilities. If ΔH and ΔS are both positive, then the enthalpy factor is unfavorable while the entropy factor is favorable. If both are negative, then the enthalpy factor is favorable, while the entropy factor is unfavorable. One simple change that illustrates the first of these possibilities is the melting of ice. $H_2O\ (s) \rightarrow H_2O\ (l)$. Melting is an endothermic process, and liquids have greater entropy than solids. Therefore, for melting, ΔH and ΔS are both positive in sign. Is the change spontaneous? You know that ice will melt only if the temperature is greater than 0°C. The change is spontaneous at high temperature, but not at low temperature. The reverse process is the freezing of water. For freezing, ΔH and ΔS are both negative. Freezing is spontaneous only at low temperatures. When only one of the two factors is favorable, the direction of the change is determined by the temperature. The table, Enthalpy, Entropy, and Spontaneous Change, on page 395 summarizes the four possibilities.

Enthalpy, Entropy, and Spontaneous Change

ΔH	ΔS	*Result*
– (exothermic	+ (increased entropy)	always spontaneous
+ (endothermic)	– (decreased entropy)	never spontaneous
+ (endothermic)	+ (increased entropy)	spontaneous at high temperature
– (exothermic)	– (decreased entropy)	spontaneous at low temperature

The Dissolving Process

You saw earlier that the dissolving of NaCl in water is endothermic. (See page 389.) At that time you could not explain why the change occurred spontaneously. The dissolving of an ionic substance in water is generally endothermic, yet many ionic substances are quite soluble in water. You can see in the table above, that an endothermic reaction can be spontaneous only if it involves increased entropy. The dissolving of NaCl is represented by the equation

$$\text{NaCl}\ (s) \rightarrow \text{Na}^+\ (aq) + \text{Cl}^-\ (aq)$$

When the solid is dissolved, the resulting ions are spread out throughout the liquid. The randomness of the system increases, so the sign of ΔS is +. Solids can dissolve spontaneously in liquids because the dissolving process increases the entropy of the system.

Free Energy Change: Entropy and Enthalpy Combined

The direction of a change depends on ΔS, ΔH, and the temperature. These can be quantitatively combined to produce a value called the **free energy**, G, a state function that predicts the spontaneity of reactions. The change in free energy is defined by the equation

$$\Delta G = \Delta H - T\Delta S.$$

If ΔG is negative, the reaction is spontaneous. If ΔG is positive, the reaction is not spontaneous. If ΔG is 0, the system is at equilibrium. Consider the situation in a change for which ΔH is negative and ΔS is positive. Based on the equation $\Delta G = \Delta H - T\Delta S$, if a positive number, ($T\Delta S$) is subtracted from a negative number, (ΔH) the result must be negative! Therefore, when ΔH is negative and ΔS is positive, ΔG is always negative, and the reaction is always spontaneous. This result is consistent with the predictions in the table, Enthalpy, Entropy, and Spontaneous Change.

PRACTICE

11.9 Air is composed mainly of nitrogen and oxygen. For the reaction $N_2\ (g) + 2\ O_2\ (g) \rightarrow 2\ NO_2\ (g)$ the enthalpy change, ΔH, is + 66.4 kJ. The entropy change, ΔS, is -121 J/K.

(a) Predict whether the reaction is spontaneous. Explain your prediction.

(b) What is the sign of ΔG for the reaction?

11.10 At room temperature, the reaction $N_2\ (g) + 3\ H_2\ (g) \rightarrow 2\ NH_3(g)$ is spontaneous. At temperatures above 310°C, the reaction is not spontaneous. Based on this information, predict the signs of ΔH, ΔS, and ΔG for the reaction at room temperature. Explain your answers.

TAKING A CLOSER LOOK

Hess's Law

Table I lists the values of ΔH for several reactions. It can also be used to find ΔH for many reactions that are not listed. You can find ΔH for a reaction from values for other reactions by applying Hess's law. **Hess's law** states that if a reaction is the exact sum of two or more other reactions, then its ΔH is the exact sum of the ΔH values of the other reactions. To illustrate the use of Hess's law, consider the reaction

$$C_2H_2\ (g) + H_2\ (g) \rightarrow C_2H_4\ (g)$$

This reaction is not listed in Table I. However, the following two reactions are listed.

$$(1)\ 2C\ (s) + H_2\ (g) \rightarrow C_2H_2\ (g) \qquad \Delta H = +227.4\ kJ$$

$$(2)\ 2C\ (s) + 2\ H_2\ (g) \rightarrow C_2H_4\ (g) \qquad \Delta H = +52.4\ kJ$$

Since in the desired reaction C_2H_2 is a reactant, rewrite reaction (1) in reverse:

(1 reversed) $C_2H_2\ (g) \rightarrow 2\ C\ (s) + H_2\ (g) \qquad \Delta H = -227.4\ kJ$

Add reaction (2) $\underline{2C\ (s) + 2\ H_2\ (g) \rightarrow C_2H_4\ (g)} \qquad \Delta H = +54.4\ kJ$

$$C_2H_2(g) + \cancel{2C\ (s)} + 2H_2\ (g) \rightarrow \cancel{2C\ (s)} + \cancel{H_2\ (g)} + C_2H_4\ (g)$$

After eliminating those substances that appear on both sides and adding the ΔHs, you get.

$$C_2H_2\ (g) + H_2\ (g) \rightarrow C_2H_4\ (g) \qquad \Delta H = -175\ kJ$$

Since you were able to add two reactions to get the exact, desired reaction, you could obtain the value of ΔH for the desired reaction using Hess's law. Hess's law can be used to find ΔS and ΔG, as well.

Standard Enthalpy and Free Energy of Formation

You were able to find the value of ΔH for a reaction by combining ΔH values for other reactions. In general, it is difficult to obtain a value for ΔH by this method. In order to more easily find ΔH values for reactions, you can use tables of ΔH_f^0, the standard molar enthalpy of formation. The **standard molar enthalpy of formation** is the enthalpy change for the formation of 1 mole of a substance from its elements in their standard state. The standard state of an element is the state in which it normally exists at 298 K and 101.3 kPa of pressure. For example, the standard state of oxygen is $O_2\ (g)$.

Standard Enthalpy and Free Energy of Formation

Substance	ΔH_f^0 (kJ/mole)	ΔG_f^0 (kJ/mole)
$CH_3OH\ (l)$	−238.6	−166.23
$CH_4\ (g)$	−74.8	−50.8
$CO\ (g)$	−110.5	−137.2
$CO_2\ (g)$	−393.5	−394.4
$H_2O\ (l)$	−285.83	−237.13
$NH_3\ (g)$	−46.19	−16.66
$NO\ (g)$	+90.37	+86.71
$NO_2\ (g)$	+33.84	+51.84

Table I in Appendix 4 lists the reaction 2 H_2 (*g*) + O_2 (*g*) → 2 H_2O (*l*), with a ΔH of −571.6 kJ. Since H_2 (*g*) and O_2 (*g*) are both elements in their standard states, this fits our definition of a formation reaction, but it forms two moles of water. Forming one mole of water would give us a ΔH of −571.6/2 = −285.8 kJ. This agrees with the heat of formation for water in the table on page 397.

To use the values in the table, Standard Enthalpy and Free Energy of Formation, to find ΔH for a given reaction, there are two rules you must bear in mind:

1. The ΔH_f^0 of any element in its standard state is zero.
2. ΔH^0 for any reaction is equal to ΔH_f^0 products ΔH_f^0 reactants. That is, to find ΔH^0, we add the total of all of the heats of formations of the products, and then subtract the sum of all of the heats of formation of the reactants.

ΔG^0 for a reaction is found from values of ΔG_f^0 in exactly the same way.

SAMPLE PROBLEM

PROBLEM

What is ΔH^0 for the reaction 2 CO (*g*) + O_2 (*g*) → 2 CO_2 (*g*)?

SOLUTION

$$\Delta H^0 = \Delta H_f^0 \text{ products} - \Delta H_f^0 \text{ reactants}$$

The product is 2 CO_2 (*g*), so ΔH_f^0 is 2 × −393.5 kJ = −787 kJ. The reactant is 2 CO (*g*) + O_2 (*g*). ΔH_f^0 of 2 CO is 2 × −110.5 kJ = −221 kJ ΔH_f^0 of O_2 (*g*) is 0. −787 kJ − (−221 kJ) = −566 kJ

PRACTICE

11.11 Find ΔH^0 for the following reactions, using the table Standard Enthalpy and Free Energy of Formation

(a) 2 NO (*g*) + O_2 (*g*) → 2 NO_2 (*g*)

(b) CH_4 (*g*) + 2 O_2 (*g*) → CO_2 (*g*) + 2 H_2O (*l*)

(c) 2 CH_3OH (*l*) + 3 O_2 (*g*) → 2 CO_2 (*g*) + 4 H_2O (*l*)

11.12 For the reaction shown in Practice 11-11, part (*a*), find ΔG^0.

11.13 For the same reaction as in the previous question, find ΔS^0 at 298 K.

Beyond their mathematical use, the values of ΔH_f^0 are valuable indicators of the relative stabilities of substances. In general, the greater the negative value of ΔH_f^0 the more stable the substance.

PRACTICE

11.14 Which of the substances listed in the table Standard Enthalpy and Free Energy of Formation is likely to be the least stable?

CHAPTER REVIEW

Data required for answering some of the questions will be found in the Table I, Heats of Reaction, in Appendix 4.

1. In which reaction do the products have a higher potential energy than the reactants?

(1) $2\ C\ (s) + H_2\ (g) \rightarrow C_2H_2\ (g)$

(2) $C\ (s) + O_2\ (g) \rightarrow CO_2\ (g)$

(3) $2\ CO\ (g) + O_2\ (g) \rightarrow 2\ CO_2\ (g)$

(4) $N_2\ (g) + 3\ H_2\ (g) \rightarrow 2\ NH_3\ (g)$

2. The greatest amount of energy is released in the combustion of 1.0 mole of

(1) $CH_4\ (g)$ (2) $CH_3OH\ (l)$ (3) $C_2H_5OH\ (l)$ (4) $Al\ (s)$

3. What happens when 1.0 mole of NH_3 (*g*) is formed from its elements?

   (1) 91.8 kJ are absorbed (3) 45.9 kJ are absorbed
   (2) 91.8 kJ are released (4) 45.9 kJ are released

4. If 2.000 moles of N_2 (*g*) and 2.000 moles of O_2 (*g*) react completely to form NO (*g*),

   (1) 182.6 kJ are absorbed (3) 365.2 kJ are absorbed
   (2) 182.6 kJ are released (4) 365.2 kJ are released.

5. ΔH for the decomposition of 2.00 moles of NH_3 (*g*) to produce N_2 (*g*) and H_2 (*g*) is

   (1) +91.8 kJ (2) −91.8 kJ (3) +183.6 kJ (4) −183.6 kJ

6. When ice melts, forming liquid water, there is

   (1) an increase in the potential energy and an increase in the entropy
   (2) a decrease in the potential energy and an increase in the entropy
   (3) an increase in the potential energy and a decrease in the entropy
   (4) a decrease in the potential energy and a decrease in the entropy

7. In what type of reaction do the products of the reaction always have more potential energy than the reactants? (1) endothermic (2) exothermic (3) spontaneous (4) decomposition

8. When one mole of a certain compound is formed from its elements under standard conditions, it absorbs 85 kJ of heat. A correct conclusion from this statement is that the reaction has a (1) ΔH_f^0 equal to −85 kJ/mole (2) ΔH_f^0 equal to +85 kJ/mole (3) ΔG_f^0 equal to −85 kJ/mole (4) ΔG_f^0 equal to +85 kJ/mole.

9. Which phrase best describes the reaction below?

$$C\ (s) + \frac{1}{2}\,O_2\ (g) \rightarrow CO\ (g) + 110.5\ \text{kJ}$$

   (1) exothermic with an increase in entropy (2) exothermic with a decrease in entropy (3) endothermic with an increase in entropy (4) endothermic with a decrease in entropy

10. Which reaction has a ΔH equal to -283 kJ at 298 K and 101.3 kPa?

(1) $C(s) + O_2(g) \rightarrow CO_2(g)$
(2) $CO(g) + \frac{1}{2} O_2(g) \rightarrow CO_2(g)$
(3) $2\,C(s) + H_2(g) \rightarrow C_2H_2(g)$
(4) $CH_4(g) + 2O_2(g) \rightarrow CO_2(g) + 2H_2O(l)$

11. As the reactants are converted to product in the reaction

$$A(g) + B(g) \rightarrow C(s)$$

the entropy of the system (1) decreases (2) increases (3) remains the same.

12. Which change is accompanied by a decrease in entropy?

(1) $H_2O(l) \rightarrow H_2O(s)$
(2) $H_2O(s) \rightarrow H_2O(g)$
(3) $H_2O(l) \rightarrow H_2O(g)$
(4) $H_2O(s) \rightarrow H_2O(l)$

13. A chemical reaction is most likely to occur spontaneously if the (1) free energy change (ΔG) is negative (2) entropy change (ΔS) is negative (3) free energy change (ΔG) is positive (4) heat of reaction (ΔH) is positive.

14. Which change results in an increase in entropy?

(1) $Br_2(l) \rightarrow Br_2(s)$
(2) $Cl_2(g) \rightarrow Cl_2(l)$
(3) $F_2(l) \rightarrow F_2(g)$
(4) $I_2(g) \rightarrow I_2(s)$

15. When formed from its elements at 298 K, which compound will be produced by an exothermic reaction? (1) nitrogen monoxide (*g*) (2) ethyne (C_2H_2) (*g*) (3) hydrogen iodide (*g*) (4) carbon dioxide (*g*)

16. The difference between the potential energy of the reactants and the potential energy of the products is (1) ΔG (2) ΔH (3) ΔS (4) ΔT.

17. Given the reaction $N_2(g) + O_2(g) + 182.6\text{ kJ} \rightarrow 2\,NO(g)$. What is the heat of formation of nitrogen monoxide in kJ/mole? (1) $\Delta H = -182.6$ (2) $\Delta H = -91.3$ (3) $\Delta H = +182.6$ (4) $\Delta H = +91.3$

18. Given the reaction

$$2\ Al\ (s) + \frac{3}{2}\ O_2\ (g) = Al_2O_3\ (s)$$

$$\Delta H = -1670\ kJ$$

The number of kilojoules of energy liberated by the oxidation of 27 g of aluminum is approximately (1) 417 (2) 835 (3) 1670 (4) 3340.

19. A chemical reaction must be spontaneous if it results in a potential energy (1) gain and an entropy increase (2) gain and an entropy decrease (3) loss and an entropy increase (4) loss and an entropy decrease.

Base your answers to questions 20 and 21 on the following information.

	Heat of Reaction ***ΔH kJ/mole***	***Free Energy of Reaction*** ***ΔG kJ/mole***
Reaction A	−94.05	−94.26
Reaction B	21.60	20.72
Reaction C	−70.96	−71.79
Reaction D	54.19	50.00

20. Which reactions are endothermic? (1) A and B (2) A and C (3) C and D (4) B and D

21. Which reactions occur spontaneously? (1) A and B (2) A and C (3) B and D (4) C and D

22. When calcium carbonate decomposes according to the equation $CaCO_3\ (s) \rightarrow CaO\ (s) + CO_2\ (g)$, the entropy of the system (1) decreases (2) increases (3) remains the same.

23. Endothermic reactions can occur spontaneously when the entropy of the system (1) decreases (2) increases (3) remains the same.

24. The entropy of a system would be greatest under conditions of

(1) high temperature and high pressure

(2) high temperature and low pressure

(3) low temperature and high pressure

(4) low temperature and low pressure.

25. Given the equation $\Delta G = \Delta H - T\Delta S$, a chemical reaction will most likely occur spontaneously if (1) ΔH is positive and ΔS is positive (2) ΔH is positive and ΔS is negative (3) ΔH is negative and ΔS is positive. (4) ΔH is negative and ΔS is negative.

26. The difference between the heat content of the products and the heat content of the reactants is (1) entropy of reaction (2) heat of reaction (3) free energy (4) activation energy.

CHEMISTRY CHALLENGE

The following questions will provide practice in answering SAT II-type questions.

For each question below, one or more of the responses given are correct. Decide which of the responses is (are) correct. Then choose

(a) if only I is correct;
(b) if only II is correct;
(c) if only I and II are correct;
(d) if only II and III are correct;
(e) if I, II, and III are correct.

1. For a system at equilibrium, which is always true?
I. $\Delta G = 0$ II. $\Delta S = 0$ III. $\Delta H = 0$.

2. Sodium chloride absorbs heat as it dissolves spontaneously in water at 25°C. For this change
I. ΔG is positive II. ΔS is positive III. ΔH is positive.

3. From the potential energy diagram for a reaction we can tell the sign of I. ΔH II. ΔS III. ΔG.

4. A spontaneous reaction is favored by
I. increasing entropy
II. decreasing enthalpy
III. a negative ΔH.

5. The sublimation of dry ice involves
   I. the breaking of bonds
   II. an increase in potential energy
   III. a decrease in entropy.
6. Given that for the reaction 2 Ca (*s*) + O_2 (*g*) → 2 CaO (*s*),

$$\Delta H^0 = -1271 \text{ kJ}, \Delta G^0 = -1208 \text{ kJ at 298 K}$$

This information is sufficient to permit us to find
   I. ΔH_f^0 of CaO   II. ΔG_f^0 of CaO   III. ΔS^0 for the given reaction at 298 K.
7. Given a system containing a mixture of gases, the entropy of the system will be increased when
   I. The volume of the system is increased
   II. The temperature of the system is increased
   III. One of the gases in the system condenses.
8. When you use a butane gas lighter, the reaction that occurs is

$$2\ C_4H_{10}\ (g) + 13\ O_2\ (g) \rightarrow 8\ CO_2\ (g) + 10\ H_2O\ (g)$$

   For this reaction, as written,

   I. ΔH^0 is negative   II. ΔG^0 is negative   III. ΔS^0 is negative.
9. For the freezing of water, H_2O (*l*) → H_2O (*s*) at room temperature
   I. ΔH^0 is positive   II. ΔG^0 is positive   III. ΔS^0 is positive.
10. ΔH^0 of formation is zero for
   I. O_2 (*g*)   II. Br_2 (*g*)   III. Hg (*s*).

The Ice-Cream Diet

Our normal body temperature is 37°C. The amount of heat needed to bring a one-gram sample of water from its freezing point (0°C) to 37°C is 4.18 j/g°C ? 1 g = 155 J. If the water starts out frozen, it requires an additional 334 J, the heat of fusion, to melt it. That is a total of 489 J absorbed for each gram of ice melted. Ice cream consists mainly of ice, so you can assume that it takes about the same amount of energy, 489 J to melt it and bring it to body temperature. The energy in food is measured in calories; there are 4.18 joules in a calorie. Therefore,

$$489 \text{ J} \times \frac{1 \text{ cal}}{4.18 \text{ J}} = 117 \text{ calories}$$

For each gram of ice cream you eat, your body must burn 117 calories.

Of course, the sugars and fats in the ice cream supply energy to the body. According to the information on the container, one gram of high-quality ice cream provides 2.5 calories. A normal portion of ice cream is 100 grams, so that would provide 250 calories. However, your body would use 11,700 to melt the 100 grams of ice cream and bring it to body temperature. Any time you use more calories than you consume, you must lose weight. Since the calories used in eating the ice cream are much greater than the calories gained from the ice cream, this diet should work. Eat ice cream and melt the pounds away!

You realize, of course, that there must be something wrong with this suggestion. You have learned that in chemical changes, such as burning the sugars in ice cream, the magnitude of ΔH is much larger than it is in physical changes, such as melting ice. Yet, every bit of information in the preceding paragraph is true; there is just one small deception. The calories that are listed on food packages are actually kilocalories. A portion of ice cream supplies 250 kilocalories while only 11.7 kilocalories (11,700 calories) were used to melt the ice cream. The energy used in eating the ice cream is insignificant compared with the energy contained in its fats and sugars. But, this author intends to keep eating it anyway!

Acids and Bases

LOOKING AHEAD

Terms such as "acid rain," "pH balanced," "litmus test" are probably familiar to you. These all deal with the properties of acids and bases. In this chapter, you will study the behavior of these important groups of compounds.

When you have completed this chapter you should be able to:

- **Explain** the behavior of electrolytes in water.
- **Define** and identify acids and bases.
- **Predict** the products of an acid-base reaction.
- **Relate** acid and base strength to the pH of a solution.
- **Predict** the behavior of acid-base indicators in aqueous solutions.
- **Perform** calculations based on acid-base titrations.

What Is an Acid? What Is a Base?

Your study of acid-base reactions will begin with definitions of acids and bases. These definitions are based on how acids and bases behave. Then, using models, you will consider definitions based on explanations for the observed behavior. You will consider two models of acids and bases.

Observed Behavior of Acids and Bases

An acid can be defined according to the properties it exhibits when in a water solution. According to this definition, an *acid:*

1. has a sour taste (*Chemicals should never be tasted unless it is absolutely certain that they are not poisonous.*)
2. neutralizes bases
3. conducts an electric current (is an electrolyte)
4. liberates hydrogen on reacting with certain metals, such as Al, Mg, and Zn
5. produces a characteristic color when reacted with certain chemicals called *indicators* (For example, acids turn a litmus solution red.)

From the properties exhibited by bases in water solutions, a *base:*

1. has a bitter taste
2. neutralizes acids
3. conducts an electric current (is an electrolyte)
4. feels slippery to the touch (*Chemicals should never be touched unless it is absolutely certain that they are not corrosive.*)
5. produces a characteristic color when reacted with certain chemicals called *indicators* (For example, bases turn a litmus solution blue.)

The Arrhenius Model

The first model to account for the behavior of acids and bases in water solutions was proposed in 1887 by the Swedish chemist Svante August Arrhenius. According to this model,

1. **Acids** ionize to produce hydrogen ions, (H^+), as their only positive ions.
2. **Bases** dissociate to produce hydroxide ions, (OH^-), as their only negative ions.
3. Electrolytes that contain positive ions other than H^+ and negative ions other than OH^- are called **salts.**

The model also explained why some acids and bases are strong and others are weak. According to the model, **strong acids** ionize almost completely, yielding a maximum quantity of H^+ ions. Similarly, **strong bases** dissociate almost completely, yielding a maximum quantity of OH^- ions. Strong acids and strong bases are strong electrolytes.

STRONG ACID: $HCl \rightarrow H^+ + Cl^-$

STRONG BASE: $NaOH \rightarrow Na^+ + OH^-$

Weak acids and **weak bases,** on the other hand, ionize or dissociate only partially and yield a small quantity of ions. Equilibrium exists between the molecules and the ions.

WEAK ACID: $CH_3COOH \leftrightarrows H^+ + CH_3COO^-$

WEAK BASE: $Mg(OH)_2 \leftrightarrows Mg^{2+} + 2\ OH^-$

In these equations, the longer of the two arrows points to the left. This means that at equilibrium, the reactants, which are molecules predominate over the products, which are ions. The small concentration of ions produces weak electrolytes.

Chemists often distinguish between the formation of ions by a dissolved ionic substance and the formation of ions by a dissolved molecular substance. $Mg(OH)_2$, for example is an ionic substance. The ions, already present in the solid, are separated or dissociated by the water. CH_3COOH, on the other hand, consists of molecules. When it dissolves in water, attractions between the water molecules and a hydrogen atom result in the formation of ions. This process is generally called *ionization.* Strictly speaking, molecular substances may ionize in water, while ionic substances may dissociate. Both processes involve the production of mobile, aqueous ions, resulting in electrical conductivity.

The hydrated proton According to Arrhenius, acids dissociate in water and produce hydrogen ions (H^+). The formula H^+ stands for an atom of H that has lost an electron—in other words, the H^+ ion is a proton. But H^+ ions do not exist alone in aqueous solution—they are surrounded by water molecules. The number of H_2O molecules attached to H^+ probably varies, but experiments lead scientists to believe that the number is most often four.

The hydrated (aquated) H^+ ion may be indicated by H^+ (*aq*) or by H_3O^+. The formula H^+ (*aq*) indicates that the H^+ ion is hydrated (aquated) but does not specify the exact number of H_2O molecules attached. The formula H_3O^+ means that one H_2O molecule is attached to one H^+ ion. This formula is not entirely accurate, but it is in common use. (You may recognize H_3O^+ as the formula for the hydronium ion.)

Naming Acids

You may already know the names of some of the common acids, such as sulfuric acid, H_2SO_4; acetic acid, $HC_2H_3O_2$, or CH_3COOH; and hydrochloric acid, HCl. These names are assigned according to rules of nomenclature, which will enable you to determine the correct names of most inorganic acids. (Organic acids are discussed in Chapter 14.)

The name of an acid is derived from the name of the negative ion formed by the loss of protons from the acid. If the acid contains more than one proton, you use the name of the ion formed when all the protons are lost. The names of several of these ions are listed in Table E in Appendix 4. For example, HNO_3 contains the NO_3^- ion, called the nitrate ion. H_3PO_4 contains PO_4^{3-}, the phosphate ion. HCl contains Cl^-, the chloride ion. The names of the acids are derived from the names of the ions, as follows.

1. If the name of the ion ends in "ide," the name of the acid is hydro . . . ic. Examples are HCl, hydro*chlor*ic acid, HI, hydro*iod*ic acid, and HBr, hydro*brom*ic acid. Acids that contain only hydrogen and another element are known as **binary acids.** Binary acids are always named hydro . . . ic.
2. If the name of the anion (negative ion) ends in "ate" the name of the acid is . . . ic. Thus, H_2SO_4, which contains the sulfate ion, is *sulfur*ic acid. HNO_3, which has a nitrate ion, is called *nitr*ic acid. $HC_2H_3O_2$ contains the acetate ion, and so is called *acet*ic acid.
3. If the name of the anion ends in "ite," the name of the acid is . . . ous. HNO_2 contains the nitrite ion, NO_2^-, and is called *nitr*ous acid. $HClO_2$ contains the chlorite ion, ClO_2^-, and is called *chlor*ous acid.

SAMPLE PROBLEM

PROBLEM

What are the names of these acids? (a) H_2SO_3 (b) $HClO_4$

SOLUTION

The name of the SO_3^{2-} ion is given in Appendix 4 as sulfite. Therefore the name of the acid containing this ion is sulfurous acid. ClO_4^- is called the perchlorate ion. The acid $HClO_4$ is called perchloric acid.

PRACTICE

12.1 Name the following acids. (a) HOCl (also often written HClO) (b) H_2S (c) $H_2C_2O_4$ (d) H_3PO_4 (e) H_3BO_3 (BO_3^{3-} is the borate ion)

Common Acids and Bases

The Arrhenius definitions tell you what to look for in an acid and a base. You will now apply your knowledge of bonding to identify the types of substances that form acids and bases, according to those definitions.

An acid forms H^+ ions; it must therefore contain hydrogen. However, not all compounds that contain hydrogen are acids. The hydrogen in the compound must be attracted by water molecules strongly enough for it to break off as a hydrated H^+, or H_3O^+, ion. Water molecules are strongly polar. In an acid, the negative end of the water molecule attracts the hydrogen on the acid molecule. The hydrogen must have a slight positive charge. Hydrogen acquires a positive charge only if it is bonded to an element of relatively high electronegativity—a nonmetal. Acids can be represented by the general formula HX, where X is a nonmetal or a group of nonmetals. The following table lists some common acids, their formulas, and relative strengths. Note that the formula of acetic acid is written two different ways. Organic acids, discussed in Chapter 14, contain "COOH" in their formulas. Organic chemists prefer to write the formulas so as to show the presence of this group of atoms. Organic chemistry has a special system of naming compounds; in this system, acetic acid can be called ethanoic acid.

Name	***Formula***	***Strength***
Hydrochloric acid	$HCl(aq)$	Very strong
Nitric acid	$HNO_3(aq)$	Very strong
Sulfuric acid	$H_2SO_4(aq)$	Strong
Phosphoric acid	$H_3PO_4(aq)$	Moderately strong
Acetic (ethanoic) acid (vinegar)	$HC_2H_3O_2(aq)$ or $CH_3COOH(aq)$	Weak
Carbonic acid	$CO_2(aq)$ or $H_2CO_3(aq)$	Weak

Bases form OH^- ions in solution. You would therefore expect bases to be compounds that contain one or more OH groups. However, not all compounds that contain an OH are bases. Alcohols, such as CH_3OH and C_2H_5OH, are not bases. A base must be able to form the negative ion OH^-. Therefore, the atom attached to the OH must form a positive ion and so is generally a metal. Bases have the general formula MOH, where M is a metal. The strongest bases contain metals that are placed in Groups 1 and 2 of the Periodic Table. Of these, the most common are NaOH, KOH, and $Ba(OH)_2$. The hydroxide compounds of metals in the remaining groups are generally insoluble in water and so do not form free OH^- ions.

If you have used a solution of ammonia as a cleanser, you may have noticed that it feels slippery to the touch. Testing it with indicators would show that it forms a basic solution. Ammonia does not contain any OH^- ions, yet it is a base in water. Ammonia can pull H^+ ions out of the water, as shown in the equation: $NH_3(g) + H_2O(l) \rightleftharpoons NH_4^+\ (aq) + OH^-\ (aq)$. The chemical equation shows that the reaction of ammonia with water does produce OH^- ions. The table below lists some common bases, their formulas, and relative strengths.

Name	***Formula***	***Strength***
Sodium hydroxide (lye)	$NaOH(aq)$	Very strong
Potassium hydroxide	$KOH(aq)$	Very strong
Barium hydroxide	$Ba(OH)_2(aq)$	Very strong
Calcium hydroxide	$Ca(OH)_2(aq)$	Moderately strong
Ammonia	$NH_3(aq)$	Weak

Although later models broadened the acid-base concept, our usual usage of the terms "acidic" and "basic" still refers to the Arrhenius definitions. Battery acid has a high concentration of H^+ ions, while lye solutions, which are basic, have a high concentration of OH^- ions.

pH

You may already be familiar with the concept of pH, as a number that indicates the acidity or basicity of a given solution. In simplest terms, at 298 K a pH of less than 7 is acidic, and the lower the pH the more acidic the solution. A pH of more than 7 is basic, and the higher the pH, the more basic the solution. A pH of 7 is neutral; pure water has a pH of 7.

The more precise definition of pH involves a mathematical function called a logarithm. The pH is defined as $-\log [H^+]$, or the negative logarithm of the molar concentration of H^+. A logarithm is a power of 10. If you wish to simply compare two different solutions, the rule is that each pH unit represents a 10-fold change in H^+ concentration. For example, a solution with a pH of 1 is 10 times more acidic than a solution with a pH of 2. Remember, the lower the pH, the more acidic the solution.

Each pH unit also represents a 10-fold change in OH^- concentration. The higher the pH, the more basic the solution. A pH of 13 is 100 times more basic than a pH of 11.

You can use pH to compare the strengths of different acids and bases. The table below lists the approximate pH values for some 0.1 *M* aqueous solutions at 298 K.

Solution (0.1 M)	***pH at 298 K***
HCl	1
HNO_3	1
$HC_2H_3O_2$	3
CO_2 (H_2CO_3)	4
$NaCl$	7
NH_3	11
$NaOH$	13

PRACTICE

12.2 If a 0.1 *M* solution of phosphoric acid, H_3PO_4, is approximately ten times more acidic than a 0.1 *M* solution of acetic acid, $HC_2H_3O_2$, what is the approximate pH of the phosphoric acid solution?

12.3 Compared with solutions of ammonia, solutions of NaOH of equal concentration are approximately

(1) 10 times more basic

(2) 10 times more acidic

(3) 100 times more basic

(4) 100 times more acidic.

12.4 Based on the table of common bases and the table of pH values, predict the pH of 0.1 *M* KOH?

12.5 Why is the pH of 0.1 *M* NaCl equal to 7?

Indicators

A convenient way to determine the pH of a solution is to add an **indicator**, a substance that changes color in a specific pH range, to the solution. Such indicators are complex molecules that behave as weak acids or as weak bases. Some of the common indicators are listed in the following table. Notice that the change of color and the pH range in which the change takes place vary with the indicator.

Indicator	***pH Range***	***Color Change***
Congo red	3–5	Blue to red
Methyl orange	3.2–4.4	Red to yellow
Methyl red	4.3–6	Red to yellow
Litmus	5.5–8.2	Red to blue
Bromthymol blue	6.0–7.6	Yellow to blue
Phenolphthalein	8.2–10	Colorless to pink
Alizarin yellow	10.1–12.1	Yellow to red

When using this table, bear in mind that the "pH range" indicates the region in which the indicator will noticeably change color. At a pH below this range, the indicator will always show its "acid color," which is listed first. Above the range, the indicator will show its "base color." For example, at a pH of 6, congo red will be red. At a pH of 5, bromthymol blue will be yellow.

PRACTICE

12.6 Use the table of indicators on page 413 to determine the colors of the following solutions:

(a) congo red in a solution of pH = 2

(b) phenolphthalein when the pH is 9.6

(c) methyl red in a solution of NaOH

(d) litmus in a solution of a strong acid

12.7 In pure water, bromthymol blue solution is green. Why do you think this is true?

12.8 Based on the pH table on page 412, which of the indicators might be used to distinguish a solution of NaOH from a solution of NH_3?

12.9 You are given two beakers, each containing a colorless solution. One of these is NaOH(*aq*), and the other is HCl(*aq*). It is not safe to taste or to touch the solutions. Other than by using an indicator, how could you determine which solution was which?

pH Calculations

The exact pH of a solution can be calculated from the H^+ concentration with the help of a scientific calculator. Before you start performing these calculations, let us take a closer look into the meaning of pH.

The ionization of water Water is a weak electrolyte. The ionization of water, $H_2O\ (l) \rightleftarrows H^+(aq) + OH^-\ (aq)$ has an equilibrium constant at 298 K, called the K_W, of 1.0×10^{-14}. In any aqueous solution, the product of the H^+ concentration and the OH^- concentration is 1.0×10^{-14}. We show this relationship with the expression $[H^+][OH^-] = 1.0 \times 10^{-14}$. In a neutral solution,

$[H^+] = [OH^-]$ so both concentrations must $= 1.0 \times 10^{-7}$ *M*. In an acidic solution, $[H^+] > [OH^-]$ while in a basic solution, $[H^+] < [OH^-]$, but in both cases, the product, $[H^+][OH^-]$ must $= 1.0 \times 10^{-14}$.

The pH is defined as $-\log [H^+]$, and a logarithm is a power of ten. Consider a solution of pure water. As shown above, the $[H^+]$ is 1.0×10^{-7} *M*. The logarithm of 1.0×10^{-7} is -7. Therefore the pH is $-(-7)$ or $+7$. When $[H^+] = [OH^-]$, the pH is 7. That is why at 298 K a pH of 7 is considered neutral.

pH from the [H⁺] With the help of a scientific calculator, you can find the pH of any aqueous solution if you know the $[H^+]$. For example, what is the pH of a solution in which the $[H^+]$ was 0.020 *M*? Since pH $= -\log [H^+]$, we must first take the logarithm of 0.020. Your calculator will tell you that the log of 0.020 is -1.7. Since the pH is the negative log, we must change the sign of our logarithm. The pH is 1.7.

SAMPLE PROBLEM

PROBLEM

What is the pH of 0.040 *M* HCl?

SOLUTION

HCl is a strong acid. We assume that it ionizes completely, producing an $[H^+]$ of 0.040 *M*. The log of 0.040 is -1.4. The pH of the solution is 1.4.

PRACTICE

12.10 Find the pH of each of the following solutions:

(a) 0.050 *M* HNO_3

(b) A solution in which the $[H^+]$ is 1.0×10^{-4} *M*

12.11 What is the concentration of $[H^+]$ in a solution that has a pH of 2.7?

pH from [OH⁻] It is also possible to find the pH of an aqueous solution if we know the $[OH^-]$. For example, what is the pH in a solution of $[OH^-] = 1.0 \times 10^{-6}$ *M*? Since the $[H^+][OH^-] = 1.0 \times 10^{-14}$, we can calculate that if $[OH^-] = 1.0 \times 10^{-6}$ *M*, then $[H^+] = 1.0 \times 10^{-8}$ *M*. From the $[H^+]$, we can now calculate that the pH is 8.0. However, there is a faster method.

To find pH from the $[OH^-]$, just take the log of the $[OH^-]$, and then add 14. In the example above, the log of 1.0×10^{-6} -6, and $-6 + 14 = 8$.

SAMPLE PROBLEM

PROBLEM

What is the pH of 0.0040 *M* NaOH?

SOLUTION

NaOH is a strong base, so the $[OH^-]$ is 0.0040 *M*. The log of 0.0040 is -2.4. Add 14 to -2.4 to get the pH of 11.6.

PRACTICE

12.12 Find the pH of each of the following solutions:

(a) 0.0020 *M* KOH

(b) A solution in which the $[OH^-]$ is 1.0×10^{-5} *M*.

(c) A solution in which the $[OH^-]$ is 5.0×10^{-12} *M*.

12.13 Find the $[OH^-]$ in a solution of pH = 6.0

Neutralization

Hydrochloric acid, HCl, and sodium hydroxide, NaOH, are two highly dangerous substances, capable of burning the skin and potentially deadly if consumed. Mixed together in equal quantities, however, they form ordinary table salt and water.

$$HCl + NaOH \rightarrow H_2O + NaCl$$

The reaction between an acid and a base to form a salt and water is called **neutralization.** The salt formed depends on the acid and base used. If nitric acid, HNO_3, is reacted with potassium hydroxide, KOH, the salt formed is potassium nitrate. Since HCl and NaOH are both strong electrolytes, their solutions contain the ions H^+, Cl^-, Na^+, and OH^-. In solution, the reaction could be written

$$H^+(aq) + Cl^-(aq) + Na^+(aq) + OH^-(aq) \rightarrow H_2O(l) + Na^+(aq) + Cl^-(aq)$$

Since the salt, NaCl, is completely dissociated in water, it is written as consisting of aqueous ions. Water is a poor electrolyte, forming few ions, and so is written as a molecule. Examining this ionic equation, you note that the Na^+ and Cl^- ions appear on both sides of the equation. In fact, these ions play no part in the neutralization reaction. Ions that are present in a solution but are not involved in the reaction are called *spectator ions*. Reactions in solution are generally written with the spectator ions omitted. In the case of the neutralization of HCl by NaOH, the *net ionic equation* is

$$H^+(aq) + OH^-(aq) \rightarrow H_2O(l)$$

The neutralization of nitric acid with potassium hydroxide produces exactly the same net equation. The K^+ ion and the NO_3^- ion are the spectator ions. As long as the acid and base are both strong electrolytes, and the salt formed is soluble, the net ionic equation for neutralization will be the same.

Acid-Base Titrations

Since the net reaction for the neutralization of a strong acid with a strong base is $H^+(aq) + OH^-(aq) = H_2O$, when the number of moles of H^+ and the number of moles of OH^- are equal, the solution will be neutral. This relationship can be used experimentally to find the concentration of an unknown acid or base, through a procedure called a titration.

A **titration** is a procedure in which measured volumes of two solutions are combined to achieve a predetermined end point. In most acid-base titrations, a measured volume of acid is added to a measured volume of base until the moles of H^+ and the moles of OH^- are equal, neutralization occurs. This particular end point is called the equivalence point.

When a strong base is added to a strong acid, the pH changes slowly until the equivalence point is approached. Then the pH

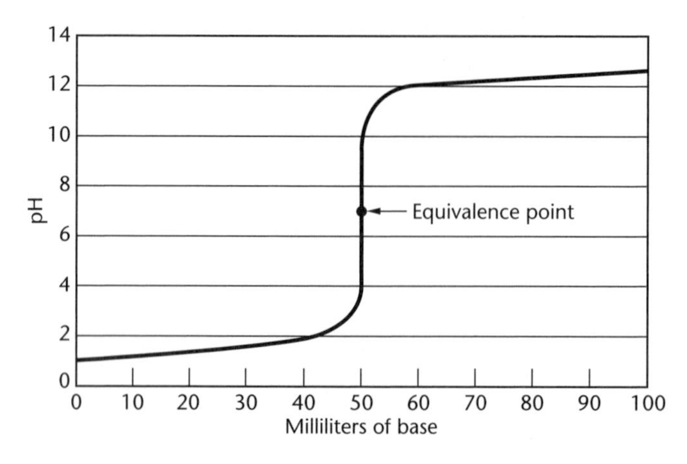

Figure 12-1 pH curve for strong acids and bases (0.1 *M* base added to 50 mL of 0.1 *M* acid)

shows a sudden, sharp increase at the equivalence point. Figure 12-1 shows the pH curve for the titration of a strong acid with a strong base.

In fact, when 49 mL of base have been added, the pH is 3; when 51 mL have been added the pH is 11. This sudden jump in pH that occurs near the equivalence point is very easy to detect with an indicator.

When weak acids or bases are used, the shape of the curve will be somewhat different. However, there will still be a sudden, large jump in pH near the equivalence point, which can be detected with an indicator.

To find the concentration of an acid or base through titration, a small, carefully measured quantity of an acid or base of unknown concentration is placed in a flask. A few drops of indicator are added. Then, using a burette, the known solution is added carefully to the flask until a single drop causes the indicator to change color. At that point, the end point, the volume of acid and base used is noted. You can then calculate the concentration of the unknown solution.

At the equivalence point, an equal number of moles of acid and base have reacted. Moles H^+ = moles OH^-. Since moles= molarity × liters, you can substitute molarity × liters for moles in the neutralization equation, which gives us, at the equivalence point,

$$\text{molarity} \times \text{liters } (H^+) = \text{molarity} \times \text{liters } (OH^-)$$

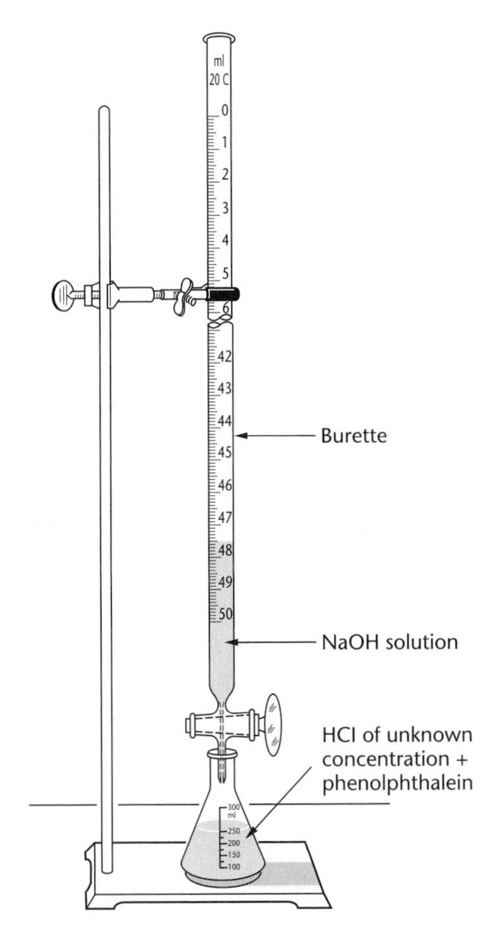

Figure 12-2 Titration

It doesn't matter what unit of volume (V) you use in this equation, as long as it is the same on both sides. The equation becomes

$$M \times V\,(\mathrm{H}^+) = M \times V\,(\mathrm{OH}^-).$$

At the equivalence point, you know the volume of the acid and the base and the molarity of the known solution, the molarity of the unknown solution can be found.

Let us consider a specific example. Suppose you wanted to find the concentration of a solution of HCl. First, you would add a carefully measured volume of the acid to a flask. In this case, let us say 25 mL of the acid. You then add a few drops of phenolphthalein indicator to the solution. (See Figure 12-2.) Phenolphthalein is colorless in acid solutions, but turns pink when there is a slight excess of OH^- ions. It is frequently used in titrations, because the color change from colorless to pink is very easy to detect.

You would then slowly add an NaOH solution of known concentration to the HCl solution. To do this, you use a burette, a piece of equipment that makes it easy to add a solution drop by drop, while measuring the volume added. The volume added is the difference between the initial reading on the burette and the final reading. When one drop of the base turns the solution from colorless to pink, you are within one drop of the equivalence point. (One extra drop is too small a quantity to significantly affect the calculations.)

You now record your final readings. Let us say that the NaOH was 0.20 *M*. The initial burette reading was 1.00 mL, and the final reading was 43.00 mL. To find the volume of NaOH added you subtract the initial reading from the final reading, so in this case, you have added 42.00 mL of base. Now you are ready to calculate the molarity of the unknown solution of HCl.

At the equivalence point, $M \times V\,(H^+) = M \times V\,(OH^-)$. The volume ($V$) of acid was 25 mL, the volume of base 42 mL, and the molarity of base was 0.2 *M*. Let x = molarity of acid.

$$x \times 25\text{ mL} = 0.20\,M \times 42\text{ mL}$$

$$x = 0.34\,M.$$

The acid was 0.34 *M*.

SAMPLE PROBLEMS

PROBLEM

1. If 40. mL of 0.20 *M* NaOH are needed to titrate 50. mL of HNO_3, what is the molarity of the HNO_3?

SOLUTION

At the equivalence point,

$$M \times V\,(H^+) = M \times V\,(OH^-)$$

The volume of acid is 50. mL. The volume of base is 40. mL, and the molarity of the base is 0.20 *M*. Let x = molarity (H^+).

$$x \times 50.\text{ mL} = 0.20\,M \times 40.\text{ mL}$$

$$x = 0.16\,M$$

PROBLEM

2. How many milliliters of 0.10 *M* HCl are needed to exactly neutralize 20. mL of 0.050 *M* NaOH?

SOLUTION

This time both molarities are known; the unknown is the volume of acid. Let x = volume (acid) in milliliters. Then, substituting into the equation

$$M \times V\,(H^+) = M \times V\,(OH^-)$$

$$0.10\ M \times x = 0.050\ M \times 20.\ \text{mL}$$

$$x = 10\ \text{mL}$$

PROBLEM

3. How many milliliters of 0.20 *M* NaOH are needed to neutralize 20. mL of 0.25 *M* H_2SO_4?

SOLUTION

At first glance, this looks the same as Problem 2. You know both molarities, and need to find the unknown volume. However, every mole of H_2SO_4 added, provides two moles of H^+ ions. When titrating with an acid that provides more than one H^+ or base that provides more than one OH^-, you multiply the molarity of that solution by the number of H^+ or OH^- ions provided by each mole of substance. Let x = volume of NaOH in mL.

$$M \times V\,(H^+) = M \times V\,(OH^-)$$

$$0.25\ M \times 20.\ \text{mL} \times 2\ H^+/\text{mole} = 0.20\ M \times x$$

$$x = 50.\ \text{mL of NaOH.}$$

PRACTICE

12.14 How many milliliters of 0.20 *M* KOH are needed to titrate 40. mL of 0.10 *M* HNO_3?

12.15 If 30. mL of 0.40 *M* HCl are titrated with 40. mL of NaOH, what is the molarity of the base?

12.16 How many milliliters of 0.40 *M* $Ba(OH)_2$ are needed to titrate 25 mL of 0.50 *M* H_2SO_4?

Other Acid-Base Theories

The Arrhenius model works well in explaining the behavior of most common acids and bases in water. Some chemists, however, prefer to broaden the acid-base concept so that it can be applied to situations where water is not present. There are two commonly applied alternatives to the Arrhenius model: the Brønsted-Lowry theory and the Lewis theory.

Brønsted-Lowry Theory

In 1923, a new set of definitions of acids and bases was proposed by Johannes Brønsted, a Danish scientist, and Martin Lowry, an English scientist. According to the **Brønsted-Lowry model,** often abbreviated as the *Brønsted model,* an *acid* is a substance that donates protons. A *base* is a substance that accepts protons. A *proton* is the same thing as a hydrogen ion, H^+. A substance therefore cannot function as an acid—donate protons—unless a base is present to accept those protons. In the Brønsted model, any reaction involving an acid must also involve a base. The reverse is also true.

You will now examine a few examples of reactions in which protons are transferred. In so doing, you will learn more details of the Brønsted-Lowry model.

HCl can donate a proton to water or to another substance that can accept protons, NH_3, for example. This ability accounts for the acidic properties of HCl. At the same time, H_2O or another substance accepts a proton from HCl and acts as a base.

$$\underset{\text{acid}}{HCl} + \underset{\text{base}}{NH_3} \rightleftharpoons NH_4^+ + Cl^-$$

$$\underset{\text{acid}}{HCl} + \underset{\text{base}}{H_2O} \rightleftharpoons H_3O^+ + Cl^-$$

(Recall that H_3O^+ is the hydronium ion; NH_4^+ is the ammonium ion.)

NH_3 can accept a proton from HCl or from other substances that can donate protons, such as H_2O. This ability accounts for the basic properties of NH_3.

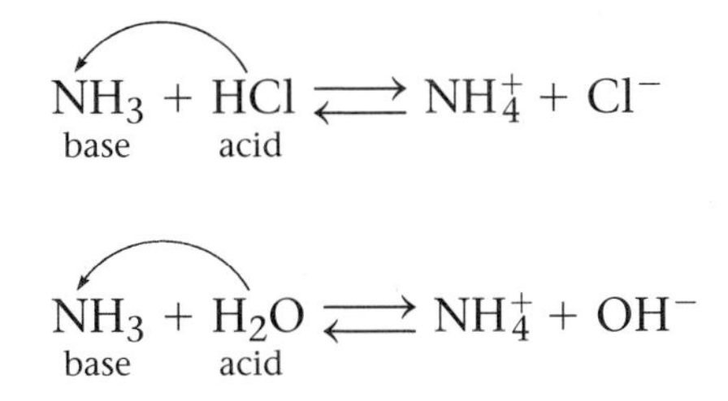

In the reaction between HCl and water, HCl forms the hydronium ion as its only positive ion. Since the HCl is also donating a proton, it is an acid in both the Arrhenius model and the Brønsted model. In the reaction between HCl and NH_3, however, neither hydronium ions nor hydroxide ions are formed. This is not an acid-base reaction according to Arrhenius, but since protons are transferred, it is an acid-base reaction according to Brønsted.

In its reaction with NH_3, shown above, water acts as an acid. The notion of water behaving as an acid may seem a little strange to you, since water has none of the properties we ordinarily associate with acids. In the Brønsted model, to say that ammonia is a base in water is the same as saying that water is an acid in ammonia. We live on a planet whose surface is covered mostly by water. It is natural for us to make water our basis for comparison, so we define water as being neutral. Those substances which, like ammonia, are bases in water are called basic, and those that are acids in water are called acidic. The planet Jupiter, however, has large amounts of ammonia in its atmosphere. If there were chemists on Jupiter, it is likely that they would consider ammonia neutral and water an acid.

Acid-Base Conjugate Pairs

In the reaction

$$HCl + H_2O \rightarrow H_3O^+ + Cl^-$$

the acid HCl donates a proton to the base H_2O. However, all chemical reactions are reversible unless they are driven to completion. The equation can thus also be read from right to left: H_3O^+ donates a proton to Cl^- and re-forms HCl and H_2O.

HCl and Cl^- are therefore related. The Cl^- ion is the base formed when HCl acts as an acid. HCl is the acid formed when Cl^-

ion acts as a base. The HCl molecule and the Cl^- ion are called a *conjugate acid-base pair.* By the same reasoning, H_2O is a base, and H_3O^+ is an acid. H_2O and H_3O^+ are a second conjugate acid-base pair. Thus in the reaction

$$\underset{\text{acid}_1}{HCl} + \underset{\text{base}_2}{H_2O} \rightleftharpoons \underset{\text{acid}_2}{H_3O^+} + \underset{\text{base}_1}{Cl^-}$$

HCl and Cl^- are conjugate pair 1, and H_2O and H_3O^+ are conjugate pair 2. In an acid-base reaction, there are two conjugate pairs of acids and bases.

If you read the next equation from left to right,

$$\underset{\text{base}_1}{NH_3} + \underset{\text{acid}_2}{H_2O} \rightleftharpoons \underset{\text{acid}_1}{NH_4^+} + \underset{\text{base}_2}{OH^-}$$

you see that NH_3 behaves as a base. It accepts a proton from H_2O. The H_2O, in this case, acts as an acid. If you read the equation in the reverse direction, an NH_4^+ ion donates a proton to an OH^- ion and re-forms NH_3 and H_2O. In the reaction, then, two conjugate acid-base pairs can again be identified. Notice that H_2O has a dual role—it can act either as an acid or as a base.

Given any acid-base reaction, it is possible to identify the acids and bases on both sides of the equation. Consider the reaction

$$H_2S + H_2O \rightleftharpoons H_3O^+ + HS^-$$

In this reaction, H_2S is donating a proton to H_2O. Therefore, H_2S is an acid, and H_2O is a base. On the product side, H_3O^+ is the acid, and HS^- the base. Confusion sometimes arises while identifying the acids and bases on the product side. One might reason that since HS^- has donated a proton, and H_3O^+ has received one, the HS^- should be the acid. The problem arises because although in ordinary speech we call a donor someone who has donated, in the Brønsted model, a proton donor is defined strictly as a particle that will donate. The term "proton donor" refers to the future, not to the past. H_3O^+ is an acid, because in the reverse reaction it will donate a proton to the base, HS^-.

There is an easy way to avoid confusion in labeling Brønsted acids and bases. A Brønsted acid always has one more H^+ than its conjugate base. In analyzing these reactions, you first find the conjugate acid-base pairs. The acid in each pair is the particle with the extra H^+.

Consider the reaction,

$$H_2S + H_2O \rightleftharpoons H_3O^+ + HS^-$$

Lines have been drawn connecting the conjugate acid-base pairs. The acid always has one more H^+ than its conjugate base. Therefore, H_2S is an acid, and HS^- is its conjugate base. H_2O is a base, and H_3O^+ is its conjugate acid.

Conjugate acid-base pairs differ by a single proton. Since protons are positive, the charge of the acid is always one higher than the charge of the conjugate base. Given any acid, it is possible to identify its conjugate base, and vice versa.

SAMPLE PROBLEMS

PROBLEM

1. What is the conjugate base of ammonia, NH_3?

SOLUTION

Since NH_3 has three Hs, its conjugate base must have two. And since NH_3 is neutral, its conjugate base must have a charge which is lower by one, or $1-$. The conjugate base of NH_3 is NH_2^-.

PROBLEM

2. What is the conjugate acid of the oxide ion, O^{2-}?

SOLUTION

Since the oxide ion has no Hs, its conjugate acid must have one. Since the charge of the oxide ion is $2-$, the charge of the conjugate acid must be one higher, or $1-$. The conjugate acid of O^{2-} is OH^-.

PRACTICE

12.17 In the following reactions, identify the conjugate acid-base pairs and label the acids and bases.

(a) $HNO_2 + H_2O \rightleftharpoons H_3O^+ + NO_2^-$

(b) $PO_4^{3-} + HF \rightleftharpoons HPO_4^{2-} + F^-$

12.18 Find the conjugate acids of each of the following:

(a) OH^-

(b) HPO_4^{2-}

(c) $C_2O_4^{2-}$

12.19 Find the conjugate bases of each of the following:

(a) H_3O^+

(b) H_3PO_4

(c) HSO_4^-

Identifying conjugate pairs is helpful in judging the strengths of acids and bases. It is also an aid in predicting the direction of acid-base reactions.

Strengths in Acid-Base Pairs

Let us use the reaction between HCl and H_2O as an example again.

$$HCl + H_2O \rightleftharpoons H_3O^+ + Cl^-$$

From experiments, it is known that HCl is a strong acid. HCl must therefore have a strong tendency to donate protons. It is reasonable to expect, then, that its conjugate base, Cl^- ion, has a weak tendency to hold or to accept protons. If you read the reaction in the opposite direction, you see that the Cl^- ion is a weak proton acceptor. Its conjugate acid (HCl) must therefore be a strong proton donor.

Here is another example.

$$\underset{acid_1}{CH_3COOH} + \underset{base_2}{H_2O} \rightleftharpoons \underset{acid_2}{H_3O^+} + \underset{base_1}{CH_3COO^-}$$

The CH_3COOH molecule is a relatively weak acid. Its conjugate base (CH_3COO^- ion) is a relatively strong base.

Recall that CH_3COOH can also be written as $HC_2H_3O_2$. The equation can be written as

$$HC_2H_3O_2 + H_2O \leftrightharpoons H_3O^+ + C_2H_3O_2^-$$

All of these examples show that for an acid to react (donate protons), a base must be present to accept protons. The extent of the overall reaction depends on the proton-donating properties of the acid and also on the proton-accepting properties of the base.

The strengths of acids can be compared by measuring their ability to donate protons to the same base, usually water. In such comparisons, only the proton donor is shown. The proton acceptor is understood to be water.

In the following table, HCl is the strongest acid, and HS^- is the weakest acid. Similarly, Cl^- is the weakest base, and S^{2-} is the strongest base. Notice that acids may be either molecules or ions. The same is true for bases.

	Acid $\longrightarrow$	***Conjugate Base***	
	Conjugate Acid $\longleftarrow$	***Base***	
Increasing acid strength (↑)	HCl	$\rightarrow H^+ + Cl^-$	Increasing base strength (↓)
	HNO_3	$\rightarrow H^+ + NO_3^-$	
	H_2SO_4	$\rightarrow H^+ + HSO_4^-$	
	H_3O^+	$\rightarrow H^+ + H_2O$	
	HSO_4^-	$\rightarrow H^+ + SO_4^{2-}$	
	CH_3COOH	$\rightarrow H^+ + CH_3COO^-$	
	H_2CO_3	$\rightarrow H^+ + HCO_3^-$	
	H_2S	$\rightarrow H^+ + HS^-$	
	$H_2PO_4^-$	$\rightarrow H^+ + HPO_4^{2-}$	
	H_3BO_3	$\rightarrow H^+ + H_2BO_3^-$	
	NH_4^+	$\rightarrow H^+ + NH_3$	
	HCO_3^-	$\rightarrow H^+ + CO_3^{2-}$	
	HPO_4^{2-}	$\rightarrow H^+ + PO_4^{3-}$	
	HS^-	$\rightarrow H^+ + S^{2-}$	

Lewis Theory

The American chemist, Gilbert Lewis, proposed another acid-base model. In the **Lewis model,** an acid is an electron pair acceptor, while a base is an electron pair donor. Consider the reaction between NH_3 and H_2O, shown on page 424. You saw that according to the Brønsted-Lowry model, ammonia is a base because it accepts a proton from the water. Lewis would say that the ammonia is a base because it donates an electron pair to the H^+ ion.

Now consider the reaction $Ag^+ (aq) + 2NH_3(aq) \rightarrow Ag(NH_3)_2{}^+(aq)$. While there is no transfer of protons, the NH_3 molecules do donate electron pairs to the silver ions. In this reaction, the ammonia is a Lewis base, while the silver ion is a Lewis acid. The Lewis model is the broadest approach to the acid-base concept.

TAKING A CLOSER LOOK

Acid-Base Equilibria

As shown in the pH table on page 412, the pH of a 0.1 *M* solution of the strong acid HCl is 1 at 298 K, while that of the weak acid $HC_2H_3O_2$ is 3. Since both solutions contain the same concentration of acid, why should the pHs be different? When an acid is dissolved in water, it forms H^+ ions. The general equation for the ionization of an acid, HA, is $HA \rightleftarrows H^+ + A^-$. The equilibrium constant for this system is called the K_a. For the acid HA,

$$K_a = \frac{[H^+][A^-]}{[HA]}.$$

For strong acids, such as HCl, the K_a is very large, indicating that virtually all of the acid is ionized. Therefore, we can assume that 0.1 *M* HCl forms 0.1 *M* H^+ ions, producing a pH of 1. However, the K_a of boric acid is 7.3×10^{-10}. This low number indicates that most of the H_3BO_3 remains un-ionized. In fact, in a 0.1 *M* boric acid solution, the H^+ concentration is only 8.5×10^{-6} *M*, and the pH is 5.1.

In the following table, acids are listed in order of strength, from strongest to weakest. This table gives us some insights into the behavior of some common acids. Sulfuric acid, sometimes called battery acid, is obviously a substance you would not wish to drink. As you can see from the table, it is a very strong acid. The vinegar many of us consume is actually dilute acetic acid. As vinegar, it is a relatively weak acid. Boric acid, H_3BO_3, has been used as an eye wash. If it is safe to put in your eyes, you would expect it to be a very weak acid, and it is. Phosphoric acid is stronger than acetic acid, yet many of us drink it frequently. Read the ingredients on a can of cola, and you will see why these beverages are very acidic.

The table below lists not only acids, but also their conjugate bases. The stronger the acid, the weaker the conjugate base. Thus you can use this table to compare base strengths as well as acid strengths. Base strengths increase as you go down the right side of the table.

The K_a of an acid also indicates the electrical conductivity of its solutions. HCl and HNO_3 are strong electrolytes, while H_2S and H_3BO_3 are weak electrolytes. Some ions and molecules can be found on both sides of the chart. These species, which include H_2O, and HSO_4^- can act both as acids and as bases. They are called **amphiprotic**, because depending on what they react with they can gain or lose protons.

Acid	*Conjugate Base*	*K_a at 298 K*
HCl	$\rightarrow H^+ + Cl^-$	Very large
HNO_3	$\rightarrow H^+ + NO_3^-$	Very large
H_2SO_4	$\rightarrow H^+ + HSO_4^-$	Very large
H_3O^+	$\rightarrow H^+ + H_2O$	1
HSO_4^-	$\rightarrow H^+ + SO_4^{2-}$	1.3×10^{-2}
H_3PO_4	$\rightarrow H^+ + H_2PO_4^-$	7.1×10^{-3}
HF	$\rightarrow H^+ + F^-$	6.7×10^{-4}
CH_3COOH	$\rightarrow H^+ + CH_3COO^-$	1.8×10^{-5}
H_2CO_3	$\rightarrow H^+ + HCO_3^-$	4.4×10^{-7}
H_2S	$\rightarrow H^+ + HS^-$	1.0×10^{-7}
$H_2PO_4^-$	$\rightarrow H^+ + HPO_4^{2-}$	6.3×10^{-8}
HSO_3^-	$\rightarrow H^+ + SO_3^{2-}$	6.2×10^{-8}
H_3BO_3	$\rightarrow H^+ + H_2BO_3^-$	7.3×10^{-10}
NH_4^+	$\rightarrow H^+ + NH_3$	5.7×10^{-10}
HCO_3^-	$\rightarrow H^+ + CO_3^{2-}$	4.7×10^{-11}
HPO_4^{2-}	$\rightarrow H^+ + PO_4^{3-}$	4.4×10^{-13}
HS^-	$\rightarrow H^+ + S^{2-}$	1.3×10^{-13}
H_2O	$\rightarrow H^+ + OH^-$	1.0×10^{-14}

PRACTICE

12.20 According to the table above, what are four examples of amphiprotic ions?

12.21 Of the bases listed on the table above, which is the weakest?

Finding the pH of Weak Acids

You have seen that the pH of strong acids can be found directly from their concentration, because you can assume that they are completely ionized. To find the pH of weak acids, you must determine exactly what concentration of H^+ ions is formed. You can do this by using the K_a value for the weak acid.

Let us consider a 0.10 *M* solution of acetic acid, CH_3COOH. Acetic acid ionizes only slightly, so at 298 K the pH of this solution is not 1. Since acetic acid is a weaker acid than HCl, you would expect the pH of the solution to be greater than 1, but less than 7. To find the pH, we must first find the $[H^+]$ resulting from the ionization of the 0.10 *M* acetic acid solution. The K_a of acetic acid is 1.8×10^{-5}. The ionization reaction can be written as

$$CH_3COOH \rightleftharpoons H^+ + CH_3COO^-$$

and the equilibrium expression as

$$\frac{[H^+][CH_3COO^-]}{[CH_3COOH]} = 1.8 \times 10^{-5}$$

The $[CH_3COOH]$ was initially 0.10 *M*, and since only a tiny fraction of the acetic acid molecules ionize, you may assume that the equilibrium concentration of CH_3COOH is still 0.10 *M*. Let $x = [H^+]$. Since the reaction forms equal numbers of moles of H^+ and CH_3COO^-, x also equals $[CH_3COO^-]$. The equation becomes

$$\frac{x^2}{0.1} = 1.8 \times 10^{-5}$$

$$x = 1.3 \times 10^{-3} = [H^+].$$

We find the pH by taking the log of the $[H^+]$ and changing its sign. The log of 1.3×10^{-3} is -2.89, so the pH is 2.89.

PRACTICE

12.22 Find the H^+ concentration and the pH in a 0.50 *M* solution of nitrous acid, HNO_2 ($K_a = 4.6 \times 10^{-4}$).

12.23 A 1.00 *M* solution of a certain acid HA has a pH of 3.00. What is the pH of a 0.100 *M* solution of the same acid?

Hydrolysis of Salts

The table on page 412 lists the pH values at 298 K of various 0.1 *M* aqueous solutions. NaCl is listed, with a pH of 7. NaCl is a neutral salt, and has no effect on the pH of the water. However, many other salts are not neutral. A 0.1 *M* solution of Na_2CO_3 has a pH of 11.6. A 0.1 *M* solution of NH_4Cl has a pH of 5.1. The process that produces acidic or basic solutions when certain salts are dissolved in water is called **hydrolysis.**

Each salt consists of a positive ion and a negative ion. The negative ion may hydrolyze by removing protons from the water. $CO_3^{2-} + H_2O \rightleftarrows HCO_3^- + OH^-$. The production of OH^- ions makes the solution basic. The hydrolysis of a negative ion can happen only if the ion is a strong enough base to pull protons from the water. Such bases must be the conjugates of weak acids. The ions CO_3^{2-}, CH_3COO^-, and PO_4^{3-} hydrolyze to form basic solutions. The ions NO_3^- and Cl^- do not hydrolyze.

Positive ions such as NH_4^+ can give protons to the water: $NH_4^+ + H_2O \rightleftarrows NH_3 + H_3O^+$. The formation of hydronium ion makes the solution acidic. In addition, metal ions can attract the negative side of water molecules, causing some of them to lose their protons: $Al^{3+} + H_2O \rightleftarrows AlOH^{2+} + H^+$ This hydrolysis produces an acidic solution. The ions of the metals in Groups 1 and 2 on the Periodic Table do not hydrolyze, while all other metal ions do hydrolyze to produce acidic solutions. The table below lists those ions that do not hydrolyze as "neutral ions," because these ions do not affect the pH of the solution.

You can use this table to predict whether a given salt is acidic, basic, or neutral. If the salt consists only of neutral ions, it is a neutral salt. If it contains one acidic ion, and one neutral ion, it is acidic. If it contains one basic ion, and one neutral ion, it is basic. When a salt contains both an acidic ion and a basic ion whether it is acidic, basic, or neutral cannot be predicted easily.

Acidic Ions	*Basic Ions*	*Neutral Ions*
NH_4^+, Al^{3+}, Cu^{2+}, Ag^+, Zn^{2+}, and all other metal ions not in Groups 1 and 2 of the Periodic Table.	CO_3^{2-}, PO_4^{3-}, S^{2-}, F^-	Na^+, K^+, Ca^{2+}, Ba^{2+}, and all other ions of Group 1 and Group 2 metals; Cl^-, Br^-, I^-, NO_3^-, SO_4^{2-}

SAMPLE PROBLEM

PROBLEM

State whether each of the following salts is acidic, basic,or neutral in aqueous solution. (a) $Ba(NO_3)_2$ (b) $ZnCl_2$ (c) Na_3PO_4

SOLUTION

(a) Ba^{2+} is formed by a Group 2 metal. It is neutral in solution. NO_3^- is the conjugate of a strong acid, and is neutral in solution. Since both ions are neutral, $Ba(NO_3)_2$ is a neutral salt.

(b) Zinc is not in Group 1 or Group 2. The Zn^{2+} ion hydrolyzes to form an acidic solution. Cl^- is a neutral ion, the conjugate base of the strong acid HCl. Since the salt contains an acidic ion and a neutral ion, $ZnCl_2$ is an acidic salt.

(c) Na^+ is a neutral ion, formed by a Group 1 metal. PO_4^{3-} is a basic ion, the conjugate base of the very weak acid, HPO_4^{2-}. Na_3PO_4 is a basic salt.

PRACTICE

12.24 Indicate whether each of the following salts is acidic, basic, or neutral in aqueous solution. (a) KCl (b) K_2CO_3 (c) CaS (d) $AgNO_3$ (e) $CrCl_3$ (f) NH_4NO_3

12.25 Write the chemical equation for the hydrolysis of the phosphate ion, PO_4^{3-}, in water.

Buffers

Some solutions can withstand the addition of moderate amounts of acid or base without undergoing a significant change in pH. Such solutions are called **buffer solutions.** As an example, consider a carbonate-bicarbonate buffer solution.

A carbonate-bicarbonate buffer system is prepared so that the concentration of bicarbonate ion HCO_3^- is equal to the concentration of the carbonate ion CO_3^{2-}. Both concentrations are 1.0 *M*. If acid is added to the solution, the following reaction takes place. All ions are hydrated.

$$CO_3^{2-} + H^+ \rightarrow HCO_3^-$$

The H^+ ions from the added acid react with the CO_3^{2-} ions in the buffer system and are used up. The pH does not change appreciably.

If base is added, the following reaction takes place:

$$HCO_3^- + OH^- \rightarrow H_2O + CO_3^{2-}$$

The OH^- ions from the added base react with the HCO_3^- ions in the buffer system and are used up. Again, the pH does not change appreciably.

A buffer system exists in our bloodstream. The pH of blood is 7.4 ± 0.05. If this pH varies by more than 0.4 unit, serious difficulties may result. The buffer system helps protect the blood from any significant changes in pH.

Acidic and Basic Oxides

The oxides of some metals react with water to form bases. These oxides are called **basic anhydrides.** The term *anhydride* is based on a Greek word that means "without water." Sodium and calcium oxides are typical basic anhydrides.

ANHYDRIDE BASE

$$Na_2O + H_2O \rightarrow 2\ NaOH$$
$$CaO + H_2O \rightarrow Ca(OH)_2$$

The oxides of some nonmetals react with water to form acids. These oxides are called **acid anhydrides.** Some relatively common acid anhydrides and the acids they form are shown in the following reactions:

ANHYDRIDE ACID

$$SO_2 + H_2O \rightarrow H_2SO_3$$
$$SO_3 + H_2O \rightarrow H_2SO_4$$
$$CO_2 + H_2O \rightarrow H_2CO_3$$
$$P_4O_{10} + 6\ H_2O \rightarrow 4\ H_3PO_4$$

(When SO_2 and CO_2 dissolve in water, they form acidic solutions. However, the acids H_2CO_3 and H_2SO_3 have never been isolated. For this reason, many chemists prefer to write the formula of carbonic acid as $CO_2 + H_2O$, and of sulfurous acid as $SO_2 + H_2O$.)

An acid anhydride can react directly with a basic anhydride, as shown in the following reactions:

BASIC ANHYDRIDE		ACID ANHYDRIDE		
CaO	+	CO_2	→	$CaCO_3$
Na_2O	+	SO_3	→	Na_2SO_4

CHAPTER REVIEW

The following questions will help you check your understanding of the material presented in the chapter.

Data required for answering some of the questions in this chapter will be found in the tables in this chapter and Tables K, L, and M in Appendix 4.

1. Which 0.1 *M* solution has the smallest pH value? (1) KOH (2) $Ca(OH)_2$ (3) HNO_3 (4) NH_3

Base your answers to questions 2 and 3 on the following information.

Beaker A contains 100 mL of 0.1 *M* H_2SO_4.

Beaker B contains 100 mL of 0.1 *M* $Ba(OH)_2$.

2. As the contents of beaker B are poured into beaker A, the pH of the resulting mixture in beaker A (1) decreases (2) increases (3) remains the same.
3. What are the products of the reaction that occurs when the two solutions are mixed? (1) BaH_2 and $HOSO_4$ (2) $BaSO_4$ and H_2O (3) $Ba(OH)_2$ and H_2O (4) SO_2 and $Ba(HSO_4)_2$

Base your answers to questions 4–7 on the table of indicators on page 413.

4. Congo red will be blue when used to test a solution with a pH of (1) 2 (2) 5 (3) 7 (4) 10

5. Which indicator will be the same color in hydrochloric acid as it is in pure water? (1) congo red (2) methyl red (3) phenolphthalein (4) methyl orange

6. Which indicator would be most useful in comparing the strengths of two different bases? (1) methyl orange (2) methyl red (3) congo red (4) alizarin yellow

7. Methyl orange would most likely be orange at a pH of (1) 5 (2) 2 (3) 3 (4) 4

8. You have 1.00 liter of 1.00 *M* NaOH. It will exactly neutralize 1.00 liter of (1) 1.00 *M* H_2SO_4 (2) 0.50 *M* H_2SO_4 (3) 2.00 *M* H_2SO_4 (4) 2.00 *M* HCl

9. How many moles of H_3O^+ ions will be required to exactly neutralize 34 g of OH^- ions? (1) 0.5 (2) 1.0 (3) 2.0 (4) 4.0

10. A conjugate acid differs from its conjugate base by one (1) ${}^{1}_{1}H$ (2) ${}^{0}_{-1}e$ (3) ${}^{1}_{0}n$ (4) ${}^{4}_{2}He$.

11. How many milliliters of 0.20 *M* H_2SO_4 are required to completely titrate 40 mL of 0.10 *M* $Ca(OH)_2$? (1) 10 (2) 20 (3) 40 (4) 80

12. Which 0.1 *M* aqueous solution will turn litmus paper blue? (1) CH_3COOH (2) CH_3OH (3) $ZnCl_2$ (4) NH_3

13. In the reaction $HC_2H_3O_2 + H_2O \rightleftharpoons H_3O^+ + C_2H_3O_2^-$, the addition of solid sodium acetate ($Na^+C_2H_3O_2^-$) results in a decrease in the concentration of (1) $C_2H_3O_2^-$ (2) H_3O^+ (3) Na^+ (4) $HC_2H_3O_2$.

14. A 2.0-mL sample of NaOH solution reacts completely with 4.0 mL of a 3.0 *M* HCl solution. What is the concentration of the NaOH solution? (1) 1.5 *M* (2) 4.5 *M* (3) 3.0 *M* (4) 6.0 *M*

15. When hydrochloric acid is neutralized by sodium hydroxide, the salt formed is sodium (1) hydrochlorate (2) chlorate (3) chloride (4) perchloride.

16. How many liters of 2.5 *M* HCl are required to exactly neutralize 1.5 L of 5.0 *M* NaOH? (1) 1.0 (2) 2.0 (3) 3.0 (4) 4.0

17. At 25°C, a solution with a pH of 7 contains (1) more H_3O^+ ions than OH^- ions (2) fewer H_3O^+ ions than OH^- ions

(3) an equal number of H_3O^+ ions and OH^- ions (4) no H_3O^+ ions or OH^- ions.

18. Which sample of HCl most readily conducts electricity? (1) HCl (*s*) (2) HCl (*l*) (3) HCl (*g*) (4) HCl (*aq*)

19. A water solution of which gas contains more OH^- ions than H_3O^+ ions? (1) HCl (2) NH_3 (3) CO_2 (4) SO_2

20. A piece of magnesium metal will react and produce hydrogen gas when it is placed in a solution of (1) NH_3 (2) NaOH (3) HCl (4) NaCl

21. Which is the strongest Brønsted base? (1) OH^- (2) $PO_4{}^{3-}$ (3) NH_3 (4) Cl^-

22. What is the conjugate base of the acid HBr? (1) H^+ (2) Br^- (3) H_3O^+ (4) OH^-

23. In which reaction does water act as a Brønsted acid?
(1) $NH_3 + H_2O \rightarrow NH_4{}^+ + OH^-$
(2) $HCl + H_2O \rightarrow H_3O^+ + Cl^-$
(3) $Ca(HCO_3)_2 \rightarrow CaCO_3 + H_2O + CO_2$
(4) $CuSO_4 + 5H_2O \rightarrow CuSO_4 \cdot 5H_2O$

24. For the reaction $HC_2H_3O_2(aq) \rightleftharpoons H^+(aq) + C_2H_3O_2{}^-(aq)$, the ionization constant, K_a, is expressed as

$$(1)\ K_a = \frac{[HC_2H_3O_2]}{[H^+][C_2H_3O_2{}^-]}$$

$$(2)\ K_a = \frac{[H^+]}{[C_2H_3O_2{}^-]}$$

$$(3)\ K_a = \frac{[C_2H_3O_2{}^-]}{[H^+]}$$

$$(4)\ K_a = \frac{[H^+][C_2H_3O_2{}^-]}{[HC_2H_3O_2]}$$

25. A Brønsted base that is stronger than HS^- is (1) $SO_4{}^{2-}$ (2) $CO_3{}^{2-}$ (3) H_2O (4) F^-

26. Phenolphthalein will be pink in a solution whose H_3O^+ ion concentration in moles per liter is (1) 1×10^{-1} (2) 1×10^{-3} (3) 1×10^{-6} (4) 1×10^{-9}.

27. What is the pH of a 0.10 *M* solution of NaOH? (1) 1 (2) 2 (3) 13 (4) 14

28. The conjugate base of NH_4^+ is (1) NH_3 (2) OH^- (3) H_2O (4) H_3O^+.

29. The ionization constant of a weak acid is 1.8×10^{-5}. A reasonable pH for a 0.1 *M* solution of this acid would be (1) 1 (2) 9 (3) 3 (4) 14.

30. What are the Brønsted bases in the reaction $H_2S + H_2O \rightleftarrows H_3O^+ + HS^-$? (1) H_2S and H_2O (2) H_2S and H_3O^+ (3) HS^- and H_2O (4) HS^- and H_3O^+

31. Which Brønsted acid has the weakest conjugate base? (1) HCl (2) H_2CO_3 (3) H_2S (4) H_3PO_4

32. Which acid is almost completely ionized in a dilute solution at 298 K? (1) CH_3COOH (2) H_2S (3) H_3PO_4 (4) HNO_3

33. A water solution of KCl has a pH closest to (1) 3 (2) 5 (3) 7 (4) 9.

Questions 34–36 are based on the following reaction, which occurs when sodium acetate is dissolved in water.

$$CH_3COO^- + H_2O \rightleftarrows CH_3COOH + OH^-$$

34. A spectator ion in this solution is (1) OH^- (2) CH_3COO^- (3) Na^+ (4) Cl^-

35. This reaction is best described as (1) neutralization (2) ionization (3) single replacement (4) hydrolysis.

36. A likely pH of the solution resulting from this reaction is (1) 5 (2) 7 (3) 9 (4) 13

37. As NaCl is dissolved in water, the pH of the solution (1) decreases (2) increases (3) remains the same.

38. As NaCl is dissolved in water, the electrical conductivity of the solution (1) decreases (2) increases (3) remains the same.

CONSTRUCTED RESPONSE

1. State two different methods you might use to distinguish a 0.1 *M* solution of nitric acid from a 0.1 *M* solution of acetic acid.

2. A chemist has a solution of nitric acid that is 0.0100 *M*. Its pH is 2.00. He dilutes 1.00 mL of the solution with water to a new total volume of 100 mL. What is the new pH of the solution?
3. Of the three compounds CH_3OH, $Ba(OH)_2$, and $C_3H_5(OH)_3$, only barium hydroxide is considered a base. Explain this observation using the Arrhenius definition of a base.
4. A student is titrating a 0.200 *M* solution of HCl with a 0.100 *M* solution of NaOH. The data so far reads

	0.200 Molar HCl	**0.100 Molar NaOH**
Inital burette reading	5.00 mL	10.00 mL
Final burette reading	20.00 mL	*(a)*
mL added	15.00 mL	*(b)*

What data should appear in the spaces labeled *(a)* and *(b)* in the table above?

Our Blood's Buffer System

Have you ever wondered what causes muscle fatigue and pain during a strenuous workout? When you exercise vigorously, your muscles begin to produce lactic acid, which enters blood vessels in the muscles and increases the acidity (that is, lowers the pH) of your blood. Human blood normally has a pH of about 7.4; when lactic acid (or any other acid) enters the bloodstream, the pH of the blood begins to drop, a condition known as acidosis. This, many scientists believe, is the main cause of muscle pain and fatigue during intense physical activities such as sprinting and lifting weights.

If the pH of the blood drops below 6.8, death occurs. Fortunately, our blood has a buffering system that keeps its pH within the fairly narrow range—6.8 to 7.8—necessary to maintain life. The primary buffer system that regulates the acid-base balance in the blood is called the carbonic acid-bicarbonate buffer, or simply the carbonate buffer. The equilibrium reactions of this system are as follows.

$$H^+ (aq) + HCO_3^- (aq) \rightleftharpoons H_2CO_3 (aq) \rightleftharpoons H_2O (l) + CO_2 (g)$$

When lactic acid produced by intense exercise enters the blood, the hydrogen ion (H^+) concentration rises, lowering the pH. This is offset, or buffered, by bicarbonate ions (HCO_3^-) present in the blood, which combine with the H^+ ions to produce carbonic acid (H_2CO_3). The carbonic acid rapidly dissociates into water (H_2O) and carbon dioxide (CO_2). In fact, part of the reason you breathe faster and harder during exercise is to get rid of the extra carbon dioxide from your lungs and keep the buffer reactions moving to the right. If the concentration of CO_2 in the blood rises faster than it can be expelled, the reactions are driven back to the left, generating excess H^+ ions and causing acidosis. Normally, this will cause muscle failure before the blood's pH can drop to a life-threatening level.

Under certain conditions, the opposite can occur. For example, when a person hyperventilates, the rapid breathing removes CO_2 from the blood too quickly, driving the reactions to the right. This reduces the H^+ concentration in the blood below normal levels, raising the blood's pH above 7.4, a condition called alkalosis. That's why a person who is hyperventilating is often told to breathe into a paper bag. By rebreathing the CO_2-rich exhaled air, the CO_2 concentration in the blood is restored to normal levels.

Who knew all this chemistry was taking place right under our noses!

13 Redox and Electrochemistry

LOOKING AHEAD

All portable electric devices—CD players, cameras, flashlights, cell phones—run on batteries. A battery is a device that drives electrons from its negative pole to its positive pole. The electrons flow because a chemical reaction is taking place in the battery, a reaction involving a transfer of electrons. Such reactions are called redox reactions. This chapter deals with redox reactions and the electricity associated with them.

When you have completed this chapter, you should be able to:

- **Define** oxidation and reduction.
- **Compare** the strengths of oxidizing and reducing agents based on the activity series.
- **Predict** the products of a redox reaction.
- **Balance** equations for reactions involving electron transfer.
- **Write** oxidation and reduction half-reactions.

Reduction and Oxidation

The chemical reaction used earliest by people was probably simple combustion. Although people made use of fire thousands of years ago, the chemical reaction involved has been understood for only the last 200 years. In 1774, the French chemist Antoine Lavoisier described burning as the combination of a material with oxygen. Burning was then understood to be rapid oxidation. The oxidation of metals was also understood to involve slow processes, such as the rusting of iron. The term "oxidation" meant "combining with oxygen."

In the early nineteenth century, when chlorine gas was studied, it was found to react with metals in a manner very similar to the way oxygen reacts. Substances that burn in oxygen generally burn in chlorine as well. The concept of oxidation was broadened to include the reaction of any metal with any nonmetal.

Let us compare the reaction of iron with oxygen and the reaction of iron with chlorine.

$$4\ Fe + 3\ O_2 \longrightarrow 2\ Fe_2O_3$$

$$2\ Fe + 3\ Cl_2 \longrightarrow 2\ FeCl_3$$

In both reactions, each iron atom loses three electrons to the nonmetal. You may recall that metals generally react with nonmetals by losing electrons. **Oxidation** was redefined to mean the loss of electrons. The nonmetal in both reactions gained electrons. The gain of electrons was defined as **reduction.** For a substance to lose electrons, some other substance must gain them. Reduction and oxidation must always occur simultaneously. Modern chemists call reactions in which there is a transfer of electrons **redox reactions.**

Redox Reactions

What happens if a coil of copper wire is placed in a solution of silver nitrate, $AgNO_3$? A deposit of silver soon appears on the coil, and the solution turns blue. The equation showing the predominant species in the reaction is

$$2\ Ag^+\ (aq) + Cu\ (s) \longrightarrow Cu^{2+}\ (aq) + 2\ Ag\ (s)$$

If you analyze this equation, you will find that two silver ions are the species that gained a total of two electrons. The copper atom is the species that lost two electrons. (The coefficients are needed to balance the equation—that is, to conserve mass and charge.) In other words, copper has been oxidized, and the silver ion has been reduced. The reaction is an oxidation-reduction reaction.

Although this discussion focuses on what happens at the atomic level, remember that real-world reactions involve quantities measured in moles. For this reaction, 2 moles of Ag^+ gain 2 moles of electrons while 1 mole of Cu loses 2 moles of electrons.

The net equation for a redox reaction can be broken down into two parts, called **half-reactions.** In the reaction between Cu (*s*) and $AgNO_3$ (*aq*), for example, the gain of electrons by Ag^+ (*aq*) and the loss of electrons by Cu (*s*) can be shown as follows:

$$2\ Ag^+\ (aq) + 2e^- \longrightarrow 2\ Ag\ (s)$$

$$Cu\ (s) \longrightarrow Cu^{2+} + 2e^-$$

The half-reaction in which electrons are lost is called the **oxidation half-reaction.** The accompanying half-reaction in which electrons are gained is called the **reduction half-reaction.**

Oxidation-reduction reactions are often called redox reactions. This shortened form of the term emphasizes the fact that oxidation and reduction always occur at the same time. The oxidation half-reaction cannot occur without the corresponding reduction half-reaction.

Oxidation Number

You will recall that in molecular compounds, electrons are shared during bond formation. Recall, too, that the atoms bonded in this way may have differing abilities to attract electrons. In other words, the atoms have different electronegativities. If atoms sharing electrons have different electronegativities, the bonding electrons are shared unequally. The electrons are closer to the atom with the higher electronegativity. The atom with the higher electronegativity becomes partially negative, and the atom with the lower electronegativity becomes partially positive.

It is easier to keep track of the transfer of electrons in chemical reactions if these shared electrons are treated as belonging to the more electronegative atom. The charge that results from assigning shared electrons in this way is called the *oxidation number*, or *oxidation state.* The oxidation state of an element will change whenever electrons are lost or gained. Therefore, every redox reaction involves changes in oxidation states. (The rules for determining the oxidation state of an element are described in Chapter 4, and you may find it useful to review them at this time.)

Electrons are transferred completely in the formation of ionic compounds. The ions that result have a positive or a negative charge. The oxidation number of a monatomic ion is the same as the charge of the ion. Note that ionic charges are indicated by a number followed by the sign, for example, 1+, and oxidation numbers are indicated by the sign followed by a number, for example +1.

Remember that the algebraic sum of all the oxidation numbers or ion charges in a compound or in a molecule is zero. That is, compounds or molecules are electrically neutral.

Oxidation

Oxidation is a reaction in which a particle loses electrons. Loss of electrons results in an increase in oxidation number. Look at the following examples of oxidation reactions:

1. Change from a free element to a positive ion

$$Cu^0 \longrightarrow Cu^{2+} + 2e^-$$

2. Change from an ion with a lower positive charge to an ion with a higher positive charge

$$\underset{\substack{\text{ferrous} \\ \text{Fe(II)}}}{Fe^{2+}} \longrightarrow \underset{\substack{\text{ferric} \\ \text{Fe(III)}}}{Fe^{3+}} + e^-$$

3. Change from a negatively charged ion to a free element

$$2\,Cl^- \longrightarrow Cl_2 + 2e^-$$

4. Change in oxidation state involving an ion made up of two or more elements (a polyatomic ion)

$$\overset{+4\ -2}{MnO_2} + 2\,H_2O \longrightarrow \overset{+7\ -2}{MnO_4{}^-} + 4\,H^+ + 3\,e^-$$

The numbers above the particles in this equation represent oxidation states. The +4 over the Mn represents the oxidation state of Mn in MnO_2. $MnO_4{}^-$ is a polyatomic ion with a charge of 1−. It contains Mn with an oxidation state of +7. Thus, the change from Mn^{+4} to Mn^{+7} represents oxidation. (To conserve charge, multiply the oxidation number of the particle by the number of particles. The total charge on both sides of the equation is zero.)

In each of the oxidations shown, the oxidation state of the oxidized element increased. Since electrons are negatively charged, a loss of electrons must always result in an increase in oxidation state.

Remember, oxidation reactions cannot take place unless a particle that can gain electrons is also present. Each of the four examples is therefore only half of a total reaction—each is a half-reaction.

Reduction

Reduction is a reaction in which a particle gains electrons. Gain of electrons results in a decrease in the oxidation number. Following are some half-reactions that illustrate reduction:

1. Change from a free element to a negative ion

$$I_2 + 2\,e^- \longrightarrow 2I^-$$

2. Change from an ion with a higher positive charge to an ion with a lower positive charge

$$Fe^{3+} + e^- \longrightarrow Fe^{2+}$$

3. Change from a positive ion to a free element

$$Ag^+ + e^- \longrightarrow Ag^0$$

4. Change in oxidation state involving an ion made up of two or more elements (a polyatomic ion)

$$\overset{+5-2}{NO_3^-} + 4\,H^+ + 3\,e^- \longrightarrow \overset{+2-2}{NO} + 2\,H_2O$$

   In this half-reaction, the oxidation state of N decreases from +5 in NO_3^- to +2 in NO, which represents reduction.

In each of these reduction half-reactions, the oxidation state of the reduced element decreased. A gain of negative electrons always results in a decrease in oxidation state. (You may have wondered why a *gain* of electrons is called reduction, since the word reduction generally refers to a decrease, not an increase. What decreases in reduction is the charge or oxidation state of the element. The term actually refers to the charge, not to the number of electrons.)

Analyzing Redox Reactions

By finding where changes in oxidation number occur you can identify the oxidation and reduction in a given redox reaction. You can also write the reduction and oxidation half-reactions. Consider the reaction

$$2\ KClO_3 \longrightarrow 2\ KCl + 3\ O_2$$

To find the oxidation and reduction half-reactions, first assign oxidation states to all of the elements in the reaction.

$$2\ \overset{+1}{K}\overset{+5}{Cl}\overset{-2}{O_3} \longrightarrow 2\ \overset{+1}{K}\overset{-1}{Cl} + 3\ \overset{0}{O_2}$$

You can see that the oxidation state of the K did not change, so potassium is neither oxidized nor reduced in this reaction. The Cl went from +5 to −1. Since the oxidation state decreased, you know that the Cl was the species reduced. The reduction half-reaction would be

$$Cl^{5+} + 6\ e^- \longrightarrow Cl^{1-}$$

A gain of six electrons was needed to reduce the oxidation state from +5 to −1. Half-reactions must be balanced for charge. In this case, the total charge on the left is 1−, as is the total charge on the right. Note that in reduction half-reactions, the electrons normally appear on the left side of the equation.

The oxygen went from −2 to 0. The oxidation state increased, so this is oxidation, and the half-reaction is

$$2\ O^{2-} \longrightarrow O_2 + 4\ e^-$$

In oxidation half-reactions the electrons normally appear on the right side of the equation. The total charge of 4− is equal on both sides of the equation. The O^{2-} was the species oxidized.

In the balanced equation above, there were actually six oxygen atoms and two chlorine atoms. Two chlorines gain 12 electrons in the reduction half-reaction. Six oxygens lose 12 electrons in the oxidation half-reaction. In any balanced redox reaction, the number of electrons lost must equal the number of electrons gained.

Consider the more complicated reaction

$$6\ \overset{+2}{Fe}SO_4 + K_2\overset{+6}{Cr_2}O_7 + 7\ H_2SO_4$$

$$\longrightarrow K_2SO_4 + \overset{+3}{Cr_2}(SO_4)_3 + 7\ H_2O + 3\ \overset{+3}{Fe_2}(SO_4)_3$$

It would be a tedious process to find the oxidation state of every element in this reaction, so look only for the ones that change. Those are identified on page 445. You can see that the Fe^{2+} was oxidized, and the Cr^{6+} was reduced.

The equation provides clues to identifying the relevant elements. You can see that the number of sulfates attached to the iron atoms changed. A change in a formula involving the same ions always indicates oxidation or reduction. In the case of the Cr, there is a polyatomic ion, $Cr_2O_7^{2-}$, that appears on the left side of the equation but not the right. A change in a polyatomic ion often signals that oxidation or reduction has occurred. The sulfur, on the other hand, always appears in this reaction as part of the SO_4^{2-} ion. When an ion stays the same, none of its atoms have been oxidized or reduced. One other sure sign of oxidation or reduction is the presence of an uncombined element on one side of the equation. Since an uncombined element has an oxidation state of 0, while a combined element almost never has an oxidation state of 0, the element was probably involved in oxidation or reduction.

The reaction appears more complicated than it really is because it includes all the spectator ions in the reaction. Spectator ions, you will recall, do not actually participate in the reaction, and can be omitted in the net ionic equation. For this reaction, the net ionic equation is

$$6\ Fe^{2+} + 14\ H^+ + Cr_2O_7^{2-} \longrightarrow 2\ Cr^{3+} + 6\ Fe^{3+} + 7\ H_2O$$

Note that in this balanced ionic equation, the total charge on the left equals the total charge on the right (24+). Ionic equations must be balanced for charge as well as for each element. When the reaction is written this way, it is much easier to see that the Fe^{2+} is oxidized and the Cr^{6+} is reduced.

SAMPLE PROBLEMS

PROBLEM

1. Identify the oxidation and reduction in the reaction

$$3\ Cu + 8\ H^+ + 2\ NO_3^- \longrightarrow 3\ Cu^{2+} + 2\ NO + 4\ H_2O$$

SOLUTION
The uncombined copper atom is going from an oxidation state of 0 to an oxidation state of +2. The copper has lost electrons and is oxidized. The NO_3^- does not appear on the right side of the reaction. A change in a polyatomic ion usually indicates reduction or oxidation. The N is going from +5 in the NO_3^- to +2 in the NO. The nitrogen has gained electrons and is reduced.

PROBLEM
2. Identify the oxidation and reduction in the reaction.

$$Zn + Cu^{2+} \longrightarrow Zn^{2+} + Cu.$$

SOLUTION
The Cu^{2+} ion is gaining two electrons in going from +2 to 0. The Cu^{2+} is reduced. However, avoid the following common error! Students often write, when solving a problem like this one, that the species that is reduced is "copper," or Cu. This is incorrect! The copper ion, Cu^{2+}, is gaining the electrons; it is reduced. The copper atom, Cu, is just the product of the reaction. In fact, metal atoms are never reduced; when metal atoms react they are oxidized. When an uncombined metal appears on the left side of a reaction, you can safely assume that it will be oxidized.

PRACTICE

13.1 Add the electrons needed to complete the following half-reactions. Be sure to add them to the correct side of the equation.

(a) $Sn^{4+} \longrightarrow Sn^{2+}$

(b) $Cl_2 \longrightarrow 2\ Cl^-$

(c) $IO_3^- + 6\ H^+ \longrightarrow 3\ H_2O + I^-$

13.2 Write the oxidation and reduction half-reactions for each of the following redox reactions.

(a) $Zn + 2\ H^+ \longrightarrow Zn^{2+} + H_2$

(b) $Mg + S \longrightarrow MgS$

(c) $Cu + 2\ H_2SO_4 \longrightarrow CuSO_4 + SO_2 + 2\ H_2O$

13.3 Complete the following half-reactions, and state whether each is an oxidation or a reduction.

(a) $Pb^{2+} \longrightarrow ? + 2e^-$

(b) $As^{5+} + 2e^- \longrightarrow ?$

(c) $P_4 \longrightarrow ? + 20e^-$

(d) $MnO_4^- + 8\ H^+ + 5e^- \longrightarrow ? + 4\ H_2O$

13.4 Identify the species reduced in each of the following reactions.

(a) $2\ Fe^{2+} + Sn^{4+} \longrightarrow 2\ Fe^{3+} + Sn^{2+}$

(b) $Cl_2 + 2\ Br^- \longrightarrow 2\ Cl^- + Br_2$

(c) $Mg + Zn^{2+} \longrightarrow Mg^{2+} + Zn$

(d) $IO_3^- + 6\ Fe^{2+} + 6\ H^+ \longrightarrow I^- + 6\ Fe^{3+} + 3H_2O$

13.5 Identify the species oxidized in each of the following reactions.

(a) $Zn + 2\ HCl \longrightarrow ZnCl_2 + H_2$

(b) $Br_2 + 2\ KI \longrightarrow 2\ KBr + I_2$

(c) $2\ MnO_4^- + 5\ SO_2 + 2\ H_2O \longrightarrow 2\ Mn^{2+} + 5\ SO_4^{2-} + 4\ H^+$

(d) $2\ Fe^{2+} + 2\ H^+ + H_2O_2 \longrightarrow 2\ Fe^{3+} + 2\ H_2O$

Predicting Oxidation-Reduction Reactions

When a piece of zinc is placed in a solution of copper(II) sulfate, the blue color of the solution gradually disappears, and pieces of copper metal form on the zinc. The reaction is

$$Zn\ (s) + Cu^{2+}\ (aq) \longrightarrow Zn^{2+}\ (aq) + Cu\ (s)$$

(The sulfate is a spectator ion.)

What would happen if a piece of copper were placed in a solution of zinc sulfate? You can write the reaction

$$Cu\ (s) + Zn^{2+}\ (aq) \longrightarrow Cu^{2+}\ (aq) + Zn\ (s)$$

However, when you actually place a piece of copper in a zinc sulfate solution, nothing at all happens. Evidently, while zinc metal can lose electrons to Cu^{2+} ions, copper metal does not lose electrons to Zn^{2+} ions. Zinc loses electrons more easily than does copper. By comparing the ability to gain or lose electrons, the activity, of the species involved, you can predict the direction of a redox reaction.

The **activity series** table that follows is used to compare the activity of metals and nonmetals. For metals, the activity series compares their tendency to lose electrons, that is to be oxidized, in an aqueous solution. In the table, lithium (Li) at the top of the table is the most active metal, and gold (Au) at the bottom is the least active. Compare Zn with Cu in the table. Zinc, which loses electrons more easily and is more active, is above Cu in the table. In a spontaneous redox reaction, the more active metal will react with the ion of the less active metal. Zinc, which is higher on the activity series than copper, can react with copper ions. Copper, which is lower in the table, will not react simultaneously with zinc ions.

For nonmetals, the activity series compares their tendency to gain electrons, that is to be reduced, in an aqueous solution. The most active nonmetal is fluorine, at the top of the table. The least

Activity Series

	Metals	***Nonmetals***	
Most	Li	F_2	Most
	Rb	Cl_2	
	K	Br_2	
	Cs	I_2	
	Ba		
	Sr		
	Ca		
	Na		
	Mg		
	Al		
	Mn		
	Zn		
	Cr		
	Fe		
	Co		
	Ni		
	Sn		
	Pb		
	H_2 (Not a metal)		
	Cu		
	Hg		
	Ag		
Least	Au		Least

active nonmetal is iodine, at the bottom of the table. The more active nonmetal will react spontaneously with the negative ion of the less active nonmetal. The activity series shows that Cl_2, near the top of the table, will react spontaneously with Br^- ions (Br_2 is below Cl_2 on the table). I_2, at the bottom of the table, will not react spontaneously with Br^- ions.

SAMPLE PROBLEMS

PROBLEM

1. Will aluminum foil react with a solution of $ZnCl_2$?

SOLUTION

Metals will react with the ions of less active metals. In this case, you must determine whether aluminum metal reacts with zinc ions. The activity series indicates that Al, which is higher in the table, is more active than zinc. The reaction will occur.

PROBLEM

2. Will iron replace zinc from solutions of its salts?

SOLUTION

Since salts are ionic substances, this question could have been worded: Does iron react with zinc ions? Since Fe is below Zn in activity, it will not replace zinc from its solutions of its salts.

PRACTICE

Use the Activity Series table to answer the following questions.

13.6 State whether each of the following pairs of atoms and ions will react spontaneously.

(a) $Pb + Ag^+$

(b) $Cu + H^+$

(c) $Cl_2 + I^-$

(d) $Mg + Fe^{2+}$

13.7 What metal will react with Zn^{2+} ion, but *not* with Al^{3+} ion?

Electrochemistry: Chemical Reactions $\longrightarrow$ Electricity

A device in which a redox reaction either produces or uses electricity is called an **electrochemical cell.** Consider again the reaction between Zn and Cu^{2+}.

$$Zn\ (s) + Cu^{2+} \longrightarrow Zn^{2+} + Cu\ (s)$$

As shown on page 447 this reaction proceeds spontaneously. The electrons flow from Zn (*s*) to Cu^{2+}. You can use a flow of electrons to perform useful work; with a voltage of 1.1, you can light a small bulb or run a radio. To make use of the electrical potential of the reaction, you must force the electrons to flow through a wire. If you simply place the piece of zinc in a solution containing copper(II) ions, the reaction will occur, but will not produce useful electricity.

An electrochemical cell that uses a chemical reaction to force a flow of electricity is called a **voltaic cell.** It is also often called a galvanic cell. (Named after the Italian scientists Luigi Galvani and Alessandro Volta.) A typical voltaic cell is shown in Figure 13-1 on page 452.

In beaker A, a strip of metallic zinc is immersed in a 1 *M* $ZnSO_4$ solution. This represents one half of the cell. In beaker B, a strip of metallic copper is immersed in a 1 *M* $CuSO_4$ solution. This represents the other half of the cell. The two strips of metal which act as conductors, are called **electrodes.** Taken individually, each solution and electrode is called a half-cell. The electrodes are connected through a voltmeter by a wire. A voltmeter measures the difference in potential, or voltage, between two points through which electrons flow. An inverted U-tube dips into both beakers. The U-tube is filled with aqueous NH_4NO_3 and is plugged with cotton or glass wool. The U-tube connects the solutions in the beakers and is called a **salt bridge.** When the two half-cells are connected as shown, and the switch is closed, the voltmeter should read 1.10 volts.

Let us further examine our sample voltaic cell. What is the direction of the electron flow? You know that electrons flow spontaneously from the species oxidized to the species reduced. In this case, the electrons flow from the zinc electrode through the wire to the copper and are finally gained by the Cu^{2+} ion.

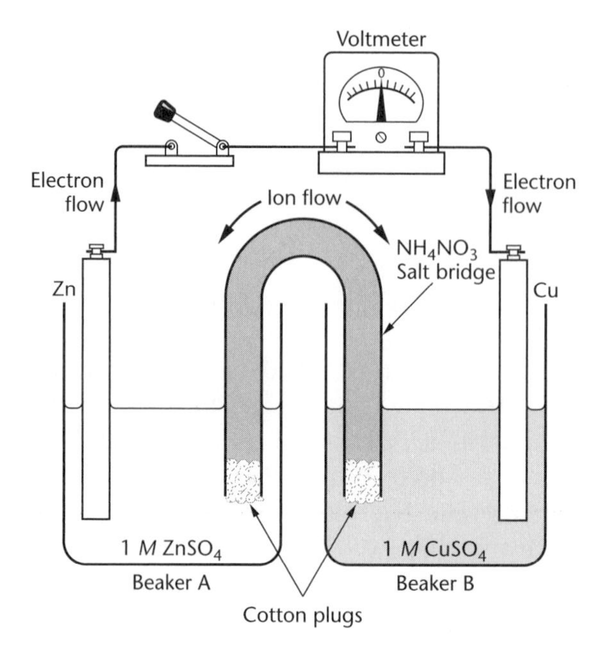

Figure 13-1 A voltaic cell

While the cell will produce 1.1 volts as shown, if the salt bridge is removed, the voltage immediately drops to zero. To better understand the role of the salt bridge, consider what would happen if the reaction had proceeded without it. At the zinc electrode, zinc is oxidized to Zn^{2+}. Since no additional negative ions are available, the solution would have excess positive charge. At the copper electrode, Cu^{2+} ions are reduced to Cu. This would produce a shortage of positive ions in the beaker, resulting in excess negative charge. Electrons will not flow from a region of positive charge to a region of negative charge, so the reaction stops.

To permit a continuous flow of electrons, both beakers must remain electrically neutral. The salt bridge contains a solution rich in ions. In our example, NH_4NO_3 was used, but any salt will do, as long as it does not interfere with the chemical reactions in the cell. The NO_3^- ions flow toward the zinc half-cell, while the NH_4^+ ions flow toward the copper half-cell. Thus both beakers remain electrically neutral, and electrons continue to flow through the wire.

Now, consider the chemical changes occurring in the two half-cells. The net chemical reaction is:

$$Zn\ (s) + Cu^{2+} \longrightarrow Zn^{2+} + Cu\ (s)$$

In the zinc half-cell, zinc metal is oxidized to Zn^{2+} ions, so the mass of the zinc electrode decreases as the reaction proceeds. The concentration of Zn^{2+} ions increases. In the copper half-cell, Cu^{2+} ions are reduced to copper metal. The mass of the copper electrode increases as copper metal is deposited on its surface. The concentration of Cu^{2+} ions decreases. These changes in concentration affect the voltage produced. When the concentrations of Zn^{2+} and Cu^{2+} are not 1 *M*, the cell will still produce electricity, but the voltage may not be 1.10 volts.

Eventually, the chemical reaction reaches equilibrium. At equilibrium, the rate of the forward and reverse reactions are equal, so electrons have an equal tendency to flow in either direction. At equilibrium, the voltmeter reads zero. The cell can no longer produce electricity. What you call a dead battery is a voltaic cell that has reached equilibrium.

Changes in concentration occur whenever electricity is produced from a voltaic cell; therefore, the voltage is constantly decreasing. However, until the reaction nears equilibrium, the change is very slight. For example, if you ran the cell shown in Figure 13-1 until half of the copper ions were used up, the voltage would drop from 1.10 volts to 1.09 volts. The change in voltage would not be noticeable.

Labeling the Electrodes

The two electrodes in a voltaic cell are called the anode and the cathode. The **anode** is defined as the electrode where oxidation takes place, while the **cathode** is the electrode where reduction takes place. In our sample voltaic cell, the zinc is oxidized and is the anode. The copper ions are reduced at the copper electrode, which is the cathode.

If you have ever put a battery into a radio, you know that you must place it correctly. The radio contains instructions that tell you where to place the + and − poles of the battery. The electrons flow from the negative pole of the battery, through the radio, to the positive pole. In this voltaic cell, you know that the electrons are flowing from the zinc to the copper. Therefore the zinc is the negative pole (electrode), and the copper is the positive pole.

In summary, a voltaic cell is made up of two half-cells. Oxidation occurs at the anode, which is the negative electrode. Reduction occurs at the cathode, which is the positive electrode.

For convenience, chemists have devised a shorthand method of representing voltaic cells. The cell shown in Figure 13-1 would be represented as

$$Zn/Zn^{2+} // Cu^{2+}/Cu.$$

Each half-cell is shown as an electrode separated by a / from the relevant ions in the cell. The two half-cells are separated by a //. The electrodes may or may not participate in the reaction. For example, in the half-cell Pt/Fe^{2+}, Fe^{3+} the reaction involves the two ions, but the platinum is an inert electrode and does not participate in the cell reaction. (If the platinum did react, its ions would have to be included in the expression.)

SAMPLE PROBLEM

PROBLEM

A cell uses the reaction $Mn + Ni^{2+} \longrightarrow Ni + Mn^{2+}$ to produce electricity.

(a) Write the half-reaction that occurs at the anode.

(b) Write the half-reaction that occurs at the cathode.

(c) Which species in this cell loses electrons?

(d) As the cell produces electricity, which ion increases in concentration?

SOLUTION

(a) Oxidation occurs at the anode. In the reaction given, the oxidation state of Mn is increasing from 0 to +2; it is being oxidized. The half reaction is $Mn \longrightarrow Mn^{2+} + 2\,e^-$

(b) In this cell the Ni^{2+} ion is reduced to Ni. The species that is reduced gains electrons. Therefore, Ni^{2+} is reduced. Note that you must include the charge, 2+, in your answer. $Ni^{2+} + 2\,e^- \longrightarrow Ni^0$

(c) We saw in part (a) that the Mn was oxidized. Therefore the Mn is losing electrons.

(d) As in any spontaneous reaction, the reactants decrease, while the products increase. Since Mn^{2+} is a product of this reaction, it increases in concentration.

PRACTICE

13.8 A voltaic cell is constructed by immersing solid cadmium in a beaker containing a solution of $Cd(NO_3)_2$. The Cd is connected with a wire to a piece of lead in a beaker containing a solution of $Pb(NO_3)_2$. A salt bridge connects the two beakers. The spontaneous reaction in the cell is $Cd + Pb^{2+} \longrightarrow Cd^{2+} + Pb$.

(a) Write a balanced equation for the half-reaction that occurs at the cathode in this cell.

(b) Write a balanced equation for the half-reaction that occurs at the anode

(c) What is the direction of electron flow in this cell?

(d) Which piece of metal would increase in mass as the cell was used?

13.9 Draw a labeled diagram of the cell represented by the expression Fe/Fe^{2+} $//Pb^{2+}$ $/Pb$. Indicate the charge of each electrode and the direction of electron flow. Label the anode and cathode.

13.10 For the chemical cell Mg/Mg^{2+} $//Zn^{2+}$ $/Zn$

(a) What is the voltage at equilibrium?

(b) Which electrode would decrease in mass as the reaction proceeds?

(c) Which ion would increase in concentration as the reaction proceeds?

Commercial Voltaic Cells

The flow of electrons from an oxidation to a reduction in a voltaic cell can be tapped and put to use. This is done in commercial electrochemical cells, such as dry cells, storage batteries, and fuel cells.

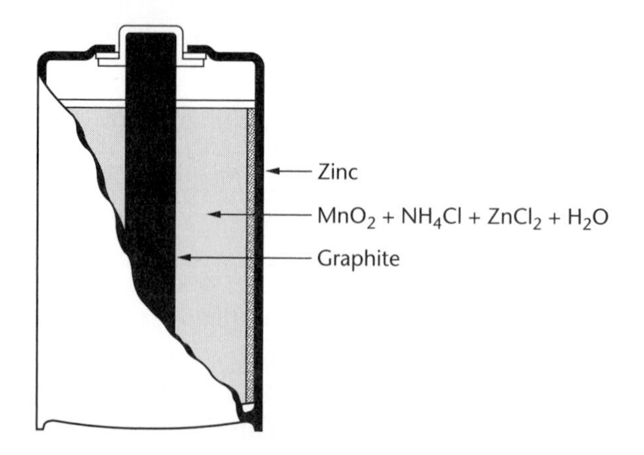

Figure 13-2 The dry cell

The common dry cell consists of a zinc container in which a rod of graphite dips into a paste made of water, ammonium chloride, zinc chloride, and manganese dioxide (see Figure 13-2). The zinc container is the anode. The graphite rod is the cathode. Oxidation takes place at the anode. Reduction takes place at the cathode. The half-reactions are as follows:

$$\text{ANODE: Zn } (s) \longrightarrow \text{Zn}^{2+} \ (aq) + 2\ e^-$$

$$\text{CATHODE: } 2\ \text{NH}_4^+ \ (aq) + 2\ \text{MnO}_2 \ (s) + 2\ e^- \longrightarrow \text{Mn}_2\text{O}_3 \ (s) + 2\ \text{NH}_3 \ (aq) + \text{H2O} \ (l)$$

Ammonia gas could interfere with the oxidation reaction. However, it is removed by the Zn^{2+}, with which it forms a stable complex ion.

$$\text{Zn}^{2+} + 2\ \text{NH}_3 \ (g) \longrightarrow \text{Zn(NH}_3)_2{}^{2+}$$

As long as the zinc remains intact and the ammonia gas is removed, the dry cell continues to function. The voltage of the dry cell is about 1.5 volts. There is no practical way of reversing the reaction in this cell, which means that the cell cannot be recharged.

Fuel cells are a recent technological development. In a fuel cell, the energy obtained by oxidizing certain gaseous fuels is converted into electricity. This conversion is very efficient (about 80 percent), and fuel cells are expected to become an important source of electrical energy. At present, they are used mainly in spacecraft. Some newer fuel cells operate with very little pollution of the atmosphere.

In a typical hydrogen fuel cell, hydrogen and oxygen are made to react at about 60°C in the presence of concentrated potassium hydroxide, the electrolyte. The half-reactions are as follows. (All ions are hydrated.)

$$\begin{array}{rl} \text{ANODE:} & 2\,H_2\,(g) + \cancel{4\,OH^-} \longrightarrow 4\,H_2O + \cancel{4e^-} \\ \text{CATHODE:} & O_2\,(g) + 2\,H_2O + \cancel{4e^-} \longrightarrow \cancel{4\,OH^-} \\ \hline \text{NET REACTION:} & 2\,H_2\,(g) + O_2\,(g) \longrightarrow 2\,H_2O + \text{energy} \end{array}$$

Electrochemistry: Electricity ⟶ Chemical Reactions

Redox reactions can be used to produce a flow of electrons through a wire in a voltaic cell. Such reactions can therefore proceed spontaneously.

The reverse process, in which electricity is used to produce chemical reactions, is used in an **electrolytic cell.** Such reactions cannot proceed spontaneously. Redox reactions that do not proceed spontaneously can be driven by an external "electron pump"—a direct-current (dc) source, such as a battery. This process is called **electrolysis.**

Electrolysis of Fused Compounds

When an ionic compound, such as NaCl, is melted (fused), its ions become mobile. Consider now what happens when two nonreactive electrodes are placed in melted NaCl. Nothing happens until the electrodes are connected to a source of direct current, as shown in Figure 13-3. Then sodium collects around the cathode. Bubbles of chlorine gas form around the anode. How can these observations be explained?

Electrons are being pumped into the electrode on the right, which makes it negatively charged, and out of the electrode on the left, which makes it positively charged. Na^+ ions are attracted to

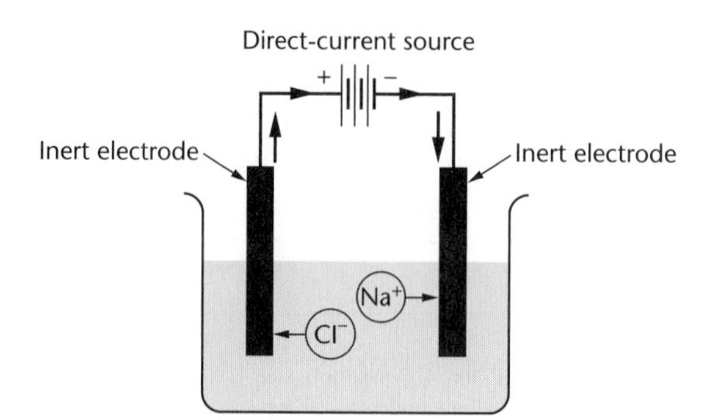

Figure 13-3 Electrolysis of a fused salt

the negatively charged electrode. There they are reduced to metallic Na.

$$Na^+ + e^- \longrightarrow Na\ (s)$$

Cl^- ions are attracted to the positively charged electrode. There they are oxidized to Cl_2 gas.

$$2\ Cl^- \longrightarrow Cl_2\ (g) + 2\ e^-$$

Reduction of Na^+ occurs at the cathode, which is the negative electrode. Oxidation of Cl^- occurs at the anode, which is the positive electrode. Recall that in the voltaic cell, the anode was negative, and the cathode was positive. This difference is the result of the electron flow. In the voltaic cell, the electrons were flowing from the negative electrode to the wire. In the electrolytic cell, the electrons are flowing from the battery or direct current source, to the negative electrode. The reaction is not spontaneous, but is driven by the "electron pump." The dc source might be a 6-volt battery, or it could be an even stronger source.

The electrolysis of any fused binary salt proceeds in much the same fashion as the electrolysis of fused NaCl. The positive electrode attracts the negative ions, which are oxidized. The positive electrode is the anode. The negative electrode attracts the positive ions, which are reduced, so the negative electrode is the cathode. Positive ions, because they are attracted by a cathode, are called **cations,** while negative ions, which are attracted to the anode, are called **anions.**

Active metals, such as sodium, potassium, magnesium, and aluminum are prepared commercially by the electrolysis of their fused salts. These elements were first isolated at the beginning of the nineteenth century, when the development of electricity made the process of electrolysis possible.

PRACTICE

13.11 Consider the electrolysis of fused KCl.

(a) Write the equation for the half-reaction that occurs at the positive electrode.

(b) Write the equation for the half-reaction that occurs at the negative electrode.

(c) Write a complete balanced equation for the overall process.

Electrolysis of Water

Since water ionizes only very slightly, electrolysis of water proceeds very slowly. Adding an electrolyte speeds up the process. The electrolyte must be chosen carefully, however, or it will be electrolyzed instead of the water.

Let's examine one possibility. Suppose sodium sulfate is used as an electrolyte. The reactions that are possible at the anode and at the cathode follow. The most probable reactions are marked by a check √. All ions in the reactions are hydrated.

CATHODE (−)

√ $2\ H_2O + 2e^- \longrightarrow H_2\ (g) + 2\ OH^-$

$Na^+ + e^- \longrightarrow Na\ (s)$

ANODE (+)

√ $2\ H_2O \longrightarrow O_2\ (g) + 4\ H^+ + 4\ e^-$

$SO_4^{2-} \longrightarrow$ not easily oxidized

If you multiply the cathode reaction by 2 and add the two reactions, you obtain

$$6\ H_2O \longrightarrow 2\ H_2\ (g) + O_2\ (g) + 4\ H^+ + 4\ OH^-$$

Since the 4 H^+ ions and the 4 OH^- ions re-form water, the net reaction is

$$2\,H_2O \longrightarrow 2\,H_2\,(g) + O_2\,(g)$$

Sulfuric acid is often used to speed up the electrolytic decomposition of water. When 0.5 *M* H_2SO_4 is the electrolyte, the half-reactions are

CATHODE (−)

$$2\,H_2O + 2e^- \longrightarrow H_2\,(g) + 2\,OH^-$$

$$\surd\ \ 2H^+ + 2e^- \longrightarrow H_2\,(g)$$

ANODE(+)

$$\surd\ \ 2\,H_2O \longrightarrow O_2(g) + 4\,H^+ + 4\,e^-$$

$$SO_4^{\ -2} \longrightarrow \text{not easily oxidized}$$

If you once again multiply the cathode reaction by 2, and add the two reactions, you get:

$$4\,H^+ + 2\,H_2O \longrightarrow 2\,H_2 + O_2 + 4\,H^+$$

Eliminating the 4 H^+ from both sides of the equation you arrive at

$$2\,H_2O \longrightarrow 2\,H_2 + O_2$$

the net reaction for the electrolysis of water. Notice, once again, that the hydrogen is produced at the negative cathode, and the oxygen is produced at the positive anode.

In electrolysis, the choice of metal electrodes is important. Platinum is frequently used because it is an excellent conductor and is not reactive. Copper is also an excellent conductor and is much less expensive than platinum. However, when electrolysis of water is attempted using copper electrodes, oxygen is not produced. Hydrogen is produced at the negative electrode, just as in the examples above, but no gas at all appears at the positive electrode. If you wait long enough, the reason becomes apparent, as a faint blue color appears in the solution. Since it is easier to oxidize Cu than H_2O, the copper electrode is oxidized, producing Cu^{2+} ions, which color the solution blue.

Electroplating

Ions of metals below H_2 in the activity series can be easily reduced. This means that metals such as copper and silver can be easily deposited on the surface of other metals such as iron. Depositing a metal coating on the surface of another metal through the use of an electric current is called **electroplating** (see Figure 13-4).

To plate copper on an iron or a steel object, the object to be plated is made the cathode of the cell. The copper then becomes the anode. The half-reactions that occur are

$$\text{ANODE } (+) \qquad Cu\ (s) \longrightarrow Cu^{2+} + 2\ e^-$$

$$\text{CATHODE } (-) \qquad Cu^{2+} + 2\ e^- \longrightarrow Cu\ (s)$$

The electrolyte is an acidified solution of $CuSO_4$. The solution is acidified to permit better plating. As Cu^{2+} ions are plated out, more Cu (s) dissolves and replaces the ions that are used up. When you add the two half-reactions, you end up with no net reaction at all! The Cu(s), the Cu^{2+}, and the 2 e^- all cancel out. During the plating process you are changing neither the concentration of Cu^{2+} ions, nor the quantity of Cu(s). You are using electricity to move copper metal from the anode to the cathode.

The same principle can be used to plate any metal on the negative cathode. The metal is placed at the positive anode, immersed in a solution of its ions. The object to be plated is placed at the negative electrode. The metal will be oxidized at the anode,

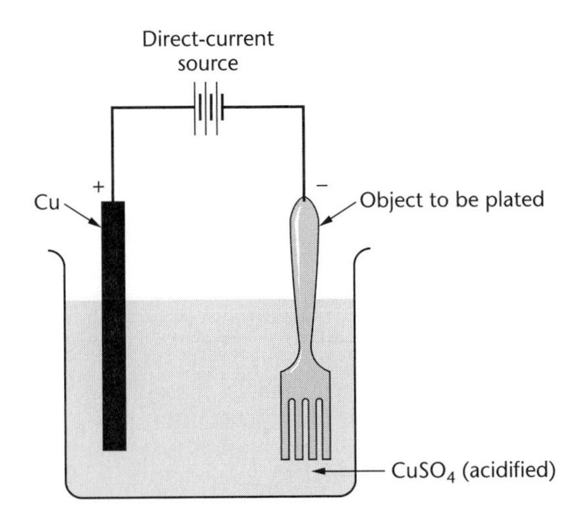

Figure 13-4 Electroplating

producing positive ions, while the same positive ions will be reduced to metal on the cathode, plating it.

Balancing Redox Reactions

All redox reactions involve the loss of electrons by one particle (oxidation) and the gain of electrons by another particle (reduction). When you write a balanced equation for a redox reaction, charge must be conserved—the number of electrons lost must equal the number of electrons gained. You should remember that even though you can write a balanced equation for a reaction, the reaction may not, in fact, occur. You can predict whether or not a redox reaction will occur from a study of the activity series, but only by experiment can you verify your prediction.

Oxidation-Number Method

Let us now consider the oxidation number method of balancing equations. Another method, called the ion-electron method, will be discussed in the Taking a Closer Look section of this chapter.

Consider the following equation:

$$Fe^{3+} + Cu \longrightarrow Cu^{2+} + Fe^{2+}$$

It may appear to you that the equation is already balanced. There is one Fe on each side of the equation, and one Cu. However, the total charge on the reactant side is 3+, while the total charge on the product side is 4+. An ionic equation must be balanced for charge, as well as for each participating element.

To balance equations using oxidation states, first write the oxidation and reduction half-reactions.

$$Fe^{3+} + e^- \longrightarrow Fe^{2+} \text{ (reduction)}$$

$$Cu \longrightarrow Cu^{2+} + 2\,e^- \text{ (oxidation)}$$

Each of these half-reactions must itself be balanced both for number of atoms and charge.

Next, use the fact that the number of electrons lost must equal the number of electrons gained. Multiply the half-reactions by the smallest whole numbers needed to make the number of electrons the same in both half-reactions. In this case, you need to multiply the reduction-half reaction by two. Now, rewriting the two half-reactions, you have $2\ Fe^{3+} + 2\ e^- \longrightarrow 2\ Fe^{2+}$ (note that every species in the half-reaction was doubled) and $Cu \longrightarrow Cu^{2+} + 2\ e^-$. Adding the two half-reactions now gives the balanced equation:

$$2\ Fe^{3+} + Cu \longrightarrow 2\ Fe^{2+} + Cu^{2+}$$

Checking, you see that there are two Fe and one Cu on each side of the equation. The total charge on each side of the equation is the same, 6+. The equation is balanced.

When the redox equation contains materials in addition to those being oxidized and reduced, these must be balanced after the process above has been completed.

Now consider the following equation, which contains several atoms and ions that do not appear in the reduction and oxidation half-reactions:

$$HCl + KMnO_4 \longrightarrow KCl + MnCl_2 + H_2O + Cl_2$$

As the first step in balancing the reaction, identify the oxidation and reduction half-reactions by finding the particles that experience a change in oxidation state. (You may wish to review the section "Analyzing Redox Reactions.") In this reaction the Mn is reduced from +7 to +2. The chlorine is oxidized from −1 to 0.

Next, write the oxidation and reduction half-reactions, making certain that each is balanced both for number of atoms and charge.

$$2\ Cl^- \longrightarrow Cl_2^{\,0} + 2\ e^-$$

$$Mn^{7+} + 5e^- \longrightarrow Mn^{2+}$$

Now, to equalize the number of electrons lost and gained, multiply both half-reactions by the appropriate whole numbers. The least common multiple of 2 and 5 is 10, so you will need 10 electrons in each half-reaction. Multiply the chlorine half-reaction by five and the manganese half-reaction by two.

$$10\ Cl^- \longrightarrow 5\ Cl_2 + 10e^-$$

$$2\ Mn^{7+} + 10e^- \longrightarrow 2\ Mn^{2+}$$

When adding the half-reactions, make certain the electrons have cancelled each other and the resulting reaction is balanced and in lowest terms.

$$2\ Mn^{7+} + 10\ Cl^- \longrightarrow 2\ Mn^{2+} + 5\ Cl_2$$

Both sides of the equation contain 2 Mn, 10 Cl, and a total charge of 4+.

Now, insert these coefficients into the original equation.

$$2\ KMnO_4 + 10\ HCl \longrightarrow KCl + 2\ MnCl_2 + 5\ Cl_2 + H_2O$$

Balance the remaining material in the reaction by inspection. Since you have 2 $KMnO_4$ there must be 2 KCl to balance the potassium. There are now 16 chlorines on the right side of the equation, so you need 6 more HCl, for a total of 16. To balance the hydrogen, you will need an 8 in front of the water. Checking the oxygens, there are now 8 on each side. The balanced equation is

$$2\ KMnO_4 + 16\ HCl \longrightarrow 2\ KCl + 2\ MnCl_2 + 5\ Cl_2 + 8\ H_2O$$

This reaction normally takes place in solution, so that the equation is more correctly written with the electrolytes shown as aqueous ions. Unbalanced, that would give us

$$K^+ + MnO_4^- + H^+ + Cl^- \longrightarrow K^+ + Cl^- + Mn^{2+} + Cl_2 + H_2O$$

The potassium ion is a spectator ion, and can be omitted. The Cl^- ion can be omitted from the product as well, giving the net equation

$$MnO_4^- + H^+ + Cl^- \longrightarrow Mn^{2+} + Cl_2 + H_2O$$

The net ionic equation can be balanced by exactly the same method used above, which would give you

$$2\ MnO_4^- + 16\ H^+ + 10\ Cl^- \longrightarrow 2\ Mn^{2+} + 5\ Cl_2 + 8\ H_2O$$

When balancing ionic equations it is important to check that the equation is balanced not only for mass but also for charge. In the ionic equation above, the total charge on the left side of the equation is 4+ and the total charge on the right side of the equation is 4+. The equation is correctly balanced.

Let us review the steps in balancing a redox reaction.

1. Find the elements that undergo a change in oxidation state.

2. Write the oxidation and reduction half-reactions, checking that they are correctly balanced for atoms and charge.
3. Multiply both half-reactions by the lowest coefficients needed to make the electrons lost in the oxidation half-reaction equal the electrons gained in the reduction half-reaction.
4. Add the two half-reactions to produce a balanced redox reaction.
5. Write the coefficients obtained in your balanced redox reaction in the appropriate places in the original equation you are balancing. Balance the remaining substances by inspection.

SAMPLE PROBLEM

PROBLEM

Balance the ionic equation

$$Fe^{2+} + ClO_3^- + H^+ \longrightarrow Fe^{3+} + Cl^- + H_2O$$

SOLUTION

The two elements that undergo a change in oxidation state are the iron and the chlorine. The two half-reactions are

$$Fe^{2+} \longrightarrow Fe^{3+} + 1e^-$$

$$Cl^{5+} + 6e^- \longrightarrow Cl^-$$

To make the electrons lost equal the electrons gained, all you have to do is multiply the oxidation equation by six. Then, adding the two half-reactions gives you

$$6\ Fe^{2+} + Cl^{5+} \longrightarrow 6\ Fe^{3+} + Cl^-$$

Placing these coefficients in the original equation, you get

$$6\ Fe^{2+} + ClO_3^- + H^+ \longrightarrow 6\ Fe^{3+} + Cl^- + H_2O$$

You can now balance the H^+ and the H_2O by inspection. Since there are three oxygens on the left, you will need three waters to balance the oxygen. Then you will need 6 H^+ to balance the hydrogen.

$$6\ Fe^{2+} + ClO_3^- + 6\ H^+ \longrightarrow 6\ Fe^{3+} + Cl^- + 3\ H_2O$$

Checking charge, you will find that there is a total charge of 17+ on the left, and a total charge of 17+ on the right. The equation is balanced.

PRACTICE

13.12 Balance the following equations:

(a) $Al + Fe^{2+} \longrightarrow Al^{3+} + Fe$

(b) $Cl_2 + Br^- \longrightarrow Br_2 + Cl^-$

(c) $Al + Sn^{4+} \longrightarrow Al^{3+} + Sn$

(d) $Na + H_2O \longrightarrow Na^+ + OH^- + H_2$

13.13 Balance the following more difficult equations.

(a) $H^+ + NO_3^- + Cu \longrightarrow Cu^{2+} + NO + H_2O$

(b) $Zn + H_2SO_4 \longrightarrow ZnSO_4 + H_2S + H_2O$

(c) $SO_2 + ClO_3^- + H_2O \longrightarrow SO_4^{2-} + Cl^- + H^+$

(d) $Fe^{2+} + H^+ + Cr_2O_7^{2-} \longrightarrow Cr^{3+} + Fe^{3+} + H_2O$

(e) $H_2O_2 + H^+ + Sn^{2+} \longrightarrow Sn^{4+} + H_2O$

TAKING A CLOSER LOOK

Oxidizing Agents

Oxidation is a reaction in which a particle loses electrons. Any particle that can cause the loss of electrons is therefore an **oxidizing agent.** The oxidizing agent is itself reduced—that is, it gains electrons.

Some typical oxidizing agents can be classified as follows:

1. Positive ions that gain electrons. A general equation for this type of reaction is

$$M^{n+} + n\,e^- \longrightarrow M^0$$

In the equation, M stands for a metal, and n stands for the charge on the metallic ion.

2. Nonmetallic elements, often halogens or oxygen, that gain electrons. A general equation for this type of reaction is

$$X_2 + 2e^- \longrightarrow 2\ X^-$$

$$Z_2 + 4e^- \longrightarrow 2\ Z^{2-}$$

X_2 stands for a halogen molecule. Z_2 stands for an oxygen molecule.

3. Negatively charged polyatomic ions that contain oxygen and another atom that is in a relatively high oxidation state. The atom tends to gain electrons and thus to change to a lower oxidation state. These oxidizing agents and typical half-reactions they undergo are shown in the examples that follow.

(a) The permanganate ion, MnO_4^-

$$MnO_4^- + 8\ H^+ + 5e^- \longrightarrow Mn^{2+} + 4\ H_2O$$

Manganese in MnO_4^- gains 5 electrons. Its oxidation state changes from +7 to +2.

(b) The nitrate ion, NO_3^-

$$NO_3^- + 4\ H^+ + 3e^- \longrightarrow NO + 2\ H_2O$$

Nitrogen in NO_3^- gains 3 electrons. Its oxidation state changes from +5 to +2.

(c) The dichromate ion, $Cr_2O_7^{2-}$

$$Cr_2O_7^{2-} + 14\ H^+ + 6\ e^- \longrightarrow 2\ Cr^{3+} + 7\ H_2O$$

Each chromium atom in $Cr_2O_7^{2-}$ gains 3 electrons and changes in oxidation state from +6 to +3

Reducing Agents

Reduction is a reaction in which a particle gains electrons. A particle that can cause the gain of electrons is a **reducing agent.** The reducing agent is itself oxidized—that is, it loses electrons.

Reducing agents can be classified as follows:

1. Active metals, such as Na, K, and Al, which tend to lose electrons readily. A general equation for the behavior of a metal as a reducing agent is

$$M^0 \longrightarrow M^{n+} + ne^-$$

In the equation, M stands for the metal and n stands for the charge on the metallic ion.

2. Halide ions that undergo the reaction

$$2\,X^- \longrightarrow X_2 + 2e^-$$

   X^- stands for a halide ion, such as Cl^-.

3. Positive ions that can lose still more electrons and change to a higher oxidation state. Two examples are

   (a) Stannous ion, tin(II), which can become stannic ion, tin(IV)

$$Sn^{2+} \longrightarrow Sn^{4+} + 2\,e^-$$

   (b) Ferrous ion, iron(II), which can become ferric ion, iron(III)

$$Fe^{2+} \longrightarrow Fe^{3+} + e^-$$

In summary, oxidizing agents are reduced, gain electrons, and decrease in charge. Reducing agents are oxidized, lose electrons, and increase in charge.

Standard Electrode Potentials

In Chapter 12, you learned that acid-base reactions proceed in the direction that produces the weaker acids and bases. You can predict the direction of a redox reaction in a similar way. Redox reactions proceed in the direction that produces the weaker oxidizing and reducing agents.

More detailed information about reduction and oxidation can be found on a table of electrode potentials, as illustrated in the following table on page 469. This table can also be used to compare the strengths of oxidizing and reducing agents. All of the half-reactions shown in the table are reduction half-reactions. Therefore, oxidizing agents appear on the left side of the table. If the reactions are considered in the reverse direction they are all oxidations, which places the reducing agents on the right side of the table. The oxidizing agents are listed in order of strength from F_2, the strongest, to Li^+, the weakest. The reducing agents *increase* in strength as we move down the right side of the table. Li (s) is the strongest reducing agent listed, and F^- the weakest. A redox

Standard Electrode Potentials

(Ionic concentrations = 1 *M* in water at 298 K. All ions are hydrated. All gases are at a partial pressure of 1 atm.)

Half-Reaction	E° (volts)
$F_2\,(g) + 2e^- \longrightarrow 2\,F^-$	+2.87
$Au^{3+} + 3e^- \longrightarrow Au\,(s)$	+1.52
$MnO_4^- + 8\,H^+ + 5e^- \longrightarrow Mn^{2+} + 4\,H_2O$	+1.51
$Cl_2\,(g) + 2e^- \longrightarrow 2\,Cl^-$	+1.36
$Cr_2O_7^{2-} + 14\,H^+ + 6e^- \longrightarrow 2\,Cr^{3+} + 7\,H_2O$	+1.33
$MnO_2\,(s) + 4H^+ + 2e^- \longrightarrow Mn^{2+} + 2\,H_2O$	+1.23
$\frac{1}{2}\,O_2\,(g) + 2\,H^+ + 2e^- \longrightarrow H_2O$	+1.23
$Br_2\,(l) + 2e^- \longrightarrow 2\,Br^-$	+1.06
$NO_3^- + 4\,H^+ + 3e^- \longrightarrow NO\,(g) + 2\,H_2O$	+0.96
$Hg^{2+} + 2e^- \longrightarrow Hg\,(l)$	+0.85
$\frac{1}{2}\,O_2\,(g) + 2\,H^+\,(10^{-7}\,M) + 2e^- \longrightarrow H_2O$	+0.82
$Ag^+ + e^- \longrightarrow Ag\,(s)$	+0.80
$\frac{1}{2}\,Hg_2^{2+} + e^- \longrightarrow Hg\,(l)$	+0.79
$NO_3^- + 2\,H^+ + e^- \longrightarrow NO_2\,(g) + H_2O$	+0.78
$Fe^{3+} + e^- \longrightarrow Fe^{2+}$	+0.77
$I_2\,(s) + 2e^- \longrightarrow 2\,I^-$	+0.53
$Cu^+ + e^- \longrightarrow Cu\,(s)$	+0.52
$Cu^{2+} + 2e^- \longrightarrow Cu\,(s)$	+0.34
$SO_4^{2+} + 4\,H^+ + 2e^- \longrightarrow SO_2\,(g) + H_2O$	+0.17
$Sn^{4+} + 2e^- \longrightarrow Sn^{2+}$	+0.15
$2\,H^+ + 2e^- \longrightarrow H_2\,(g)$	0.00
$Pb^{2+} + 2e^- \longrightarrow Pb\,(s)$	−0.13
$Sn^{2+} + 2e^- \longrightarrow Sn\,(s)$	−0.14
$Ni^{2+} + 2e^- \longrightarrow Ni\,(s)$	−0.26
$Co^{2+} + 2e^- \longrightarrow Co\,(s)$	−0.28
$2\,H^+\,(10^{-7}\,M) + 2e^- \longrightarrow H_2\,(g)$	−0.41
$Fe^{2+} + 2e^- \longrightarrow Fe\,(s)$	−0.44
$Cr^{3+} + 3e^- \longrightarrow Cr\,(s)$	−0.74
$Zn^{2+} + 2e^- \longrightarrow Zn\,(s)$	−0.76
$2\,H_2O + 2e^- \longrightarrow 2\,OH^- + H_2\,(g)$	−0.83
$Mn^{2+} + 2e^- \longrightarrow Mn\,(s)$	−1.18
$Al^{3+} + 3e^- \longrightarrow Al\,(s)$	−1.66
$Mg^{2+} + 2e^- \longrightarrow Mg\,(s)$	−2.37
$Na^+ + e^- \longrightarrow Na\,(s)$	−2.71
$Ca^{2+} + 2e^- \longrightarrow Ca\,(s)$	−2.87
$Sr^{2+} + 2e^- \longrightarrow Sr\,(s)$	−2.89
$Ba^{2+} + 2e^- \longrightarrow Ba\,(s)$	−2.92
$Cs^+ + e^- \longrightarrow Cs\,(g)$	−2.92
$K^+ + e^- \longrightarrow K\,(s)$	−2.92
$Rb^+ + e^- \longrightarrow Rb\,(s)$	−2.93
$Li^+ + e^- \longrightarrow Li\,(s)$	−3.00

reaction will proceed spontaneously only if the stronger reducing agent is reacting with the stronger oxidizing agent.

All the half-reactions shown in the table are occurring in water, and all ions are hydrated. Comparisons using this table are completely valid only in aqueous solution.

Consider again the reaction between metallic zinc and the copper(II) ion.

$$Zn\ (s) + Cu^{2+} \longrightarrow Zn^{2+} + Cu\ (s)$$

Does this reaction proceed as shown? Using the table of electrode potentials, find the four species shown in the equation. Zn (*s*) appears below Cu (*s*) on the right side of the table, so Zn (*s*) is a stronger reducing agent than Cu (*s*). Cu^{2+} appears above Zn^{2+} on the left side of the table, so Cu^{2+} is a stronger oxidizing agent than Zn^{2+}. Since the stronger oxidizing and reducing agents are on the left side of the chemical equation, this reaction does proceed as written.

Reducing agents lose electrons, while oxidizing agents gain them. Electrons will flow spontaneously from the stronger reducing agent to the stronger oxidizing agent. You can use this tendency to predict whether a given reaction will occur.

You have seen that the direction of a redox reaction can be predicted using the Activity Series, or the Electrode Potentials. However, the Activity Series cannot be used to predict a reaction between ions, such as the reaction

$$Sn^{2+} + 2\ Fe^{3+} \longrightarrow Sn^{4+} + 2\ Fe^{2+}.$$

SAMPLE PROBLEM

PROBLEM

1. Will metallic nickel react spontaneously with a solution containing Zn^{2+}?

SOLUTION

Find Ni (*s*) and Zn^{2+} on the table of electrode potentials. Ni (*s*) is a weaker reducing agent than Zn (*s*). Zn^{2+} is a weaker oxidizing agent than Ni^{2+}. For a redox reaction to occur, the reactants must be the stronger reducing and oxidizing agents. The reaction does not occur.

PROBLEM

2. Will Cl_2 (*g*) react spontaneously with a solution of KBr?

SOLUTION

KBr contains the ions K^+ and Br^-. K^+ appears on the left side of the table; it is an oxidizing agent. Br^- is a reducing agent, and Cl_2 is an oxidizing agent. Since the oxidizing agent, Cl_2, can react only with a reducing agent, it cannot react with K^+. The K^+ is a spectator ion in this reaction. Finding Cl_2 and Br^- on the table, you note that the chlorine is a stronger oxidizing agent than bromine, while Br^- is a stronger reducing agent than Cl^-. The reaction will occur.

PROBLEM

3. Will aluminum spontaneously replace zinc from its salts?

SOLUTION

Zinc salts contain the zinc ion, Zn^{2+}. Finding Al (*s*) and Zn^{2+} on the table, we see that Al is a stronger reducing agent than Zn, and Zn^{2+} is a stronger oxidizing agent than Al^{3+}. Aluminum does replace zinc from its salts, which means that Al (*s*) does react with Zn^{2+}.

PRACTICE

(Use the table of electrode potentials.)

13.14 Indicate whether the following reactions proceed spontaneously.

(a) $Fe\ (s) + Ni^{2+} \longrightarrow Fe^{2+} + Ni\ (s)$

(b) $2\ F^- + Cl_2 \longrightarrow F_2 + 2\ Cl^-$

(c) $Cr_2O_7^{2-} + 14\ H^+ + 3\ Sn^{2+} \longrightarrow 2\ Cr^{3+} + 3\ Sn^{4+} + 7\ H_2O$

13.15 Indicate whether the following pairs of materials react with each other spontaneously.

(a) Cu (*s*) and Ag^+
(b) Br_2 (*l*) and I^-
(c) Cr (*s*) and a solution of NiF_2
(d) Al (*s*) and a solution of NaCl

Reaction of Acids with Metals

Acids produce H^+ ions in aqueous solution. Acids will therefore react with those metals that can reduce H^+ ions. On the table of electrode potentials, all metals listed below H_2 should react with H^+ ions, while metals listed above H_2 should not. The reaction $Zn + 2\,H^+ \longrightarrow H_2 + Zn^{2+}$ occurs spontaneously. Weaker reducing agents such as Cu, Ag, and Au will not react with the H^+ ion. In general, the stronger the reducing agent, the faster the reaction.

HNO_3 reacts vigorously with zinc, but the reaction does not produce hydrogen. HNO_3 also reacts with copper, which should not be able to reduce the H^+ ions. The explanation for these observations is found in the table of electrode potentials. Nitric acid produces the nitrate ion, NO_3^-, which is a much stronger oxidizing agent than H^+. While the reaction of Cu with H^+ is not spontaneous, the reaction of Cu with NO_3^- is spontaneous.

$$3\,Cu + 2\,NO_3^- + 8\,H^+ \longrightarrow Cu^{2+} + 2\,NO + 4\,H_2O.$$

Because the nitrate ion is more easily reduced than the H^+ ion, reactions of metals with nitric acid produce oxides of nitrogen instead of hydrogen gas.

E⁰

Since redox reactions involve a transfer of electrons, these reactions can be used as sources of electrical energy. The tendency for electrons to flow in a given direction depends upon the relative strengths of the reducing and oxidizing agents involved. A quantitative measure of the relative strengths of oxidizing and reducing agents is provided by the standard electrode potentials.

The standard electrode potentials listed on page 469 are also called reduction potentials, since all of the half-reactions are listed as reductions. The unit of electrical potential is the volt, and the symbol for electrical potential is *E*. The **standard electrode potential,** E^0, is the electrical potential when all ions are 1 *M* concentration and all gases are at 1 atm pressure.

Reduction half-reactions cannot take place alone. They must be accompanied by oxidation half-reactions. To assign an E^0 value to each half-reaction, a standard half-cell was needed as the basis for comparison. The standard chosen is the half-cell involving H_2 and H^+.

$$2\,H^+ + 2\,e^- \longrightarrow H_2\,(g)$$

By definition, E^0 for this half-reaction is 0.00 volts, in either direction.

The value of E expresses the relative tendency of the reaction to occur spontaneously. If E is positive, the reaction is spontaneous under the given conditions. If E is negative, the reaction is not spontaneous, but the reverse reaction is spontaneous. If $E = 0$, the system is at equilibrium. You may recall using the value of ΔG to predict whether a reaction is spontaneous. ΔG and E are closely related, but always opposite in sign (except, when both are 0).

Notice that all of the reductions listed above the standard hydrogen half-reaction have positive values of E^0. These reductions will proceed spontaneously, when H_2 is oxidized to H^+. All of the reductions listed below the hydrogen half-reaction have negative values of E^0. These reductions will not proceed spontaneously to oxidize H_2. However, the reverse reactions, the oxidations, will proceed spontaneously when H^+ is reduced. The E^0 of an oxidation half-reaction, (right to left on the table) is called the oxidation potential. It is found by changing the sign of the given reduction potential. The E^0 for the oxidation of Zn to Zn^{2+}, for example, is $+0.76$ volts.

Determination of E^0 for a Redox Reaction

The net E^0 for any redox reaction is the sum of the E^0 values of the two half-reactions. Remember, to obtain the correct value of E^0 for the oxidation half-reaction, you must change the sign of the value given on the table of reduction potentials.

Consider the reaction between Zn (s) and Cu^{2+}.

$$Zn\,(s) + Cu^{2+} \longrightarrow Zn^{2+} + Cu\,(s)$$

From the table on page 469, you can see that the standard electrode potential for the reduction of Cu^{2+} to Cu is $+0.34$ volts. The Zn is oxidized in this reaction. The table tells us that E^0 for

$Zn^{2+} + 2\,e^- \longrightarrow Zn$ is -0.76 volts, but this is a reduction. The zinc half-reaction is going in the opposite direction, with the zinc being oxidized, so you obtain E^0 by changing the sign of the given reduction potential. The E^0 for the oxidation of Zn is $+0.76$ volts. The net E^0 for the redox reaction is

$$0.76 \text{ volts} + 0.34 \text{ volts} = 1.10 \text{ volts}$$

Since the voltage is positive, you know that the reaction is spontaneous as written.

You learned previously that Zn does react spontaneously with Cu^{2+}; you were able to predict whether a redox reaction is spontaneous without actually calculating the net E^0 for the reaction. You can use the value of E^0 to check your predictions.

In Sample Problem 1, on page 470 it was predicted that Ni would not react spontaneously with Zn^{2+}. Let us calculate the net E^0 for this reaction. Zn^{2+} is reduced to Zn, with an E^0 of -0.76 volts. Ni appears on the right side of the table. It is oxidized to Ni^{2+} with an E^0 of $+0.26$. Because this is an oxidation, you change the sign of the E^0 given in the table. The net E^0 for the reaction between Ni and Zn^{2+} is the sum of the voltages of the two half reactions. The negative E^0 value indicates that the reaction is not spontaneous.

$$Zn^{2+} + 2e^- \longrightarrow Zn \qquad E^0 = -0.76 \text{ volts}$$

$$Ni \longrightarrow Ni^{2+} + 2e^- \qquad E^0 = +0.26 \text{ volts}$$

$$\text{net } E^0 = -0.50 \text{ volts}$$

PRACTICE

13.16 Find the net E^0 for each of the following redox reactions and indicate whether each reaction occurs spontaneously.

(a) $Cl_2\,(g) + 2\,Br^- \longrightarrow 2\,Cl^- + Br_2\,(l)$

(b) $Cu\,(s) + 2\,Ag^+ \longrightarrow Cu^{2+} + 2\,Ag\,(s)$

(c) $Zn\,(s) + Mg^{2+} \longrightarrow Zn^{2+} + Mg\,(s)$

(d) $Cu\,(s) + Cl_2\,(g) \longrightarrow Cu^{2+} + 2\,Cl^-$

(e) $Cr_2O_7{}^{2-} + 14\,H^+ + 3\,Sn^{2+} \longrightarrow 2\,Cr^{3+} + 3\,Sn^{4+} + 7\,H_2O$

Balancing Equations by the Ion-Electron Method

When redox reactions occur in aqueous solution, the solutions are often made acidic. This prevents the precipitation of insoluble metal hydroxides. Reactions in acidic media are often balanced by the ion-electron method. This method does not use oxidation states; it uses the actual charges of the ions. To illustrate the ion-electron method, let us use the following reaction, which takes place in an acid medium.

$$Fe^{2+} + MnO_4^- \longrightarrow Fe^{3+} + Mn^{2+}$$

1. Write a half-reaction for each ion that undergoes a change in charge.

$$Fe^{2+} \longrightarrow Fe^{3+}$$

$$MnO_4^- \longrightarrow Mn^{2+}$$

2. Add H_2O where necessary to balance the oxygens.

$$MnO_4^- \longrightarrow Mn^{2+} + 4\ H_2O$$

(There are no oxygens in the iron half-reaction)

3. Add H^+ ions to balance the hydrogens.

$$MnO_4^- + 8\ H^+ \longrightarrow Mn^{2+} + 4\ H_2O$$

(There are no hydrogens in the iron half-reaction.)

4. Add electrons to balance the charge.

$$Fe^{2+} \longrightarrow Fe^{3+} + e^-$$

$$MnO_4^- + 8\ H^+ + 5e^- \longrightarrow Mn^{2+} + 4\ H_2O$$

5. Multiply the half-reactions by coefficients that make the electrons lost equal electrons gained. In this case, you need to multiply the iron half-reaction by five.
6. Add the two half-reactions.

$$5\ Fe^{2+} + MnO_4^- + 8\ H^+ \longrightarrow 5\ Fe^{3+} + Mn^{2+} + 4\ H_2O$$

The equation is now balanced. When a reaction contains additional materials, they can be balanced by inspection.

The ion-electron method can be adapted for use in basic media as well, although you are less likely to encounter such reactions.

In basic media, only steps 2 and 3 are different. In step 2, you still add water to balance the oxygen, but you add it to the side with the extra oxygen.

$$MnO_4^- + 4\,H_2O \longrightarrow Mn^{2+}$$

In step 3, you add to the opposite side twice as many OH^- ions as you added water in step 2.

$$MnO_4^- + 4\,H_2O \longrightarrow Mn^{2+} + 8\,OH^-$$

You then need the same number of electrons to balance the charge as you did in acidic media. The rest of the process is the same.

This reaction does not actually occur in basic media, but the balanced equation would be

$$MnO_4^- + 4\,H_2O + 5\,Fe^{2+} \longrightarrow 5\,Fe^{3+} + Mn^{2+} + 8\,OH^-$$

PRACTICE

13.17 Balance in acidic media.

(a) $NO_3^- + Cu \longrightarrow Cu^{2+} + NO$

(b) $SO_2 + ClO_3^- \longrightarrow SO_4^{2-} + Cl^-$

(c) $Fe^{2+} + Cr_2O_7^{2-} \longrightarrow Cr^{3+} + Fe^{3+}$

Disproportionation Reactions

When chlorine gas is bubbled into a hot solution of sodium hydroxide, the following reaction occurs:

$$3\,Cl_2 + 6\,OH^- \longrightarrow 3\,H_2O + ClO_3^- + 5\,Cl^-$$

The oxidation state of chlorine in chlorine gas is zero. In the ClO_3^- ion the oxidation state of Cl is +5, while in the Cl^- ion the Cl is −1. The chlorine was both oxidized and reduced. Reactions in which the same element is both oxidized and reduced are called **disproportionation reactions**, or **auto-oxidations.** One very common reaction of this type is the decomposition of hydrogen peroxide, a common household disinfectant.

$$2\,H_2O_2 \longrightarrow 2\,H_2O + O_2$$

The oxidation state of oxygen in the peroxide is −1. The oxidation state of oxygen is reduced to −2 in water and oxidized to 0 in oxygen gas. This reaction is very slow in the absence of a catalyst and in the absence of light. To inhibit the decomposition, hydrogen peroxide is stored in dark bottles.

In the table of electrode potentials on page 469, for the half-reaction

$$Cu^{+} + e^{-} \longrightarrow Cu \qquad E^{0} = +0.52 \text{ volts}$$

For the oxidation of the copper(I) ion,

$$Cu^{+} \longrightarrow Cu^{2+} + e^{-} \qquad E^{0} = -0.16 \text{ volts.}$$

(It is not listed in the table.) From these two values, you can predict that a 1 *M* solution of Cu^{+} ion cannot be prepared. Adding the two half-reactions together gives us

$$2\ Cu^{+} \longrightarrow Cu + Cu^{2+} \qquad E^{0} = +0.36 \text{ volts.}$$

Since the E^0 is positive, the reaction is spontaneous. The Cu^{+} ion disproportionates, auto-oxidizes, in aqueous solution.

Recharging a Chemical Cell

When a car battery goes dead, you generally do not throw it away. It can be recharged several times. Certain smaller batteries, called "Nicads," containing nickel and cadmium, can also be recharged. You recharge a cell by driving the cell reaction in the reverse direction. Let us see how the zinc-copper cell discussed earlier could be recharged. The cell reaction was

$$Zn(s) + Cu^{2+} \longrightarrow Zn^{2+} + Cu(s) \qquad E^{0} = 1.1 \text{ volts}$$

The voltage drops to zero at equilibrium. At this point there is almost no Cu^{2+} ion left in the cell, and twice as much Zn^{2+} ion as there was initially. To recharge the cell, you must drive the reaction to the left, producing Cu^{2+} ions and removing Zn^{2+} ions. The voltage of a battery depends upon the ion concentrations, but not on the quantity of solid metal. (That is why the larger "D" batteries have exactly the same voltage as the much smaller "AA" batteries. The difference is that the "D" cells last longer.)

Since E^0 is positive for the forward reaction, it must be negative for the reverse reaction. To run a reaction that has a negative E^0, you must supply electricity. This means you must turn the chemical cell into an electrolytic cell by using a source of direct current that supplies a voltage greater than 1.1 volts. Attach the power source to the chemical cell negative to negative and positive to positive. Recall that in the chemical cell the zinc was the negative electrode, and the copper the positive electrode. By attaching the negative pole of our power source to the zinc, we force electrons into the zinc. The electrons are then gained by the Zn^{2+} ions forming solid zinc at the cathode. Attaching the positive pole of the power source to the copper, pulls electrons from the copper, producing Cu^{2+} ions. Cu^{2+} ions form and Zn^{2+} ions are removed. If you continue the process long enough, you can get back to the 1 *M* concentrations of the original cell. The cell is then recharged, and will once again supply 1.1 volts of electricity.

How long a given cell lasts depends not upon its electrical potential, but upon the quantity of current it supplies. Current, which is measured in amperes, expresses the rate at which electrons are passing through the circuit. Most battery-operated devices use very small amounts of current, which is why even tiny batteries can last several hours.

CHAPTER REVIEW

The following questions will help you check your understanding of the material presented in the chapter.

Write the number of the word or expression that best completes the statement or answers the question.

1. In the reaction $Cu + 2\,Ag^+ \longrightarrow Cu^{2+} + 2\,Ag$, which species gains electrons? (1) Cu (2) Ag (3) Cu^{2+} (4) Ag^+
2. In the reaction $Zn + 2\,HCl \longrightarrow ZnCl_2 + H_2$, which species loses electrons? (1) Zn (2) H^+ (3) Cl^- (4) H_2
3. The reduction of chlorine, Cl_2, results in the formation of (1) Cl^+ (2) Cl^{3+} (3) Cl^{5+} (4) Cl^-
4. Which half reaction correctly represents oxidation?
   (1) $Zn + 2\,e^- \longrightarrow Zn^{2+}$ (3) $Zn^{2+} + 2\,e^- \longrightarrow Zn$
   (2) $Cl_2 \longrightarrow 2\,Cl^- + 2\,e^-$ (4) $2\,Cl^- \longrightarrow Cl_2 + 2e^-$

5. The species reduced in a redox reaction always (1) increases in charge (2) is oxidized (3) gains electrons (4) loses protons

6. Which of the following reactions is not a redox reaction?

   (1) $Ba + Cl_2 \longrightarrow BaCl_2$ (2) $Cl_2 + 2\ HBr \longrightarrow Br_2 + 2\ HCl$

   (3) $H_2SO_4 + BaCl_2 \longrightarrow 2\ HCl + BaSO_4$

   (4) $3\ Cl_2 + 6\ NaOH \longrightarrow 5\ NaCl + NaClO_3 + 3\ H_2O$

Questions 7 to 11 are based on the activity series for the six elements listed below. They are listed in order, from most active to least active.

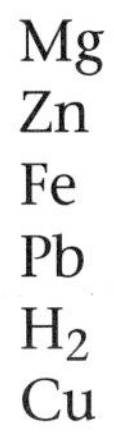

7. Which element will react with Pb^{2+} ions but not with Zn^{2+} ions? (1) Mg (2) Fe (3) Cu (4) H_2

8. A solution containing Al^{3+} ions reacts spontaneously with only one of the metals listed above. This metal must be (1) Mg (2) Fe (3) Pb (4) Cu

9. Which of the following reactions is spontaneous?

   (1) $Fe + Zn^{2+} \longrightarrow Fe^{2+} + Zn$ (3) $Pb + Cu^{2+} \longrightarrow Pb^{2+} + Cu$

   (2) $Cu + 2\ H^+ \longrightarrow Cu^{2+} + H_2$ (4) $Pb + Mg^{2+} \longrightarrow Pb^{2+} + Mg$

10. Mercury metal, Hg, does not react with strong acids. Based on this information, and the activity series above, one could conclude that (1) Hg is less active than Pb (2) Hg is more active than Mg (3) Hg is less active than Cu (4) Hg is more active than Pb

11. When copper metal is placed in a solution of silver nitrate, gray crystals of silver form on the piece of copper, and the solution gradually turns blue. What would be the most likely result of placing pieces of metallic silver in a solution of zinc nitrate? (1) pieces of zinc will form on the silver (2) silver ions will be produced in the solution (3) zinc ions will be oxidized (4) no reaction will occur

12. The electronic equation $Mg^{2+} + 2\ e^- \longrightarrow Mg^0$ indicates that the (1) magnesium ion is being oxidized (2) magnesium ion

is being reduced (3) magnesium atom is being oxidized (4) magnesium atom is being reduced.

13. As the number of particles oxidized during a chemical reaction increases, the number of particles reduced (1) decreases (2) increases (3) remains the same.

14. Which is the equation for the reduction that takes place in the reaction below?

$$2\ KBr + 3\ H_2SO_4 + MnO_2 \longrightarrow 2\ KHSO_4 + MnSO_4 + 2\ H_2O + Br_2$$

(1) $Mn^{4+} + 2\ e^- \longrightarrow Mn^{2+}$
(2) $2\ Br^- \longrightarrow Br_2^0 + 2\ e^-$
(3) $Mn^{2+} + 2\ e^- \longrightarrow Mn^{4+}$
(4) $2\ Br^- + 2\ e^- \longrightarrow 2\ Br^{2-}$

15. Which reaction occurs at the negative electrode during the electrolysis of fused (molten) calcium fluoride?
(1) $CaF_2 \longrightarrow Ca^{2+} + F_2 \uparrow$
(2) $Ca^{2+} + 2\ e^- \longrightarrow Ca^0$
(3) $2\ F^- \longrightarrow F_2 + 2\ e^-$
(4) $Ca^0 \longrightarrow Ca^{2+} + 2\ e^-$

16. An aluminum-nickel cell produces the reaction:

$$2\ Al\ (s) + 3\ Ni^{2+}\ (aq) \longrightarrow 2\ Al^{3+}\ (aq) + 3\ Ni\ (s)$$

As an aluminum-nickel chemical cell approaches equilibrium, the mass of the nickel electrode (1) decreases (2) increases (3) remains the same.

17. What is the coefficient of MnO_4^- when the following redox reaction is balanced?

$$\_\_MnO_4^- + 5\ SO_2 + 2\ H_2O \longrightarrow \_\_Mn^{2+} + 5\ SO_4^{2-} + 4\ H^+$$

(1) 1 (2) 2 (3) 3 (4) 4

18. How many moles of electrons are needed to reduce 1 mole of Fe^{3+} to Fe^{2+}? (1) 1 (2) 2 (3) 3 (4) 5

19. As a sodium atom is oxidized, the number of protons in its nucleus (1) decreases (2) increases (3) remains the same.

20. Given the equation

$$2\ Al(s) + 3\ Cu^{2+}\ (aq) \longrightarrow 2\ Al^{3+}\ (aq) + 3\ Cu\ (s)$$

The total number of moles of electrons transferred from 2 Al (*s*) to 3 Cu^{2+} (*aq*) is (1) 9 (2) 2 (3) 3 (4) 6.

21. As the reaction in a chemical cell approaches equilibrium, the potential of the cell (1) decreases (2) increases (3) remains the same.

22. What is the coefficient of Fe^{2+} when the redox equation below is correctly balanced?

$$\_\_\_\ Fe^{2+} + \_\_\_\ NO_3^- + \_\_\_\ H^+ \longrightarrow \_\_\_\ Fe^{3+} + \_\_\_\ NO + 2\ H_2O$$

(1) 1 (2) 2 (3) 3 (4) 4

23. In the reaction $MnO_2 + 4\ HCl \longrightarrow MnCl_2 + Cl_2 + 2\ H_2O$, the oxidation number of oxygen (1) decreases (2) increases (3) remains the same.

24. Which will occur when Sr is oxidized? (1) It will form an isotope. (2) It will attain a positive oxidation number. (3) It will become a negative ion. (4) It will become radioactive.

25. How many moles of Ag^+ will be reduced when 1 mole of Fe^{2+} changes to Fe^{3+}? (1) 1 (2) 2 (3) 3 (4) 4

26. In the reaction $Zn + CuSO_4 \longrightarrow ZnSO_4 + Cu\ (s)$, 1 mole of zinc will (1) lose 1 mole of electrons (2) gain 1 mole of electrons (3) lose 2 moles of electrons (4) gain 2 moles of electrons.

27. A voltaic cell differs from an electrolytic cell in that a voltaic cell (1) produces an electric current by means of a chemical reaction (2) produces a chemical reaction by means of an electric current (3) has oxidation and reduction occurring at the electrodes (4) has ions migrating between the electrodes.

28. In the reaction $2\ KClO_3 \longrightarrow 2\ KCl + 3\ O_2$, the oxidation number of chlorine (1) decreases (2) increases (3) remains the same.

29. When 1 mole of Ni^{3+} changes to Ni^{2+}, the Ni^{3+} undergoes (1) oxidation by losing electrons (2) oxidation by gaining electrons (3) reduction by losing electrons (4) reduction by gaining electrons.

Base your answers to questions 30 and 31 on the following reaction:

$$2\ Cr\ (s) + 3\ Cu^{2+}\ (aq) \longrightarrow 2\ Cr^{3+}\ (aq) + 3\ Cu\ (s)$$

30. The electronic equation that represents the oxidation reaction that occurs is

(1) $2\ Cr^0 - 6\ e^- \longrightarrow 2\ Cr^{3+}$ (3) $2\ Cr^{3+} - 6\ e^- \longrightarrow 2\ Cr^0$

(2) $2\ Cr^0 + 6\ e^- \longrightarrow 2\ Cr^{3+}$ (4) $2\ Cr^{3+} + 6\ e^- \longrightarrow 2\ Cr^0$

31. If 3 moles of Cu react according to the equation above, the total number of moles of electrons transferred will be (1) 1 (2) 2 (3) 3 (4) 6.

Base your answers to questions 32 through 39 on the following diagram and equation.

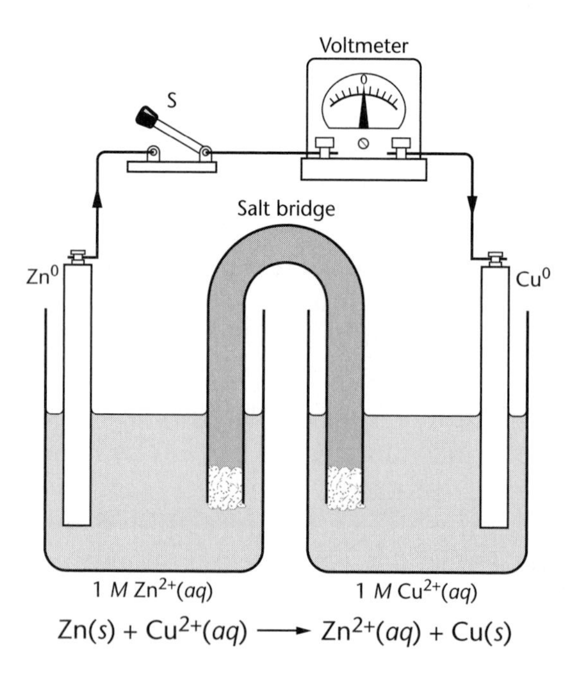

32. The flow of electrons in this cell will be from (1) Cu to Zn (2) Zn to Cu (3) Zn to Zn^{2+} (4) Cu^{2+} to Cu.

33. The anode in this cell is the (1) Zn, and is the negative pole (2) Zn, and is the positive pole (3) Cu, and is the negative pole (4) Cu, and is the positive pole.

34. Which statement about the cell is most accurate?
   (1) Electrons flow through both the wire and the salt bridge.
   (2) Ions flow through both the wire and the salt bridge.
   (3) Electrons flow through the wire, and ions through the salt bridge.
   (4) Ions flow through the wire, and electrons through the salt bridge.

35. In the spontaneous reaction that occurs in this cell, the species reduced is the (1) Cu (2) Zn (3) Cu^{2+} (4) Zn^{2+}.

36. If the salt bridge were removed and switch S closed, the voltage would be (1) 0.00 volts (2) 0.34 volt (3) 0.76 volt (4) 1.10 volts.

37. When the reaction reaches chemical equilibrium, the cell voltage will be (1) 0.00 volts (2) 0.34 volt (3) 0.76 volt (4) 1.10 volts.

38. In this reaction, the substance oxidized is (1) Cu (2) Zn (3) Cu^{2+} (4) Zn^{2+}

39. Based on the table of electrode potentials on page 469, what will be the maximum voltage in the cell when switch S is closed? (1) 0.00 volts (2) 0.34 volt (3) 0.76 volt (4) 1.10 volts

In answering questions 40 through 50 refer to the table of electrode potentials on page 469.

40. Which half-reaction is the standard for the measurement of E^0 voltages?

(1) $Na = Na^+ + e^-$	(3) $Li = Li^+ + e^-$
(2) $H_2 = 2\ H^+ + 2\ e^-$	(4) $2\ F^- = F_2 + 2\ e^-$

41. Which ion can act as both an oxidizing agent and a reducing agent? (1) Na^+ (2) Sn^0 (3) Sn^{2+} (4) Zn^{2+}

42. Which metal will react with 1 *M* HCl? (1) Au (2) Ag (3) Cu (4) Zn

43. In the reaction $Cl_2\ (g) + 2\ Br^-\ (aq) \longrightarrow 2\ Cl^-\ (aq) + Br_2\ (l)$, the E^0 value is equal to (1) −2.42 volt (2) −0.30 volt (3) 0.30 volt (4) 2.42 volts.

44. Given the reaction $Ca\ (s) + 2\ H^+\ (aq) \longrightarrow Ca^{2+}\ (aq) + H_2\ (g)$, the potential ($E^0$) for the overall reaction is (1) −2.87 volts (2) 0.00 volts (3) 2.87 volts (4) 5.74 volts.

45. Which pair of half-cells will produce a cell with the highest potential (E^0)?
(1) Zn, Zn^{2+} (1 *M*) and H_2, H^+ (1 *M*)
(2) Mg, Mg^{2+} (1 *M*) and H_2, H^+ (1 *M*)
(3) Fe, Fe^{2+} (1 *M*) and H_2, H^+ (1 *M*)
(4) Al, Al^{3+} (1 *M*) and H_2, H^+ (1 *M*)

Base your answers to questions 46 through 48 on the following diagrams of the half-cells and on the information in the table of standard electrode potentials on page 469.

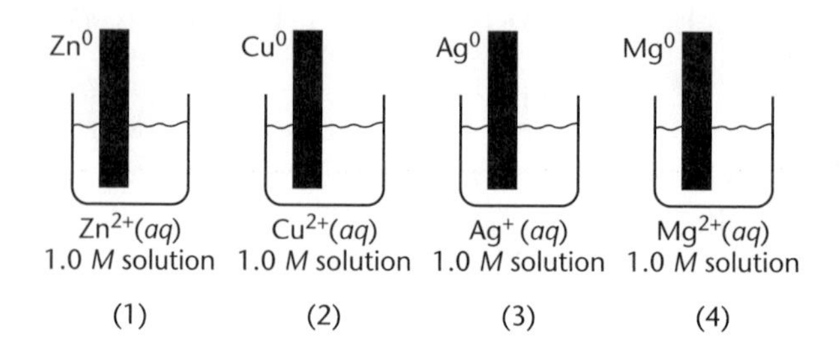

46. Which half-cell contains a metal electrode that is the strongest reducing agent? (1) 1 (2) 2 (3) 3 (4) 4

47. Half-cell 2 is connected to half-cell 3 by means of a wire and a salt bridge. When the cell reaches equilibrium, the voltage will be (1) 1.14 v (2) 0.80 v (3) 0.46 v (4) 0.00 v.

48. Half-cell 3 is connected to half-cell 4 by means of a wire and a salt bridge. Which electronic equation represents the oxidation reaction that occurs?
(1) $Mg^0 + 2e^- \longrightarrow Mg^{2+}$
(2) $Mg^0 \longrightarrow Mg^{2+} + 2e^-$
(3) $Mg^{2+} + 2e^- \longrightarrow Mg^0$
(4) $Mg^{2+} \longrightarrow Mg^0 + 2e^-$

49. Which metal will react with dilute hydrochloric acid to liberate hydrogen gas? (1) Ag (2) Au (3) Cu (4) Mg

50. A standard zinc half-cell is connected to a standard copper half-cell by means of a wire and a salt bridge. Which electronic equation represents the oxidation reaction that takes place?
(1) $Cu^0 - 2e^- \longrightarrow Cu^{2+}$
(2) $Cu^{2+} + 2e^- \longrightarrow Cu^0$
(3) $Zn^0 - 2e^- \longrightarrow Zn^{2+}$
(4) $Zn^{2+} + 2e^- \longrightarrow Zn^0$

51. Which reaction will occur spontaneously at room temperature?

(1) $Co^{2+} + Pb^0 \longrightarrow$ (3) $Sr^{2+} + Ag^0 \longrightarrow$

(2) $2\,Cl^- + I_2{}^0 \longrightarrow$ (4) $2\,I^- + Br_2{}^0 \longrightarrow$

52. In the reaction $Cl_2 + H_2O \longrightarrow HClO + HCl$, chlorine is (1) oxidized only (2) reduced only (3) both oxidized and reduced (4) neither oxidized nor reduced.

CONSTRUCTED RESPONSE

Base your answers to questions 1 to 3 on the following information.

A student has a piece of nickel, a solution of $Ni(NO_3)_2$, a piece of copper, and a solution of $Cu(NO_3)_2$. In order to set up a voltaic cell, he places the strips of metal into the beakers of solution as shown below.

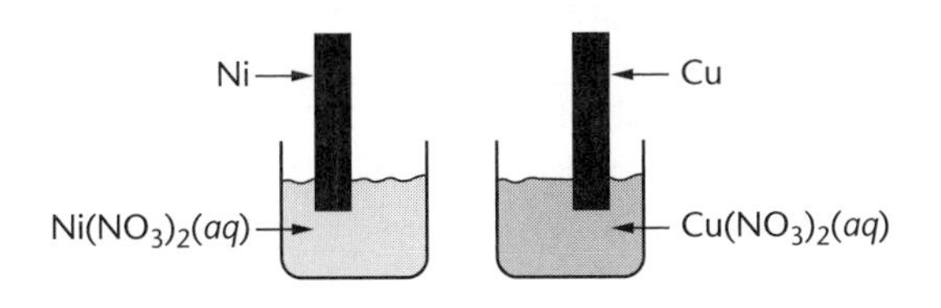

1. What additional equipment is needed to set up a working voltaic cell? Draw the diagram of the cell above, with the necessary additional equipment.
2. The reaction in the cell is $Ni + Cu^{2+} \longrightarrow Ni^{2+} + Cu$.
   (a) In your diagram of the cell, indicate the direction of electron flow.
   (b) What species is reduced in this reaction?
   (c) Write the oxidation half reaction occurring in this cell.
3. The $Cu(NO_3)_2$ solution is blue, due to the presence of Cu^{2+} ions, while the $Ni(NO_3)_3$ solution is green due to the presence of Ni^{2+} ions. The intensity of each color depends on the concentration of each ion in the solution. Predict what color changes would occur in each of these solutions, as the reaction proceeds in the direction indicated in question 2.

4. Given the reaction: $4\ Fe\ (s) + 3\ O_2\ (g) \longrightarrow 2\ Fe_2O_3\ (s)$
   (a) What is the oxidation state of the iron in $Fe\ (s)$?
   (b) What is the oxidation state of the oxygen in Fe_2O_3?
   (c) Write the balanced oxidation half-reaction for this redox reaction.
5. The following redox reaction occurs spontaneously in a voltaic cell:

$$Al + Zn^{2+} \longrightarrow Al^{3+} + Zn \text{ (not balanced)}$$

   (a) Write the half-reaction for the reduction that occurs
   (b) Write the half-reaction for the oxidation that occurs.
   (c) Write the balanced equation for the redox reaction above.
   (d) Name the half-reaction that occurs at the cathode.

Fuel Cells: Energy for the Future

If you have not heard about fuel cells yet, chances are you will start hearing a lot about them soon. A fuel cell is an energy conversion device that uses a chemical reaction between hydrogen and oxygen to generate electricity, with water and heat as the only by-products.

Invented in 1839, fuel cells were initially used mainly in laboratories and, later, in the United States' space program. (Fuel cells were used on Gemini and Apollo flights in the 1960s and are still used on the space shuttle.) But because fuel cells produce "clean" energy (and because of their portability), there is great interest today in using them for applications ranging from cellular phones and laptop computers to large-scale power production. Perhaps most promising is the fuel cell's potential to power cars, replacing the internal combustion engine, which emits harmful air pollutants and greenhouse gases.

A fuel cell, like a battery, consists of electrodes (solid electrical conductors) in an electrolyte (an electrically conductive medium). Many fuel cells use a thin plastic film called a *proton exchange membrane,* or PEM, as the electrolytic medium. Hydrogen gas is fed to the fuel cell's negative electrode, or anode, while oxygen from the air is directed to the cathode on the opposite side of the PEM. At the anode, hydrogen molecules are oxidized, yielding positively charged H^+ ions and negatively charged electrons. The H^+ ions, or protons, can pass through the PEM to the cathode, but to reach the cathode, the electrons must travel around the PEM through an external circuit, generating an electrical current. Once at the cathode, the electrons and H^+ ions combine with oxygen to form water. Essentially, the process is the opposite of electrolysis, in which an electric current is used to split water molecules into oxygen and hydrogen.

An electric car powered by a fuel cell using pure hydrogen as its fuel would be a true ZEV, zero-emission vehicle. However, special high-pressure tanks must be developed that can store enough hydrogen gas to give such a car a useful traveling range. And of course, a network of hydrogen fueling stations does not yet exist. But these hurdles will most likely be overcome in time. Already, auto makers Honda and Toyota have built first-generation fuel cell vehicles; Ford and DaimlerChrysler plan to have similar cars in limited production by 2004. Don't be surprised if, somewhere down the road, you find yourself driving a car powered by a hydrogen fuel cell!

Organic Chemistry

LOOKING AHEAD

Medicines, cosmetics, fossil fuels, plastics, man-made fabrics, and flavorings—these are some of the products developed by organic chemists. Organic chemists study the compounds of one element, carbon, an element common to all known forms of life.

When you have completed this chapter, you should be able to:

- **Provide** formulas and molecular structures for hydrocarbons consisting of up to 10 carbon atoms.
- **Use** the IUPAC naming system to identify a compound from its structure.
- **Define and identify** isomers.
- **Recognize and name** compounds based on a list of common functional groups.
- **Identify and predict** several of the important reactions involving organic substances.

Organic Compounds

In the past, organic compounds were thought to be related only to living things. All other compounds were called inorganic compounds. Although the terms *organic* and *inorganic* still are used, it is now known that organic compounds can be produced outside of living things. In fact, most of the 10 million or more known organic compounds can be and are made in the laboratory.

Organic compounds are compounds that contain carbon. (A few carbon-containing compounds are classified as inorganic compounds. These include carbon dioxide, CO_2, carbon monoxide, CO, and sodium carbonate, Na_2CO_3.) Since carbon occurs in such a huge number of compounds, it is not surprising that carbon compounds are of enormous importance in everyday life. For example, coal, natural gas, and petroleum are naturally occurring substances that contain carbon. Carbon compounds are used in the manufacture of medicines, dyes, and plastics. And, above all, carbon compounds are found in all living things. An understanding of the chemistry of organic compounds is vital to an understanding of the world around you.

Characteristics of Organic Compounds

The vast majority of organic compounds are molecular. Bonds within these molecules are covalent. Bonds between the molecules are intermolecular attractions. As a result of these weak attractions, many organic compounds have relatively low melting points and low solubilities in water. Molecular compounds tend to react more slowly than do ionic compounds. Organic reactions are therefore slow, on the whole, and take more time to occur than is required for inorganic reactions.

The Bonding Behavior of Carbon

The unique bonding behavior of carbon is mainly responsible for the large number of organic compounds that exist. Let us examine this bonding behavior.

Carbon has four valence electrons. Its electronegativity, 2.6, is large enough to prevent carbon from losing electrons to other nonmetals, but not large enough to permit it to take electrons from most metals. Carbon achieves an octet, eight valence electrons, by forming four covalent bonds. As a result, nearly all carbon compounds are molecular.

Carbon atoms also bond readily with other carbon atoms. Organic compounds can contain chains of carbons of enormous length, with molecular masses in the millions of amu. Because of the complexity and variety of organic compounds, chemists have agreed on certain methods of describing and representing them.

Structural Formulas

A **structural formula** shows the arrangement of bonds in a molecule. Some kinds of structural formulas show the structure of a molecule in three dimensions. For example, a molecule of C_2H_6 (ethane) can be shown this way:

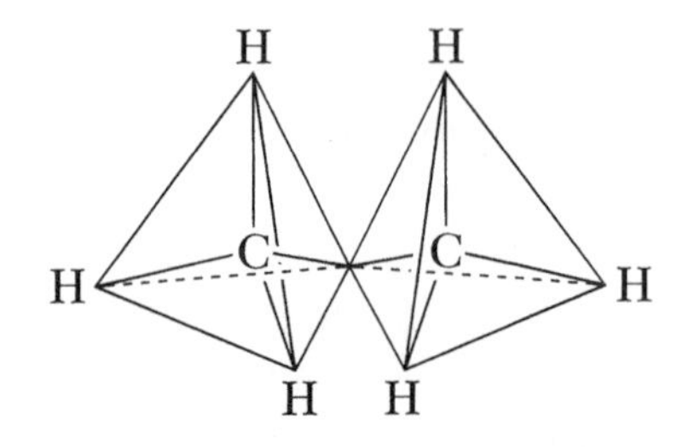

Three-dimensional formulas give more information about molecules than do other kinds of structural formulas. Three-dimensional formulas show the shapes of molecules, for example. However, two-dimensional formulas are commonly used. A two-dimensional formula for ethane is

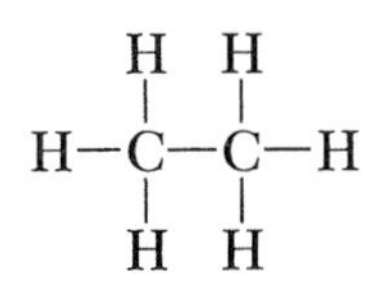

Condensed structural formulas are also often used. The compound pentane

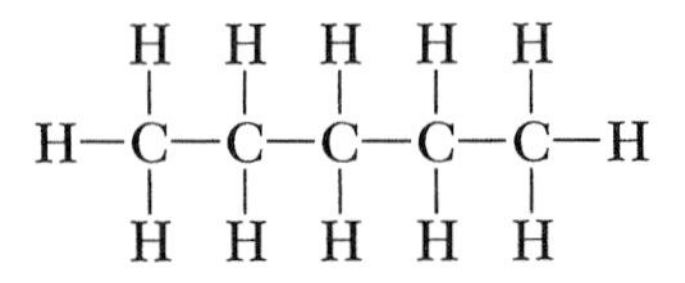

may be written in condensed form as $CH_3(CH_2)_3CH_3$. The condensed formula for the structure

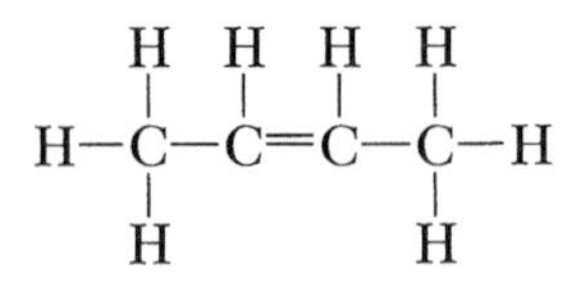

is $CH_3CHCHCH_3$.

Structural formulas are of special importance in organic chemistry. Suppose that you want to write the structural formula for C_2H_6O. According to the bonding rules, C has four bonds, O has two, and H has one. If these rules are followed, two entirely different formulas can be drawn for a molecule of C_2H_6O.

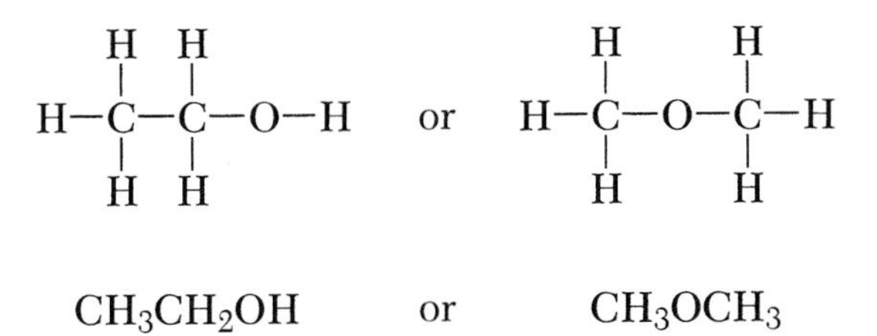

These two compounds have the same number and kinds of atoms but very different structures. Their physical, chemical, and biological properties are vastly different. Compounds that have the same number and kinds of atoms but different structures are called **isomers.** The existence of isomers is one of the reasons that there are so many carbon compounds.

Hydrocarbons

An enormous number of compounds are made up of only carbon and hydrogen. These compounds are called **hydrocarbons.** Hydrocarbons are divided into groups, called **homologous series,** that follow the same general formula. In the three homologous series that follow, each member of the series differs from the next by one carbon atom and two hydrogen atoms.

The Alkanes

The **alkanes** have single bonds between carbon atoms. Typical alkanes are

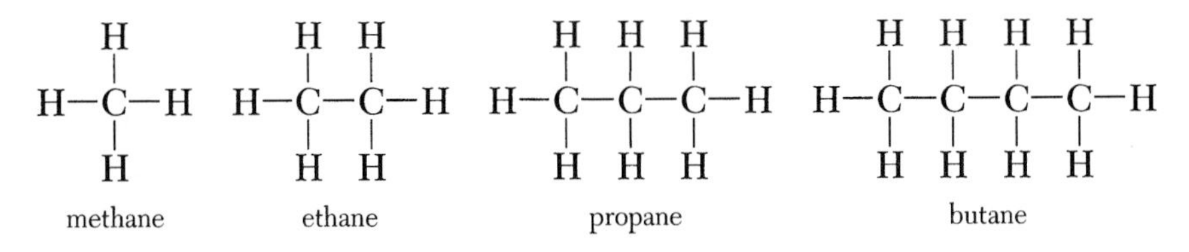

Alkanes have the general formula C_nH_{2n+2}. The general formula shows the relationship between the number of carbon and hydrogen atoms. Following butane in the homologous series above are pentane, C_5H_{12}; hexane, C_6H_{14}; heptane, C_7H_{16}; octane, C_8H_{18};

and so on. Compounds in this family of hydrocarbons contain the suffix *-ane* in their names.

The prefixes, meth-, eth-, prop-, but-, etc., indicate the number of carbons in the chain. You should become familiar with the first ten prefixes, shown in the table below.

Prefix	***Number of Carbon Atoms***
Meth-	1
Eth-	2
Prop-	3
But-	4
Pent-	5
Hex-	6
Hept-	7
Oct-	8
Non-	9
Dec-	10

The Alkenes

The **alkenes** have one double bond between carbon atoms. Alkenes are also called *olefins*. An olefin that has more than one double bond is called a *polyolefin*. Alkenes follow the general formula C_nH_{2n}. The simplest member of this homologous series is ethene, or ethylene. Ethene can be shown by any of the following formulas:

$$C_2H_4 \quad \text{or} \quad \begin{matrix} H & & H \\ & C{=}C & \\ H & & H \end{matrix} \quad \text{or} \quad \begin{matrix} & H & H & \\ & | & | & \\ H{-} & C & {=}C & {-}H \end{matrix}$$

Other members of this homologous series are propene, C_3H_6; butene, C_4H_8; and so on. Compounds in this family of hydrocarbons have the suffix *-ene* in their names.

An example of a polyolefin is 1,3-butadiene. The numbers in the name indicate the positions of the double bonds. The naming of organic compounds will be considered in more detail later in the chapter.

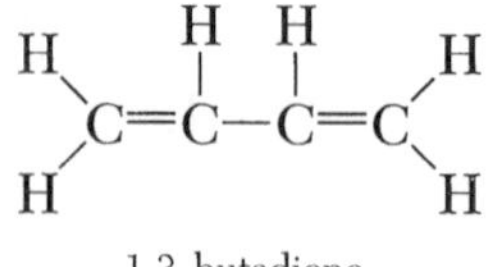

1,3-butadiene

The hydrocarbons that have two double bonds between carbon atoms have the suffix *-diene* in their names.

The Alkynes

The **alkynes** have one triple bond between carbon atoms. Alkynes form a homologous series with the general formula C_nH_{2n-2}. A common alkyne is ethyne, more commonly known as acetylene.

$$C_2H_2 \quad \text{or} \quad H{-}C{\equiv}C{-}H$$

ethyne

Other alkynes are

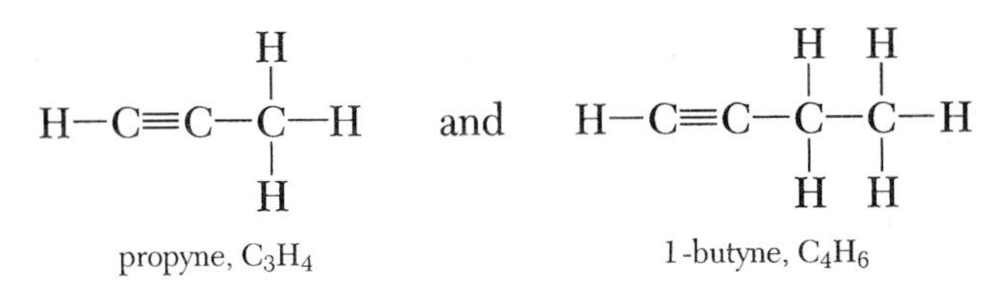

propyne, C_3H_4 and 1-butyne, C_4H_6

The compounds in this family of hydrocarbons have the suffix *-yne* in their names. The 1 in 1-butyne indicates the position of the triple bond.

The Homologous Series of Hydrocarbons table below summarizes the names and general formulas of the three homologous series discussed above. We can use the table of prefixes and the Homologous Series of Hydrocarbons table together to obtain the formulas of many hydrocarbons.

Homologous Series of Hydrocarbons

Name	General Formula	Examples: Name	Examples: Structural Formula
alkanes	C_nH_{2n+2}	ethane	$H_3C{-}CH_3$ (H—C—C—H with H above and below each C)
alkenes	C_nH_{2n}	ethene	$H_2C{=}CH_2$ (C=C with two H on each C)
alkynes	C_nH_{2n-2}	ethyne	$H{-}C{\equiv}C{-}H$

n = number of carbon atoms

SAMPLE PROBLEM

PROBLEM

What is the formula for pentene?

SOLUTION

The table of prefixes tells us that the prefix *pent-* means that there are five carbons. Since pentene ends with the letters *ene* it is an alkene. The Homologous Series of Hydrocarbons table tells us that the general formula for an alkene is C_nH_{2n}. Therefore, pentene is C_5H_{10}.

PRACTICE

14.1 Give the chemical formula of each of the following:

(a) butene

(b) octane

(c) hexyne

(d) decane

14.2 State whether each of the following is an alkane, alkene, or alkyne:

(a) C_7H_{16}

(b) C_2H_2

(c) C_5H_{10}

Derivatives of Hydrocarbons

Hydrocarbons are the simplest kind of organic compounds in that they are made up only of hydrogen and carbon. More complex compounds contain atoms from other elements in addition to hydrogen and carbon atoms. These additional atoms replace one or more hydrogen atoms in a hydrocarbon.

Hydrocarbon Radicals

When a single hydrogen is removed from a hydrocarbon, the remaining structure is called a hydrocarbon radical, often represented by the letter *R*. When these are derived from alkanes, they are called **alkyl groups**, and are named by substituting the letters *yl* for the *ane*. Thus, CH_4 is methane, while CH_3 is called a methyl group.

Functional Groups

When one or more hydrogen atoms of a hydrocarbon are replaced by atoms or groups of atoms of different elements, the replacement atoms are called **functional groups.** Usually the functional group of an organic compound determines the chief characteristics of the compound. Thus different compounds with the same functional group usually have important properties in common. Examples of some functional groups are halogens, alcohols, ethers, ketones, acids, aldehydes, and amines.

The table below lists some of the most important functional groups and their formulas.

Organic Functional Groups

Class of Compound	Functional Group	General Formula	Example
Halide (halocarbon)	—F (fluoro-) —Cl (chloro-) —Br (bromo-) —I (iodo-)	*R*—X (X represents any halogen)	$CH_3CHClCH_3$ 2-chloropropane
Alcohol	—OH	*R*—OH	$CH_3CH_2CH_2OH$ 1-propanol
Ether	—O—	*R*—O—*R′*	$CH_3OCH_2CH_3$ methoxyethane (methyl ethyl ether)
Aldehyde	$-\overset{O}{\overset{\|\|}{C}}-H$	$R-\overset{O}{\overset{\|\|}{C}}-H$	$CH_3CH_2\overset{O}{\overset{\|\|}{C}}-H$ propanal
Ketone	$-\overset{O}{\overset{\|\|}{C}}-$	$R-\overset{O}{\overset{\|\|}{C}}-R'$	$CH_3\overset{O}{\overset{\|\|}{C}}CH_2CH_2CH_3$ 2-pentanone
Organic acid	$-\overset{O}{\overset{\|\|}{C}}-OH$	$R-\overset{O}{\overset{\|\|}{C}}-OH$	$CH_3CH_2\overset{O}{\overset{\|\|}{C}}-OH$ propanoic acid
Ester	$-\overset{O}{\overset{\|\|}{C}}-O-$	$R-\overset{O}{\overset{\|\|}{C}}-O-R'$	$CH_3CH_2\overset{O}{\overset{\|\|}{C}}OCH_3$ methyl propanoate
Amine	$-\overset{\|}{N}-$	$R-\overset{R'}{\overset{\|}{N}}-R''$	$CH_3CH_2CH_2NH_2$ 1-propanamine
Amide	$-\overset{O}{\overset{\|\|}{C}}-\overset{\|}{N}H$	$R-\overset{O}{\overset{\|\|}{C}}-\overset{R'}{\overset{\|}{N}}H$	$CH_3CH_2\overset{O}{\overset{\|\|}{C}}-NH_2$ propanamide

R represents a bonded atom or group or atoms.

The names of some compounds containing these functional groups appear in the "example" column of the table. These names are determined according to a system designed by an international committee of chemists. The names are called the IUPAC names. (The letters stand for *I*nternational *U*nion of *P*ure and *A*pplied Chemistry.)

Naming Organic Compounds

Several different systems have been used to name organic compounds. Isopropyl alcohol, 2-propanol, and dimethylcarbinol are three names for the same substance. This text will focus on the IUPAC names, but will also include some examples of "common names," names that were in use before the IUPAC system was developed.

Naming Hydrocarbons

You are already familiar with the prefixes indicating the number of carbons in a chain and the suffixes indicating the bond type. In Practice 14.1 you were asked to write the formula of butene. Butene is an alkene with one double bond and the formula C_4H_8. However, when you draw the structure of butene, you see that there are two possibilities.

Formula A	Formula B
H H H H | | | | H—C—C=C—C—H | | H H	H H H H | | | | H—C=C—C—C—H | | H H
or	or
$CH_3–CH=CH–CH_3$	$CH_2=CH–CH_2–CH_3$

Both of these molecules have the formula C_4H_8, but they are different substances. Different substances that have the same chemical formula are called isomers. One of the rules of the

IUPAC system is that different substances must have different names. To make clear which structure is intended, the carbon atoms in the molecule are numbered and a number is included in the name of the molecule. The molecule in Formula A is 2-butene, and the molecule in Formula B is 1-butene. The number refers to the location of the carbon atom just before the double bond.

You could also draw the structure of C_4H_8 this way:

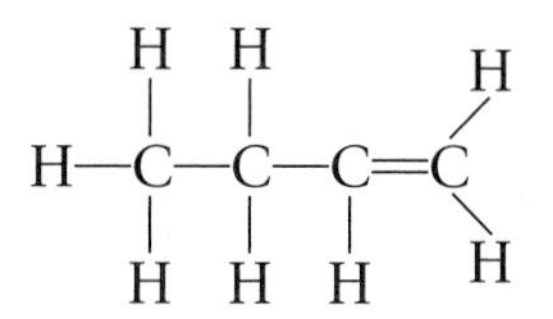

However, if you take this molecule "out of the paper," turn it around, and put it back, it is exactly the same as Formula B. It would still be called 1-butene, and not 3-butene. When you number the carbons in a chain, begin from the end of the molecule that is closest to the double or triple bond. In the two examples that follow, the molecule is named right to left, since that method gives lower numbers in the name.

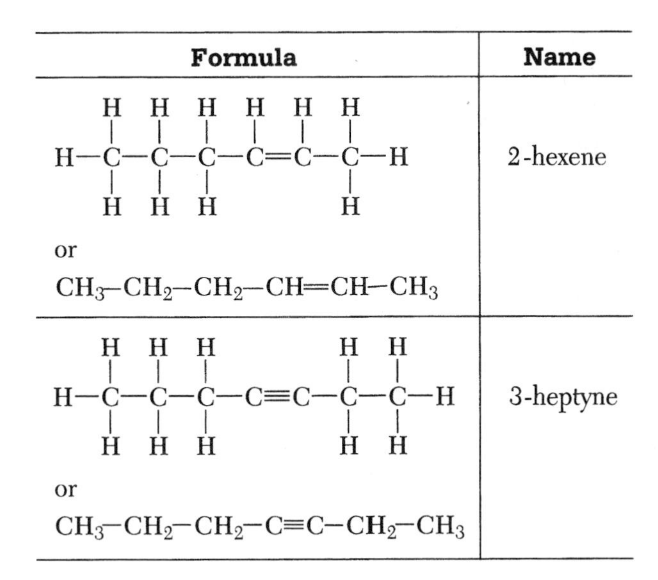

Formula	Name
or $CH_3{-}CH_2{-}CH_2{-}CH{=}CH{-}CH_3$	2-hexene
or $CH_3{-}CH_2{-}CH_2{-}C{\equiv}C{-}CH_2{-}CH_3$	3-heptyne

Some hydrocarbons contain carbon atoms that are not part of the main chain. The main chain is the longest continuous chain of carbon atoms and is the basis of the name. Consider the molecule

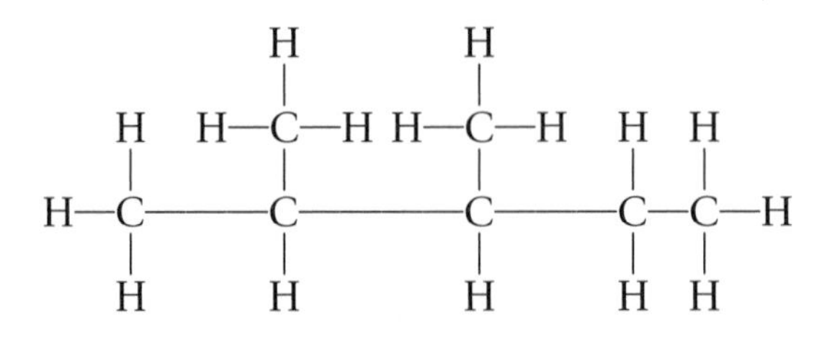

The longest chain of carbon atoms is 5, but there are two additional carbon groups that are not part of that chain. Carbon groups that are not part of the main chain are called side chains. Side chains are named as hydrocarbon radicals; their names contain the suffix *-yl.* Hydrocarbon radicals are also called alkyl groups, which are formed by removing one hydrogen from an alkane. The molecule shown is called 2,3-dimethylpentane. You obtain this name by applying the following rules.

1. Identify the longest continuous chain of carbons, and name that chain. Five single-bonded carbons give us the name *pentane.* (You may wish to refer again to the table of prefixes on page 492.)

2. Identify anything that is attached to the main chain, other than hydrogen, and name it. Use the numerical prefixes *di-, tri-, tetra-, penta-,* and *hexa-* to indicate how many of each species occur in the molecule. (These prefixes are discussed on page 162 in Chapter 4.) In this case, there are two CH_3 groups attached to our main chain. Since CH_3 is called a methyl group, the name becomes dimethylpentane.

3. Use numbers to indicate the location of the attached species. Number the carbons from the side that produces lower numbers in the name. In this case, the compound is called 2,3-dimethylpentane.

Note that there can be more than one group attached to a given carbon. Each group must still be numbered separately, so the name of the compound with the structure

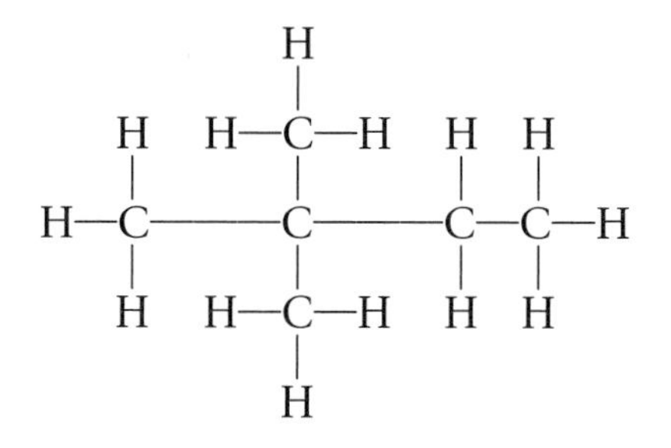

is 2,2-dimethylbutane.

It is important to bear in mind the rule that the names of organic compounds are based upon the longest continuous chain of carbon atoms. For example, there is no organic compound called 2-ethylbutane. If you try to draw such a compound you get

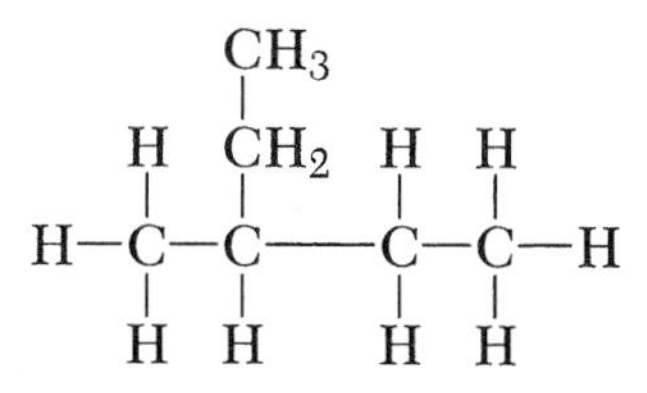

Examining this structure closely, you see that the longest continuous chain of carbon atoms is five. The correct name of the compound is 3-methylpentane.

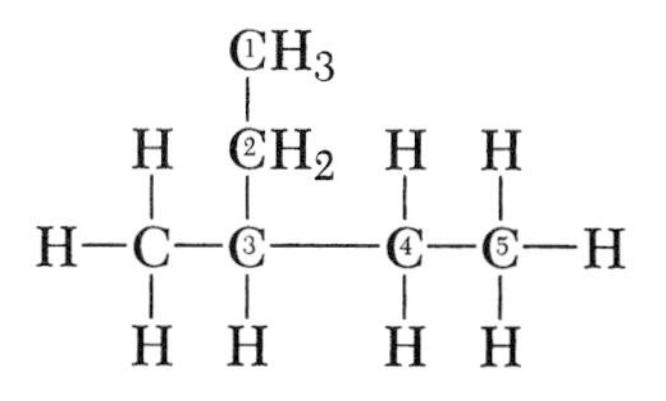

From Name to Molecular Structure

Just as you can systematically obtain the name of a hydrocarbon when given its molecular structure, you can draw the structure of a hydrocarbon when given its IUPAC name.

SAMPLE PROBLEMS

PROBLEM

1. Draw the structure of 2,3-dimethylbutane.

SOLUTION

We know that butane consists of a chain of four carbons, with single bonds.

We begin our structure as C—C—C—C.

There are two methyl groups, each containing one carbon, attached to the chain. One group on the second and one on the third carbon in the chain. The structure would now look like this:

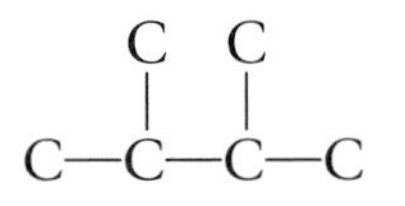

Finally, recalling that a carbon atom forms four bonds, fill in the hydrogens:

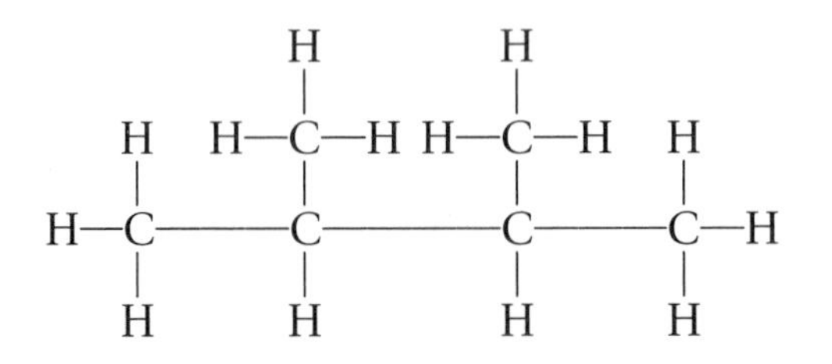

PROBLEM

2. Draw the structure of 1-pentyne.

SOLUTION

The name tells us that there are five carbons, with a triple bond in the first bonding position. Begin by writing

C≡C—C—C—C

Fill in the hydrogens, again being sure that each carbon has exactly four bonds.

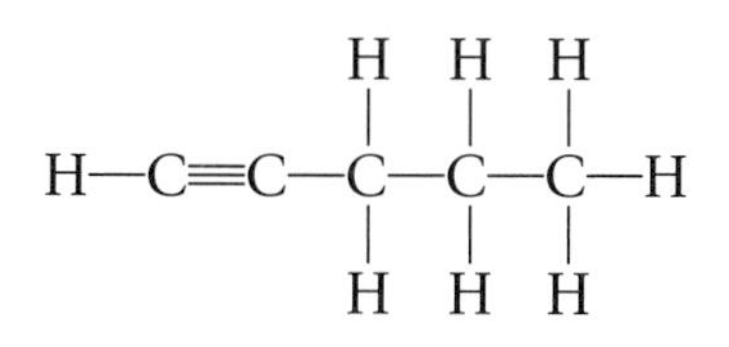

PRACTICE

14.3 Draw molecular structures for each of the following compounds:

(a) 2-pentene

(b) 3-methyl-1-pentene

(c) 2,2,3-trimethylhexane

(d) 2-pentyne

14.4 Name the following hydrocarbons:

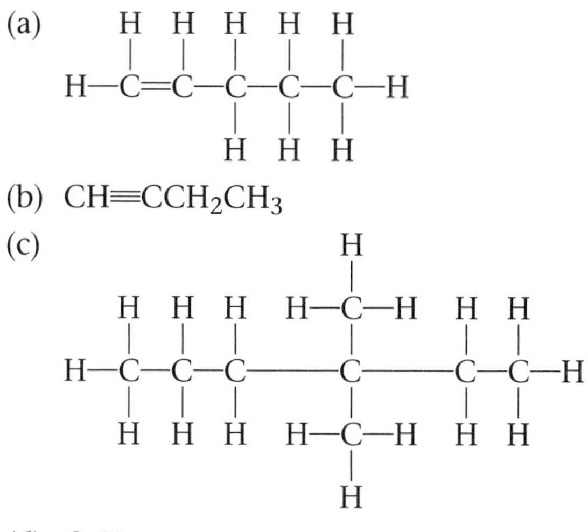

(a)

(b) $CH{\equiv}CCH_2CH_3$

(c)

(d) C_2H_2

14.5 Provide the name of an isomer of each of the following compounds:

(a) hexane

(b) 2,2-dimethylpropane

(c) 1-butene

Naming Compounds with Functional Groups

Each of the functional groups in the table on page 495 has its own system of IUPAC names. The following rules will help you to correctly name a large number of common organic compounds:

1. *Halides (halocarbons).* Organic compounds that contain one or more atoms of the Group 17 elements attached to a hydrocarbon chain are called **halides,** or **halocarbons.** The general formula of the halocarbons is

   $$RX$$

   Compounds that contain fluorine, for example, are called fluorocarbons. Halocarbons are named by replacing the *-ine* of the name of the halogen with the letter *o* and using this as a prefix before the name of the parent hydrocarbon. CH_3F is called fluoromethane. As was the case with alkyl groups, numerical prefixes indicate the number of each type of halogen atom, and numbers indicate the location on the hydrocarbon chain. Note the correct names of the compounds below.

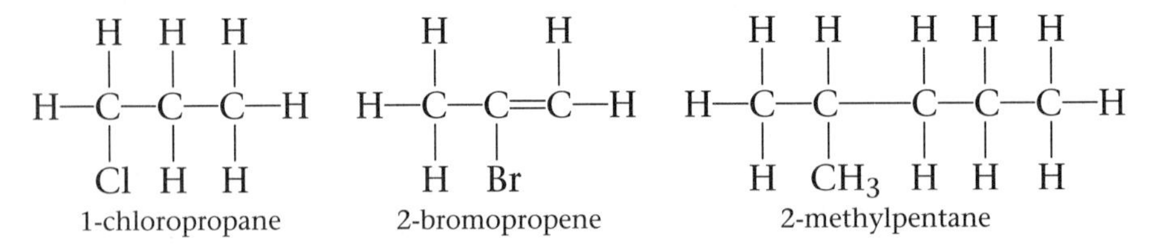

2. *Alcohols.* An organic compound in which one or more hydrogen atoms on a hydrocarbon chain have been replaced by a hydroxyl group, —OH, is called an **alcohol.** Alcohols have the general formula

   $$R\text{—OH}.$$

   (Recall that *R* represents a hydrocarbon radical, a chain of carbons and hydrogen of any length) To name an alcohol, add *-ol* to the hydrocarbon name, and identify the position of the OH group.

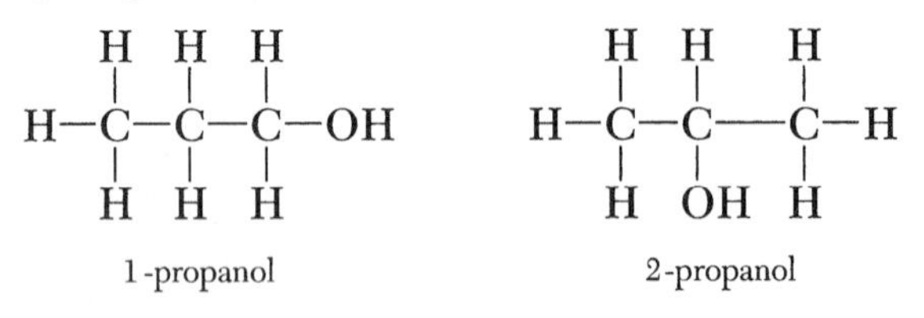

These two molecules are also isomers.

3. *Aldehydes.* An organic compound in which a carbonyl group, —CHO, is located at the end of a hydrocarbon chain is called an **aldehyde.** Aldehydes have the general formula

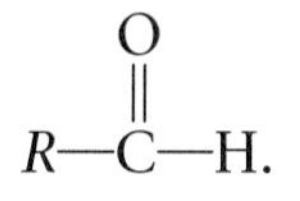

In an aldehyde, the C=O bond is always on an end carbon. Add the suffix *-al* to the root name; common names appear in parentheses.

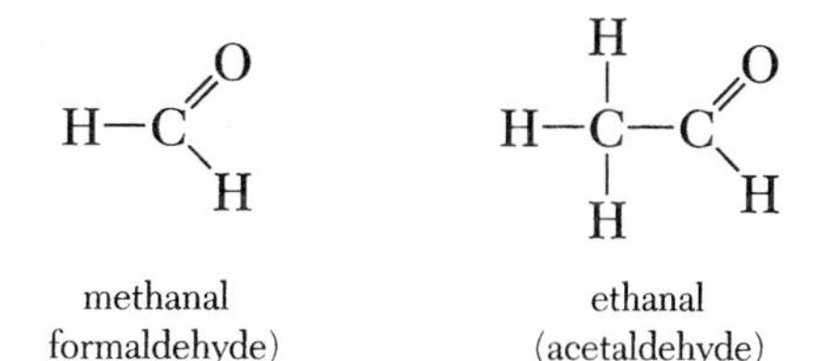

methanal (formaldehyde) — ethanal (acetaldehyde)

4. *Ketones.* An organic compound in which the carbonyl group, —CO—, is located within a hydrocarbon chain is called a **ketone.** The general formula of a ketone is

$$R{-}\overset{\displaystyle O}{\overset{\|}{C}}{-}R'.$$

(*R′* is used to indicate that the two groups on either side of the C=O need not be the same.) Note that the difference between a ketone and an aldehyde is that the C=O bond in a ketone is not on an end carbon. To name ketones, use a number to indicate the location of the oxygen, and add the suffix *-one* to the hydrocarbon root.

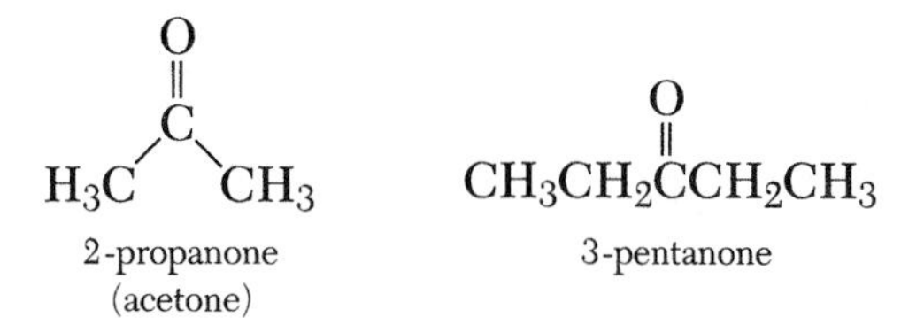

2-propanone (acetone) — 3-pentanone

5. *Ethers.* An organic compound in which an oxygen atom is positioned between two carbon atoms in a hydrocarbon

chain is called an **ether.** The oxygen is attached to carbons on both sides, so the general formula for an ether is

$$R—O—R'.$$

Two methods of naming ethers are still commonly used. One method is to name the two *R* groups in the formula *R*—O—*R'* and add the word *ether.* The method recommended by the IUPAC is to name the shorter of the two groups followed by *-oxy,* followed by the name of the longer group. Both names are given for the ethers shown below.

$CH_3CH_2—O—CH_2CH_3$ $CH_3—O—CH_2CH_3$

diethyl ether
ethoxyethane

methyl ethyl ether
methoxyethane

6. *Acids.* An organic compound that contains a carboxyl group, —COOH, at the end of a hydrocarbon chain is called a **carboxylic acid,** or an **organic acid.** The general formula for an organic acid is

$$R—\overset{\overset{\displaystyle O}{\|}}{C}—OH$$

or *R*COOH. Add the suffix *-oic* and the word *acid* to the root. The prefix describes the total number of carbon atoms.

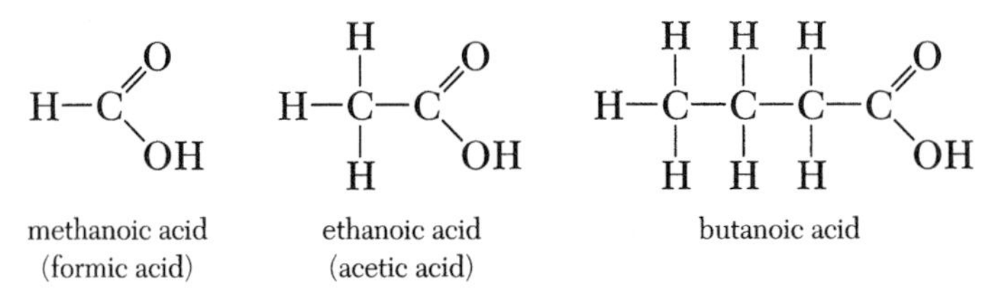

methanoic acid
(formic acid)

ethanoic acid
(acetic acid)

butanoic acid

7. *Amines.* An organic compound derived from ammonia (NH_3) by the replacement of one or more hydrogen atoms by organic radicals is called an **amine.** In an amine, as many as three hydrocarbon radicals can be attached to a nitrogen atom, giving the general formula

$$R—\underset{}{\overset{\overset{\displaystyle R'}{|}}{N}}—R''.$$

The simplest amines are those that contain only one group of carbon atoms; these would have the general formula RNH_2. These amines are named by adding the suffix *-amine* to the name of the base hydrocarbon, with the *e* in the hydrocarbon name dropped. $C_2H_5NH_2$ would be ethanamine. It is also often called ethylamine.

8. *Amides.* An organic compound that contains the carboxyl group, —CO—, and an amine group is called an **amide.** An amide looks like a combination ketone and amine. Its general formula is

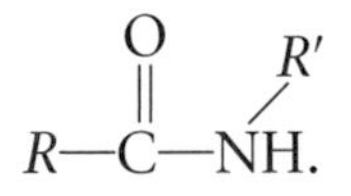

The simplest amides, like the simplest amines, contain only one carbon group, and would then have the formula $RCONH_2$. As with amines, amides are named by dropping the *e* from the name of the base hydrocarbon and adding *-amine* to the end of the name. CH_3CONH_2 is called ethanamide.

The naming system for esters is derived from the reaction used to form them, and will be discussed shortly, in the "Reactions" section of this chapter.

SAMPLE PROBLEM

PROBLEM

Give the correct IUPAC names for these organic compounds:

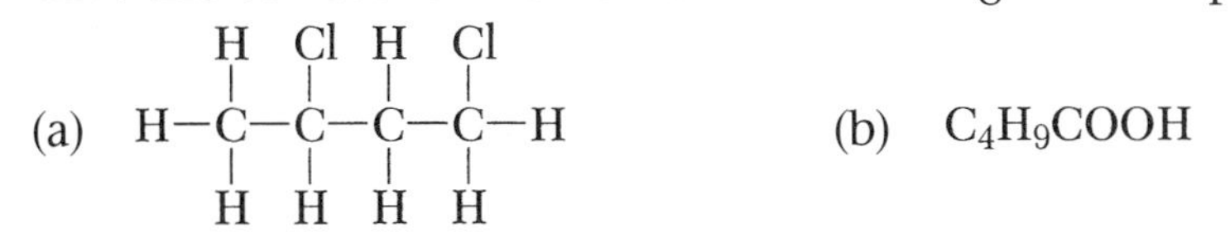

```
        H  CH3  CH3  H  H
        |   |    |   |  |
(c)  H—C—C————C————C—C—H
        |   |    |   |  |
        H  CH3  CH3  H  H
```

(d) $C_2H_5COCH_3$

(e)

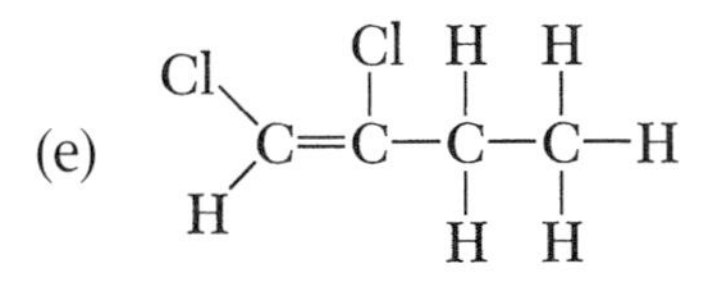

SOLUTION

(a) There are four carbons in the longest continuous chain, and all of the carbon-carbon bonds are single, so the base name is butane. There are two chlorines attached, one to the first carbon and the other on the third. (Numbering right to left produces lower numbers.) The correct name is 1,3-dichlorobutane.

(b) The COOH group identifies this compound as an acid. Since it has a total of 5 carbons, it is pentanoic acid.

(c) The longest chain has five carbons. There are four methyl groups attached, two to the second carbon and two to the third carbon. The name is 2,2,3,3-tetramethylpentane.

(d) The compound fits the general formula *RCOR′*, so it is a ketone. Since it has a total of four carbons, it is called 2-butanone.

(e) The double bond makes the compound an alkene. The name must indicate the locations of the two chlorines as well as the location of the double bond. The correct name is 1,2-dichloro -1-butene.

PRACTICE

14.6 Provide the IUPAC name for the following organic compounds. You may wish to refer to the table of Organic Functional Groups on page 495.

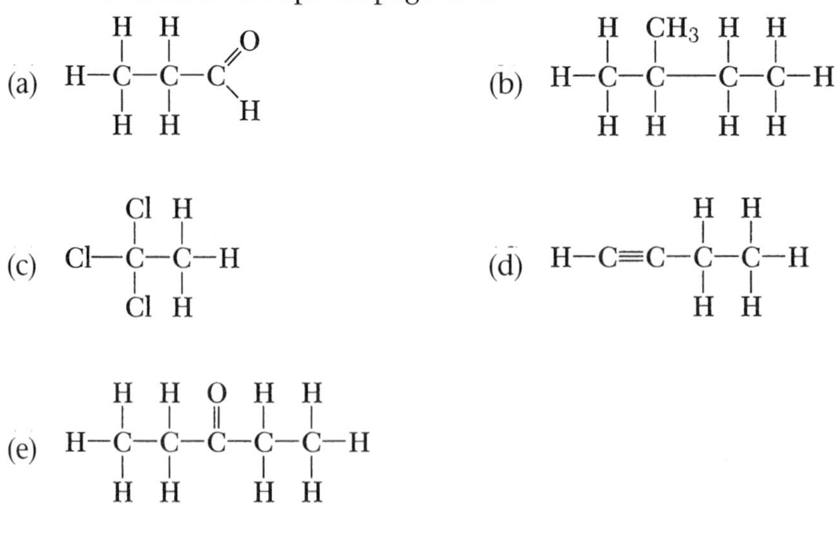

14.7 Draw the structure of each of the following compounds.

(a) butanoic acid
(b) butanal
(c) 2,3-dimethyl pentane
(d) 2-methyl-3-pentanol
(e) 2,2,3,3-tetrabromohexane
(f) 2-pentanone
(g) methoxypropane (methyl propyl ether)

14.8 Why is there no compound called 1-propanone?

Isomers

Examples of isomers have been pointed out in preceding sections. Isomers will now be considered in more detail.

Recall that isomers are molecules that have the same number and kinds of atoms but differ in structure, that is, in the arrangement of their bonds and atoms. In hydrocarbons, different sequences of carbon atoms produce different isomeric forms. Ethane, C_2H_6, has no isomers because only one C—C arrangement is possible.

```
   H  H
   |  |
H—C—C—H
   |  |
   H  H
```

Propane, C_3H_8, also has no isomers because only one C—C—C arrangement is possible.

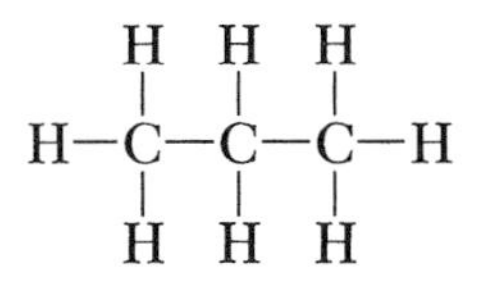

In butane, C_4H_{10}, two sequences of carbon atoms are possible: a continuous chain of carbon atoms, C—C—C—C, and a branched chain,

```
C—C—C
  |
  C
```

The compound with the continuous chain is called the *normal* (*n*-) compound, and the compound with the branched chain is called the *iso*- compound. The IUPAC names are in parentheses.

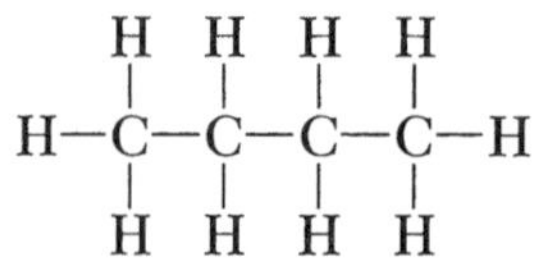

butane
(*n*-butane)

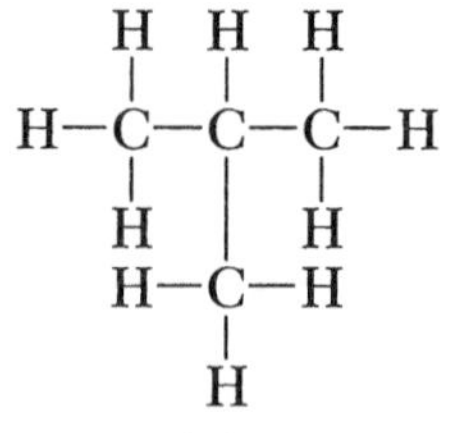

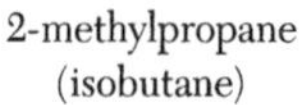

2-methylpropane
(isobutane)

Be careful, though.

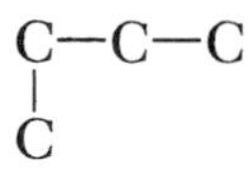

and

C—C—C
    |
    C

are the same as C—C—C—C.

The greater the number of carbon atoms, the greater the number of chain isomers. The compound $C_{20}H_{42}$ has 366,319 possible isomers. The compound $C_{40}H_{82}$ has 62,491,178,805,831 possible isomers.

Structural isomers may also involve the position of functional groups and the position of a double or triple bond.

1. *Position of a functional group.* The molecular formula for both of the following molecules is C_3H_7OH:

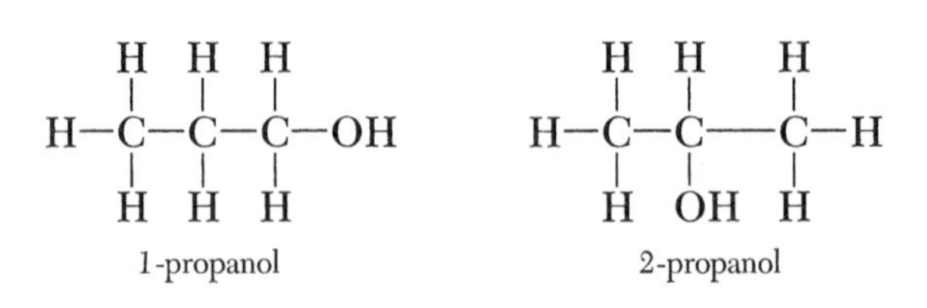

1-propanol 2-propanol

The position of the OH functional group is identified by the number of the carbon atom to which it is bonded.

2. *Position of a double bond.* The molecular formula for both of the following molecules is C_4H_8:

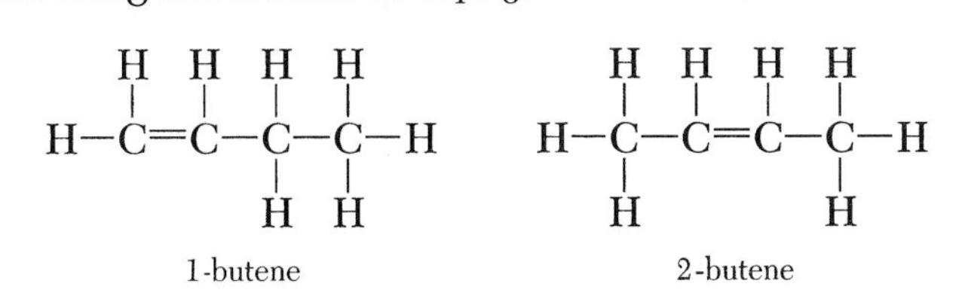

The position of a multiple bond is identified by the number of the carbon atom after which it appears.

3. *Position of a triple bond.* The molecular formula for both of the following molecules is C_4H_6:

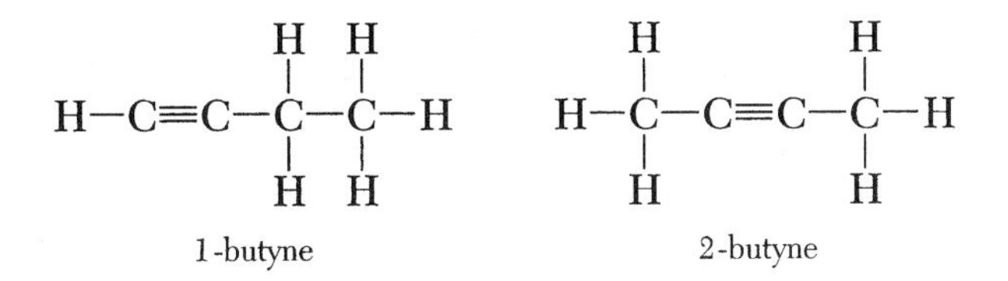

4. *Differences in functional groups or bond arrangements.* Compounds with two double carbon-carbon bonds are called *dienes.* 1,3-Butadiene has four carbons and six hydrogens.

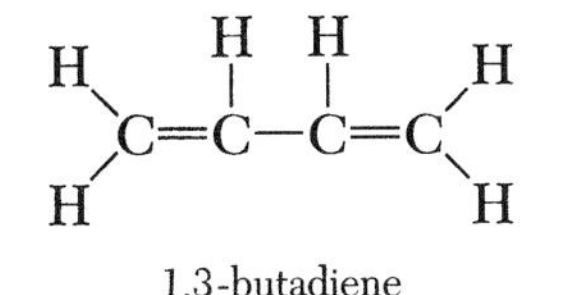

Its molecular formula, C_4H_6, is the same as that of the two butyne molecules. Every diene has an alkyne isomer.

Recall that the difference between an aldehyde and a ketone is the placement of the carbonyl group. Acetone, CH_3COCH_3, is an isomer of the aldehyde propanal, CH_3CH_2CHO. Both substances have the molecular formula C_3H_6O. Every ketone has an aldehyde isomer.

Every ether has an alcohol isomer. For example, dimethyl ether, CH_3OCH_3, is an isomer of ethanol, C_2H_5OH. These isomers, despite having the same molecular formulas, are quite different from each other in their chemical and physical properties.

Some Types of Organic Reactions

Organic compounds take part in a large number of reactions that are often very complex. However, many reactions fall into certain classes, or types. Understanding these types of reactions is of great importance in the analysis, control, and synthesis of organic compounds. The types of reactions you will now consider include substitution, addition, polymerization, esterification, saponification, fermentation, and combustion.

Substitution and Addition

Organic compounds containing molecules in which carbon atoms are bonded by single bonds are called **saturated compounds.** There is no room on such molecules for more atoms or groups of atoms. If one or more atoms (usually H atoms) in the original molecule are removed, however, new atoms or groups of atoms can be substituted for the original atoms. This kind of reaction is called **substitution.**

A reaction in which a halogen is substituted for a hydrogen atom is as follows:

$$CH_4 + Cl_2 \longrightarrow \underset{\text{methyl chloride}}{CH_3Cl} + HCl$$

This halogen substitution reaction takes place when exposed to light or at temperatures of 300°C or greater.

Unsaturated compounds have one or more double or triple bonds between carbon atoms. Two electrons (one single bond or one electron pair) are enough to hold carbon atoms together. The other electrons of double or triple bonds can therefore be made available for bonding to other atoms. It is not necessary to remove H atoms from the original molecule—the new atoms or groups of atoms can be added to the molecule as the double (or triple) bond opens. A reaction of this kind is called **addition.** Two types of addition reactions are halogenation and hydrogenation.

1. *Halogenation.* In **halogenation**, two halogen atoms from the same molecule are added to an opened double or triple bond. The general reaction for the addition of a halogen is

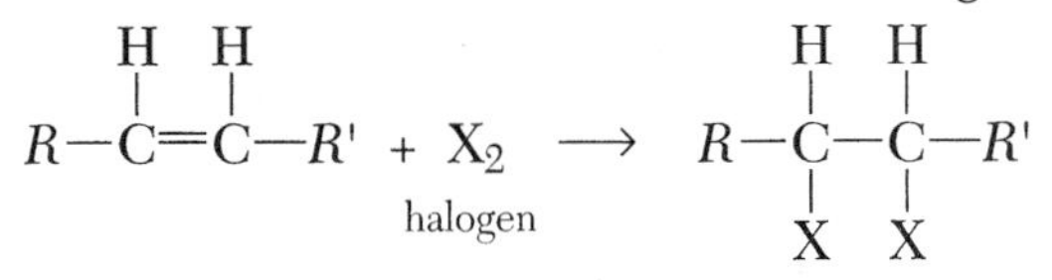

An example of halogenation with bromine is

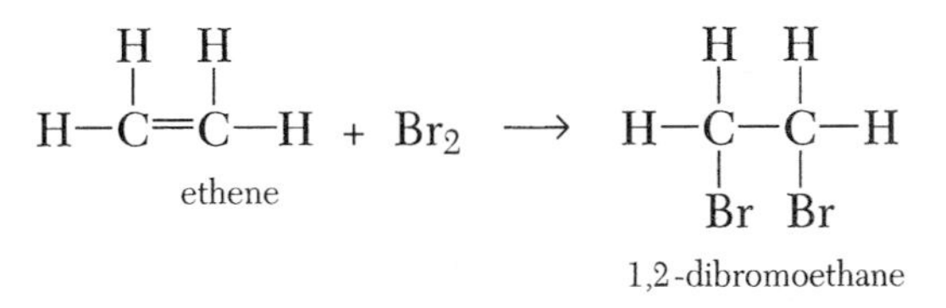

2. *Hydrogenation.* In the **hydrogenation** reaction, the double or triple bond of an unsaturated molecule is saturated with hydrogen atoms. Palladium can be used as a catalyst in the reaction. An example of a hydrogenation reaction is

$$H_3C-CH_2-CH_2-CH=CH-CH_3 \xrightarrow[Pd]{H_2} H_3C-CH_2-CH_2-CH_2-CH_2-CH_3$$

C_6H_{12} (2-hexene) $\qquad$ C_6H_{14} (hexane)

Molecules with double or triple bonds readily enter into many reactions. Such molecules are sometimes hydrogenated to reduce their level of reactivity. Many foods, for example, contain unsaturated molecules and may be hydrogenated to make them less susceptible to spoilage. Solid shortenings and margarines contain hydrogenated vegetable oils.

SAMPLE PROBLEM

PROBLEM

Complete the following chemical reactions and give the correct structure and IUPAC name for each product:

(a) C_3H_6 (propene) + $Br_2 \longrightarrow$

(b) $C_2H_6 + Cl_2 \longrightarrow$

SOLUTION

(a) C_3H_6 is unsaturated and so will react by addition. The double bond opens, and the bromine atoms bond to the available carbon atoms. The product is $C_3H_6Br_2$ and is called 1,2-dibromopropane.

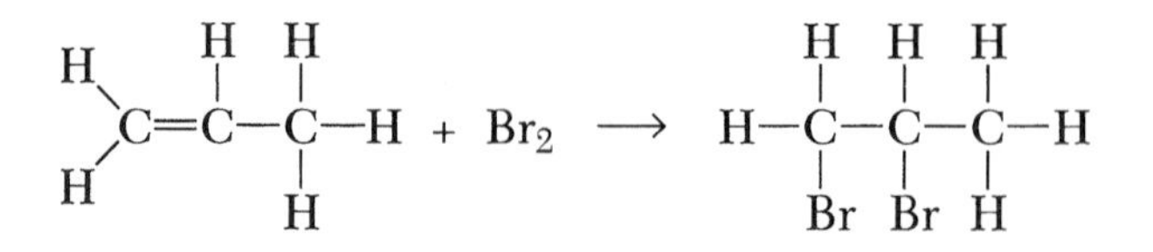

(b) Ethane, C_2H_6, is saturated and reacts with chlorine by substitution. A chlorine atom replaces one of the hydrogen atoms. The other hydrogen atom combines with a remaining chlorine atom to form HCl. The products are HCl, hydrogen chloride, and C_2H_5Cl, chloroethane.

```
   H  H                   H  H
   |  |                   |  |
H—C—C—H  +  Cl2  →  H—C—C—H  +  HCl
   |  |                   |  |
   H  H                  Cl  H
```

PRACTICE

14.9 Complete the chemical reaction and give the correct IUPAC name for each product.

(a) $CH_4 + Cl_2 \longrightarrow$

(b) C_4H_8 (2-butene) $+ Br_2 \longrightarrow$

(c) $C_2H_6 + Cl_2 \longrightarrow$

(d) $C_3H_6 + H_2 \longrightarrow$

(e) $C_2H_2 + 2\ Br_2 \longrightarrow$

Polymerization

Molecules that contain an $—NH_2$ group and a —COOH group separated by a single carbon are called amino acids. Their reactions are extremely important in biochemistry, the branch of chemistry that studies living things. Consider what happens when a two-carbon amino acid, labeled A below, reacts with a three-carbon amino acid, labeled B.

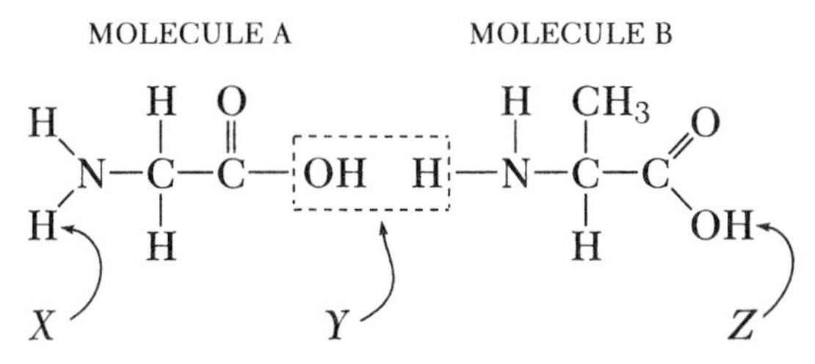

Water splits out between the two molecules, at the site marked *Y* in the diagram. A new bond forms between one carbon atom of molecule A and the nitrogen atom of the other molecule. The new molecule that results is

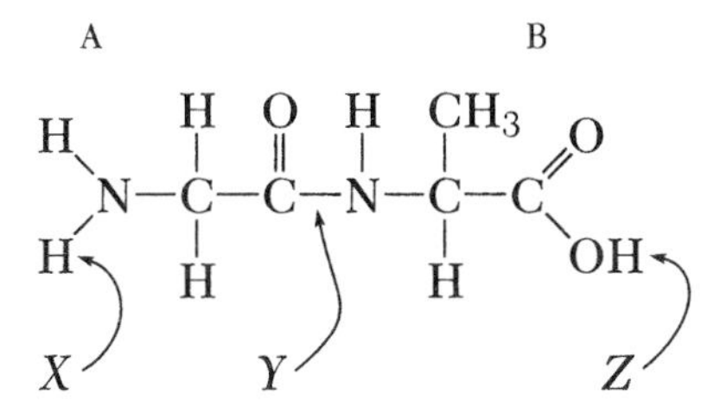

This process can be repeated. The H at site *X* may combine with the OH from another B molecule. Or the OH at site *Z* may combine with an H from another A molecule. A very long chain of alternating A and B units can be formed in this way.

The simple, repeating units of a chain of this type are called **monomers.** Both molecule A and molecule B are monomers. The chain—the molecule formed from the repeating monomers—is called a **polymer.** The process by which a polymer is formed is called **polymerization.**

Polymers are very large molecules with molecular weights in the thousands. Many thousands of monomers may make up a polymer. The polymer just described is natural silk, made by the silkworm. The structural formula for the molecule is as follows. (The letter *n* stands for the number of repeating units.)

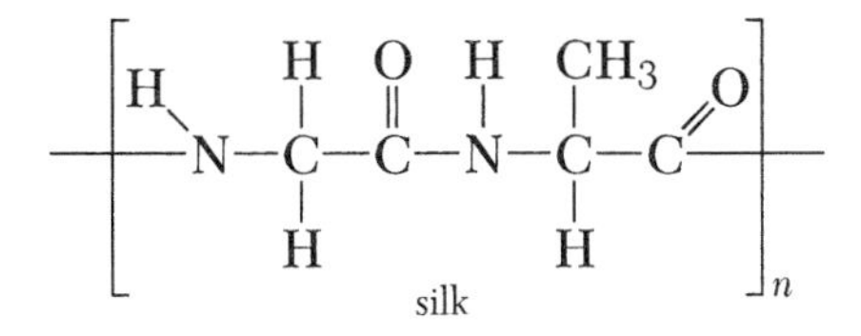

In organic chemistry, the word *condensation* is sometimes used when water splits out between two molecules. Thus the silk molecule is called a *condensation polymer.* The process by which it is formed is called *condensation polymerization.* Many important synthetic products are condensation polymers. Dacron and nylon are examples, as shown by the equations below.

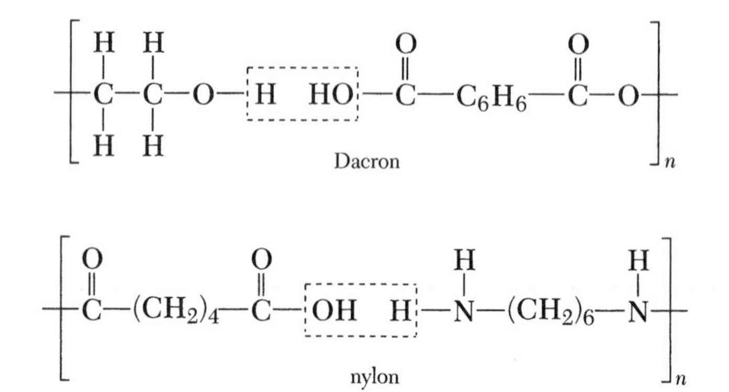

Different types of Dacron and nylon are produced by varying the length of the chain and by various other processes.

Another polymerization process is called *addition polymerization.* In this process, molecules of a single monomer are polymerized. The monomer units must have double bonds, which permit the molecules to link.

An example of this process is the polymerization of ethylene (ethene) to form polyethylene. The structure of an ethylene molecule is

```
H.    .H
 :C::C:
H'    'H
```

Two ethylene molecules may join as follows:

```
  H H H H
  .. .. .. ..
H:C:C:C::C
  .. ..    ..
  H H     H
```

If the process is repeated, polyethylene results.

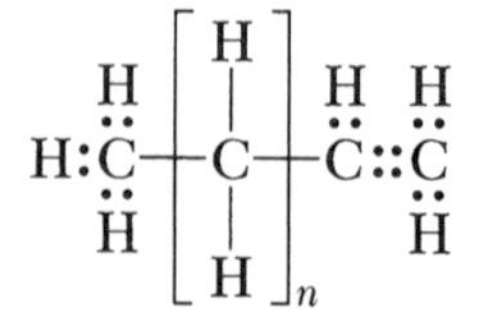

Teflon® and saran are also produced by addition polymerization. If fluorine atoms are substituted in the repeating

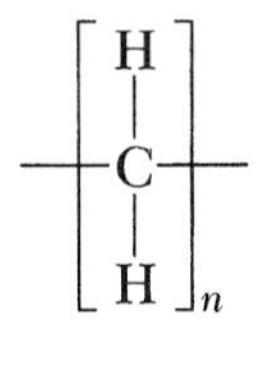

structure, Teflon®

$$\left[\begin{array}{c} F \\ | \\ -C- \\ | \\ F \end{array}\right]_n$$

results. If chlorine atoms are substituted, saran is formed.

Esterification

When an alcohol reacts with an acid, water splits out between the alcohol molecule and the acid molecule. A product called an **ester** forms. The reaction is called **esterification.** The general equation for esterification is

$$\underset{\text{acid}}{R{-}\overset{\overset{\displaystyle O}{\|}}{C}{-}OH} + \underset{\text{alcohol}}{H OR'} \xrightarrow{H_2SO_4} \underset{\text{ester}}{R{-}\overset{\overset{\displaystyle O}{\|}}{C}{-}O{-}R'} + H_2O$$

Sulfuric acid is generally used as a catalyst in esterification reactions. Esters have the general formula *RCOOR'*, or

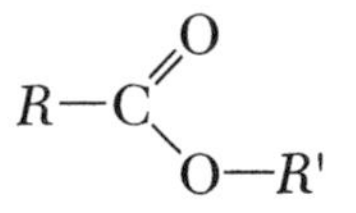

An example of esterification is the reaction between ethanoic acid (acetic acid) and methanol.

$$CH_3COOH + CH_3OH \xrightarrow{H_2SO_4} CH_3COOCH_3 + H_2O$$

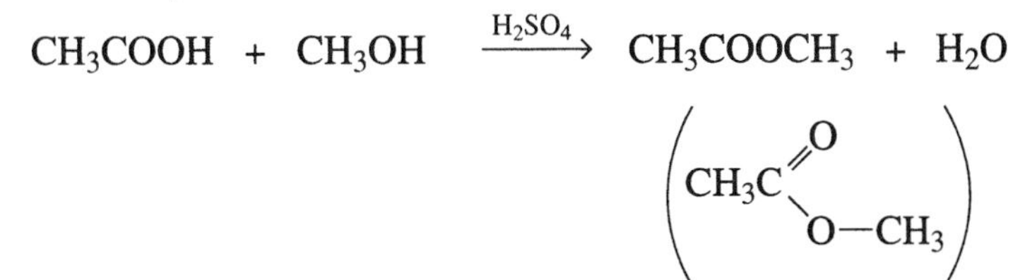

The product of this reaction is called methyl ethanoate. The portion of the ester that came from the alcohol is named first, and the portion that came from the acid is named second. The *-ic* in the acid name is changed to *-ate* in the ester. Another way to look at it is that the chain of carbons attached to a single oxygen is named first with the ending *-yl,* followed by the chain of carbons that is attached to both oxygens with the ending *-ate.* For example, the ester $C_2H_5COOC_2H_5$, which has the structure

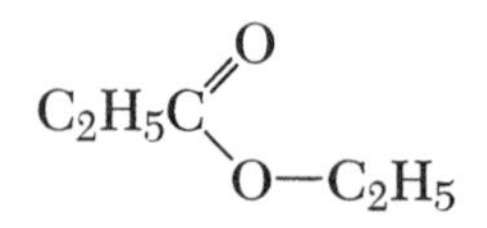

is called ethyl propanoate.

Some esters are responsible for the flavors of fruits. Most of these are still known by their common names; the IUPAC names are included in parentheses. Amyl acetate (pentyl ethanoate) is responsible for the flavor of bananas. Octyl acetate (octyl ethanoate) gives oranges their flavor, and amyl butyrate (pentyl butanoate) accounts for the flavor of apples.

Many animals fats are tri-esters of glycerol (triglycerides). Glycerol is a trihydroxy alcohol; it contains three OH groups. The formula of glycerol is $C_3H_5(OH)_3$. The esterification of glycerol produces fats with the following structure:

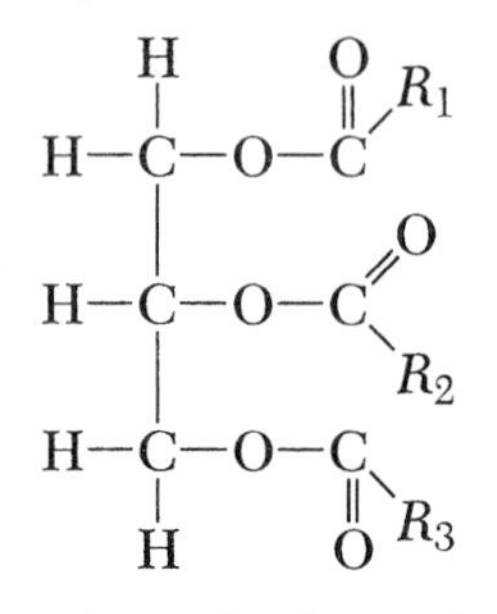

All the R groups are long-chain-hydrocarbon groups from fatty acids, such as stearic acid, $C_{17}H_{35}COOH$. Thus beef fat, glyceryl tri-stearate, has the formula $(C_{17}H_{35}COO)_3C_3H_5$.

Saponification

Tri-esters of glycerol may be broken down by a hydrolysis reaction with NaOH. One of the products of such a reaction—sodium stearate—is a soap. The process is therefore called **saponification** (the prefix *sapon-* means "soap"). Sodium stearate can be produced from beef fat, glyceryl tri-stearate, for example.

$$(C_{17}H_{35}COO)_3C_3H_5 + 3\,NaOH \longrightarrow \underset{\text{sodium stearate}}{3\,C_{17}H_{35}COONa} + \underset{\text{glycerol}}{C_3H_5(OH)_3}$$

The other product of this reaction is glycerol, which, as you know, is also an important material in the home and in industry.

Soaps and detergents The sodium stearate molecule consists of a chain of carbon atoms. One end of the molecule, containing the hydrocarbon portion, is nonpolar. The other end, containing —COONa, is polar.

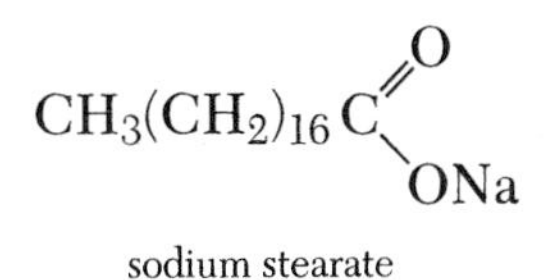

sodium stearate

This structure of the molecule explains the cleansing action of soap. The polar end of the soap molecule is soluble in water. The nonpolar end is soluble in oil. Normally, water and oils cannot mix because of high surface tension between the oil and water. When soap is added to an oil-water system, the polar end of the soap molecule dissolves in the water. The nonpolar end dissolves in the oil. The soap molecule thus lowers the surface tension between the oil and water and permits water to penetrate into the oil film.

Molecules that have both polar and nonpolar properties are classified as *detergents. Soaps* are detergents made up of a metal, such as sodium, linked to a fatty acid radical, such as stearate. Synthetic detergents, called *syndets* or *soapless soaps,* are made without fats. In place of a fat, a long-chain alkyl sulfate group is substituted.

RCOONa	R—O—SO_2—ONa
soap	syndet
R is a long-chain alkyl radical	R is an alkyl radical varying from C_7H_{15} to $C_{18}H_{37}$

Some detergents are *biodegradable.* This means that the large detergent molecules can be broken down by bacteria. The bacteria secrete organic compounds, called *enzymes,* that catalyze the breakdown reaction. The *R* group in biodegradable detergents is a straight chain. An example is

$$CH_3CH_2CH_2CH_2CH_2CH_2CH_2CH_2CH_2CH_2CH_2CH_2—O—SO_2—ONa$$

The enzymes of bacteria cannot catalyze the breakdown of detergents in which *R* is a branched hydrocarbon, as in the following compound:

$$\underset{\displaystyle CH_3}{CH_3–\underset{|}{CH}}–CH_2–\underset{\underset{\displaystyle CH_3}{|}}{CH}–CH_2–\underset{\underset{\displaystyle CH_3}{|}}{CH}–CH_2–\underset{\underset{\displaystyle CH_3}{|}}{CH}–C_6H_4–SO_2–ONa$$

Detergents with branched chains are therefore not biodegradable. Nonbiodegradable detergents add to water pollution.

Fermentation

Enzymes secreted by living organisms are involved in the breakdown of other organic compounds. An enzyme of yeast, for example, catalyzes a **fermentation** process in which a simple sugar is broken down into ethanol and carbon dioxide. Ethanol, you recall, is the alcohol present in all alcoholic beverages.

$$C_6H_{12}O_6\ (aq) \longrightarrow 2\ CH_3CH_2OH\ (l) + 2\ CO_2\ (g)$$

Combustion

Combustion, or burning, is the process in which a compound combines with oxygen. Nearly all organic compounds are combustible. Most of the compounds discussed in this chapter contain only the elements carbon, hydrogen, and oxygen. When these compounds burn in sufficient oxygen, they produce carbon dioxide and water. The complete combustion of ethanol, for example, is

$$C_2H_5OH + 3\ O_2 \rightarrow 2\ CO_2 + 3\ H_2O.$$

The complete combustion of butane is

$$2\ C_4H_{10} + 13\ O_2 \rightarrow 8\ CO_2 + 10\ H_2O.$$

Table I in Appendix 4 on page 594 lists several additional organic combustion reactions. Note that they are all highly exothermic.

SAMPLE PROBLEMS

PROBLEM

1. Write the formula for the missing product (X) in the following reaction:

$$CH_3COOH + C_2H_5OH \rightarrow H_2O + X$$

SOLUTION

First, recognize that this is a reaction between an acid, RCOOH, and an alcohol, ROH. It is esterification. Esters have the general formula RCOOR'. In this case, the ester would be $CH_3COOC_2H_5$.

PROBLEM

2. Write the formula and coefficient for the missing reactant (X) in the following reaction:

$$(C_{17}H_{35}COO)_3C_3H_5 + X \rightarrow C_3H_5(OH)_3 + 3\ C_{17}H_{35}COONa$$

SOLUTION

The products of the reaction are glycerol and a soap, sodium stearate. Saponification is the reaction of a tri-ester with NaOH, sodium hydroxide. Since the product contains 3 sodiums and 3 OH groups, the reactant must be 3 NaOH.

PRACTICE

14.10 Write the formula and give the name of the missing product in each of the following reactions:

(a) $C_3H_6 + Cl_2 \rightarrow X$

(b) $CH_4 + 2\ O_2 \rightarrow CO_2 + X$

(c) $C_6H_{12}O_6 \rightarrow 2\ C_2H_5OH + X$

(d) $C_2H_6 + Cl_2 \rightarrow HCl + X$

(e) $(C_{17}H_{35}COO)_3C_3H_5 + 3\ NaOH \rightarrow 3\ C_{17}H_{35}COONa + X$

14.11 Write the formula and give the name of the missing reactant in each of the following reactions:

(a) $C_2H_5COOH + X \rightarrow H_2O + C_2H_5COOCH_3$

(b) $X + 2\ Cl_2 \rightarrow C_2H_2Cl_4$

(c) $X + Cl_2 \rightarrow HCl + C_4H_9Cl$

(d) $X + 5\ O_2 \rightarrow 3\ CO_2 + 4\ H_2O$

14.12 Draw the structural formulas of the following:

(a) The ester formed when ethanol reacts with methanoic acid

(b) ethyl butanoate

(c) glycerol

(d) the product formed when Br_2 is reacted with 2-butene

TAKING A CLOSER LOOK

The Chemistry of the Functional Groups

The functional group concept is important in the study of organic chemistry. The reactions of a given organic compound depend to a very large extent on the functional group(s) contained in that compound. You will now briefly investigate the chemistry of some of the most important functional groups.

Alcohols

The functional group in an *alcohol* is —OH. Alcohols, like other organic compounds but unlike inorganic bases, are generally nonionic. Like water, alcohols react with some metals to form hydrogen.

$$ROH + Na \rightarrow H_2\ (g) + RONa$$

Examples of some common alcohols are shown in the table below. Ethanol is present in all alcoholic beverages. Phenol is an important aromatic alcohol used in the preparation of many dyes and medicines. Alcohols that contain two —OH groups are called dihydroxy alcohols. An example is ethylene glycol, which is used as a permanent antifreeze in automobiles and as an important starting material for the manufacture of certain polymers, such as Dacron. Alcohols with three —OH groups are called trihydroxy alcohols. Glycerol, used for medicinal products, explosives, and plastics, is an example.

Name	***Formula***
Methyl alcohol (methanol)	CH_3OH or H—O attached to tetrahedral C with three H
Ethyl alcohol (ethanol)	CH_3CH_2OH (C_2H_5OH) or H—C(H)(H)—C(H)(H)—O—H
Phenol	C_6H_5OH or benzene ring—O—H
Ethylene glycol (1,2-ethanediol)	$C_2H_4(OH)_2$ or H—C(H)(OH)—C(H)(OH)—H
Glycerol (glycerin) (1,2,3-propanetriol)	$C_3H_5(OH)_3$ or H—C(H)(OH)—C(H)(OH)—C(H)(OH)—H

Primary, secondary, and tertiary alcohols Alcohols are often divided into three categories: *primary, secondary,* and *tertiary.* The category is determined by the other atoms attached to the carbon with the OH. In a primary alcohol, the OH is attached to an end carbon that has at least two hydrogens also attached to it. The general formula for a primary alcohol is RCH_2OH. Methanol and ethanol are both primary alcohols. 1-Propanol, shown below, is a primary alcohol, but 2-propanol is not.

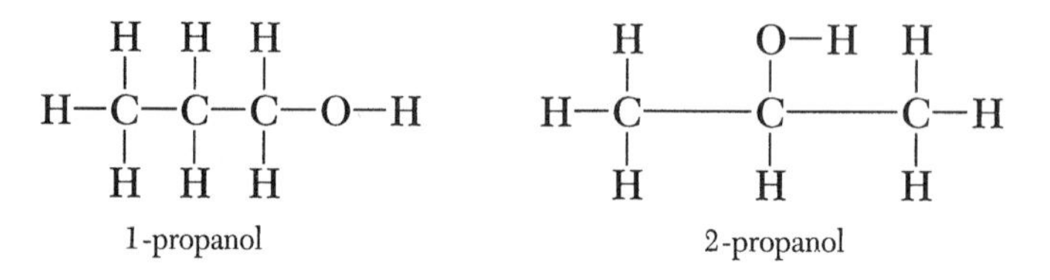
1-propanol 2-propanol

The compound 2-propanol is a secondary alcohol. In a secondary alcohol, the OH is attached to a carbon that has only one hydrogen attached to it. The carbon is generally located in the middle of the chain rather than at the end, so that it has two other carbons attached to it. The general formula for a secondary alcohol is $RCHOHR'$, or

$$R-\underset{\displaystyle H}{\overset{\displaystyle OH}{\overset{|}{\underset{|}{C}}}}-R'$$

R' is used to call attention to the fact that the two functional groups need not be the same. As shown below, 2-butanol is a secondary alcohol, but 2-methyl-2-propanol is not.

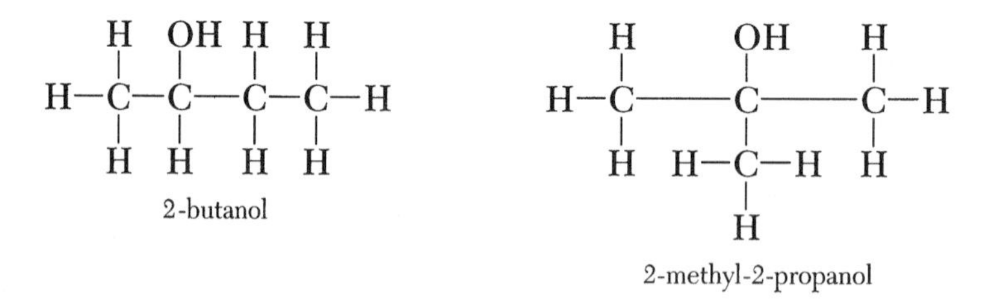
2-butanol 2-methyl-2-propanol

When the carbon attached to the OH has no hydrogens on it, the alcohol is tertiary. The carbon with the OH has three functional groups attached, so that the general formula for a tertiary alcohol is

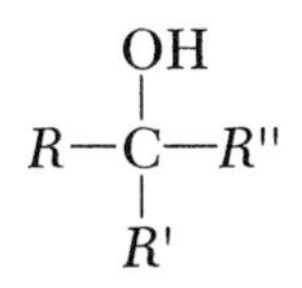

Tert butyl alcohol, or 2-methyl-2-propanol, is a tertiary alcohol. The numbers in the name indicate the locations of the functional groups.

Aldehydes

A carbon atom with a double bond to an oxygen atom has the formula

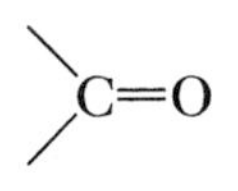

and is called a *carbonyl* group. In an organic compound called an *aldehyde,* the functional group is a derivative of a carbonyl.

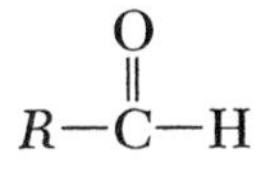

The simplest aldehyde has a carbonyl group and two H atoms bonded to the C atom—the first compound in the table. Other aldehydes have an H atom on one side of the carbonyl and an organic *R* group on the other side.

Name	**Formula**
Methaldehyde, or methanal (formaldehyde)	$HCHO$ or H—C(=O)—H
Acetaldehyde (ethanal)	CH_3CHO or H_3C—C(=O)—H
Cinnamaldehyde	$C_6H_5(CH)_2CHO$ or C_6H_5—CH=CH—C(=O)—H

Notice that in an aldehyde, the oxygen is always attached to an end carbon in a chain. Aldehydes are prepared through the oxidation of primary alcohols.

Some aldehydes have distinctive odors and flavors. Some are used in perfumes. Formaldehyde is used as a preservative. In 40 percent solution, it is called formalin. The cinnamaldehyde molecule is responsible for the flavor of cinnamon.

Ketones

Ketones, like aldehydes, contain the carbonyl group,

```
   O
   ‖
 —C—
```

However, in ketones, the carbonyl group is located in the interior of the carbon-carbon chain, not at the end. A common ketone is acetone, also called 2-propanone. Acetone has the formula

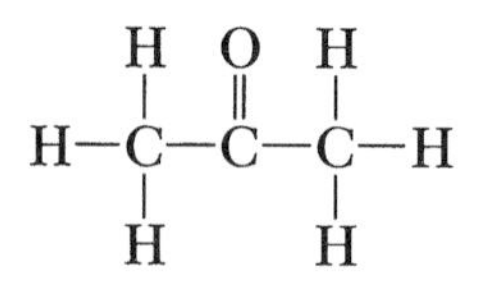

Acetone is an important commercial solvent. Ketones are prepared through the oxidation of secondary alcohols.

PRACTICE

14.13 Identify each of the following compounds as an alkane, alkene, alkyne, ketone, aldehyde, primary alcohol, or secondary alcohol.

```
(a)  H          H
      \         |
       C=C—C—H
      /   |  |
     H    H  H

(b)    H  H  H  OH  H
       |  |  |  |   |
     H—C—C—C—C——C—H
       |  |  |  |   |
       H  H  H  H   H

(c)    H  O  H  H
       |  ‖  |  |
     H—C—C—C—C—H
       |     |  |
       H     H  H

(d)     O  H  H
        ‖  |  |
        C—C—C—H
       /   |  |
      H    H  H
```

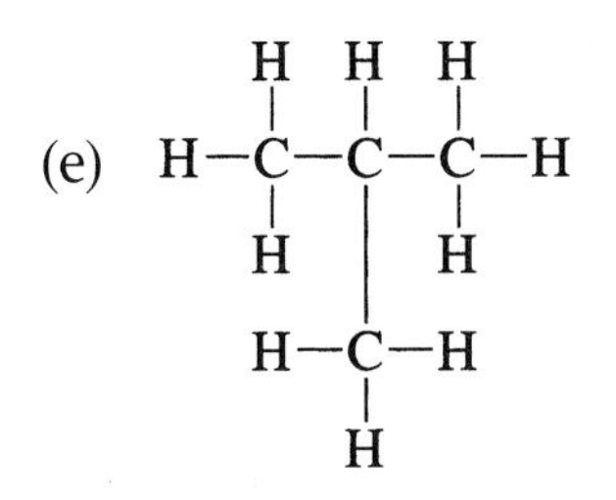

14.14 Draw structural formulas for each of the following compounds:

(a) butane (b) propyne

(c) 1-butanol (d) 1-pentene

Ethers

Ethers correspond to oxides in the inorganic compounds. Ethers have two *R* groups, which may be the same or different. The ether used in hospitals, called diethyl ether, has the formula

$$(CH_3CH_2)_2O \quad \text{or} \quad \begin{array}{ccccccccccc} & & H & & H & & & & H & & H & & \\ & & | & & | & & & & | & & | & & \\ H & - & C & - & C & - & O & - & C & - & C & - & H. \\ & & | & & | & & & & | & & | & & \\ & & H & & H & & & & H & & H & & \end{array}$$

In the vanillin molecule, present in vanilla, there are three functional groups around a benzene ring: an aldehyde, an alcohol, and an ether.

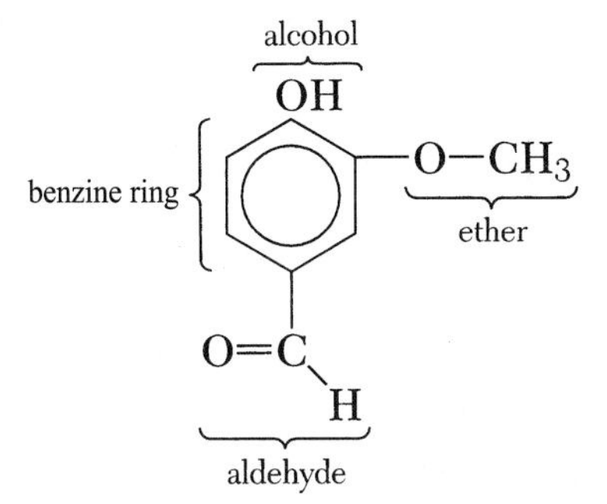

Ethers may be prepared through the dehydration of alcohols. As shown below, the dehydration of ethanol produces diethyl ether.

$$2\ C_2H_5OH \longrightarrow C_2H_5OC_2H_5 + H_2O$$

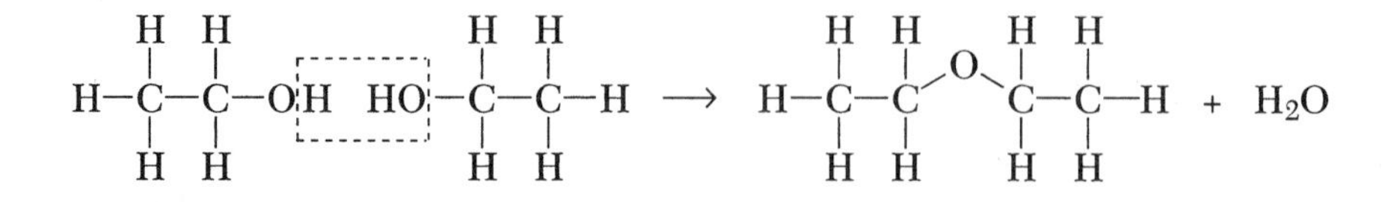

Acids

The functional group in *organic acids,* which are also called carboxylic acids, is the *carboxyl* group.

$$-C(=O)OH$$

An organic acid may have one or more carboxyl groups, as in oxalic acid.

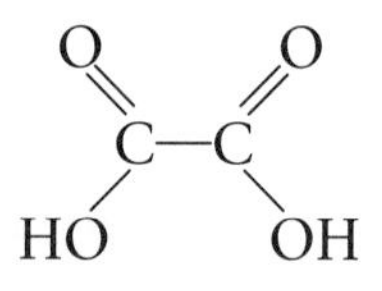

Acids may be produced through the oxidation of primary alcohols, or the oxidation of aldehydes. The oxidation of ethanol produces acetic acid.

$$C_2H_5OH + O_2 \longrightarrow CH_3COOH + H_2O$$

This reaction proceeds slowly at room temperature. However, fine wines are often stored for several years. If the bottles are not tightly sealed, the ethanol in the wine may be oxidized to acetic acid, which ruins the taste of the wine. Vinegar is a dilute solution of acetic acid. When a wine tastes vinegary it is due to the oxidation of ethanol to acetic acid.

Most organic compound are nonelectrolytes. Many are insoluble in water, and most of the soluble compounds, such as the smaller alcohols and ketones, do not form ions in water. However, organic acids are indeed acids; they do ionize in water to produce H_3O^+ ions. For example, $CH_3COOH + H_2O \rightleftharpoons H_3O^+ + CH_3COO^-$.

Organic acids are generally weak acids, and ionize only slightly. They are considered weak electrolytes.

Amines

The functional group in *amines* is $—NH_2$. Amines can be pictured as derivatives of ammonia, NH_3, in which one of the hydrogen atoms has been replaced by an organic *R* group.

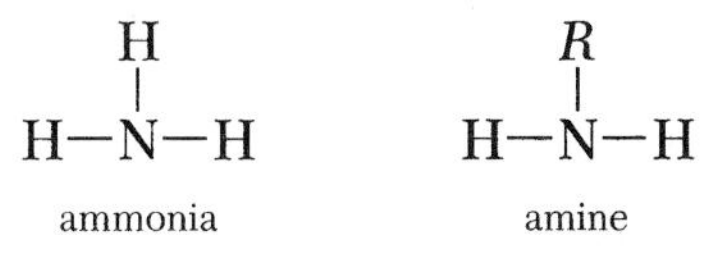

Examples of amines are methylamine, CH_3NH_2, and phenylamine, $C_6H_5NH_2$, which is also called aniline.

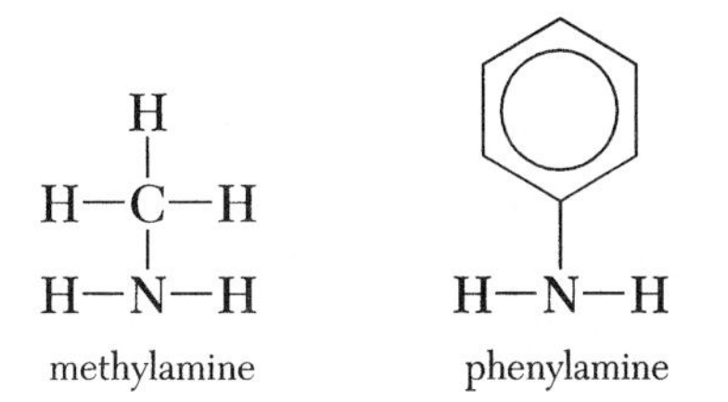

Amines are used in the manufacture of dyes, medicines, perfumes, and other useful products.

Amino Acids and Proteins

Amino acids have the general formula

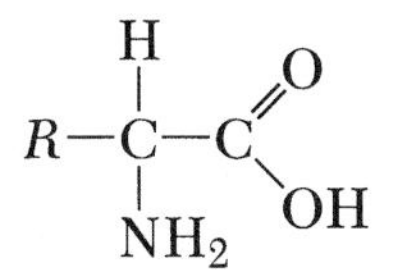

Amino acids are the building blocks of proteins. Proteins are complex organic compounds that contain carbon, oxygen, hydrogen, nitrogen, and, often, sulfur and phosphorus.

Proteins are derived from amino acids in the following type of reaction:

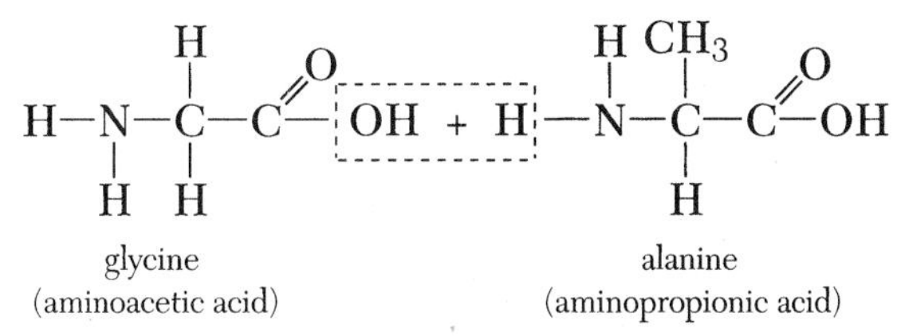

In this reaction, water splits off between the two amino acids and a new bond forms between them. The single molecule that results from the joining of the two amino acids is called a *dipeptide.*

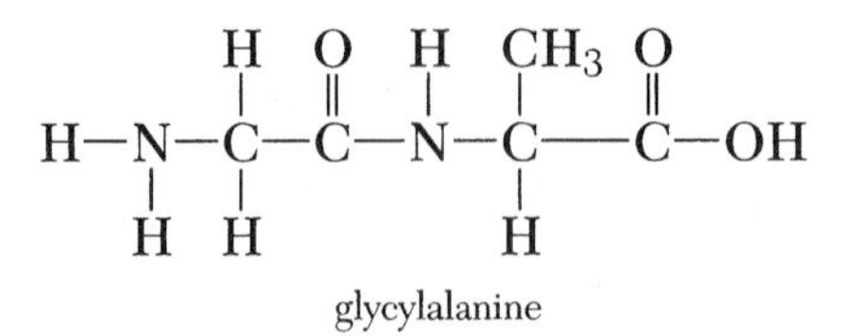

glycylalanine

The bond formed by the reaction that links the two amino acids is called a *peptide linkage.*

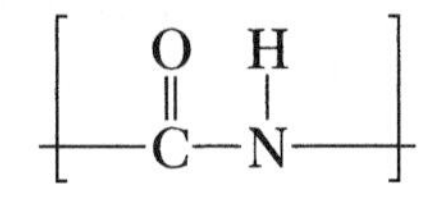

When several amino acids join in this way, the resulting compound is called a *polypeptide.* Proteins are polypeptides. They have molecular weights that range from 10,000 to many millions. A section of a protein can be represented like this:

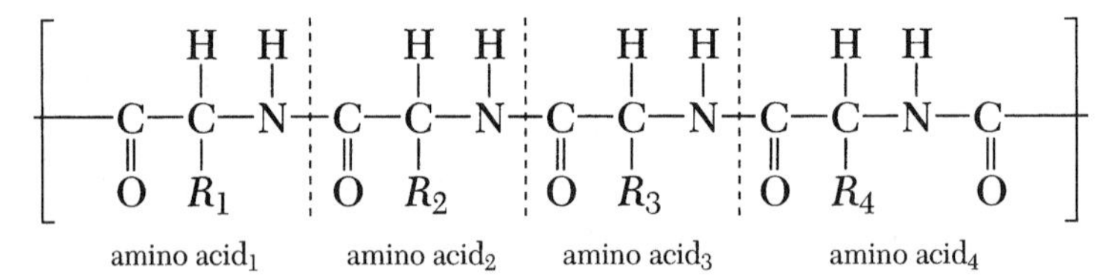

amino acid$_1$ amino acid$_2$ amino acid$_3$ amino acid$_4$

The *R* groups can stand for parts of different amino acids. As you can see, a protein molecule may be very complex.

There are 20 or so amino acids that may enter into the formation of human proteins. The kinds of amino acids present and their arrangement in a molecule determine the particular kind of protein. A human insulin molecule, for example, has two long, connected chains of amino acids. The total number of amino acids in the chains is 51; 16 different kinds of amino acids are represented in this total. Amino acids may combine in an almost endless variety of ways to form an enormous number of proteins.

Cis–Trans Isomerism

For convenience, structural formulas have thus far been represented as two-dimensional. However, recall that the carbon atom is tetrahedral. The structures of molecules of carbon compounds are therefore three-dimensional. Since this is so, two molecules may have the same sequence of carbon atoms, but the atoms in each molecule may have a different orientation in space. For example, the following molecules have the same composition ($C_2H_2Cl_2$), but they have different properties. Molecule A is a polar molecule, and molecule B is nonpolar.

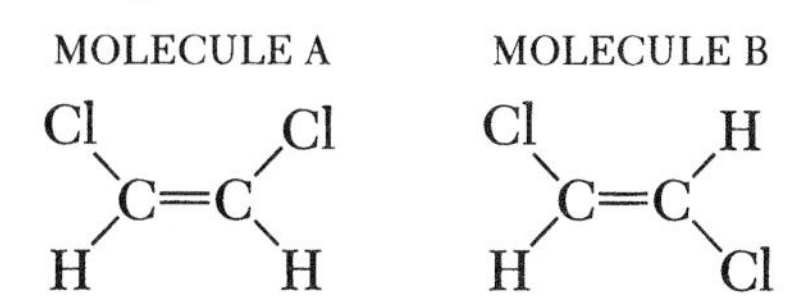

These two molecules are examples of a different kind of isomerism: They are *cis–trans* isomers. Notice that each molecule has a double bond. Also, each molecule has two chlorine atoms substituted for two hydrogen atoms. When the two substituted atoms or groups of atoms are on the same side of the double bond, they are in the *cis* position. When they are on opposite sides of the double bond, they are in the *trans* position. Thus molecule A is called cis-dichloroethylene, and molecule B is called trans-dichloroethylene.

Cis–trans isomerism exists only when double or triple bonds are present. It occurs because rotation around double and triple bonds is restricted. Cis–trans isomerism cannot exist in a molecule that has only single bonds between carbon atoms. In such molecules, the atoms rotate freely around the single bond. According to the laws of probability, therefore, any substituted atoms spend an equal amount of time on the same side and on opposite sides of the molecule. The following structures, for example, are equivalent. Any of these structural formulas may be used to represent 1,2-dichloroethane.

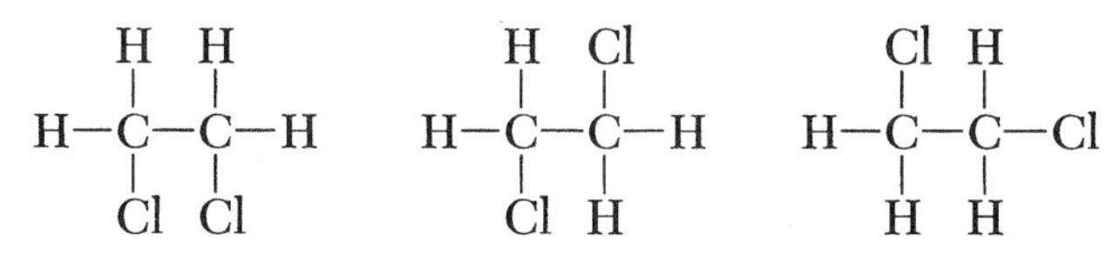

PRACTICE

14.15 Draw and name five different compounds with the formula C_6H_{14}.

14.16 Draw and name one isomer of 2-butanone.

CHAPTER REVIEW

The following questions will help you check your understanding of the material presented in the chapter.

1. Which represents the functional group of an organic acid? (1) —COOH (2) —O*R* (3) —CHO (4) —NH_2
2. Which formula represents a member of the same homologous series as C_8H_{14}? (1) C_3H_4 (2) C_3H_5 (3) C_3H_6 (4) C_3H_8
3. Which is an isomer of

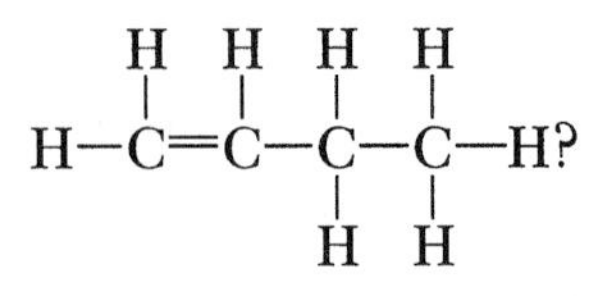

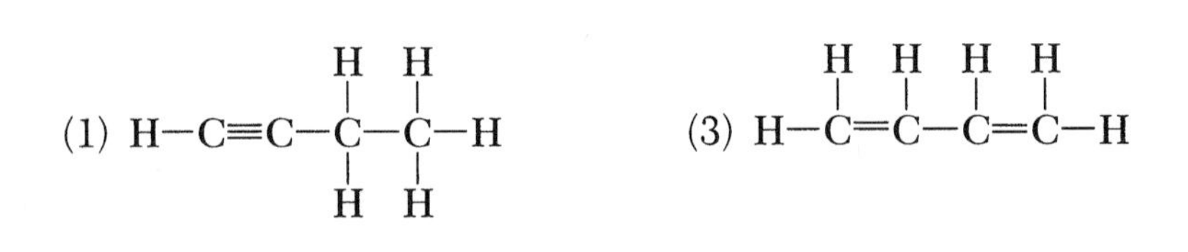

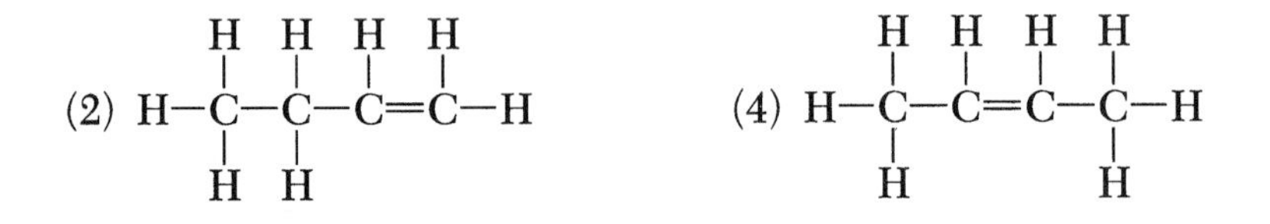

4. One of the products produced by the reaction between CH_3COOH and CH_3OH is (1) HOH (2) H_2SO_4 (3) HCOOH (4) CH_3CH_2OH.
5. Which molecule contains four carbon atoms? (1) ethane (2) butane (3) methane (4) propane
6. Which formula may represent an unsaturated hydrocarbon? (1) C_2H_6 (2) C_3H_6 (3) C_4H_{10} (4) C_5H_{12}

7. Which is the correct structural formula for methanol?

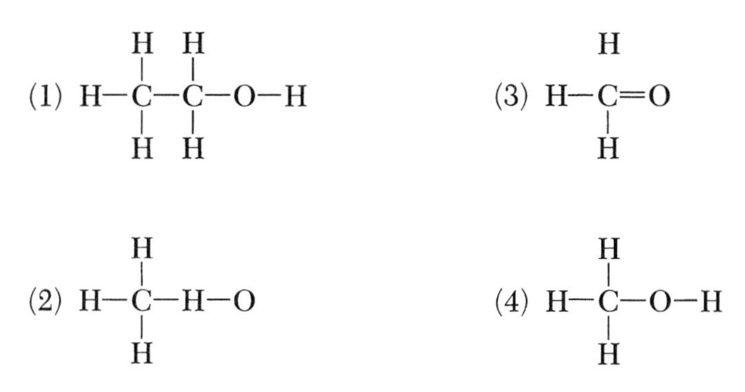

8. What is the correct IUPAC name of the compound represented by the following structural formula?

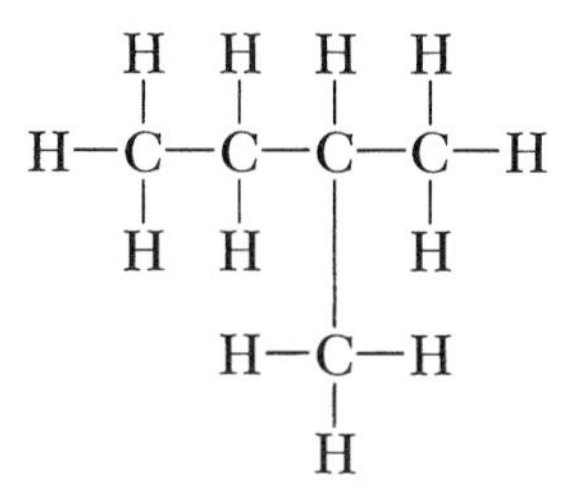

(1) pentane (2) isobutane (3) 2-methylbutane (4) butane

9. How many carbon atoms are contained in an ethyl group?
(1) 1 (2) 2 (3) 3 (4) 4

10. What is the correct formula of 1,1-dibromoethane?

```
        Br  H                      H   H
        |   |                      |   |
(1) H—C—C—H              (3) Br—C—C—B
        |   |                      |   |
        H   Br                     H   H

        Br  Br                     Br  H
        |   |                      |   |
(2) H—C—C—H              (4) H—C—C—H
        |   |                      |   |
        H   H                      Br  H
```

11. Which structural formula represents an alcohol?

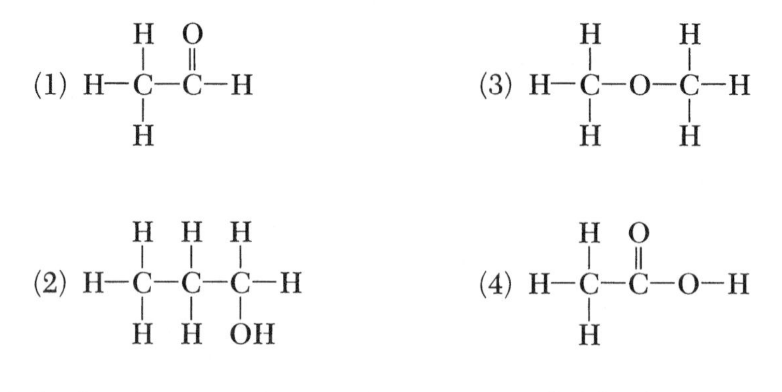

12. Which compound can have isomers? (1) C_2H_4 (2) C_2H_2 (3) C_2H_6 (4) C_4H_8

13. In the alkane series, the formula of each member of the series differs from its preceding member by (1) CH (2) CH_2 (3) CH_3 (4) CH_4

14. Which organic compound is a product of a saponification reaction? (1) CCl_4 (2) $C_3H_5(OH)_3$ (3) C_6H_6 (4) $C_6H_{12}O_6$

15. Which is an example of a homologous series? (1) CH_4, C_2H_4, C_3H_8 (2) CH_4, C_2H_2, C_3H_6 (3) CH_4, C_2H_6, C_3H_6 (4) CH_4, C_2H_6, C_3H_8

16. Which hydrocarbon is saturated? (1) C_2H_6 (2) C_3H_4 (3) C_4H_6 (4) C_2H_2

17. The structural formula of 1,3-dichlorobutane is

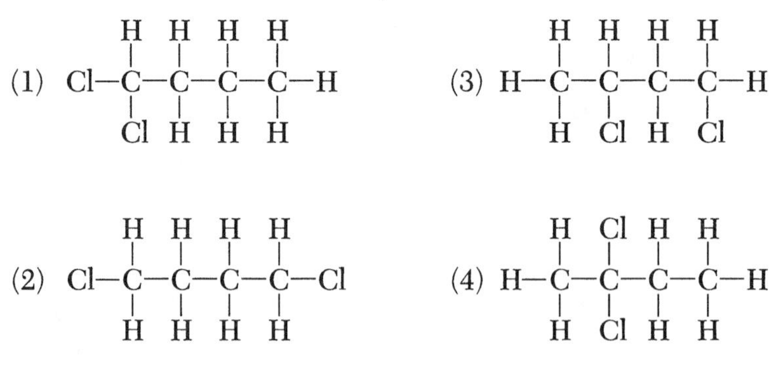

18. A fermentation reaction and a saponification reaction are similar in that they both can produce (1) an ester (2) an alcohol (3) an acid (4) a soap.

19. The general formula of organic acids can be represented as

$$\text{(1) } R{-}C\begin{smallmatrix}\diagup\!\!\diagup O\\ \diagdown H\end{smallmatrix} \quad \text{(2) } R{-}\overset{\diagup\!\!\diagup O}{C}{-}O{-}R' \quad \text{(3) } R{-}OH \quad \text{(4) } R{-}C\begin{smallmatrix}\diagup\!\!\diagup O\\ \diagdown OH\end{smallmatrix}$$

20. Which is an isomer of 2,2-dimethylpropane? (1) ethane (2) propane (3) pentane (4) butane

21. The molecule with the structural formula

$$CH_3{-}CH_2{-}CH_2{-}\overset{\overset{\displaystyle O}{\|}}{C}{-}CH_3$$

is classified as a(n) (1) aldehyde (2) alcohol (3) ketone (4) ether.

22. Which is an isomer of the compound propanoic acid, CH_3CH_2COOH?

(1) $CH_2{=}CHCOOH$	(3) $CH_3CH_2CH_2COOH$
(2) $CH_3CH(OH)CH_2OH$	(4) $HCOOCH_2CH_3$

23. Which molecular formula represents pentene? (1) C_4H_8 (2) C_4H_{10} (3) C_5H_{10} (4) C_5H_{12}

24. The compound CH_3CONH_2 is classified as (1) a ketone (2) an acid (3) an amide (4) an amino acid.

25. The reaction $C_3H_6 + H_2 \rightarrow C_3H_8$ is an example of (1) substitution (2) addition (3) polymerization (4) esterification.

26. Which formula represents a nonpolar compound?

$$\text{(1) } \begin{array}{c} H{:}\overset{\cdot\cdot}{\underset{\cdot\cdot}{O}}{:} \\ \;\;\;H \end{array} \quad \text{(2) } H{:}\overset{\cdot\cdot}{\underset{\cdot\cdot}{Cl}}{:} \quad \text{(3) } \begin{array}{c} H{:}\overset{\cdot\cdot}{\underset{\cdot\cdot}{N}}{:}H \\ H \end{array} \quad \text{(4) } \begin{array}{c} H \\ H{:}\overset{\cdot\cdot}{\underset{\cdot\cdot}{C}}{:}H \\ H \end{array}$$

27. Which is the third member of the alkene series? (1) propane (2) propene (3) butane (4) butene

28. Which is an isomer of *n*-butane?

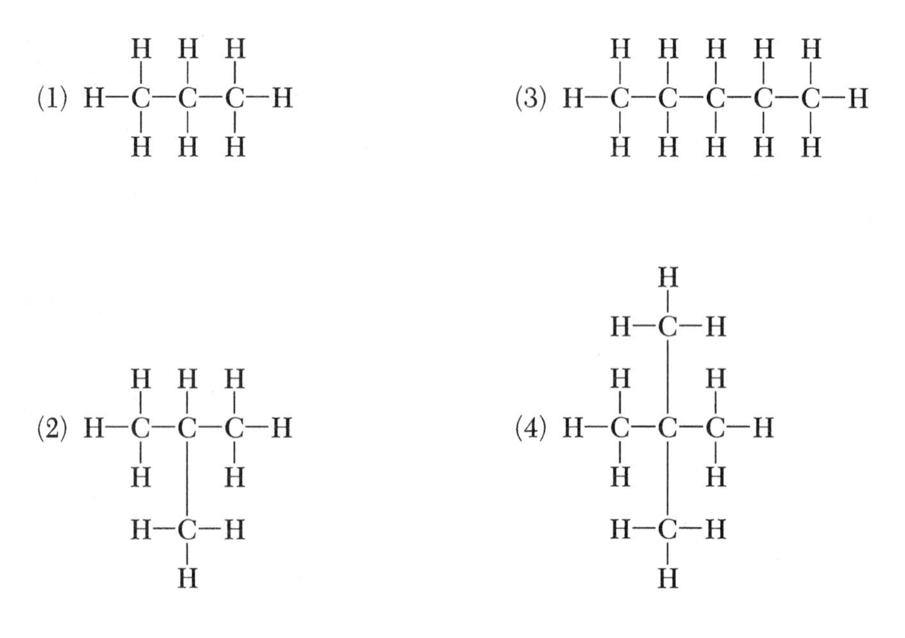

29. The product of a reaction between a hydrocarbon and chlorine is 1,2-dichloropropane. The hydrocarbon must have been (1) C_5H_{10} (2) C_2H_4 (3) C_3H_6 (4) C_4H_8.

30. A solution of methanol differs from a solution of acetic acid in that the solution of acetic acid (1) contains molecules only (2) has a pH of 7 (3) turns red litmus to blue (4) conducts electricity.

31–35. (1) pentanoic acid (2) butanal (3) 2-propanol (4) pentane (5) 2-methyl -2-propanol.

For each substance below, choose the substance from the list above that is its isomer.

31. 2-butanol

32. ethyl propanoate

33. 2-methylbutane

34. 2-butanone

35. ethyl methyl ether (methoxyethane)

CONSTRUCTED RESPONSE

1. Draw the structures of the following organic compounds:
   (a) methyl propanoate
   (b) 1,1-dichloroethane
   (c) 3-hexyne
   (d) 1-propanol
   (e) butanal
   (f) 2-pentanone
2. For each of the four compounds named below, (a) draw the structural formula of the compound, (b) draw the structural formula of an isomer of the compound and (c) name the isomer.
   (1) 2-butanol
   (2) propanal
   (3) ethoxyethane (diethyl ether)
   (4) methyl ethanoate (methyl acetate)

Kekulé: Lord of the Rings

Michael Faraday discovered benzene, a hydrocarbon with the formula C_6H_6, in 1826. However, its structure remained a mystery for the next 40 years. Chemists could draw several possible structures for the formula C_6H_6. One possibility contains two triple bonds as shown below.

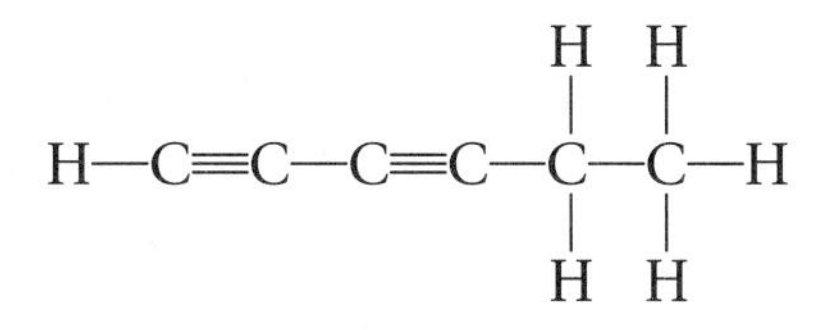

It is also possible for a molecule of C_6H_6 to contain four double bonds, or one triple bond and two double bonds. However, all these structures indicate an unsaturated hydrocarbon, which means that benzene should react with halogens, such as bromine, by addition. In fact, benzene reacts with bromine by substitution.

$$C_6H_6 + Br_2 \longrightarrow C_6H_5Br + HBr$$

Why would an unsaturated hydrocarbon react in this way? Chemists were unable to explain it until Kekulé had his dream.

Friedrich August von Kekulé was a German chemist. One evening in 1865 after having worked on the benzene problem for some time, he dozed off in front of his fireplace. He dreamed of hydrocarbon molecules, which danced before him as if they were taunting him. The molecules then turned into snakes. Eventually, the snakes began to bite their own tails, forming rings. According to Kekulé's description of the event, he awakened "as if by a flash of lightning." Based on his dream, he proposed the following structure for the elusive molecule.

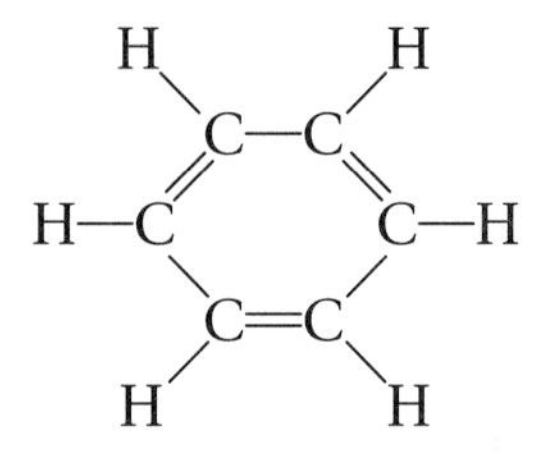

This molecule has the correct formula, C_6H_6, and each carbon atom has the required four covalent bonds. But how does this "circular reasoning," explain why benzene reacts as if it were a saturated hydrocarbon? It still has three double bonds. Kekulé suggested that the double bonds could change places, or resonate, like this.

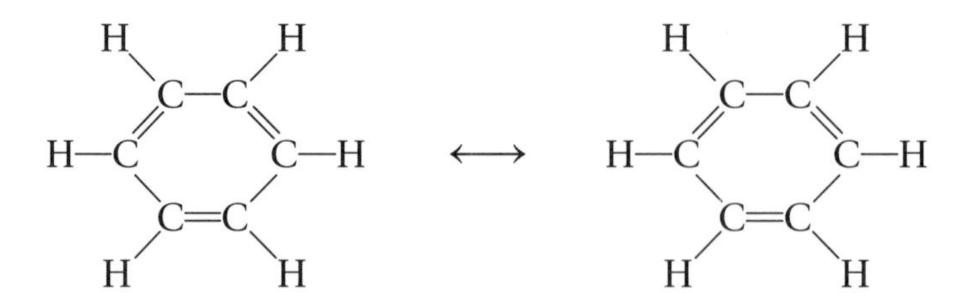

This resonating double bond would be more stable than an ordinary double bond, and so would not break as easily.

The C —H bond would break instead, just as in saturated hydrocarbons.

Our concept of resonance has changed. (See pages 147 to 148.) Modern chemists often represent the benzene molecule simply as a hexagon surrounding a circle. The circle represents the three double bonds, which are shared equally by the six carbon atoms.

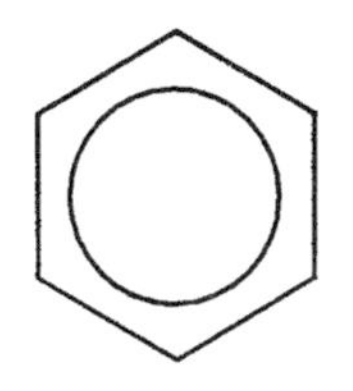

The structure above, called a benzene ring, is found in an enormous number of organic compounds, from pain relievers, such as aspirin, to explosives like TNT. Kekulé's dream led to our understanding of the structure of these molecules. Fellow chemists, let us continue to dream.

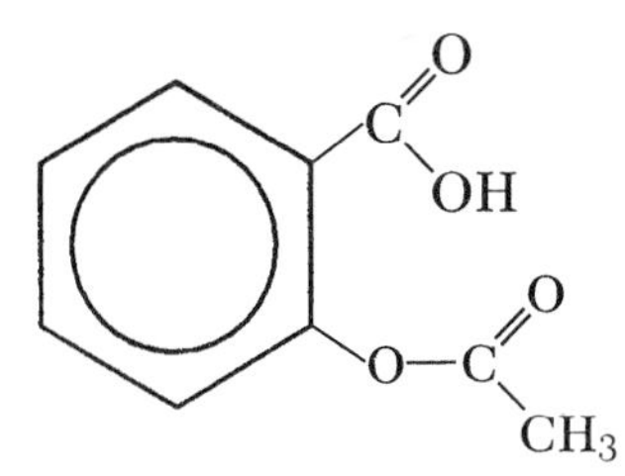

Aspirin

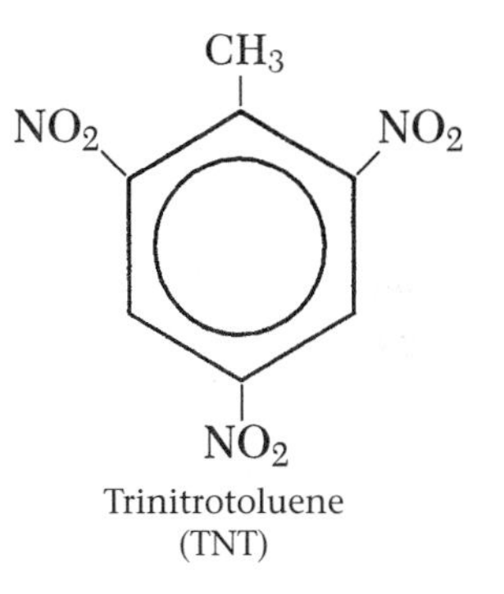

Trinitrotoluene (TNT)

Nuclear Chemistry

LOOKING AHEAD

Chemical reactions generally involve the bonding electrons in atoms. We judge the stabilities of elements based upon their tendencies to combine with other elements, by gaining, losing, or sharing electrons. Radon is a noble gas element, which does not readily combine with other elements. However, while the electron configuration of radon is very stable, the nucleus is very unstable. Radon participates in nuclear reactions, which involve changes in the protons and neutrons. In this chapter, you will study nuclear reactions and the energy associated with them.

When you have completed this chapter, you should be able to:

- **Describe** the most common types of nuclear decay.
- **Write** balanced nuclear equations.
- **Solve** problems involving half-lives.
- **Understand** the significance of, and problems associated with fusion and fission reactions.
- **Describe** some of the important uses of radioactive isotopes.
- **Recognize** several different types of nuclear reactions.

Radioactivity

The French chemist Antoine-Henri Becquerel is credited with the discovery of radioactivity. In 1896, he noticed that some photographic film that had been stored near a uranium sample had black streaks on it, as if it had been exposed to light. With the help of Pierre and Marie Curie, he was able to show that some form of invisible radiation, coming from the uranium, was affecting the film. The Curies called the process that produces these invisible rays *radioactivity.* **Radioactivity** is the spontaneous breakdown of an unstable atomic nucleus, yielding particles and/or radiant energy.

With the use of charged plates, as shown in Figure 15-1, scientists were able to identify three distinct types of radiation. They named them for the first three letters of the Greek alphabet. The positively charged rays were called alpha rays (α), the negatively charged rays were called beta rays (β), and the neutral rays were called gamma rays (γ).

Radioisotopes

Recall that all elements have *isotopes*—that is, the elements exist in different forms, the atoms of which have the same number of

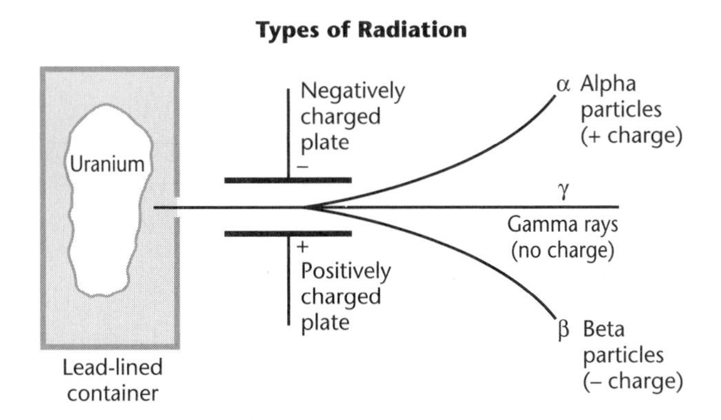

Figure 15-1

protons but a different number of neutrons. The mass numbers of the isotopes of an element therefore differ, since mass number is equal to the number of nucleons—the neutrons and the protons. Consider hydrogen, for example. There are three isotopes of hydrogen. The most common isotope has one proton and no neutrons in its nucleus. Another isotope, called *deuterium,* has 1 proton and 1 neutron. The third isotope, *tritium,* has 1 proton and 2 neutrons. Each of these isotopes has a different mass number, but all have atomic number 1. The isotopes may be represented as ${}^{1}_{1}H$, ${}^{2}_{1}H$, and ${}^{3}_{1}H$, where the superscript is the mass number and the subscript is the atomic number. These and other isotopes are generally referred to as *nuclides,* a term that describes all atoms without regard to mass number or atomic number.

Stable and Unstable Nuclei

The nuclei of certain atoms are stable. Under ordinary circumstances, stable nuclei do not undergo change. The nuclei of other atoms are unstable. These nuclei undergo change spontaneously, that is, without outside help. When unstable nuclei undergo change, they also give off radiation. Elements whose atoms have unstable nuclei are radioactive and are called **radioisotopes,** or **radionuclides.**

One source of nuclear instability is overcrowding. No stable nucleus has ever been found that has more than 83 protons. Therefore, every element with an atomic number 84 and greater is radioactive. Experiments have also shown that stability is related to the ratio of neutrons to protons in the nucleus. This ratio is symbolized as n/p.

Each of the points plotted on the graph in Figure 15-2 represents the n/p ratio of a nuclide. Notice that most of the many points on the graph can be connected by the shaded band. The nuclides whose n/p ratios fall within this band are stable. This part of the graph is called the *stability zone,* or *stability belt.*

Now, study the dashed line on the graph. Any point on this line represents an equal number of protons and neutrons, or an n/p ratio of 1. Notice that the beginning segment of each curve on the graph—that is, the curve for the stability zone and the curve for an n/p ratio of 1—overlap. This means that the light elements—elements up to about atomic number 20—are stable and have an

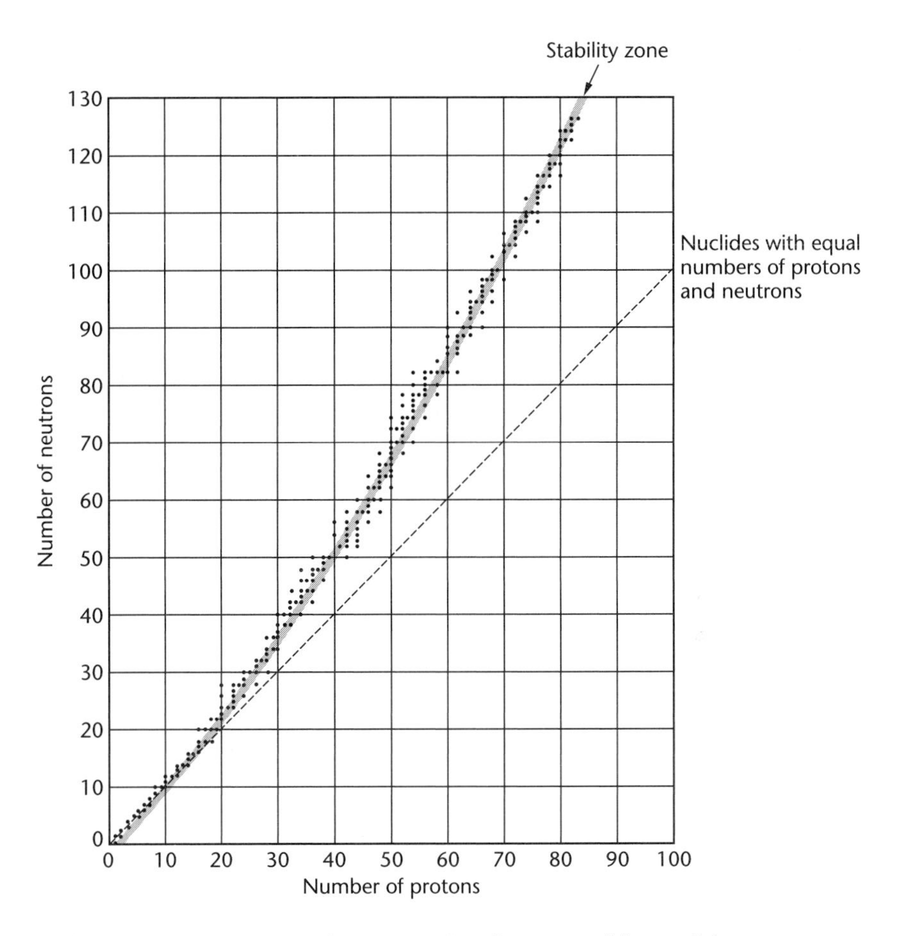

Figure 15-2 Neutron/proton ratio of some stable nuclides

n/*p* ratio of 1. Heavier elements—elements with atomic numbers higher than 20—have *n*/*p* ratios greater than 1. For example, an element with 40 protons (this is the element with atomic number 40, remember) has about 50 neutrons ($n/p = 1.25$). The heavier elements have more neutrons than protons. On the graph, the curve connecting the *n*/*p* ratios for the heavier elements slopes toward the neutron axis.

The *n*/*p* ratio of the most unstable—the radioactive—nuclides is outside the zone of stability. When any of these nuclides undergoes spontaneous change, a new nuclide is formed. The new nuclide has a more stable ratio of neutrons to protons.

Nuclear Emissions

The change that an unstable nucleus undergoes is called *disintegration,* or *decay.* When unstable nuclei disintegrate, or decay, certain particles—alpha or beta particles—or bundles of electromagnetic energy, called gamma radiation, are emitted. (Gamma radiation may be symbolized as γ.) Some elements, such as radium, give off alpha and beta particles, and gamma radiation. Other elements emit only one or two types of emissions.

Alpha (α) particles have a charge of 2+ and a mass four times that of a hydrogen atom. An alpha particle has the same charge and mass as a helium nucleus and is considered to be a helium nucleus. Beta particles are electrons. [There are actually two types of beta particles. Beta(−) particles (β^-) are electrons. Beta (+) particles (β^+) have the same mass as an electron but an opposite charge. These particles are called *positrons.*] Gamma radiation has no charge and no mass. The symbols and properties of the common types of emissions are summarized in the table that follows.

Name	**Symbol**	**Charge**	**Mass**	**Relative Penetrating Power** *(approx.)*	**Relative Ionizing Power** *(approx.)*
Alpha	α or $^{4}_{2}He$	2+	4+	1	10,000
Beta (−) (electron)	β^- or $^{\ 0}_{-1}e$	1−	0	100	100
Positron [beta (+)]	β^+ or $^{\ 0}_{+1}e$	1+	0	100	100
Gamma	γ	0	0	10,000	1

Notice that gamma radiation has the greatest penetrating power of all the types of emissions listed. This means that gamma radiation is the most dangerous of the emissions—it can attack the skin and bones of the human body.

The **ionizing power** of nuclear emissions refers to their ability to remove one or more electrons from an atom or molecule.

Separating Nuclear Emissions

Nuclear emissions can be separated by a magnetic or an electric field. Alpha particles and positrons, because of their positive charge, are deflected toward the negative pole, or electrode. Electrons, because of their negative charge, are deflected toward the positive pole, or electrode. Gamma radiation has no charge and is not influenced by magnetic or electric fields.

Spontaneous Nuclear Reactions—Natural Radioactivity

Four major types of nuclear changes occur spontaneously in nature and are referred to as **natural radioactivity**, or sometimes as **natural transmutation.** In each type of change, a new nuclide with a more stable n/p ratio is formed.

Beta (−) decay Nuclides with an n/p ratio that is too high for stability appear at the left of the stability belt shown in the graph on page 541. These unstable nuclei undergo spontaneous change that favors a lower number of neutrons and a higher number of protons. Such a change comes about when a beta (−) particle is emitted from the nucleus. But the beta (−) particle is an electron, and, as you know, electrons do not exist in nuclei. How, then, can an electron be emitted from the nucleus? Some scientists believe that during beta (−) decay, a neutron breaks up forming a proton and an electron. The electron is emitted.

Beta (−) decay is shown in the following equation, called a nuclear equation. The subscripts in the equation show charge. The superscripts show mass numbers. In a balanced nuclear equation, the sum of the subscripts on one side of the equation must be equal to the sum of the subscripts on the other side of the equation. Similarly, the sums of the superscripts must be equal on both sides of the equation.

$$^{1}_{0}n \longrightarrow {}^{1}_{1}p + {}^{0}_{-1}e$$

The equation shows that in beta (−) decay, a neutron in the nucleus disappears, producing a new proton and an electron. The n/p ratio decreases. The electron produced is emitted from the nucleus with great force.

A nuclear equation for a reaction in which an element undergoes beta (−) decay can be written in either of two ways. In the equations that follow, the superscripts are the mass numbers and the subscripts are the atomic numbers of the elements or the charge of the particle that is emitted.

	PARENT (ORIGINAL) NUCLEUS		DAUGHTER (NEW) NUCLEUS		EMISSION
	${}^{14}_{6}\text{C}$	$\longrightarrow$	${}^{14}_{7}\text{N}$	$+$	${}^{0}_{-1}e$
or					
	${}^{14}_{6}\text{C}$	$\longrightarrow$	${}^{14}_{7}\text{N}$	$+$	β^-
n/p RATIO	$8/6 = 1.33$		$7/7 = 1.00$		

The carbon-14 nucleus loses a neutron and gains a proton. An electron is emitted. The new nucleus is nitrogen-14.

If these equations are representative, several principles for nuclear changes involving beta (−) decay can be stated.

1. The atomic number increases by 1.
2. The mass number remains the same.
3. The sums of the subscripts on the two sides of the equation are equal. (This is true also for alpha decay.)
4. The sums of the superscripts on the two sides of the equation are equal. (This is true also for alpha decay.)
5. The n/p ratio is decreased.

Positron or Beta (+) decay Nuclides with an n/p ratio that is too low for stability appear at the right of the stability belt shown in the graph on page 541. These unstable nuclei undergo spontaneous change that favors a higher number of neutrons and a lower number of protons. Such a change comes about when a beta (+) particle (positron) is emitted from a nucleus. The mass of a positron is the same as the mass of an electron, but it has an opposite charge. The name positron is derived from "positive electron." Scientists believe that when a proton decays, it forms a neutron and a positron [beta (+) particle], which is emitted.

$$ {}^{1}_{1}p \longrightarrow {}^{1}_{0}n + {}^{0}_{+1}e $$

A nuclear equation for a reaction in which an element undergoes positron decay may be written in either of the following ways:

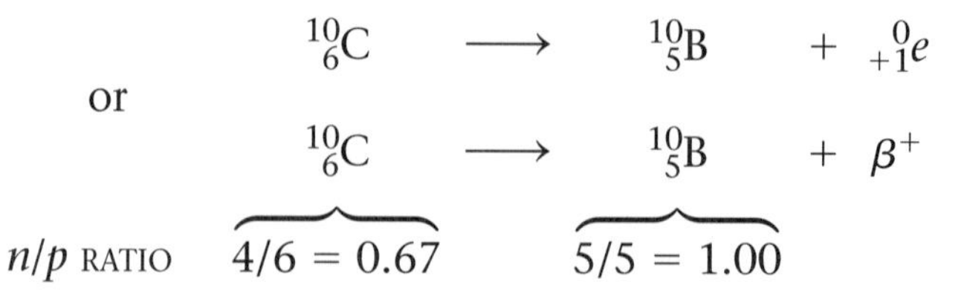

$$ {}^{10}_{6}\text{C} \longrightarrow {}^{10}_{5}\text{B} + {}^{0}_{+1}e $$

or

$$ {}^{10}_{6}\text{C} \longrightarrow {}^{10}_{5}\text{B} + \beta^{+} $$

n/p RATIO $4/6 = 0.67$ $5/5 = 1.00$

In this example, carbon-10 is the parent, or original, nucleus and boron-10 is the daughter, or new, nucleus.

In beta (+) decay, atomic number decreases by one, mass number remains the same, and the n/p ratio increases.

Alpha decay Repulsions occur between protons in the nucleus of any atom. In atoms with high atomic numbers (from about 84 and higher), the repulsions between the large numbers of protons contribute to the instability of the nuclei. Such atoms undergo spontaneous change that lowers the number of protons. Such a change comes about when an alpha particle (helium nucleus) is emitted from an atom. An example of alpha decay is

$$^{238}_{92}\text{U} \longrightarrow {}^{234}_{90}\text{Th} + {}^{4}_{2}\text{He}$$

$$\text{or} \quad ^{238}_{92}\text{U} \longrightarrow {}^{234}_{90}\text{Th} + \alpha$$

In alpha decay, the atomic number decreases by 2, and the mass number decreases by 4. The n/p ratio increases. An increase in the n/p ratio tends to offset the effect of the large proton–proton repulsions in the nucleus.

Gamma radiation Gamma radiation has neither charge nor appreciable mass. It consists of high-energy electromagnetic waves. These waves resemble X rays, but they have a higher frequency and therefore a greater energy. This high energy accounts for the penetrating power of gamma radiation. Gamma radiation, you will recall, can cause serious damage to living tissue.

Gamma radiation is often associated with emission of alpha and beta particles. If gamma radiation alone takes place, then neither atomic number nor mass number changes—parent and daughter nuclei are identical.

Summary The following table summarizes the nuclear changes that occur in spontaneous nuclear reactions:

Type of Emission	***Symbol***	***Change in Atomic Number***	***Change in Mass Number***
Alpha	${}^{4}_{2}\text{He}$	Decreases by 2	Decreases by 4
Beta (−) (electron)	${}^{0}_{-1}e$	Increases by 1	None
Positron [beta(+)]	${}^{0}_{+1}e$	Decreases by 1	None
Gamma	γ	None	None

Writing Nuclear Equations

All nuclear equations must be balanced both for charge and for mass. The subscripts must balance, and the superscripts must balance. In the decay of C-14, for example, the equation is

$$^{14}_{6}C \longrightarrow {}^{14}_{7}N + {}^{0}_{-1}e.$$

The total charge, or the total of the subscripts, is 6 on the left, and 7 − 1, or 6, on the right. The particle formed has an atomic number of 7; it is nitrogen. The atomic number determines the identity of the element. The total mass on both sides of the equation is 14.

If you know the type of radiation emitted by a given nuclide, you can identify the new nuclide formed in the nuclear reaction. The table on page 549 lists the particles emitted by several nuclides. Consider francium-220. What new element is formed by the radioactive decay of this nuclide? The table tells us that Fr-220 emits an alpha particle. The Periodic Table tells you that Fr has an atomic number of 87. You can write the equation for the decay of Fr-220:

$$^{220}_{87}Fr \longrightarrow {}^{4}_{2}He + X$$

You know that the equation must be balanced both for charge and for mass. Therefore, *X* must have an atomic number of 85, and a mass of 216. The Periodic Table tells us that the element with an atomic number of 85 is At (astatine). Therefore, the element formed in this decay, element *X* in your equation, is At-216. The nuclear equation for the decay of Fr-220 is

$$^{220}_{87}Fr \longrightarrow {}^{4}_{2}He + {}^{216}_{85}At$$

PRACTICE

15.1 Write the nuclear equation for the radioactive decay of the following nuclides and identify the new element formed. Use the table on page 549, or Table N in Appendix 4, to identify the particle emitted

(a) Co-60

(b) U-238

(c) K-42

(d) K-37

15.2 Fill in the missing particle in each of the following nuclear equations.

(a) ${}^{235}_{92}\text{U} \longrightarrow {}^{231}_{90}\text{Th} + ?$

(b) ${}^{14}_{6}\text{C} \longrightarrow {}^{14}_{7}\text{N} + ?$

(c) ${}^{232}_{90}\text{Th} \longrightarrow {}^{4}_{2}\text{He} + ?$

(d) ${}^{99}_{43}\text{Tc} \longrightarrow {}^{0}_{-1}e + ?$

Artificial Transmutation

In 1934 Frédéric and Irène Joliot-Curie discovered that a stable nucleus can be made unstable (radioactive) by bombardment with a high-energy particle, or "bullet." Irène Joliot-Curie was the daughter of Marie Curie, the first woman scientist to win two Nobel prizes. The Joliot-Curies bombarded stable boron-10 atoms with alpha particles from naturally radioactive radium. They produced artificially radioactive nitrogen as a result of this bombardment.

$${}^{10}_{5}\text{B} + {}^{4}_{2}\text{He} \longrightarrow {}^{13}_{7}\text{N} + {}^{1}_{0}n$$

Since 1934, a large number of artificial radioactive elements have been produced by nuclear bombardment. Changing of one element into another by artificial means such as this is called **artificial transmutation.** In artificial transmutation, as in natural radioactive changes, the sum of the superscripts on one side of the equation is equal to the sum of the superscripts on the other side. The same is true for the subscripts. Note though, that in artificial transmutation there will always be two particles on the reactant side, while in natural radiation there will be only one particle on the reactant side.

The nuclear projectiles, or "bullets," used to bombard nuclei are usually positively charged particles. These particles may be protons, alpha particles, or deuterons. (A *deuteron* is the nucleus of the hydrogen isotope deuterium. The symbol for a deuteron is ${}^{2}_{1}\text{H}$.) Such particles must have a very high energy, or they will be repelled by the positively charged nucleus. Special machines called *particle accelerators* are used to obtain nuclear particles with sufficiently high energy to penetrate the nucleus.

Neutrons can also be used as "bullets" for many nuclear changes. Because neutrons do not have a charge, they are not repelled by the positive charge of a nucleus. Neutrons therefore do not have to be accelerated to higher energies.

PRACTICE

15.3 Fill in the missing particle in each of the following nuclear equations:

(a) $^{238}_{92}\text{U} + ^{1}_{0}n \longrightarrow ^{0}_{-1}e + ?$

(b) $^{14}_{6}\text{C} + ^{4}_{2}\text{He} \longrightarrow ^{1}_{0}n + ?$

(c) $^{239}_{93}\text{Np} \longrightarrow ^{0}_{-1}e + ?$

(d) $^{235}_{92}\text{U} + ^{1}_{0}n \longrightarrow ^{143}_{56}\text{Ba} + ? + 3\,^{1}_{0}n$

15.4 Which of the equations above shows natural radiation?

Half-Life

The rate of radioactive decay is not affected by the chemical or physical makeup of the radioactive atom or by the environment. Scientists describe the rate of decay of radioactive atoms in terms of half-life. **Half-life** is the time required for one-half of any given mass of radioactive nuclei to decay. The half-life for the beta (−) decay of thorium-234, for example, is 24 days. If you start with 10 grams of thorium-234, after 24 days, 5 grams are left; after 48 days, 2.5 grams are left; after 72 days, 1.25 grams are left; and so on.

The half-lives of different atoms vary enormously. The half-life for the alpha decay of U-238 is 4.5 billion years. In extreme contrast, the half-life for the alpha decay of Po-214 is 1.6×10^{-4} second. Half-lives of some radioisotopes are given in the table on the next page.

To calculate the fraction of the original mass of nuclide remaining, use the following equation

$$\text{fraction remaining} = \left(\frac{1}{2}\right)^{\frac{t}{T}}$$

In the formula, t is the time elapsed and T is the half-life. The exponent $\frac{t}{T}$ is the number of half-life periods.

$$\frac{t}{T} = \text{number of half-life periods}$$

The equation is easy to use when the value of $\frac{t}{T}$ is a whole number. To find the fraction of nuclide remaining after a fractional num ber of half-life periods requires the use of a scientific calculator or table of logarithms. For now, our discussion will be confined to half-life problems in which the number of half-life periods is a whole number.

Selected Radioisotopes

Nuclide	***Half-Life***	***Decay Mode***	***Nuclide Name***
^{198}Au	2.69 d	β^-	gold-198
^{14}C	5730 y	β^-	carbon-14
^{37}Ca	175 ms	β^+	calcium-37
^{60}Co	5.26 y	β^-	cobalt-60
^{137}Cs	30.23 y	β^-	cesium-137
^{53}Fe	8.51 min	β^+	iron-53
^{220}Fr	27.5 s	α	francium-220
^{3}H	12.26 y	β^-	hydrogen-3
^{131}I	8.07 d	β^-	iodine-131
^{37}K	1.23 s	β^+	potassium-37
^{42}K	12.4 h	β^-	potassium-42
^{85}Kr	10.76 y	β^-	krypton-85
^{16}N	7.2 s	β^-	nitrogen-16
^{19}Ne	17.2 s	β^+	neon-19
^{32}P	14.3 d	β^-	phosphorus-32
^{239}Pu	2.44×10^4 y	α	plutonium-239
^{226}Ra	1600 y	α	radium-226
^{222}Rn	3.82 d	α	radon-222
^{90}Sr	28.1 y	β^-	strontium-90
^{99}Tc	2.13×10^5 y	β^-	technetium-99
^{232}Th	1.4×10^{10} y	α	thorium-232
^{233}U	1.62×10^5 y	α	uranium-233
^{235}U	7.1×10^8 y	α	uranium-235
^{238}U	4.51×10^9 y	α	uranium-238

ms = milliseconds; s = seconds; min = minutes; h = hours; d = days; y = years

SAMPLE PROBLEMS

PROBLEM

1. A sample of a radioactive substance with an original mass of 16 g was studied for 8 hours. When the study was completed, only 4 g of the substance remained. What is the half-life of the substance?

SOLUTION

In this problem, you are given enough information to find the fraction remaining

$$\frac{4\text{g}}{16\text{g}} = \frac{1}{4}$$

Now you must determine the number of half-life periods that have passed. Substitute the values into the equation.

$$\text{fraction remaining} = \left(\frac{1}{2}\right)^{\frac{t}{T}}$$

$$\frac{1}{4} = \left(\frac{1}{2}\right)^{\frac{t}{T}}, \frac{1}{4} = \left(\frac{1}{2}\right)^{2}$$

Now that you know the value of $\frac{t}{T}$ is 2, you can find the value of T.

$$\frac{t}{T} = \text{number of half-life periods}$$

$$\frac{8}{T} = 2, T=4$$

The problem can also be solved logically, without the use of the equations. The passage of each half-life reduces the mass of the sample by one half. During the first half-life, the mass of the substance decreased by one half, from 16 g to 8 g. During the next half-life, the mass of the substance decreased from 8 g to 4 g. That makes two half-lives passed in 8 hours; therefore, one half-life must be 4 hours.

PROBLEM

2. A sample of I-131 had an original mass of 16 g, how much will remain after 24 days?

SOLUTION

The table gives the half-life of I-131 as 8 days. The number of half-life periods that have passed in 24 days is

$$\frac{t}{T} = \text{number of half-life periods}$$

$$\frac{24}{8} = 3$$

Solving the equation for the fraction remaining, you get

$$\text{fraction remaining} = \left(\frac{1}{2}\right)^{\frac{t}{T}}$$

$$\left(\frac{1}{2}\right)^3 = \frac{1}{8}$$

You are not finished yet. At this point you know that $\frac{1}{8}$ of the original 16-g sample remains.

$$\frac{1}{8} \times 16g = 2g$$

Again, you can solve this problem without using the equation. Since the half-life is 8 days, one half of the sample decays every 8 days. You began with 16 g. After 8 days, half of 16 g, or 8 g, remains. After 8 more days, for a total of 16 days, 4 g remains. After another 8 days, for a total of 24 days, 2 g of I-131 remains.

PROBLEM

3. A sample was found to contain 2.00 mg of C-14. How many grams of C-14 were there in the sample 11,460 years ago?

SOLUTION

The half-life (T) of C-14 is 5730 years. You are given the time elapsed (t) as 11,460 years.

First find the number of half-life periods.

$$\frac{t}{T} = \frac{11{,}460 \text{ years}}{5730 \text{ years}} = 2 \text{ half-life periods}$$

Next calculate the fraction remaining after 2 half-life periods.

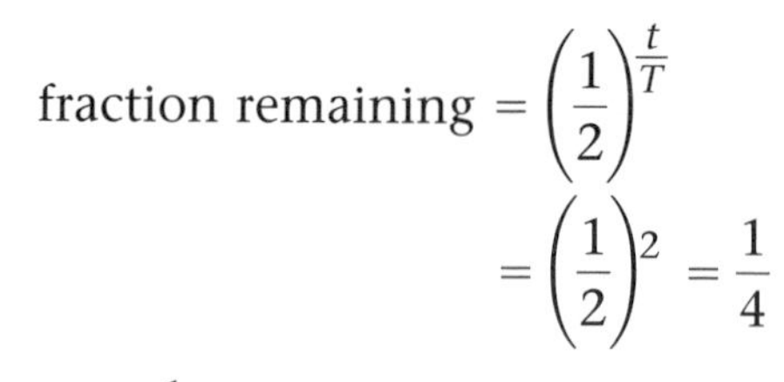

$$\text{fraction remaining} = \left(\frac{1}{2}\right)^{\frac{t}{T}} = \left(\frac{1}{2}\right)^{2} = \frac{1}{4}$$

Therefore, 2.00 mg is $\frac{1}{4}$ of the original sample. The original sample must have been

$$4 \times 2.00 \text{ mg, or } 8.00 \text{ mg.}$$

To solve this problem without using the equation, you can reason as follows: Since the amount of the nuclide becomes halved as you go forward one half-life, the amount must double if you go back one half-life. In this problem, you are going back two half-lives, so the quantity must double twice. Doubling 2.00 mg twice gives you 8.00 mg.

PRACTICE

(Use the table on page 549)

15.5 How long would it take 16 g of Ra-226 to break down until only 1.0 g remained?

15.6 How much of a sample of Co-60 with an original mass of 200 g remains after 15.78 years?

15.7 The half-life of Ra-221 is not listed on the table. In 2.0 minutes, a 64 mg sample of Ra-221 decays to a mass of 4.0 mg. What is the half-life of Ra-221?

Radioactive Decay Series

Recall that uranium-238 changes spontaneously into thorium-234 by losing an alpha particle (helium nucleus).

$$^{238}_{92}\text{U} \longrightarrow {}^{234}_{90}\text{Th} + {}^{4}_{2}\text{He} \qquad T = 4.5 \times 10^{9} \text{ years}$$

Thorium-234, in turn, changes into protactinium-234 by beta (−) (electron) emission.

$$^{234}_{90}\text{Th} \longrightarrow {}^{234}_{91}\text{Pa} + {}^{0}_{-1}e \qquad T = 24 \text{ days}$$

Protactinium also decays. After 12 spontaneous changes, uranium-238 becomes lead-206. Each change has its own half-life, and the entire set of changes is called a *radioactive decay series.* The total change from uranium-238 to lead-206 is determined by the slowest step—the rate-determining step. This is the step that has the longest half-life—4.5 billion years, the half-life of uranium-238.

Every sample of naturally occurring uranium-238 contains some lead-206. From the known half-lives of the elements in the series and from the proportion of uranium-238 to lead-206, it is possible to calculate the age of a rock that contains uranium. In this way, the age of Earth is estimated to be nearly 4.5 billion years.

Radiodating

Knowledge of the half-life of radioactive elements makes it possible to date many ancient objects. For example, carbon-14, which has a half-life of 5730 years, can be used to date objects that are up to 80,000 years old. Let's see how a "radiocarbon clock" works.

The atmosphere always contains a certain amount of C-14. This radioactive carbon is formed when nitrogen in the atmosphere is bombarded by neutrons.

$$^{14}_{7}N + ^{1}_{0}n \longrightarrow ^{14}_{6}C + ^{1}_{1}H$$

The neutrons are produced as a result of collisions between air molecules and protons that are part of the radiation that falls on Earth's upper atmosphere from outer space. Since the amount of nitrogen in the atmosphere and the intensity of radiation from outer space are fairly constant, the amount of C-14 in the atmosphere is also constant.

C-14 combines with oxygen and forms radioactive carbon dioxide. Carbon dioxide, as you probably know, is taken in by green plants in the food-making process. As a result of this process, the carbon becomes part of the plants' tissues. Since part of the carbon dioxide in the atmosphere is radioactive, some C-14 is present in the tissues of all green plants. When an animal feeds on green plants or on other animals that feed on green plants, C-14 becomes part of the animal's tissues also. All living things contain some radioactive carbon.

As long as a living thing is alive, it has a constant amount of C-14 in its tissues. Like all radionuclides, the C-14 in the tissues undergoes decay. However, C-14 is continually being taken in, and the concentration in the tissues remains constant.

When a plant or animal dies, it stops taking in C-14. The C-14 present in the tissues at the time of death continues to decay at a constant rate. The concentration of C-14 in the tissues therefore decreases as time goes by. If the radioactivity of the tissues is measured, the date of death can be determined from this measurement and the half-life of carbon.

Let us suppose that you want to date a wooden carving. You know that the wood contains C-14 because it was once part of a living tree. You also know that C-14 has a half-life of 5730 years. You can measure the **decay rate,** the amount of radiation emitted by a given sample, in counts per minute, or cpm. Let us say that the carbon in your carving has a decay rate of 7.65 cpm. The same quantity of carbon in a fresh sample of wood from a living tree has a decay rate of 15.3 cpm. The carbon in the wood carving is emitting half as much radiation as the carbon in the fresh sample, so it is reasonable to conclude that the wood carving contains only half as much C-14 as a fresh sample. Half of the C-14 in the carving has decayed since the wood stopped taking in carbon dioxide. One half-life, or 5730 years, must have passed.

Different radionuclides must be used to date objects older than 80,000 years. As you have learned, uranium-238, with a half-life of 4.5 billion years, can be studied to determine the age of ancient uranium-bearing rocks. This is the method used to estimate the age of Earth. More recently formed rocks, which do not contain uranium, can be dated by studying their content of potassium-40, which has a half-life of about 1.3 million years.

Energy of Nuclear Changes

That matter can be converted into energy and energy can be converted into matter is expressed mathematically by Einstein's equation $E = mc^2$. In the equation, E stands for energy, m for mass, and c for the speed of light. The speed of light is 3×10^8 meters per second. The square of that number is 9×10^{16}. Multiplied by such a tremendously large number, the change of even a small quantity of mass into energy results in a very large quantity of energy. For example, if one gram of matter were converted completely into energy, about 9×10^{13} J—nearly 23 billion kilocalories—of heat would be liberated. This quantity of energy is nearly equal to the energy released by the explosion of some 40,000 tons of TNT. It is about 100 million times the quantity of energy involved in common chemical changes.

In an ordinary chemical reaction, a small amount of mass is converted into energy. However, the change in mass is so very tiny it is ignored. But mass–energy conversion is of great importance in nuclear changes, particularly in fission and fusion reactions. **Fission** is a reaction in which large nuclei split into two or more smaller nuclei. **Fusion** is a reaction in which small nuclei unite into larger nuclei. You will now consider these reactions.

Nuclear Fission

A heavy nucleus, such as U-238, can become more stable if it undergoes fission. This occurs during natural radioactivity, when U-238 undergoes a series of decays and changes to Pb-206. In this process, a large nucleus is split into more stable nuclei, and energy is released.

Fission may also be brought about artificially. An example of this type of artificial transmutation is the change brought about by the bombardment of U-235 with a neutron. One of the many possible reactions that may result from this bombardment is

$$^{235}_{92}\text{U} + {}^{1}_{0}n \longrightarrow {}^{143}_{56}\text{Ba} + {}^{90}_{36}\text{Kr} + 3\,{}^{1}_{0}n + 1.9 \times 10^{10}\ \text{kJ/mole}$$

The energy released by this reaction is enormous—19 billion kilojoules per mole of uranium.

Notice that bombardment of 1 uranium atom with 1 neutron results in the release of 3 neutrons. These 3 neutrons can, in turn, bombard 3 other uranium atoms, resulting in the release of 9 more neutrons. These 9 neutrons can bombard 9 more uranium atoms, releasing 27 neutrons, and so on. A reaction such as this is called a *chain reaction.* Once started, a chain reaction keeps on going by itself. Not only is a chain reaction self-sustaining but it also increases in magnitude as it goes on (see Figure 15-3 on page 556). If a chain reaction is not controlled, it results in an enormous release of energy. Such a release of energy occurred in the explosions of the first nuclear bombs.

Chain reactions can be controlled, primarily by slowing down the neutrons in devices called *reactors.* The energy released by the

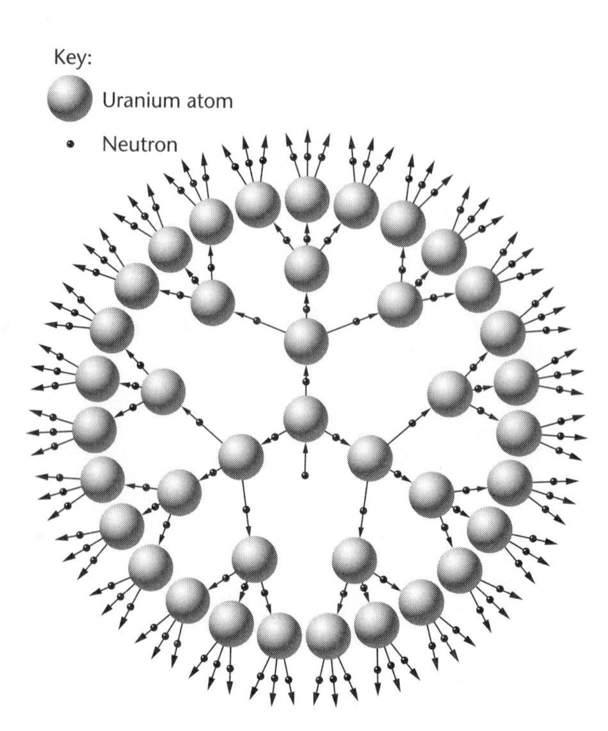

Figure 15-3 A chain reaction

reactions can thus be regulated and used for many peaceful purposes, including the production of electricity. One of the advantages of nuclear fission as a means of producing energy is that only a small quantity of fuel is needed to obtain a large quantity of energy. On the other hand, the reactors produce poisonous radioactive wastes and are potential sources of danger to living things. It is of utmost importance that reactors be designed and built to ensure maximum safety. Accidents that have occurred in nuclear reactors have focused attention on the seriousness of this problem.

While there are problems, nuclear power also has several advantages. Nuclear power plants produce no air pollution. Power plants using coal, oil, or natural gas all increase the amount of carbon dioxide in the atmosphere, while nuclear power plants do not. Increased use of nuclear power will decrease our dependence on fossil fuels. The future of nuclear power is widely debated throughout the world, and the debate seems likely to continue for a long time.

PRACTICE

15.8 Below you will find the steps in the production of Pu-239 from U-238. Identify the materials in the equations that are represented with the letters W, X, and Y.

1. $^{238}U + W \longrightarrow X$
2. $X \longrightarrow {}_{-1}^{0}e + Y$
3. $Y \longrightarrow {}_{-1}^{0}e + {}^{239}Pu$

Nuclear Fusion

Even greater quantities of energy are released by fusion of small nuclei into larger nuclei than by fission. One example of a fusion reaction is the union of the nuclei of two isotopes of hydrogen—deuterium and tritium. The fusion of a deuteron (the nucleus of a deuterium atom) and a triton (the nucleus of a tritium atom) yields helium and a neutron.

$${}_{1}^{2}H + {}_{1}^{3}H \longrightarrow {}_{2}^{4}He + {}_{0}^{1}n + \text{energy}$$

Fusion reactions like this one are believed to be the source of the sun's energy. Fusion reactions have also been artificially produced in hydrogen bombs. Fusion reactions are generally cleaner (they produce less poisonous wastes) than fission reactions. However, to start fusion reactions, enormously high temperatures—on the order of the temperatures of the sun—are required. As yet, no practical way of developing these temperatures and controlling the reactions has been found. If and when these problems are solved, fusion reactions may become a major source of energy for peaceful purposes. Our dependence on oil and other fuels will be lessened, and energy crises will be avoided.

Other Uses of Radioisotopes

Radioactive materials are used to generate power and in radioactive dating, as you have seen. They are also used extensively in the field of medicine.

Radiation can kill cells or can cause them to become malignant. However, carefully directed beams of gamma rays can be used to kill cancerous cells without extensively damaging healthy tissue. Radiation therapy has become one of the most commonly used treatments for cancer. Cobalt-60 is one nuclide that is frequently used as a source of the radiation. Radiation can shrink cancers that are too large to be operable so they can be safely removed. After surgery, the area that surrounded the cancer is often treated with radiation to destroy any malignant cells that may have been left behind.

Ordinary iodine contains just one isotope, I-127, which is not radioactive. Iodine is normally absorbed by the thyroid gland and is necessary for the gland to function properly. Isotopes of an element are chemically indistinguishable from each other. If a small amount of the radioactive isotope I-131 is introduced into the body, it too will be absorbed by the thyroid. Because I-131 is radioactive, doctors can easily trace its path through the body to the thyroid and use I-131 to diagnose thyroid disorders. Similarly, Tc-99 (technetium-99) is used to locate brain tumors and to visualize blood flow patterns in the heart. The radioactive isotopes used in medical diagnoses generally have short half lives and are quickly eliminated from the body.

Radioactive Tracers

Radioisotopes are chemically identical to stable isotopes of the same element and will participate in exactly the same chemical reactions. Because they are radioactive, they are easily identified and can be used to trace the course of a chemical reaction. The paths of many organic reactions are studied by using carbon-14 as a tracer. One of the reactants is prepared using carbon-14, which can then be located at various stages of the reaction, providing valuable information about the path or mechanism of the reaction.

Irradiation of Food

Radiation may also be used to kill bacteria in foods. The treatment of meat with radiation could help prevent food poisoning. However, many people are afraid that while the radiation is killing the bacteria, it might cause changes in the meat. The debate continues. Some supermarkets now sell irradiated foods.

Detection of Radioactivity

Atoms and molecules in the path of radiation are ionized—that is, they are stripped of electrons. In other words, alpha particles,

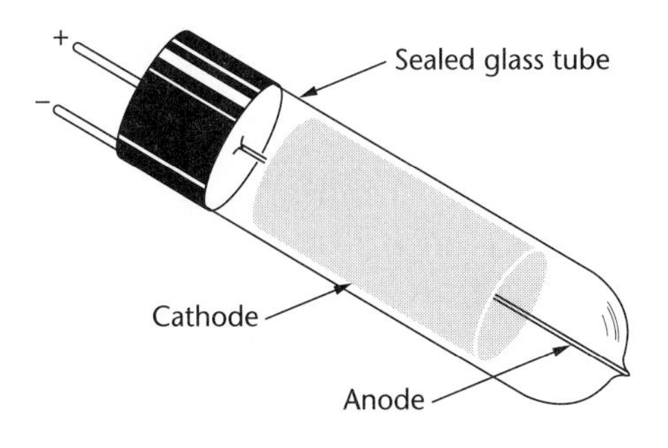

Figure 15-4 The Geiger-Müller tube

beta particles, gamma radiation, and other nuclear emissions have enough energy to remove some electrons from atoms or molecules they collide with. Positively charged particles and free electrons are left behind after the collisions.

Devices that are used to detect radioactivity are based on this ionizing ability of radiation. One such device is the Geiger counter. This device houses a sealed glass tube that contains argon gas at low pressure. Within the glass tube is a metallic cylinder with a wire running through the center (see Figure 15-4). The metallic cylinder is the cathode, and the wire is the anode. A high difference of potential (voltage)—just a little less than the voltage needed to ionize the argon gas—is maintained between the cathode and the anode. When gamma radiation enters the tube, it ionizes many gas atoms. Electrons are attracted to the anode, and argon ions are attracted to the cathode. The process goes on continuously until large numbers of electrons are traveling toward the anode. This produces a surge of current, which can be detected as a flash of light or a clicking noise, depending on how the counter is constructed. Some detection devices are designed so that a recording, such as a graph, is made of the electrical pulses.

Nuclear Binding Energy

If 8 protons unite with 8 neutrons, a nucleus of O-16 is formed. A proton has a mass of 1.008142 amu. A neutron has a mass of 1.008982 amu. The sum of the mass of 8 protons and the mass of 8 neutrons is 16.136992 amu.

$$8(1.008142) + 8(1.008982) = 16.136992 \text{ amu}$$

However, when the mass of O-16 is measured, it is found to be 16.000000 amu. There has been a loss of mass of 0.136992 amu. The mass lost when nucleons unite and form a nucleus is called the **mass defect.**

When 8 protons and 8 neutrons combine and form a nucleus of oxygen, there is a loss of mass of 0.136992 amu. This mass is converted into energy according to the $E = mc^2$ relationship. The energy resulting from the conversion is called **nuclear binding energy.** Nuclear binding energy is the energy released when nucleons come together and form a nucleus. Nuclear binding energy is also the energy absorbed when a nucleus is split into its individual nucleons. The higher the binding energy per nucleon, the more stable is the nucleus.

To compare the nuclear binding energy of different nuclides, scientists divide the mass defect by the number of nucleons in each nuclide. In the case of O-16, the mass defect per nucleon in amu is

$$\frac{0.136992 \text{ amu}}{16},$$

or 0.00856 amu per nucleon. The greater the mass defect per nucleon, the greater the binding energy per nucleon.

Nuclear Binding Energy vs. Mass Number

A graph of the binding energy per nucleon in relation to the mass number for various atoms is shown in Figure 15-5. The graph shows the following information:

1. The most stable nuclei are in the region of mass number 56—approximately the atomic mass of iron—because the values for binding energy per nucleon are greatest in this region.
2. The very small nuclei and the very large nuclei have low binding energies per nucleon compared with nuclei in the middle range. The very small and very large nuclei are less stable than the nuclei in the middle range.
3. As mass numbers increase up to about 56, binding energies increase. There is also an increase in stability. If small nuclei combine to form larger nuclei (fusion), there is a release of energy equal to the increased binding energies.

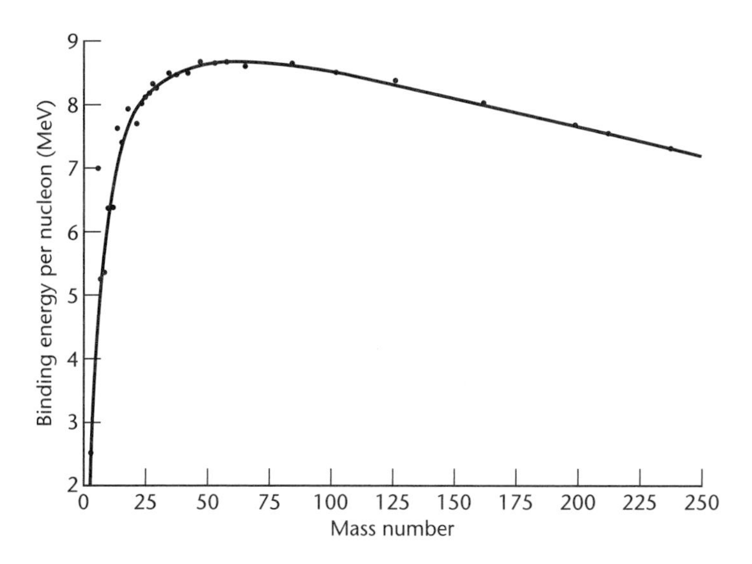

Figure 15-5 Binding energy vs. mass number

4. If large nuclei, such as uranium, are split into smaller nuclei (fission), there is an increase in binding energy. A release of energy also occurs during this type of change.
5. The changes in binding energy are shown by the slope of the curve. The change in binding energy is sharper with increasing mass number up to about 56 than with decreasing mass number to the same point. There may be a greater increase in binding energy—and therefore a greater release of energy—with fusion changes than with fission changes.

A Summary of Nuclear Binding Energy

Some of the important points to remember about nuclear binding energy are as follows:

1. When nucleons unite into a nucleus, there is a loss of mass. The lost mass, called the mass defect, is converted into energy according to the equation $E = mc^2$. The energy released when protons and neutrons unite and form a nucleus is called the binding energy of the nucleus. The higher the binding energy per nucleon, the more stable is the nucleus.
2. A graph of binding energy per nucleon plotted against mass number shows that greater nuclear stability is attained when:

(a) certain light nuclei, such as the nuclei of hydrogen and lithium, combine and form a heavier nucleus. (This is an example of fusion.)

(b) certain heavy nuclei, such as the nuclei of uranium, split into smaller nuclei. (This is an example of fission.)

TAKING A CLOSER LOOK

More Advanced Half-Life Problems

Radium-226 has a half-life of 1600 years. From your study of half-life, you know that a 100.-gram sample will decay to 50.0 grams in 1600 years; after 3200 years only 25 grams will remain. Each additional 1600 years results in the decay of another half of the remaining sample. However, we have not yet investigated what happens when the time is *less* than the half-life. How much of a 100. gram sample of the ^{226}Ra would remain after 800 years? Recall that we used the equation

$$\text{fraction remaining} = \left(\frac{1}{2}\right)^{\frac{t}{T}}$$

where $\frac{t}{T}$ is the number of half-life periods, equal to the time, t, divided by the half-life, T. In this case, $\frac{t}{T} = 0.5$. If you have a scientific calculator, you can raise any number to any power, so you can find the value of the expression.

$$\text{fraction remaining} = \left(\frac{1}{2}\right)^{\frac{t}{T}}$$

$$= \left(\frac{1}{2}\right)^{0.5} = 0.707$$

$$0.707 \times 100.\text{g} = 70.7\text{g}$$

Similarly, you could determine the quantity of radium remaining after *any* time interval. How much of a 100. gram sample would remain after 2000 years? $\frac{t}{T}$, the number of half-life periods would be 2000/1600, or 1.25.

$$\text{fraction remaining} = \left(\frac{1}{2}\right)^{\frac{t}{T}}$$

$$= \left(\frac{1}{2}\right)^{1.25} = 0.420$$

$$0.420 \times 100.\text{g} = 42.0\text{g}$$

PRACTICE

15.9 If the initial mass of a sample of ^{42}K (half-life = 12.4 hours) is 10.0 grams, calculate how much mass remains after

(a) 3.10 hours

(b) 18.6 hours

This chapter about nuclear chemistry concludes your introduction to the study of chemistry. Where will you go from here? Will you take other courses in chemistry? Or is this course, perhaps, all that you feel you need or want to know about chemistry? Whatever you choose, intelligent citizenship requires that you make decisions as you try to solve the many problems that confront you—the energy crisis and water and air pollution, to name only a few. In the years ahead, the information you have learned in this course will be useful to you. But far more valuable will be your ability to approach the solving of problems by the methods emphasized in science—gathering information, formulating and testing hypotheses, and, above all, keeping an open mind. In short, all that you have learned in your study of chemistry will enable you to better understand the world around you and to make a suitable adjustment to it.

CHAPTER REVIEW

The following questions will help you check your understanding of the material presented in the chapter.

Data for answering questions in this chapter appear in the table of selected half-lives of radioisotopes on page 549.

1. Three types of changes are illustrated below:

   Change 1: $CO_2\ (g) \longrightarrow CO_2\ (s)$
   Change 2: $C\ (s) + O_2\ (g) \longrightarrow CO_2\ (g)$
   Change 3: $6\ {}^{1}_{0}n + 6\ {}^{1}_{1}H \longrightarrow {}^{12}_{6}C$

   Based on this information, which of the following statements is true? (1) All three changes represent the formation of chemical bonds. (2) Nuclear changes occur in all. (3) The size of particles increases from change 1 to change 3. (4) The energy released per mole of carbon increases from change 1 to change 3.

2. The energy released during nuclear reactions is (1) slightly less than the energy released during chemical reactions (2) much less than the energy released during chemical reactions (3) about the same as the energy released during chemical reactions (4) much greater than the energy released during chemical reactions.

3. When a radioactive element emits a beta (−) particle, (1) the positive charge of the nucleus increases by 1 (2) the number of neutrons in the nucleus increases (3) oxidation takes place (4) the atomic mass of the element increases by 1.

4. In the nuclear change ${}^{239}_{93}Np \longrightarrow {}^{239}_{94}Pu + X$, the particle X is (1) a positron (2) a neutron (3) an alpha particle (4) an electron.

5. In the nuclear transmutation ${}^{40}_{18}Ar + {}^{4}_{2}He \longrightarrow Y + {}^{1}_{0}n$, the element Y is (1) ${}^{44}_{19}K$ (2) ${}^{43}_{20}Ca$ (3) ${}^{44}_{22}Ti$ (4) ${}^{43}_{19}K$.

6. From electron to proton to neutron, the masses of the particles (1) decrease (2) increase (3) remain the same.

7. Based on the information in the table on page 549 what is the symbol for the nuclide formed by the radioactive decay of ${}^{226}Ra$? (1) ${}^{226}Ac$ (2) ${}^{230}Th$ (3) ${}^{222}Rn$ (4) ${}^{222}Po$

8. When a radioactive element releases an alpha particle, the number of neutrons in the nucleus (1) decreases (2) increases (3) remains the same.

9. As the atomic masses of a series of isotopes of a given element increase, the number of protons in each nucleus (1) decreases (2) increases (3) remains the same.

10. Nuclide X is radioactive because its neutron to proton ratio is too large. Nuclide X is most likely to release (1) positrons (2) alpha particles (3) beta particles (4) protons.

11. If a radioactive substance has a half-life of 9 days, what fraction of its original mass will remain after 27 days? (1) $\frac{1}{8}$ (2) $\frac{1}{4}$ (3) $\frac{1}{2}$ (4) $\frac{3}{4}$

12. When an atom emits a beta particle, the total number of nucleons (1) decreases (2) increases (3) remains the same.

13. A 40.0-milligram sample of ^{33}P decays to 10.0 mg in 50.0 days. What is the half-life of ^{33}P? (1) 12.5 days (2) 25.0 days (3) 37.5 days (4) 75.0 days

14. In the equation $^{226}_{88}Ra \longrightarrow {}^{222}_{86}Rn + X$, X represents (1) a neutron (2) a proton (3) a beta particle (4) an alpha particle.

15. A sample of radioactive strontium-90 has a mass of 1 g after 112.4 years. What was the mass of the original sample? (1) 16 g (2) 12 g (3) 8 g (4) 4 g

16. When a radioactive element forms a chemical bond with another element, its half-life (1) decreases (2) increases (3) remains the same.

17. An originally pure radioactive sample of $^{232}_{90}Th$ now contains atoms of $^{232}_{91}Pa$. This results because some atoms of $^{232}_{90}Th$ each emitted (1) a neutron (2) a beta particle (3) an alpha particle (4) a gamma ray.

18. A sample of two naturally occurring isotopes contains 4×10^{23} atoms of isotope ^{24}X and 2×10^{23} atoms of isotope ^{25}X. The average atomic mass of the element is equal to

    (1) $(24 \times 4) + (25 \times 2)$

    (2) $(24 \times 2) + (25 \times 4)$

    (3) $\dfrac{(24 \times 2) + (25 \times 4)}{6}$

    (4) $\dfrac{(24 \times 4) + (25 \times 2)}{6}$

19. A sample contains 100.0 mg of iodine-131. At the end of 32 days, the number of milligrams of iodine-131 that will remain will be (1) 25.00 (2) 12.50 (3) 6.250 (4) 3.125.

20. Nuclear fusion is not currently available as an energy source because
    (1) we are unable to safely dispose of the resulting radioactive wastes
    (2) the amount of energy produced is too small to make the process practical
    (3) the temperatures at which the reaction takes place are too high to make the process feasible
    (4) it produces greenhouse gases that would produce global warming.

21. The radioactive isotope often used to study organic reaction mechanisms is (1) carbon-12 (2) carbon-14 (3) uranium-235 (4) uranium-238.

22. Which nuclear reaction shows the process called natural transmutation?
    (1) $^{238}_{92}U \longrightarrow {}^{4}_{2}He + {}^{234}_{90}Th$
    (2) $^{235}_{92}U + {}^{1}_{0}n \longrightarrow {}^{143}_{56}Ba + {}^{90}_{36}Kr + 3\,{}^{1}_{0}n$
    (3) $^{9}_{4}Be + {}^{4}_{2}He \longrightarrow {}^{12}_{6}C + {}^{1}_{0}n$
    (4) $^{14}_{7}N + {}^{4}_{2}He \longrightarrow {}^{18}_{9}F$

23. In a fusion reaction, the major problem related to causing the nuclei to fuse into a single nucleus is the (1) small mass of the nucleus (2) large mass of the nucleus (3) attractions of the nuclei (4) repulsions of the nuclei.

24. Which radioisotope is used in diagnosing thyroid disorders? (1) cobalt-60 (2) uranium-235 (3) iodine-131 (4) lead-206

25. A nuclide frequently used in finding the age of the remains of a living organism is (1) ^{238}U (2) ^{14}C (3) ^{60}Co (4) ^{131}I.

26. Which type of radiation could not have its movement altered through the use of a charged plate? (1) gamma rays (2) alpha particles (3) beta particles (4) positron emission

27. During a fusion reaction, the total mass of the materials in the system (1) decreases (2) increases (3) remains the same.

28. During a fission reaction, the total mass of the materials in the system (1) decreases (2) increases (3) remains the same.

CONSTRUCTED RESPONSE

1. A scientist finds that a sample of a radioisotope that initially has a mass of 16 grams has decayed down to a mass of 1.0 gram in 15.28 days.
   (a) What is the half-life of the isotope?
   (b) Based on Table N on page 595, what is the likely identity of this nuclide?
2. $^{238}_{92}U$ is an alpha emitter.
   (a) Write the nuclear reaction for the decay of $^{238}_{92}U$.
   (b) After four additional decay steps, the nuclide $^{226}_{88}Ra$ is produced. Write the four nuclear reactions leading to the formation of this nuclide. (Several different answers are possible.)
3. There is still much debate over whether we should continue to build nuclear power plants.
   (a) Give one scientific argument against the use of nuclear power.
   (b) State one advantage of nuclear power over power generated from fossil fuels.

Lise Meitner: The Greatest Unknown Scientist

Her friends included such towering figures of science as Niels Bohr and Max Planck. Albert Einstein once called her "the German Madame Curie," yet you've probably never heard of Lise Meitner. Nevertheless, her work as a scientist helped lead to a development that really made a lot of noise—the atomic bomb!

Born in Austria in 1878, Meitner entered the University of Vienna in 1901 and in 1906 became the first woman to receive a doctorate in physics from that institution. She moved to Berlin in 1907, where she met an organic chemist named Otto Hahn, with whom she collaborated professionally throughout much of her life. Together, they began working in the emerging field of radioactivity. Their research was interrupted by World War I, during which Meitner volunteered as a radiologist in an Austrian field hospital. In 1917 she and Hahn isolated the most stable isotope of the element protactinium, the radioactive "parent" element of actinium.

Studying radiation and nuclear physics, Meitner pioneered the use of the Geiger counter in this research. After the discovery of the neutron in 1932, Meitner, Hahn, and another colleague began experiments that involved bombarding heavy elements with neutrons. However, by 1938 the social climate in Germany was changing, and Meitner, who was Jewish, fled to Sweden with the help of Niels Bohr.

Even from afar, her involvement in the research remained critical, and she instructed Hahn to bombard uranium with neutrons. When he did so, the experiment yielded puzzling results, which Hahn and other contemporary scientists were at a loss to explain. Yet Meitner, upon reviewing the findings, immediately realized that the uranium nucleus had split, making her the first to recognize the process of nuclear fission. Using Einstein's famous equation $E = mc^2$, she and her colleagues were able to calculate the huge amount of energy released in this process.

Meitner was invited to join the Manhattan Project, which developed the first atomic bomb, but being a committed pacifist, Meitner declined. She was later horrified by the devastation caused by the bomb in Japan. In 1944, Otto Hahn was awarded the Nobel Prize in chemistry for his research on fission, but Meitner was snubbed by the Nobel committee. She moved to England in 1958, and died in 1968, just a few days before her 90th birthday. In 1992, element 109 was named "Meitnerium" in her honor.

Appendix 1
Measurements

As often as possible, chemists use numbers to express their observations. The quantitative approach helps chemists to recognize relationships that might not otherwise be evident and to make important generalizations.

Units of Measurement

In chemistry, as in other sciences, measurements are expressed in units of the metric system or of a modified version of the metric system, the International System of Units (Système Internationale d'Unités), abbreviated SI.

The metric system and the SI are decimal systems of weights and measurements—that is, the units of the systems are related by factors of ten. Prefixes in the names of the units denote the size of the units. The decimal relationship between units and the prefixes are shown in the table of units of length on page 558.

The decimal relationship between units makes conversion from one unit to another an easy operation. Compare the conversion from kilometers to centimeters with the conversion from similar units in the customary system—that is, from miles to inches! This ease in calculations is one reason that the metric system—or SI—is in use in everyday life in most parts of the world today and is coming into use in the United States.

Units in the customary system can be converted into corresponding metric units. The following table lists conversion factors for some units commonly used.

Conversion Factors	
1 inch	= 2.54 centimeters
39.37 inches	= 1 meter
1 pound	= 453.6 grams
1.06 quarts	= 1 liter
(1 milliliter	= 1.000027 cubic centimeters, or approximately 1 cm^3)

Length

The standard unit of length in the metric system and SI is the meter, equal to 39.37 inches, or slightly more than one yard. The centimeter (10^{-2} meter), millimeter (10^{-3} meter), and picometer (10^{-12} meter) are commonly used units of length. The Angstrom unit (Å) (not part of the metric system) is used to express very small measurements, such as the dimensions of atoms and the bond distances within molecules. One Angstrom unit (Å) is 10^{-10} meter, or 10^{-8} centimeter.

Prefix	Meaning	Example
Deci-	One-tenth (10^{-1})	1 *deci*meter = 0.1 meter (10^{-1} m)
Centi-	One-hundredth (10^{-2})	1 *centi*meter = 0.01 meter (10^{-2} m)
Milli-	One-thousandth (10^{-3})	1 *milli*meter = 0.001 meter (10^{-3} m)
Micro-	One-millionth (10^{-6})	1 *micro*meter = 0.000001 meter (10^{-6} m)
Nano-	One-billionth (10^{-9})	1 *nano*meter = 0.000000001 meter (10^{-9} m)
Pico-	One-trillionth (10^{-12})	1 *pico*meter = 0.000000000001 meter (10^{-12} m)
Deka-	Ten times (10^{1})	1 *deka*meter = 10 meters (10^{1} m)
Hecto-	One hundred times (10^{2})	1 *hecto*meter = 100 meters (10^{2} m)
Kilo-	One thousand times (10^{3})	1 *kilo*meter = 1000 meters (10^{3} m)
Mega-	One million times (10^{6})	1 *mega*meter = 1,000,000 meters (10^{6} m)
Giga-	One billion times (10^{9})	1 *giga*meter = 1,000,000,000 meters (10^{9} m)
Tera-	One trillion times (10^{12})	1 *tera*meter = 1,000,000,000,000 meters (10^{12} m)

Volume

The metric unit of volume is the liter, which is slightly larger than the quart (1 liter = 1.06 quarts). The SI unit of volume is the cubic meter, m^3. For common laboratory measurements of volume, the liter and the milliliter are commonly used. The milliliter can be considered equivalent to one cubic centimeter, or 1 cm^3.

0.5 mL = 0.0005 L
1.0 mL = 0.001 L
20.0 mL = 0.020 L
150 mL = 0.150 L

Mass

Newton's first law of motion states that a body at rest tends to remain at rest, and a body in motion tends to remain in motion unless acted on by some outside force. This property of matter is called *inertia*. Inertia is proportional to the amount of matter, or mass, that is present in a body. The mass of a body remains the same anywhere in the universe.

Mass cannot be measured directly. Instead, the gravitational force acting on the mass of a body is measured. In other words, the *weight* of the body is measured. The weight of a body depends on the distance of the body from the center of Earth or from some other object in space, such as the sun.

The weight of a body is proportional to its mass. This means that equal masses have equal weights at equal distances from the center of Earth. The masses of bodies can therefore be compared by weighing the masses.

Metric and SI units of mass in common use are the gram, milligram, and kilogram.

5 mg = 0.005 g	0.5 kg = 500 g
250 mg = 0.250 g	1.0 kg = 1000 g

Temperature

Temperature is a measure of the average kinetic energy of a system. Temperature may be expressed on the Celsius (centigrade) scale and on the Kelvin (absolute) scale.

On the Celsius scale, the freezing point of water at 1 atm is 0°C. The boiling point of water at 1 atmosphere is 100°C.

On the Kelvin scale, zero represents the absence of any molecular motion, or zero kinetic energy. The freezing point of water on the Kelvin scale is 273 K, and the boiling point of water is 373 K. The size of one kelvin is thus the same as one Celsius degree. Temperature can be converted from one scale to the other according to the following relationship:

$$K = degrees\ C + 273$$

It is important to remember that Kelvin scale truly measure average kinetic energy. This is not true of other temperature scales. When a gas at constant pressure is heated from 20°C to 40°C, the kinetic energy of the molecules does not double—it increases from 293 K (273 + 20°C) to 313 K (273 + 40°C). When a gas at constant pressure is heated from 293 K (20°C) to 586 K (313°C), the kinetic energy of the molecules doubles (and the volume doubles).

Heat Energy

Heat is conveniently measured by transferring it to a liquid, usually water. The heat transfer is usually carried out in a calorimeter, which contains a known mass of water at a known temperature. The quantity of heat needed to raise the temperature of 1 gram of water 1 K is 4.18 joules. This number, called the specific heat capacity of water, is used to calculate the amount of heat absorbed by a given sample of water.

$$\text{Heat} = 4.18\text{J/g} \cdot \text{K} \times \text{g}_{\text{water}} \times \Delta\text{t}$$

Suppose that a calorimeter contains 200. grams of water at a temperature of 294.0 K. As a heat producing, or exothermic, reaction takes place in the calorimeter, the water temperature rises to 314.0 K. This means that the heat given off by the reaction caused 200. grams of water to undergo a temperature change of 20.0 K. The quantity of heat transferred is

$$200.\ \text{grams} \times 4.18\ \text{J/g} \cdot \text{K} \times 20.0\ \text{K} = 16{,}700\ \text{joules, or } 16.7\ \text{kilojoules.}$$

(Rounded to 3 significant figures.)

Although the SI unit of heat is the joule, heat is still often expressed in calories.

$$1.00 \text{ calorie} = 4.18 \text{ joules.}$$

Pressure

The SI unit of pressure is the pascal, which is based on the metric units of force (the newton) and area (the square meter). The pascal is a very small amount of pressure; standard atmospheric pressure is 101,300 pascals. Because the pascal is such a small unit, pressure is often expressed in kilopascals, abbreviated kPa.

Other units of pressure still in popular use include the atmosphere and the torr. One atmosphere is defined as standard atmospheric pressure, which is equal to 101.3 kPa. The torr is a unit based on the mercury barometer. In a barometer, a column of mercury 1-millimeter high is defined as one torr. Standard atmospheric pressure is 760 torr.

$$101.3 \text{ kPa} = 1.000 \text{ atm} = 760.0 \text{ torr}$$

Since the prefixes used in the metric system (see page 558) are the same for all metric units, you can easily convert between metric units.

SAMPLE PROBLEMS

PROBLEM

1. Convert a heat of 14 joules to kilojoules.

SOLUTION

The prefix *kilo* means 1000, therefore, 1 kJ = 1000 J.

$$14 \text{ J} \times \frac{1 \text{ kJ}}{1000 \text{ J}} = 0.014 \text{ kJ}$$

The expression $\frac{1 \text{ kJ}}{1000 \text{ J}}$ is called a conversion factor. A conversion factor is the ratio of the two units. When converting from one unit to another, multiply by the conversion factor that has the desired unit in the numerator.

PROBLEM
2. Convert 21.6 liters to milliliters.

SOLUTION
There are 1000 mL in 1 L. Since the conversion is to mL, use the ratio with the mL in the numerator.

$$21.6 \text{ L} \times \frac{1000 \text{ mL}}{1 \text{ L}} = 21{,}600 \text{ mL}$$

PRACTICE

A.1 How many grams are in 125 milligrams?
A.2 How many liters are in 250 milliliters?
A.3 If you are 183 cm tall, what is your height in meters?
A.4 How many millimeters are in 25.3 centimeters?
A.5 Given that 2.2 pounds = 1 kilogram. If your mass is 62,000 grams, what is your weight in pounds?

Uncertainty In Measurement

Suppose that you mass a block of metal on three different centigram balances and obtain the following masses: 14.61 grams, 14.62 grams, and 14.63 grams. What is the mass of the block? In the three massings, you obtained only the first three digits each time. In other words, the last digit in each mass is uncertain. You can accurately describe the mass of the block only as lying between 14.61 grams, 14.62 grams, and 14.63 grams. You could mass the block on any number of centigram balances, and you would continue to obtain values that agree only in the first three digits. This situation and many similar ones suggest that it is impossible to reproduce a series of measurements without error.

The limitation of measurement caused by errors is called *uncertainty*. Uncertainty results from shortcomings of the experimenter and of the equipment used. All experimental observations, which we call facts, are therefore uncertain to some degree. Uncertainty can be reduced by increasing the skills and powers of observation of the experimenter and by refining equipment, but it can never be eliminated.

Precision and Accuracy

Measurements can be described as *precise* or as *accurate.* Although these terms are often used interchangeably, they do not have the same meaning. *Precision* refers to the reproducibility of a measurement. *Accuracy* refers to the closeness of a measurement to the accepted value of the measurement. For example, a scientist finds the molar mass of oxygen to be 37.1, 37.4, and 36.9. The measurements were closely reproduced in each determination and are therefore precise. However, the accepted value for the molar mass of oxygen is 32.0. The measurements are therefore not accurate.

Whenever possible, you should indicate the precision or accuracy of the measurements you make in your work in chemistry. The metal block that weighs between 14.61 grams and 14.63 grams can be described as weighing 14.62 grams $\pm$ 0.01 gram. The symbol $\pm$ (plus or minus) expresses the range of uncertainty in this measurement.

If a measurement you obtain can be compared with a known or accepted value for this measurement, you should indicate your *experimental error*—the difference between the observed value and the accepted value. Suppose that in a gas law experiment, you determine the molar mass of oxygen, O_2, to be 30.0 grams. The accepted value for this measurement, which you can find in a reference book, is 32.0 grams. The error in your observation is 30.0 g − 32.0 g = −2.0 grams. The minus sign indicates that your measurement is lower than the accepted value. Scientists often express error as an absolute value; a statement that the measurement was off by two grams does not indicate whether the experimental value was too high or too low.

Experimental error is often expressed as *percentage error* or *percent error,* which is calculated as follows:

$$\text{Percent error} = \frac{(\text{observed value} - \text{accepted value})}{\text{accepted value}} \times 100\%$$

The results you obtained in measuring the molar mass of oxygen involved a percent error of

$$\frac{30.0\text{ g} - 32.0\text{ g}}{32.0\text{ g}} \times 100\% = -6.3\%$$

Again, the negative value indicates that the observed value was lower than the accepted value. However, percent error may be expressed as an absolute value. The student might have reported the percent error as 6.3% without meaning to imply whether the result was too high or too low.

PRACTICE

A.6 In a laboratory experiment, a student determines the atomic mass of an oxygen atom to be 18.0 amu. What is the percent error in the experiment?

A.7 If my scale tells me that I weigh 189 pounds, when I actually weigh 180. pounds, what is the percent error?

A.8 The percent water in the hydrate $CuSO_4 \cdot 5H_2O$ is 36%. Two students performing an experiment to determine the percent water in that substance obtain two different answers, but both report percent errors of 10.%. What were the two values obtained by the students?

Expressing Uncertainty with Significant Figures

The measurement of 14.62 grams that you found for the block of metal contains three certain digits—1, 4, and 6—and one doubtful digit—2. This measurement is said to contain four significant figures, or sig figs. The numbers that express a measurement (including the last digit, which is doubtful) are called *significant figures.*

The number of significant figures in a measurement is determined by the calibration of the measuring instrument. On instruments with digital readouts, the last digit is the doubtful place. If a digital balance records a mass as 34.62 grams, the 2 in the hundredths place is the doubtful digit. On instruments such as the

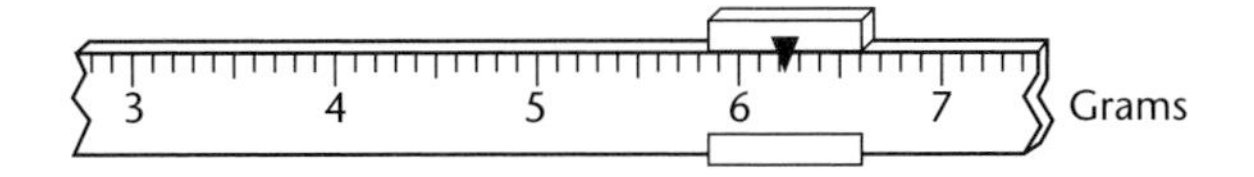

triple-beam balance, graduated cylinder, or burette, a correct reading should include one estimated place. The diagram above shows a section of a triple-beam balance, with the rider, used to read from 0 to 10 grams, in the position indicated. How should the mass be correctly expressed? If you said 6.2 grams, you left out the estimated place. The correct reading could be 6.21 grams, or 6.22 grams; it is difficult to tell. That is why it is referred to as an estimated place!

Rules for Working with Significant Figures

Rules for using and interpreting significant figures follow:

1. All nonzero digits are significant figures. The measurement 6.45 grams has 3 significant figures.
2. Zeros to the left of all nonzero digits are not significant figures. When the above measurement is converted to kilograms it becomes 0.00645 kg. It still has 3 significant figures.
3. Zeros to the right of all nonzero digits are assumed to be significant figures if, and only if, the number contains a written decimal point. The measurement 6.45 grams is 6450 milligrams. It still has three significant figures. The zero is not a significant figure, because the number does not contain a decimal point. Consider the measurement 65.0 grams. This measurement has three significant figures. The final zero is a significant figure, because the number has a written decimal point. If you convert the 65.0 grams to kilograms, you would write it as 0.0650 kg. The zeros to the left are not significant, (see rule 2) but the one to the right is, since the number has a decimal point. There are still 3 significant figures in the measurement.
4. When you add or subtract measurements, your result should contain the same number of decimal places as the quantity with the fewest decimal places. For example, if you were adding 3.5 cm + 0.12 cm your answer should be rounded off to 3.6 cm. You cannot write 3.62 cm, because one of the measurements has no value in the hundredths place. If you

write the numbers to be added or subtracted one under the other, you must round off to the last complete column.

$$\begin{array}{r} 3.5 \text{ cm} \\ + \underline{0.12} \text{ cm} \\ 3.62 \text{ cm} \end{array}$$

This must be rounded to 3.6 cm. The hundredths column is incomplete; there is no value there in the first measurement. Therefore you cannot include the hundredths in your answer.

5. Multiplying or dividing a measurement by a definite number does not change the number of significant figures. For example, if you are asked to double the mass of an object whose mass is 3.45g, you would keep the three significant figures, and write 6.90 grams.
6. In multiplication and division of measurements, the result can contain no more significant figures than are contained in the least certain measurement.

   The product obtained by multiplying 4.12 meters by 2.1 meters can contain only two significant figures. The product (8.652) must be rounded off to contain the proper number of significant figures. Thus 4.12 meters × 2.1 meters equals 8.7 square meters. (Use the preceding rules for rounding off 8.652 to two significant figures.)

PRACTICE

A.9 Indicate the number of significant figures in each of the following measurements.

(a) 6.30 grams

(b) 0.00453 liters

(c) 85,000 grams

(d) 0.003030 kg

A.10 Perform the following operations and express your answers to the correct number of significant figures.

(a) A mass of 5.51 grams divided by a volume of 4.2 liters

(b) A volume of 49.2 mL minus a volume of 48 mL

(c) A pressure of 101.3 kPa multiplied by a volume of 15.0 liters.

A.11 To the correct number of significant figures, find the volume of a cube with a side of 3.5 cm.

A.12 In an experiment to find the density of a metal, a student obtains the following data.

Volume of water in cylinder:	41.4 mL
Volume of water + metal sample:	45.4 mL
Mass of metal sample:	24.3 grams

Find the density of the metal to the correct number of significant figures.

A.13 A small piece of metal is correctly placed on the pan of a triple-beam balance. The riders are all at the zero mark except for the rider on the 0–10-gram beam, which is located at the position shown below. What is the mass of the metal?

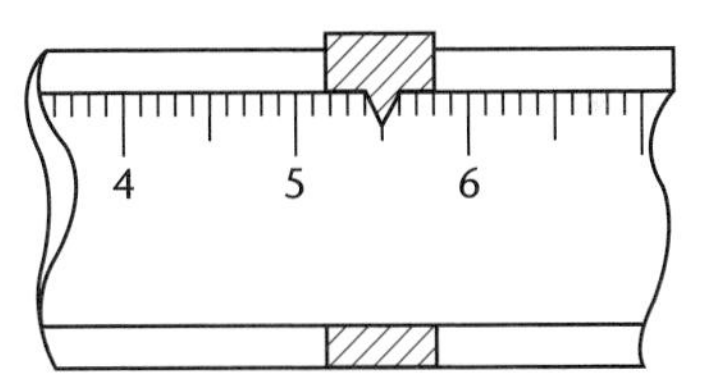

Appendix 2 Scientific Notation and Calculators

Expressing very small or very large numbers with zeros is awkward and increases the chances for making errors in calculations. Instead, very small or very large numbers are written as decimal numbers between 1 and 10 multiplied by 10 raised to an appropriate power. For example, 2,000,000,000 kilograms (2 gigagrams) is written as 2×10^9 kilograms. Similarly, 0.000001 gram (1 microgram) is written as 10^{-6} gram. This system of expressing numbers is called *scientific notation.*

The use of scientific notation helps to clarify the number of significant figures in a measurement. When the number 0.0072 gram is written as 7.2×10^{-3} gram, only the significant figures appear in the number. Suppose you wanted to express the measurement 6.70 kilograms in grams. It would be 6700 grams. However, while 6.70 has three significant figures, 6700 would be assumed to have only two. The problem is resolved if you use scientific notation. The measurement is 6.70×10^3 grams, and has three significant figures.

To express a number in scientific notation, move the decimal point until it is directly after the first nonzero digit. The number of places the point is moved gives the power of 10. In the number 36,400, you must move the decimal 4 places to the left to get 3.64×10^4. In the number 0.0024, you must move the decimal 3 places to the right, to get 2.4×10^{-3}. Notice that for large numbers, where the decimal is moved to the left, the exponent is positive, and for small numbers, where the decimal is moved to the right, the exponent is negative.

Scientific Calculators

Scientific calculators are approved for use on most standard chemistry examinations, and it is important to become familiar with them. They are particularly helpful when working with numbers in scientific notation.

The calculator key used in scientific notation is usually identified either with the letters EE or Exp. If you wish to enter the number 4.6×10^5, for example, you must enter 4.6 EE 5. Most calculators will now read 4.6 05. Note that when entering scientific notation, you do not use the "×" key. The EE function should be thought of as meaning "times ten to the . . . " Entry of numbers with negative exponents varies with the calculator. On some calculators, you enter the number the same way you would for a positive exponent, and then hit the +/− key. To enter 4.6×10^{-5} you would enter 4.6 EE 5 +/−. On other calculators, you press the EE key and hit the − before the exponent number (4.6 EE −5).

Most calculators also provide a way to convert numbers into and out of scientific notation. These vary considerably from one calculator to another, but it is worth your while to read the instructions that came with your calculator, and learn how to do this.

Beware of the following common error, when using a scientific calculator. Suppose you wish to find the value of the expression $\frac{15 \times 13}{12 \times 11}$. If you enter "15 × 13 ÷ 12 × 11," your calculator will read 178.75. Obviously, this is not the correct answer! The calculator is programmed to take the results of the first three actions and multiply them by 11, while you need to multiply only the denominator by 11. To avoid this error, make it a practice to use the times key for all multiplication in the numerator, but use the ÷ key for all multiplication in the denominator. If you try 15 × 13 ÷ 12 ÷ 11, your calculator will read 1.477, the correct answer.

Calculator Practice: Perform the following calculations on your scientific calculator.

1. $\frac{3.4 \times 10^4}{2 \times 10^{-2}}$

2. $5.3 \times 10^4 \times 6.2 \times 10^5$

3. $\dfrac{545 \times 3.4 \times 10^5}{21 \times 6.2 \times 10^{-4}}$

4. The combined gas law can be written: $V_2 = \dfrac{P_1V_1T_2}{P_2T_1}$

Use your calculator to find the new volume of a gas, V_2, when V_1 = 20.0 mL, P_1 = 120. kPa, T_1 = 300. K, T_2 = 273 K, and P_2 = 185 kPa.

Answers

1. 1.7×10^6
2. 3.3×10^{10}
3. 1.4×10^{10}
4. 11.8 liters

Appendix 3 Laboratory Preparations

To study the properties of elements and compounds, it is necessary, when possible, to prepare and collect small samples of the substances. A sample is prepared by use of a reaction in which the element or compound is formed. To insure that a sample of sufficient quantity will be obtained, the reaction should be one that goes to completion.

The reaction takes place in a vessel, such as a test tube or a flask, which is called the *generator.* The method by which the sample is collected depends on the physical properties of the substance. If the sample is a gas, it can be collected in one of two ways, depending on the solubility and density of the gas. Gases that have limited solubility in water are collected by *water displacement.* This method is used to collect oxygen and hydrogen, for example. Gases that are soluble in water are collected by *air displacement.* Ammonia is one of the gases collected by this method. A third method, *condensation,* is used in some special cases for gases, such as bromine, that can be condensed easily.

Caution: Special methods of preparation and collection are important for certain substances because they are dangerous to handle. Students should not perform any of the following preparations without the approval and supervision of their teacher.

Preparation and Collection of Gases

Diagrams for the preparation and collection of the following gases are included: Oxygen, hydrogen, chlorine, carbon dioxide, sulfur dioxide, ammonia, and hydrogen chloride. (Following the section on the collection of gases you will find diagrams for the Preparation and Collection of Volatile Liquids and Solids.)

Oxygen

$$2\ H_2O_2\ (l) \longrightarrow 2\ H_2O + O_2\ (g)$$

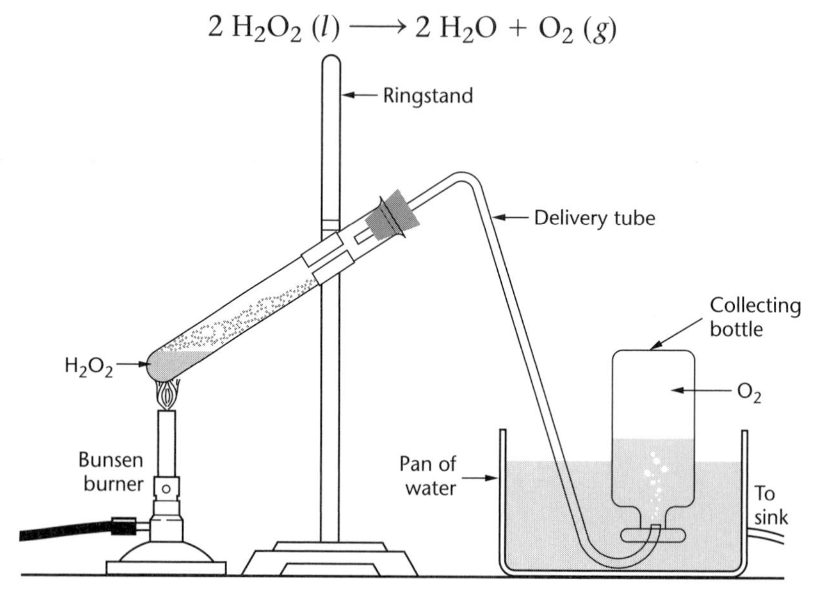

Hydrogen

$$(1)\ Zn\ (s) + H_2SO_4\ (aq) \longrightarrow ZnSO_4\ (aq) + H_2\ (g)$$

$$(2)\ Mg\ (s) + 2\ HCl\ (aq) \longrightarrow MgCl_2\ (aq) + H_2\ (g)$$

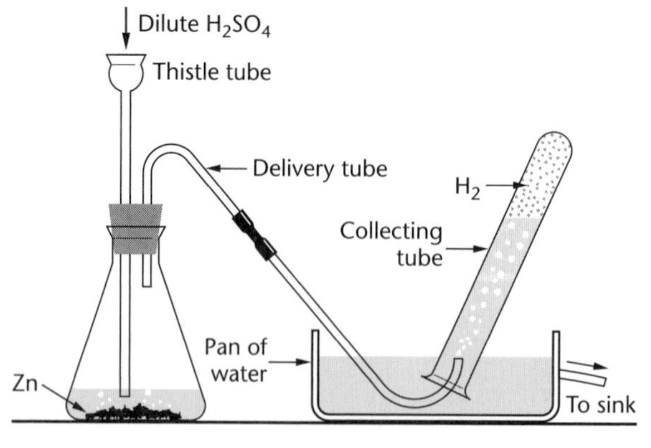

Chlorine

(1) $MnO_2\ (s) + 4\ HCl\ (aq) \longrightarrow MnCl_2\ (aq) + 2\ H_2O + Cl_2\ (g)$

(2) $2\ KMnO_4\ (s) + 16\ HCl\ (aq) \longrightarrow$
$2KCl\ (aq) + 2\ MnCl_2\ (aq) + 8\ H_2O + 5\ Cl_2\ (g)$

(3) $2\ NaCl\ (s) + 2\ H_2SO_4\ (aq) + MnO_2\ (s) \longrightarrow$
$Na_2SO_4\ (aq) + MnSO_4\ (aq) + 2\ H_2O + Cl_2\ (g)$

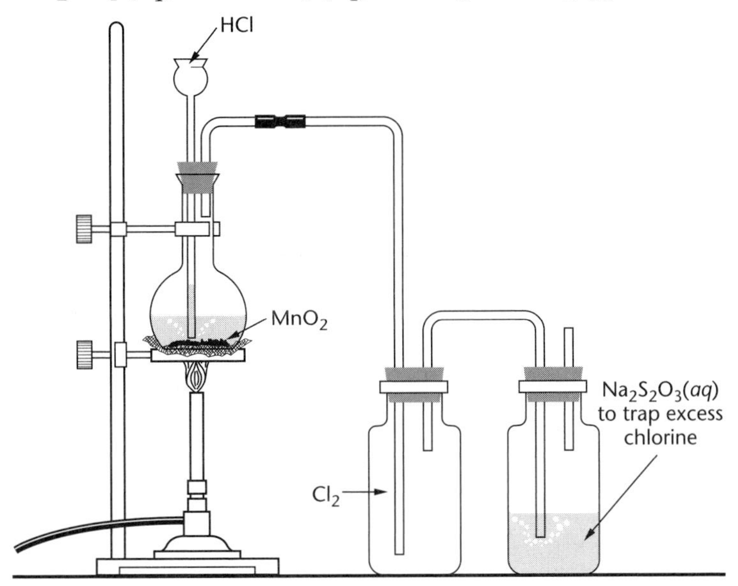

Carbon Dioxide

(1) $CaCO_3\ (s) + 2\ HCl\ (aq) \longrightarrow CaCl_2\ (aq) + H_2O + CO_2\ (g)$

(2) $NaHCO_3\ (s) + HCl\ (aq) \longrightarrow NaCl\ (aq) + H_2O + CO_2\ (g)$

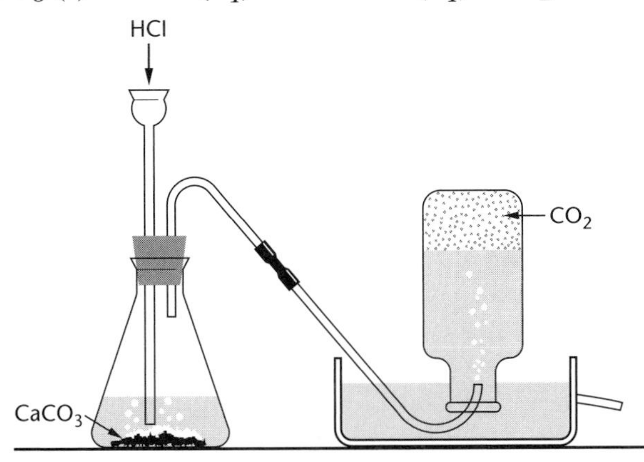

Sulfur Dioxide

(1) $Na_2SO_3\ (s) + 2\ HCl\ (aq) \longrightarrow 2\ NaCl\ (aq) + H_2O + SO_2\ (g)$

(2) $NaHSO_3\ (s) + HCl\ (aq) \longrightarrow NaCl\ (aq) + H_2O + SO_2\ (g)$

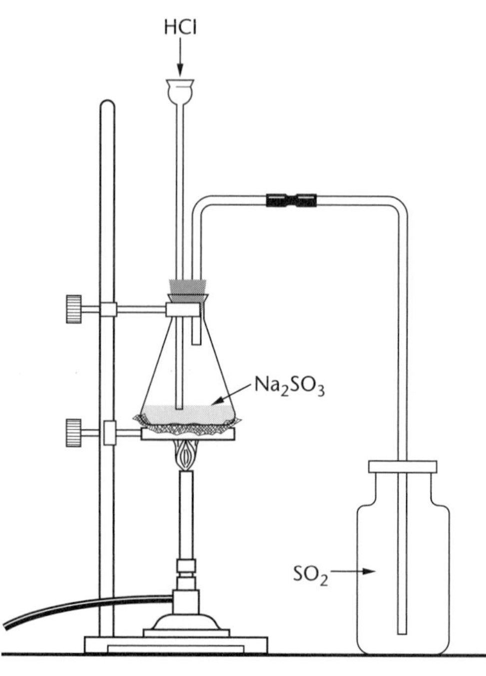

Ammonia

$2\ NH_4Cl\ (s) + Ca(OH)_2\ (s) \longrightarrow CaCl_2\ (s) + 2\ H_2O + 2\ NH_3\ (g)$

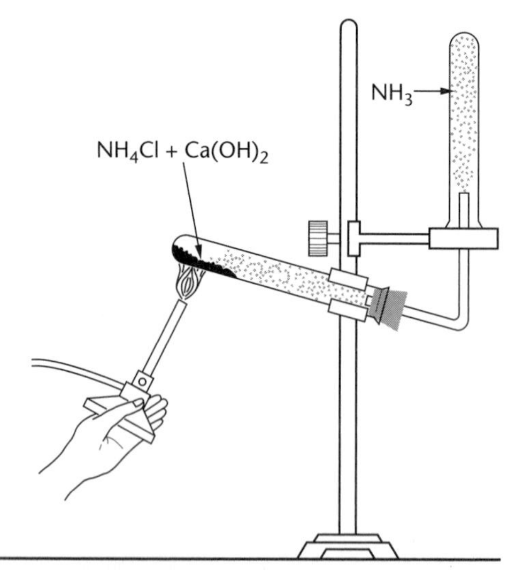

Hydrogen Chloride

$$2\ \mathrm{NaCl}\ (s) + \mathrm{H_2SO_4}\ (aq) \longrightarrow \mathrm{Na_2SO_4}\ (aq) + 2\ \mathrm{HCl}\ (g)$$

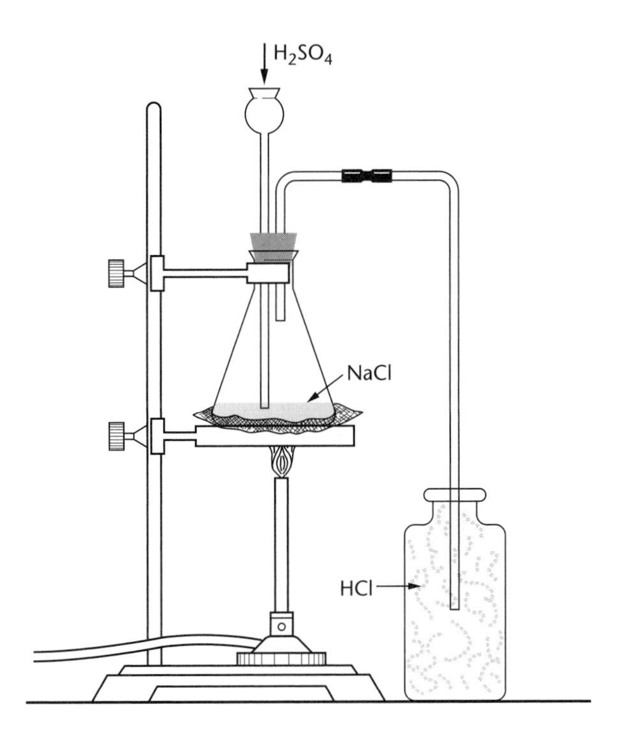

Preparation and Collection of Volatile Liquids

Bromine

$$2\ KBr\ (s) + 2\ H_2SO_4\ (aq) + MnO_2\ (s) \longrightarrow K_2SO_4\ (aq) + MnSO_4\ (aq) + 2\ H_2O + Br_2\ (l)$$

Nitric Acid

$$2\ NaNO_3\ (s) + H_2SO_4\ (l) \longrightarrow Na_2SO_4\ (s) + 2\ HNO_3\ (l)$$

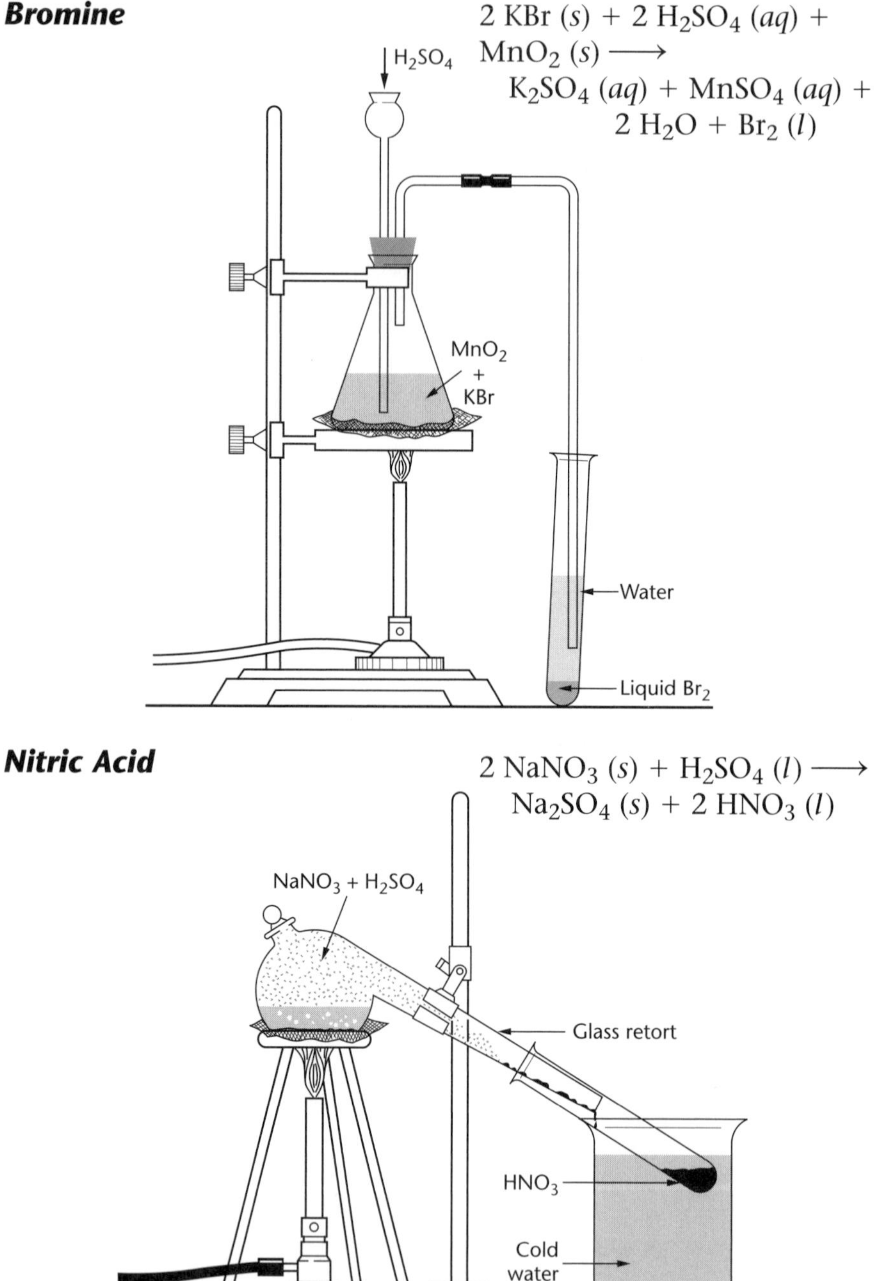

Preparation and Collection of a Volatile Solid

Iodine

$$2\ KI\ (s) + 2\ H_2SO_4\ (aq) + MnO_2\ (s) \longrightarrow K_2SO_4\ (aq) + MnSO_4\ (aq) + 2\ H_2O + I_2\ (s)$$

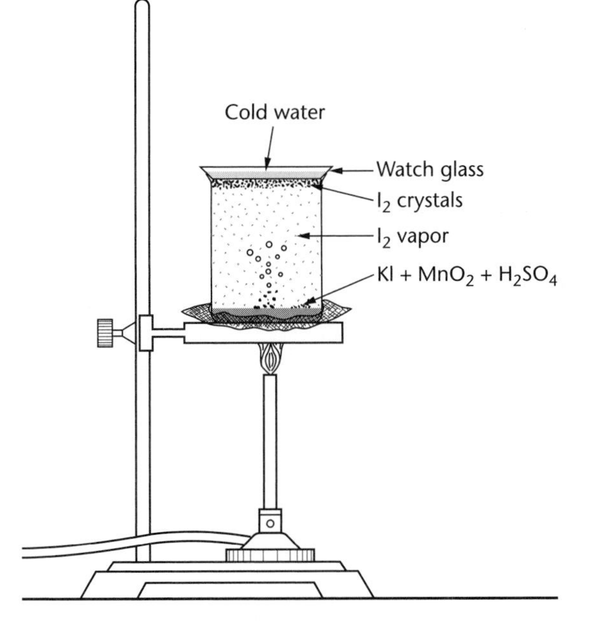

Appendix 4
Reference Tables

The following pages contain the Reference Tables for the Physical Setting—Chemistry. Students should become familiar with these tables and their uses. Being able to properly apply the information provided in the Reference Tables will help you answer many of the questions that you encounter on examinations.

C Reference Tables for Physical Setting/CHEMISTRY

2002 Edition

Table A
Standard Temperature and Pressure

Name	Value	Unit
Standard Pressure	101.3 kPa 1 atm	kilopascal atmosphere
Standard Temperature	273 K 0°C	kelvin degree Celsius

Table B
Physical Constants for Water

Heat of Fusion	334 J/g
Heat of Vaporization	2260 J/g
Specific Heat Capacity of H_2O (ℓ)	4.18 J/g•K

Table C
Selected Prefixes

Factor	Prefix	Symbol
10^3	kilo-	k
10^{-1}	deci-	d
10^{-2}	centi-	c
10^{-3}	milli-	m
10^{-6}	micro-	μ
10^{-9}	nano-	n
10^{-12}	pico-	p

Table D
Selected Units

Symbol	Name	Quantity
m	meter	length
g	gram	mass
Pa	pascal	pressure
K	kelvin	temperature
mol	mole	amount of substance
J	joule	energy, work, quantity of heat
s	second	time
L	liter	volume
ppm	part per million	concentration
M	molarity	solution concentration

Table E
Selected Polyatomic Ions

H_3O^+	hydronium	CrO_4^{2-}	chromate
Hg_2^{2+}	dimercury (I)	$Cr_2O_7^{2-}$	dichromate
NH_4^+	ammonium	MnO_4^-	permanganate
$C_2H_3O_2^-$ CH_3COO^-	acetate	NO_2^-	nitrite
		NO_3^-	nitrate
CN^-	cyanide	O_2^{2-}	peroxide
CO_3^{2-}	carbonate	OH^-	hydroxide
HCO_3^-	hydrogen carbonate	PO_4^{3-}	phosphate
$C_2O_4^{2-}$	oxalate	SCN^-	thiocyanate
ClO^-	hypochlorite	SO_3^{2-}	sulfite
ClO_2^-	chlorite	SO_4^{2-}	sulfate
ClO_3^-	chlorate	HSO_4^-	hydrogen sulfate
ClO_4^-	perchlorate	$S_2O_3^{2-}$	thiosulfate

Table F
Solubility Guidelines for Aqueous Solutions

Ions That Form *Soluble* Compounds	Exceptions
Group 1 ions (Li^+, Na^+, etc.)	
ammonium (NH_4^+)	
nitrate (NO_3^-)	
acetate ($C_2H_3O_2^-$ or CH_3COO^-)	
hydrogen carbonate (HCO_3^-)	
chlorate (ClO_3^-)	
perchlorate (ClO_4^-)	
halides (Cl^-, Br^-, I^-)	when combined with Ag^+, Pb^{2+}, and Hg_2^{2+}
sulfates (SO_4^{2-})	when combined with Ag^+, Ca^{2+}, Sr^{2+}, Ba^{2+}, and Pb^{2+}

Ions That Form *Insoluble* Compounds	Exceptions
carbonate (CO_3^{2-})	when combined with Group 1 ions or ammonium (NH_4^+)
chromate (CrO_4^{2-})	when combined with Group 1 ions, Ca^{2+}, Mg^{2+}, or ammonium (NH_4^+)
phosphate (PO_4^{3-})	when combined with Group 1 ions or ammonium (NH_4^+)
sulfide (S^{2-})	when combined with Group 1 ions or ammonium (NH_4^+)
hydroxide (OH^-)	when combined with Group 1 ions, Ca^{2+}, Ba^{2+}, Sr^{2+}, or ammonium (NH_4^+)

Table G Solubility Curves

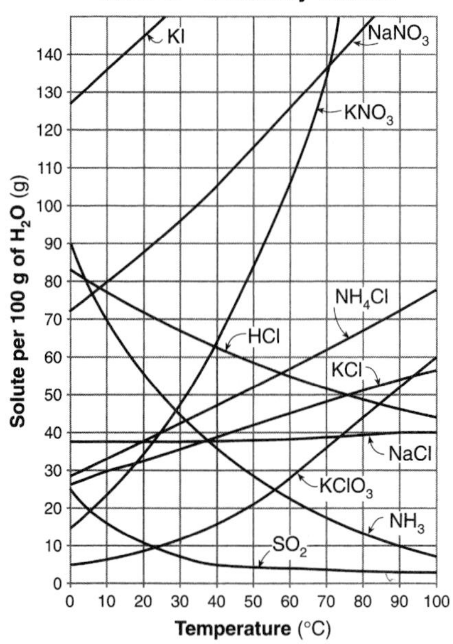

Table H
Vapor Pressure of Four Liquids

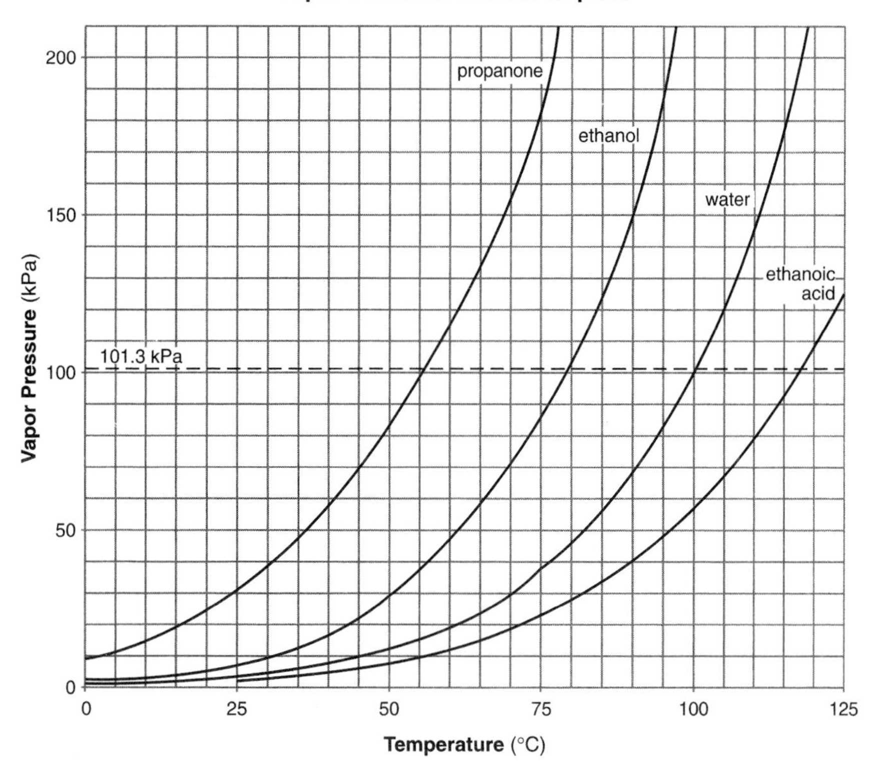

Table I
Heats of Reaction at 101.3 kPa and 298 K

Reaction	ΔH (kJ)*
$CH_4(g) + 2O_2(g) \longrightarrow CO_2(g) + 2H_2O(\ell)$	–890.4
$C_3H_8(g) + 5O_2(g) \longrightarrow 3CO_2(g) + 4H_2O(\ell)$	–2219.2
$2C_8H_{18}(\ell) + 25O_2(g) \longrightarrow 16CO_2(g) + 18H_2O(\ell)$	–10943
$2CH_3OH(\ell) + 3O_2(g) \longrightarrow 2CO_2(g) + 4H_2O(\ell)$	–1452
$C_2H_5OH(\ell) + 3O_2(g) \longrightarrow 2CO_2(g) + 3H_2O(\ell)$	–1367
$C_6H_{12}O_6(s) + 6O_2(g) \longrightarrow 6CO_2(g) + 6H_2O(\ell)$	–2804
$2CO(g) + O_2(g) \longrightarrow 2CO_2(g)$	–566.0
$C(s) + O_2(g) \longrightarrow CO_2(g)$	–393.5
$4Al(s) + 3O_2(g) \longrightarrow 2Al_2O_3(s)$	–3351
$N_2(g) + O_2(g) \longrightarrow 2NO(g)$	+182.6
$N_2(g) + 2O_2(g) \longrightarrow 2NO_2(g)$	+66.4
$2H_2(g) + O_2(g) \longrightarrow 2H_2O(g)$	–483.6
$2H_2(g) + O_2(g) \longrightarrow 2H_2O(\ell)$	–571.6
$N_2(g) + 3H_2(g) \longrightarrow 2NH_3(g)$	–91.8
$2C(s) + 3H_2(g) \longrightarrow C_2H_6(g)$	–84.0
$2C(s) + 2H_2(g) \longrightarrow C_2H_4(g)$	+52.4
$2C(s) + H_2(g) \longrightarrow C_2H_2(g)$	+227.4
$H_2(g) + I_2(g) \longrightarrow 2HI(g)$	+53.0
$KNO_3(s) \xrightarrow{H_2O} K^+(aq) + NO_3^-(aq)$	+34.89
$NaOH(s) \xrightarrow{H_2O} Na^+(aq) + OH^-(aq)$	–44.51
$NH_4Cl(s) \xrightarrow{H_2O} NH_4^+(aq) + Cl^-(aq)$	+14.78
$NH_4NO_3(s) \xrightarrow{H_2O} NH_4^+(aq) + NO_3^-(aq)$	+25.69
$NaCl(s) \xrightarrow{H_2O} Na^+(aq) + Cl^-(aq)$	+3.88
$LiBr(s) \xrightarrow{H_2O} Li^+(aq) + Br^-(aq)$	–48.83
$H^+(aq) + OH^-(aq) \longrightarrow H_2O(\ell)$	–55.8

*Minus sign indicates an exothermic reaction.

Table J
Activity Series**

Most ↓ Least	Metals	Nonmetals	Most ↓ Least
	Li	F_2	
	Rb	Cl_2	
	K	Br_2	
	Cs	I_2	
	Ba		
	Sr		
	Ca		
	Na		
	Mg		
	Al		
	Ti		
	Mn		
	Zn		
	Cr		
	Fe		
	Co		
	Ni		
	Sn		
	Pb		
	**H_2		
	Cu		
	Ag		
	Au		

**Activity Series based on hydrogen standard

Note: H_2 is *not* a metal

Table K
Common Acids

Formula	Name
$HCl(aq)$	hydrochloric acid
$HNO_3(aq)$	nitric acid
$H_2SO_4(aq)$	sulfuric acid
$H_3PO_4(aq)$	phosphoric acid
$H_2CO_3(aq)$ or $CO_2(aq)$	carbonic acid
$CH_3COOH(aq)$ or $HC_2H_3O_2(aq)$	ethanoic acid (acetic acid)

Table L
Common Bases

Formula	Name
$NaOH(aq)$	sodium hydroxide
$KOH(aq)$	potassium hydroxide
$Ca(OH)_2(aq)$	calcium hydroxide
$NH_3(aq)$	aqueous ammonia

Table M
Common Acid–Base Indicators

Indicator	Approximate pH Range for Color Change	Color Change
methyl orange	3.2–4.4	red to yellow
bromthymol blue	6.0–7.6	yellow to blue
phenolphthalein	8.2–10	colorless to pink
litmus	5.5–8.2	red to blue
bromcresol green	3.8–5.4	yellow to blue
thymol blue	8.0–9.6	yellow to blue

Table N
Selected Radioisotopes

Nuclide	Half-Life	Decay Mode	Nuclide Name
^{198}Au	2.69 d	β^-	gold-198
^{14}C	5730 y	β^-	carbon-14
^{37}Ca	175 ms	β^+	calcium-37
^{60}Co	5.26 y	β^-	cobalt-60
^{137}Cs	30.23 y	β^-	cesium-137
^{53}Fe	8.51 min	β^+	iron-53
^{220}Fr	27.5 s	α	francium-220
^{3}H	12.26 y	β^-	hydrogen-3
^{131}I	8.07 d	β^-	iodine-131
^{37}K	1.23 s	β^+	potassium-37
^{42}K	12.4 h	β^-	potassium-42
^{85}Kr	10.76 y	β^-	krypton-85
^{16}N	7.2 s	β^-	nitrogen-16
^{19}Ne	17.2 s	β^+	neon-19
^{32}P	14.3 d	β^-	phosphorus-32
^{239}Pu	2.44×10^4 y	α	plutonium-239
^{226}Ra	1600 y	α	radium-226
^{222}Rn	3.82 d	α	radon-222
^{90}Sr	28.1 y	β^-	strontium-90
^{99}Tc	2.13×10^5 y	β^-	technetium-99
^{232}Th	1.4×10^{10} y	α	thorium-232
^{233}U	1.62×10^5 y	α	uranium-233
^{235}U	7.1×10^8 y	α	uranium-235
^{238}U	4.51×10^9 y	α	uranium-238

ms = milliseconds; s = seconds; min = minutes;
h = hours; d = days; y = years

Table O
Symbols Used in Nuclear Chemistry

Name	Notation	Symbol
alpha particle	$^{4}_{2}He$ or $^{4}_{2}\alpha$	α
beta particle (electron)	$^{0}_{-1}e$ or $^{0}_{-1}\beta$	β^-
gamma radiation	$^{0}_{0}\gamma$	γ
neutron	$^{1}_{0}n$	n
proton	$^{1}_{1}H$ or $^{1}_{1}p$	p
positron	$^{0}_{+1}e$ or $^{0}_{+1}\beta$	β^+

Table P
Organic Prefixes

Prefix	Number of Carbon Atoms
meth-	1
eth-	2
prop-	3
but-	4
pent-	5
hex-	6
hept-	7
oct-	8
non-	9
dec-	10

Table Q
Homologous Series of Hydrocarbons

Name	General Formula	Examples	
		Name	Structural Formula
alkanes	C_nH_{2n+2}	ethane	H H | | H–C–C–H | | H H
alkenes	C_nH_{2n}	ethene	H H \ / C=C / \ H H
alkynes	C_nH_{2n-2}	ethyne	H–C≡C–H

n = number of carbon atoms

of the Elements

18
4.00260 0 **He** 2 2

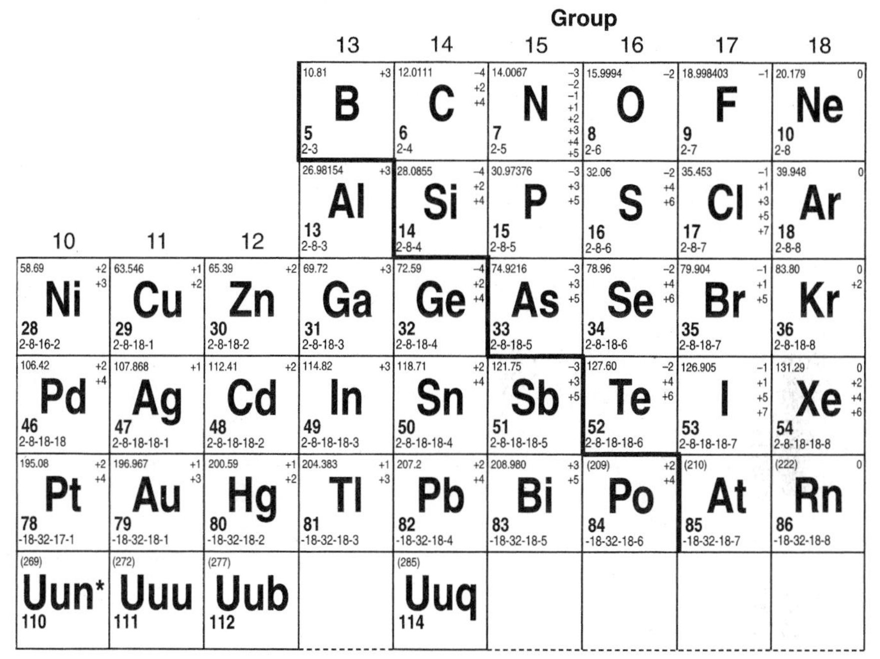

Group

10	11	12	13	14	15	16	17	18
			10.81 +3 **B** 5 2-3	12.0111 −4 +2 +4 **C** 6 2-4	14.0067 −3 −2 −1 +1 +2 +3 +4 +5 **N** 7 2-5	15.9994 −2 **O** 8 2-6	18.998403 −1 **F** 9 2-7	20.179 0 **Ne** 10 2-8
			26.98154 +3 **Al** 13 2-8-3	28.0855 −4 +2 +4 **Si** 14 2-8-4	30.97376 −3 +3 +5 **P** 15 2-8-5	32.06 −2 +4 +6 **S** 16 2-8-6	35.453 −1 +1 +3 +5 +7 **Cl** 17 2-8-7	39.948 0 **Ar** 18 2-8-8
58.69 +2 +3 **Ni** 28 2-8-16-2	63.546 +1 +2 **Cu** 29 2-8-18-1	65.39 +2 **Zn** 30 2-8-18-2	69.72 +3 **Ga** 31 2-8-18-3	72.59 −4 +2 +4 **Ge** 32 2-8-18-4	74.9216 −3 +3 +5 **As** 33 2-8-18-5	78.96 −2 +4 +6 **Se** 34 2-8-18-6	79.904 −1 +1 +5 **Br** 35 2-8-18-7	83.80 0 +2 **Kr** 36 2-8-18-8
106.42 +2 +4 **Pd** 46 2-8-18-18	107.868 +1 **Ag** 47 2-8-18-18-1	112.41 +2 **Cd** 48 2-8-18-18-2	114.82 +3 **In** 49 2-8-18-18-3	118.71 +2 +4 **Sn** 50 2-8-18-18-4	121.75 −3 +3 +5 **Sb** 51 2-8-18-18-5	127.60 −2 +4 +6 **Te** 52 2-8-18-18-6	126.905 −1 +1 +5 +7 **I** 53 2-8-18-18-7	131.29 0 +2 +4 +6 **Xe** 54 2-8-18-18-8
195.08 +2 +4 **Pt** 78 -18-32-17-1	196.967 +1 +3 **Au** 79 -18-32-18-1	200.59 +1 +2 **Hg** 80 -18-32-18-2	204.383 +1 +3 **Tl** 81 -18-32-18-3	207.2 +2 +4 **Pb** 82 -18-32-18-4	208.980 +3 +5 **Bi** 83 -18-32-18-5	(209) +2 +4 **Po** 84 -18-32-18-6	(210) **At** 85 -18-32-18-7	(222) 0 **Rn** 86 -18-32-18-8
(269) **Uun*** 110	(272) **Uuu** 111	(277) **Uub** 112		(285) **Uuq** 114				

*The systematic names and symbols for elements of atomic numbers above 109 will be used until the approval of trivial names by IUPAC.

151.96 +2 +3 **Eu** 63	157.25 +3 **Gd** 64	158.925 +3 **Tb** 65	162.50 +3 **Dy** 66	164.930 +3 **Ho** 67	167.26 +3 **Er** 68	168.934 +3 **Tm** 69	173.04 +2 +3 **Yb** 70	174.967 +3 **Lu** 71
(243) +3 +4 +5 +6 **Am** 95	(247) +3 **Cm** 96	(247) +3 +4 **Bk** 97	(251) +3 **Cf** 98	(252) **Es** 99	(257) **Fm** 100	(258) **Md** 101	(259) **No** 102	(260) **Lr** 103

Table S
Properties of Selected Elements

Atomic Number	Symbol	Name	First Ionization Energy (kJ/mol)	Electro-negativity	Melting Point (K)	Boiling* Point (K)	Density** (g/cm^3)	Atomic Radius (pm)
1	H	hydrogen	1312	2.1	14	20	0.00009	37
2	He	helium	2372	—	1	4	0.000179	32
3	Li	lithium	520	1.0	454	1620	0.534	155
4	Be	beryllium	900	1.6	1551	3243	1.8477	112
5	B	boron	801	2.0	2573	3931	2.340	98
6	C	carbon	1086	2.6	3820	5100	3.513	91
7	N	nitrogen	1402	3.0	63	77	0.00125	92
8	O	oxygen	1314	3.5	55	90	0.001429	65
9	F	fluorine	1681	4.0	54	85	0.001696	57
10	Ne	neon	2081	—	24	27	0.0009	51
11	Na	sodium	496	0.9	371	1156	0.971	190
12	Mg	magnesium	736	1.3	922	1363	1.738	160
13	Al	aluminum	578	1.6	934	2740	2.698	143
14	Si	silicon	787	1.9	1683	2628	2.329	132
15	P	phosphorus	1012	2.2	317	553	1.820	128
16	S	sulfur	1000	2.6	386	718	2.070	127
17	Cl	chlorine	1251	3.2	172	239	0.003214	97
18	Ar	argon	1521	—	84	87	0.001783	88
19	K	potassium	419	0.8	337	1047	0.862	235
20	Ca	calcium	590	1.0	1112	1757	1.550	197
21	Sc	scandium	633	1.4	1814	3104	2.989	162
22	Ti	titanium	659	1.5	1933	3580	4.540	145
23	V	vanadium	651	1.6	2160	3650	6.100	134
24	Cr	chromium	653	1.7	2130	2945	7.190	130
25	Mn	manganese	717	1.6	1517	2235	7.440	135
26	Fe	iron	762	1.8	1808	3023	7.874	126
27	Co	cobalt	760	1.9	1768	3143	8.900	125
28	Ni	nickel	737	1.9	1726	3005	8.902	124
29	Cu	copper	745	1.9	1357	2840	8.960	128
30	Zn	zinc	906	1.7	693	1180	7.133	138
31	Ga	gallium	579	1.8	303	2676	5.907	141
32	Ge	germanium	762	2.0	1211	3103	5.323	137
33	As	arsenic	944	2.2	1090	889	5.780	139
34	Se	selenium	941	2.6	490	958	4.790	140
35	Br	bromine	1140	3.0	266	332	3.122	112
36	Kr	krypton	1351	—	117	121	0.00375	103
37	Rb	rubidium	403	0.8	312	961	1.532	248
38	Sr	strontium	549	1.0	1042	1657	2.540	215
39	Y	yttrium	600	1.2	1795	3611	4.469	178
40	Zr	zirconium	640	1.3	2125	4650	6.506	160

Atomic Number	Symbol	Name	First Ionization Energy (kJ/mol)	Electro-negativity	Melting Point (K)	Boiling* Point (K)	Density** (g/cm^3)	Atomic Radius (pm)
41	Nb	niobium	652	1.6	2741	5015	8.570	146
42	Mo	molybdenum	684	2.2	2890	4885	10.220	139
43	Tc	technetium	702	1.9	2445	5150	11.500	136
44	Ru	ruthenium	710	2.2	2583	4173	12.370	134
45	Rh	rhodium	720	2.3	2239	4000	12.410	134
46	Pd	palladium	804	2.2	1825	3413	12.020	137
47	Ag	silver	731	1.9	1235	2485	10.500	144
48	Cd	cadmium	868	1.7	594	1038	8.650	171
49	In	indium	558	1.8	429	2353	7.310	166
50	Sn	tin	709	2.0	505	2543	7.310	162
51	Sb	antimony	831	2.1	904	1908	6.691	159
52	Te	tellurium	869	2.1	723	1263	6.240	142
53	I	iodine	1008	2.7	387	458	4.930	132
54	Xe	xenon	1170	2.6	161	166	0.0059	124
55	Cs	cesium	376	0.8	302	952	1.873	267
56	Ba	barium	503	0.9	1002	1910	3.594	222
57	La	lanthanum	538	1.1	1194	3730	6.145	138
Elements 58–71 have been omitted.								
72	Hf	hafnium	659	1.3	2503	5470	13.310	167
73	Ta	tantalum	728	1.5	3269	5698	16.654	149
74	W	tungsten	759	2.4	3680	5930	19.300	141
75	Re	rhenium	756	1.9	3453	5900	21.020	137
76	Os	osmium	814	2.2	3327	5300	22.590	135
77	Ir	iridium	865	2.2	2683	4403	22.560	136
78	Pt	platinum	864	2.3	2045	4100	21.450	139
79	Au	gold	890	2.5	1338	3080	19.320	146
80	Hg	mercury	1007	2.0	234	630	13.546	160
81	Tl	thallium	589	2.0	577	1730	11.850	171
82	Pb	lead	716	2.3	601	2013	11.350	175
83	Bi	bismuth	703	2.0	545	1833	9.747	170
84	Po	polonium	812	2.0	527	1235	9.320	167
85	At	astatine	—	2.2	575	610	—	145
86	Rn	radon	1037	—	202	211	0.00973	134
87	Fr	francium	393	0.7	300	950	—	270
88	Ra	radium	—	0.9	973	1413	5.000	233
89	Ac	actinium	499	1.1	1320	3470	10.060	—
Elements 90 and above have been omitted.								

*Boiling point at standard pressure
**Density at STP

Table T
Important Formulas and Equations

Density	$d = \frac{m}{V}$	d = density m = mass V = volume
Mole Calculations	number of moles = $\frac{\text{given mass (g)}}{\text{gram-formula mass}}$	
Percent Error	% error = $\frac{\text{measured value} - \text{accepted value}}{\text{accepted value}} \times 100$	
Percent Composition	% composition by mass = $\frac{\text{mass of part}}{\text{mass of whole}} \times 100$	
Concentration	parts per million = $\frac{\text{grams of solute}}{\text{grams of solution}} \times 1\,000\,000$	
	molarity = $\frac{\text{moles of solute}}{\text{liters of solution}}$	
Combined Gas Law	$\frac{P_1V_1}{T_1} = \frac{P_2V_2}{T_2}$	P = pressure V = volume T = temperature (K)
Titration	$M_AV_A = M_BV_B$	M_A = molarity of H^+ M_B = molarity of OH^- V_A = volume of acid V_B = volume of base
Heat	$q = mC\Delta T$ $q = mH_f$ $q = mH_v$	q = heat H_f = heat of fusion m = mass H_v = heat of vaporization C = specific heat capacity ΔT = change in temperature
Temperature	K = °C + 273	K = kelvin °C = degrees Celsius
Radioactive Decay	fraction remaining = $\left(\frac{1}{2}\right)^{\frac{t}{T}}$ number of half-life periods = $\frac{t}{T}$	t = total time elapsed T = half-life

Glossary

activated complex: A temporary, high-energy, transitional structure, somewhere between reactants and products (In an energy profile, the activated complex is at the top of the crest that forms the potential energy barrier)

acid anhydrides: Oxides of some nonmetals that react with water to form acids

acids: Substances that ionize to produce hydrogen ions (H^+) as their only positive ions; proton donors

activation energy: The energy needed to weaken or break bonds before new bonds can be formed; depends on the nature of the reacting particles

addition reaction: A reaction of unsaturated compounds during which atoms or groups of atoms are added to the molecule as double or triple bonds open

alcohol: An organic compound in which one or more hydrogen atoms of a hydrocarbon have been replaced by a hydroxyl group, –OH, (*R*–OH)

aldehyde: An organic compound in which the carbonyl group –CHO is located at the end of a hydrocarbon chain (*R*–CHO)

alkali metal family: The metals of Group 1: lithium, sodium, potassium, rubidium, cesium, and francium

alkaline earth metal family: The metals of Group 2: beryllium, magnesium, calcium, strontium, barium, and radium

alkanes: Hydrocarbons that have single bonds between carbon atoms

alkenes: Hydrocarbons that have a double bond between two of its carbon atoms

alkyl groups: Hydrocarbon radicals derived from the alkanes

alkynes: Hydrocarbons that have a triple bond between two of its carbon atoms

allotropic: Describes elements or compounds that exist in two or more forms in the same physical state; each form has its own chemical and physical properties, for example red and white phosphorus

alpha particles: Particles emitted from a radioisotope that have a charge of 2+ and a mass number of four, considered to be a helium nucleus (α)

amide: An organic compound that contains the carbonyl group –CO– and an amine group

amine: An organic compound derived from ammonia (NH_3) by the replacement of one or more hydrogen atoms by alkyl groups

amino acids: Organic compounds that contain an amine group ($-NH_2$) and a carboxyl group (–COOH)

amphiprotic: Species that, depending on what they react with, can either gain or lose protons

amphoteric hydroxides: Hydroxides of metalloids that act as either acids or bases, depending on the chemical environment

analysis reaction: (See **decomposition reaction.**)

anions: Negative ions, which are attracted by a positive anode

anode: The electrode where oxidation takes place; the negative electrode of the voltaic cell and the positive electrode of the electrolytic cell

aqueous solution: A solution in which water is the solvent

Arrhenius model: Acids ionize to produce hydrogen ions, (H^+), as their only positive ions; bases dissociate to produce hydroxide ions, (OH^-), as their only negative ions; electrolytes that contain positive ions other than H^+ and negative ions other than OH^- are called salts

artificial transmutation: The changing of one element into another by artificial means

atmosphere: A non-SI unit of pressure (atm)

atom: The smallest part of an element that retains the properties of the element

atomic mass unit: A unit on the arbitrary scale for the mass of elements equal to 1/12 the mass of an atom of the most abundant form of carbon (carbon-12)

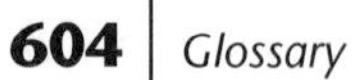

atomic mass: The weighted average mass of all the isotopes in a naturally occurring sample of an element

atomic number: The number of protons in the nucleus of an atom, it describes a particular atom and the element of which it is a part

auto-oxidations: (See **disproportionation reactions.**)

Avogadro's number: 6.02×10^{23} particles

balanced equations: Equations that show the proportions in which materials react and products are formed in accordance with the law of conservation of mass; in a properly balanced equation, there must be the same number of atoms of each element on the left side of the equation as on the right side

bases: Substances that dissociate to produce hydroxide ions (OH^-) as their only negative ions; proton acceptors

basic anhydrides: Oxides of some metals that react with water to form bases

beta particles: Particles emitted from the nucleus of a radioisotope that are electrons (β^-)

binary acids: Acids that contain only hydrogen and another element

boiling point: The temperature at which the vapor pressure of a liquid equals the pressure of the gas acting on the liquid

bonds: The attractive forces between atoms

Boyle's law: A constant temperature, the volume of a fixed mass of gas varies inversely with the pressure

Brønsted-Lowry model: An acid is a substance that donates protons; a base is a substance that accepts protons

buffer solutions: Solutions that can withstand the addition of moderate amounts of acid or base without undergoing a significant change in pH

carboxylic acid: An organic compound that contains a carboxyl group, –COOH, at the end of a hydrocarbon chain (*R*–COOH); also called an organic acid

catalyst: A substance that changes the rate of a chemical reaction without being permanently changed itself

cathode: The electrode where reduction takes place; the positive electrode of the voltaic cell and the negative electrode of the electrolytic cell

cations: Positive ions, which are attracted by a negative cathode

Charles' law: The volume of a fixed mass of gas at constant pressure varies directly with its absolute (Kelvin) temperature

chemical bonds: The attractive forces that tend to pull atoms together when atoms unite; the forces resulting from the simultaneous attraction of two nuclei for one or more pairs of electrons

chemical changes: Changes in which a new material or substance is formed

chemical properties: Describe how matter behaves when it changes into another kind of matter

chemistry: The science that seeks to explain how and why matter behaves as it does

colligative properties: Properties, such as vapor pressure, boiling point, and freezing point, that depend on the number of particles of solute in a solution, not on the chemical nature of the particles

combination reaction: A reaction in which two simple substances combine to form a single substance

combustion: The reaction in which an element or compound reacts with oxygen

common ion effect: Describes the decrease in solubility of a solute caused by the presence in solution of another solute with which it has an ion in common

composition reaction: (See **combination reaction.**)

compound: A substance formed when two or more elements combine in fixed proportions by mass

concentrated: Describes a solution that contains a relatively high proportion of solute

coordinate covalent bond: A covalent bond in which both electrons in a shared pair are supplied by one atom

covalent bonds: The chemical bonds formed by the sharing of one or more pairs of electrons

critical temperature: The temperature at which a given substance cannot exist in the liquid phase

crystal lattice energy: In the solution process, the energy needed (absorbed) to separate the solid particles, whether they are ions or molecules

Dalton's law of partial pressures: At constant temperature, the pressure of a mixture of gases, which do not react, equals the sum of the partial pressures of the gases in the mixture

decay rate: the amount of radiation emitted by a given sample, measured in counts per minute

decomposition reaction: A reaction in which a single substance produces two or more simpler substances

dehydrating agent: A substance that removes water or the elements that make up water from matter with which the substance is in contact

density: A measure of the mass per unit volume of a given sample of matter

diatomic: Describes molecules that have two atoms

dilute: Describes a solution that contains a much smaller proportion of solute than solvent

dipole-dipole forces: The attractive forces, or bonds, that occur between polar molecules

dipoles: molecules that contain two, oppositely charged centers of charge

dispersion forces: The attractive forces, or bonds, that hold together nonpolar molecules; they result from the interaction of the electron clouds of neighboring molecules

disproportionation reactions: Reactions in which the same element is both oxidized and reduced

double bond: A covalent bond in which two atoms share two pairs of electrons

double replacement reaction: A reaction that occurs between ionic compounds in solution in which the positive ions simply "change partners"

electrochemical cell: A device in which a redox reaction either produces or uses electricity

electrodes: The metallic conductors in an electrochemical cell

electrolysis: The process by which redox reactions that do not proceed spontaneously can be driven by an external "electron pump"—a direct current (dc) source, such as a battery; a process that uses electricity to separate compounds into their elements

electrolytes: Compounds that conduct electricity in solution

electrolytic cell: An electrochemical cell that uses electricity to produce an otherwise nonspontaneous chemical reaction

electromagnetic radiation: Energy that travels through space as waves at the speed of 3×10^8 meters per second

electron affinity: The amount of energy released when an atom receives an electron and becomes a negatively charged ion

electronegativity: The power of an atom in a molecule to attract electrons to itself

electrons: Negatively charged particles that are located outside the nucleus of atoms, they are the smallest of the three fundamental particles Each electron has a unit charge of −1
electroplating: The process of depositing a metal coating on the surface of another metal through the use of an electric current
element: A substance that contains only one kind of atom
emission spectroscopy: The study of the energy—electromagnetic radiation—emitted by excited atoms
empirical formula: A formula that expresses the smallest whole number ratio among its elements
endothermic: Describes reactions that absorb, or remove, heat from their surroundings
energy of activation: (See **activation energy.**)
energy: The ability, or capacity, to do work
enthalpy: (See **heat content.**)
entropy: Describes randomness (*S*)
equilibrium constant: The ratio of the product concentration(s) to the reactant concentration(s) at equilibrium (K_{EQ})
equilibrium: The situation that exists when two opposing processes go on at the same time and at the same rate
ester: An organic compound formed in the reaction between an organic acid and an alcohol ($RCOOR'$)
esterification: The reaction between an alcohol and an organic acid forming an ester and water
ether: An organic compound in which an oxygen atom is positioned between two carbon atoms ($R–O–R'$)
excited state: The state of an atom that has absorbed energy that has moved electrons to higher levels
exothermic: Describes reactions that release, or give off, heat to their surroundings

families: (See **groups.**)
fermentation: The process, catalyzed by enzymes, in which a simple sugar is broken down into ethanol and carbon dioxide
first ionization energy: The amount of energy needed to remove the most loosely bound electron from an atom
fission: A reaction in which large nuclei split into smaller nuclei
formula mass: The sum of the atomic masses of the elements in an ionic compound represented by a certain formula

formula: A statement of the number of atoms of each element in a compound

free energy: A state function that predicts the spontaneity of reactions (*G*)

freezing point: The temperature at which the liquid and its solid form have the same vapor pressure

frequency: The number of complete pulses, or cycles, that pass a given point in one second

functional group: An atom or a group of atoms that replace one or more hydrogen atoms of a hydrocarbon and determine the characteristics of the compound

fusion: A reaction in which small nuclei unite into larger nuclei

gamma radiation: radiation that has no charge and no mass; high-energy electromagnetic waves (γ)

Gay-Lussac's law: The pressure of a fixed mass of gas at constant volume varies directly with the Kelvin (absolute), temperature

gram-atomic mass: The atomic mass of an element expressed in grams

gram-formula mass: The formula mass of an ionic compound expressed in grams

gram-molecular mass: The molecular mass of a substance expressed in grams

ground state: The lowest energy state of the electrons in an atom

groups: The vertical columns of elements on the Periodic Table

half-life: The time required for one-half of any given mass of radioactive nuclei to decay

half-reactions: The two parts of a redox reaction

halides: Organic compounds that contain one or more atoms of the Group 17 elements attached to a hydrocarbon chain ($R–X$)

halocarbon: (See **halides.**)

halogen family: The elements of Group 17: fluorine, chlorine, bromine, iodine, and astatine

halogenation: A reaction in which the two halogen atoms from a halogen molecule are added to an opened double or triple bond in an organic molecule

hard water: Describes water that contains significant concentrations of calcium and/or magnesium ions

heat content: Describes the energy, expressed in joules or kilojoules, stored in a sample of matter; enthalpy (*H*)
heat of fusion: The quantity of heat required to melt a given quantity of a solid
heat of reaction: The difference in energy between the products and the reactants (ΔH)
heat of vaporization: The quantity of heat required to boil a given quantity of liquid
heat: A measure of the total energy in a system
Heisenberg uncertainty principle: It is impossible to simultaneously measure both the position and the momentum of an object exactly
Hess's law: Law that states if a reaction is the exact sum of two or more other reactions, then its ΔH is the exact sum of the ΔH values of the other reactions
heterogeneous mixture: A mixture in which equal volumes have different compositions
homogeneous: Describes a sample of matter that has the same properties throughout
homogeneous mixture: A mixture in which equal volumes have the same composition
homologous series: A series of organic compounds that follow the same general formula
Hund's rule: No orbital in a sublevel may contain two electrons unless all the orbitals in that sublevel contain at least one electron; all unpaired electrons in a given atom have the same spin
hydrated: Surrounded by water molecules
hydration energy: In the solution process, the energy released as a result of the attractions between solute particles and solvent particles
hydrocarbons: Binary compounds made up only of carbon and hydrogen
hydrogen bond: An attractive force between molecules that occurs when an atom of hydrogen acts as a bridge between two small, highly electronegative elements, such as the oxygens in H_2O or the fluorines in HF
hydrogenation: A reaction in which the double or triple bond of an unsaturated molecule undergoes a reaction with hydrogen; the product is always more saturated than the reactant molecule

hydrolysis: The process that produces acidic or basic solutions when certain salts are dissolved in water

hydronium ion: A hydrated proton, the H_3O^+ ion

indicator: A substance that changes color in a specific pH range

inorganic: Describes compounds that do not contain carbon, also applied to carbon dioxide and carbonate compounds

ionic bond: A bond between elements of such greatly differing electronegativity that electrons are displaced completely and a transfer of electrons takes place forming positive and negative ions

ionic solids: Solids that form as a result of ionic bonding; they have high melting points, do not conduct electricity in the solid state, conduct electricity when melted or dissolved in water

ionization energy: The energy required to remove the most loosely held electron from an isolated, neutral, gaseous atom

ionizing power: The ability of radiation to remove one or more electrons from an atom or molecule

isomers: Compounds that have the same number and kinds of atoms but have different structural formulas

isotopes: Different forms of the same element that have the same number of protons (atomic number), but a different number of neutrons in their nucleus (different mass number)

ketone: An organic compound in which the carbonyl group –CO– is located within a hydrocarbon chain (*R*–CO–*R*′)

kinetic energy: Energy of motion

law of conservation of energy: Under ordinary conditions, energy can be neither created nor destroyed, but it can be converted from one form to another

law of conservation of matter (mass): Matter (mass) may be neither created nor destroyed

law of octaves: Law developed by John Newlands stating that if the elements are arranged in order of increasing atomic weight, the first and eighth elements have related properties

Le Chatelier's Principle: The principle that if a system in equilibrium is subjected to a stress (a change in concentration, pressure of a gas, or temperature), the system adjusts to partially relieve the stress and restore equilibrium

Lewis model: An acid is an electron pair acceptor; a base is an electron pair donor
limiting reactant: The reactant used up in a chemical reaction
lone pair: A pair of electrons not involved in bond formation

macromolecule: (See **network solid.**)
mass defect: The mass lost when nucleons unite and form a nucleus
mass: A measure of the quantity of matter contained in a body
mass number: The sum of the protons and neutrons in the nucleus of an atom
mechanism of reaction: The sequence of steps involved in a reaction
metallic bonds: Bonds formed by a sea of mobile valence electrons surrounding relatively stationary positive ions
metallic solids: Solids formed as a result of metallic bonding; they are excellent conductors of electricity in solid and liquid state
metalloids: Elements whose properties lie somewhere between those of metals and those of nonmetals
metals: Most elements in the Periodic Table are metals
molality: Concentration expressed as moles of solute dissolved in 1000 grams, or 1 kilogram, of solvent (m)
molar gas volume: The volume of one mole of any gas at STP is 22.4 liters
molar mass: The mass in grams of one mole of a substance
molarity: Concentration expressed as moles of solute per liter of solution (M)
mole: The amount of any substance that contains Avogadro's number of particles, 6.02×10^{23} particles
molecular mass: The sum of the atomic masses of all the atoms in a molecule
molecular solids: Polar and nonpolar substances held in the solid state by intermolecular attractions; they have low melting and boiling points, do not conduct electricity in the solid or liquid state
monatomic: describes molecules that have only one atom
monomers: The simple, repeating units of a polymer chain
multiple covalent bond: A covalent bond in which atoms share two or more pairs of electrons

natural radioactivity: (See **natural transmutation.**)
natural transmutation: Nuclear changes that occur spontaneously in nature, forming a new nuclide with a more stable n/p ratio
net ionic equation: A chemical equation that omits the spectator ions
network bond: The covalent bonds found in network solids
network solid: A structure in which all the atoms in the crystals are covalently joined in one large network, these substances have exceptionally high melting points
neutralization: The reaction between an acid and a base to form a salt and water
neutron: An electrically neutral particle, its mass is only slightly greater than the mass of a proton
nitrogen fixation: The changing of nitrogen in the air into useable nitrogen compounds
noble gases: Elements whose outermost energy level has completed *s* and *p* sublevels; they do not readily form chemical bonds with other elements: helium, neon, argon, krypton, xenon, and radon
nonelectrolytes: Compounds that do not conduct electricity in solution
nonmetals: The elements on the right side of the Periodic Table
nonpolar covalent bond: A covalent bond between atoms that have the same attraction for electrons; results from the equal sharing of electron pairs
normal boiling point: The temperature at which the vapor pressure of a liquid reaches 101.3 kPa
nuclear binding energy: The energy released when nucleons come together to form a nucleus; the energy absorbed when a nucleus is split into its individual nucleons
nucleons: Protons or neutrons, the particles in the nucleus
nucleus: The center of the atom, which contains the mass

octet rule: Elements generally bond so as to attain a structure containing eight valence electrons
orbital: Describes the shape of the region where an electron may be found
organic acid: (See **carboxylic acid.**)
organic compound: Compounds that contain carbon (Carbon dioxide and compounds containing the carbonate ion are not considered to be organic compounds.)

oxidation half-reaction: The half-reaction in which electrons are lost

oxidation number: The charge an atom would acquire if all its bonds were treated as ionic bonds

oxidation state: See (**oxidation number.**)

oxidation: The loss of electrons; an algebraic increase in oxidation number

oxidizing agent: A particle that can cause the loss of electrons, oxidation, and by doing so, undergoes reduction

partial pressure: The pressure exerted by gas in a mixture of gases

parts per million: Concentration expressed as parts of solute by mass per one million parts by mass of total solution (ppm)

pascal: The SI unit of pressure (P)

Pauli exclusion principle: In an atom, only two electrons, which have opposite spins, can occupy the same orbital

percentage concentration: Concentration expressed as parts by mass of solute per 100 parts by mass of total solution

periodic law, revised: Law stating that the properties of the elements are periodic functions of their atomic number

periodic law: Law stating that if the elements are arranged in order by increasing atomic weight (mass), the physical and chemical properties repeat regularly

periods: The horizontal rows of elements on the Periodic Table

phase diagram: A graph that shows the phases present at various temperatures and pressures

phase: One of the states of matter; a clearly separate region of a sample that is different from the rest of the sample

photosynthesis: The process by which green plants use carbon dioxide to manufacture food

physical changes: Changes in which the form but not the identity of the matter changes

physical properties: Describe the qualities of a substance that can be demonstrated without changing the composition of the substance

polar covalent bond: A covalent bond between two atoms that have some difference in their ability to attract the electron pair in the bond, resulting in an unequal sharing of the electron pair; a bond of partial ionic character with one side partially positive and the other side partially negative

polar molecules: (See **dipoles.**)
polyatomic ions: Ions that contain two or more atoms; most polyatomic ions contain coordinate covalent bonds
polymer: The chain, or molecule, formed by repeating monomers
polymerization: The process by which a polymer is formed from repeating monomers
positive ions: The positively charged particles that result from the loss of one or more electrons from a neutral atom
positrons: Particles emitted from the nucleus of a radioisotope that have the same mass as an electron but an opposite charge (β^+)
potential energy: The energy a body possesses because of its position with respect to another body
precipitate: An insoluble product that forms in a chemical reaction
pressure: The force per unit area
product(s): The material(s) to the right of the yields arrow in a chemical equation
protons: Positively charged particles that are located in the nucleus of atoms, each has a unit charge of +1

quantized: Describes electron energy that can have only certain values within a given range

radioactivity: The spontaneous breakdown on an unstable atomic nucleus, yielding particles and/or radiant energy
radioisotopes: Atoms that have an unstable nucleus
rate of reaction: The rate at which the reactants are consumed and the products are formed
rate-determining step: The slowest of the steps in a reaction mechanism
reactant(s): The material(s) to the left of the yields arrow in a chemical equation
redox reactions: Reactions that involve the transfer of electrons and/or changes in oxidation numbers
reducing agent: A particle that can cause the gain of electrons, reduction, and by doing so, undergoes oxidation
reduction half-reaction: The half-reaction in which electrons are gained
reduction: The gain of electrons; an algebraic decrease in oxidation number

salts: Electrolytes that contain positive ions other than H^+ and negative ions other than OH^-

saponification: The hydrolysis of a fat with a base to produce a soap and glycerin

saturated compounds: Organic compounds containing molecules in which carbon atoms are bonded by single bonds

saturated: Describes a solution that contains the maximum amount of a given solute that can be dissolved at a given temperature and pressure

semimetals: (See **metalloids.**)

single replacement reaction: A chemical reaction in which an uncombined element replaces a different element from a compound producing a new compound and a different uncombined element

skeleton equation: An equation that shows reactants and products, but does not indicate in what proportions they are reacting

solubility product constant: The product of the concentration of the ions in a solution (K_{SP})

solute: In a solution the substance that is dissolved

solution: A homogeneous mixture consisting of a solvent and a solute

solvent: In a solution, the substance that does the dissolving

specific heat capacity: The quantity of heat required to change the temperature of one gram of substance by one degree

spectator ions: Ions that do not participate in a reaction

spectrometer: A device that reads the amount of light of a given frequency that is absorbed by a solute

spectroscopy: The study of electromagnetic radiation

standard electrode potential (E^0): The electrical potential when all ions are at 1M concentration and all gases are at 1 atm pressure

standard molar enthalpy of formation: The enthalpy change for the formation of 1 mole of a substance from its elements in their standard state ($\Delta H^0{}_f$)

standard molar entropy: Entropy measured for 1 mole at 1 atmosphere for gases, and at 1 molar for solutes (S^0)

state function: A variable that depends only on the initial and final conditions

strong acids: Acids that ionize almost completely, yielding a maximum quantity of H^+ ions

strong bases: Bases that dissociate almost completely, yielding a maximum quantity of OH^- ions

structural formula: A formula that shows the arrangement of bonds between atoms in a molecule

sublimation: The direct change of phase from solid to gas

substance: Any variety of matter that has a definite, unvarying set of properties and for which all samples have identical composition

substitution reaction: A reaction in which one or more hydrogen atoms are removed from a saturated hydrocarbon and replaced with new atoms or groups

supersaturated: Describes a solution that contains more of a given solute than will normally dissolve at a given temperature and pressure

synthesis reaction: (See **combination reaction.**)

system: A convenient part of the environment that can be isolated for study

temperature: A measure of the average kinetic energy of the particles in a system

titration: A procedure in which measured volumes of two solutions are combined to achieve a predetermined end point

transition elements: Elements that complete inner, or lower, energy levels before completing outer, or higher, energy levels

triple bond: A covalent bond in which two atoms share three pairs of electrons

triple point: The conditions of temperature and pressure at which a given substance can exist in all three phases at the same time

unsaturated: Describes a solution that can dissolve more of a given solute at a given temperature and pressure

unsaturated compounds: Organic compounds that have one or more double or triple bonds between carbon atoms

valence electrons: The electrons in the outer energy levels of an atom that are likely to become involved in the formation of chemical bonds

vapor pressure: In a closed container, the pressure exerted by a vapor that exists above its liquid form

voltaic cell: An electrochemical cell that uses a spontaneous chemical reaction to produce an electric current

volume: A measure of the space that matter takes up

wavelength: The distance between peaks or troughs on a wave

weak acids: Acids that ionize only partially, yielding small quantities of H^+ ions

weak bases: Bases that ionize only partially, yielding small quantities of OH^- ions

weight: A measure of the force with which gravity pulls a body toward the center of Earth

Index